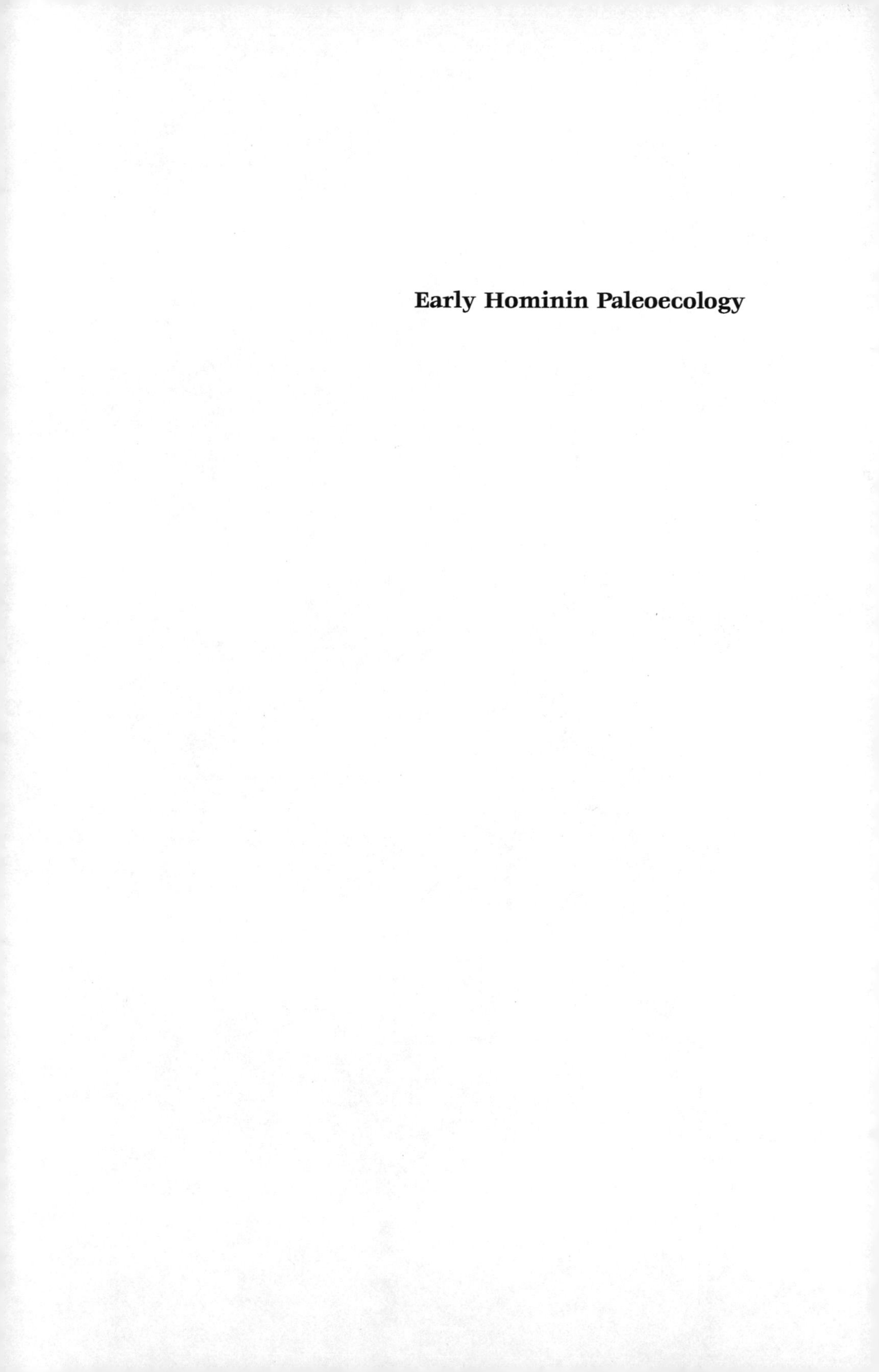

Early Hominin Paleoecology

Early Hominin Paleoecology

EDITED BY
Matt Sponheimer, Julia A. Lee-Thorp, Kaye E. Reed, and Peter S. Ungar

UNIVERSITY PRESS OF COLORADO
Boulder

Published by University Press of Colorado
5589 Arapahoe Avenue, Suite 206C
Boulder, Colorado 80303

Printed in the United States of America

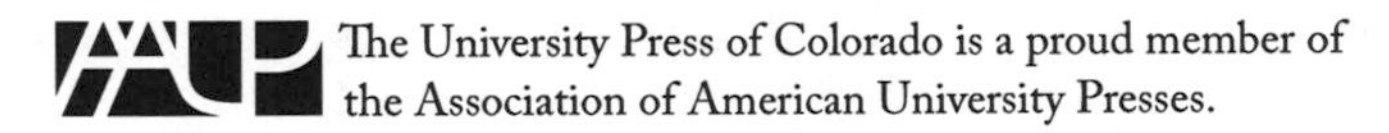
The University Press of Colorado is a proud member of the Association of American University Presses.

The University Press of Colorado is a cooperative publishing enterprise supported, in part, by Adams State University, Colorado State University, Fort Lewis College, Metropolitan State University of Denver, Regis University, University of Colorado, University of Northern Colorado, Utah State University, and Western State Colorado University.

∞ This paper meets the requirements of the ANSI/NISO Z39.48-1992 (Permanence of Paper).

Library of Congress Cataloging-in-Publication Data

Early hominin paleoecology / edited by Matt Sponheimer, Julia A. Lee-Thorp, Kaye E. Reed, and Peter S. Ungar.
pages cm
Includes bibliographical references and index.
ISBN 978-1-60732-224-5 (hardcover : alk. paper) — ISBN 978-1-60732-225-2 (ebook)
1. Fossil hominids. 2. Paleoecology. 3. Paleontology. I. Sponheimer, Matt.
GN282.E37 2013
569.9—dc23
2013002837

Design by Daniel Pratt

22 21 20 19 18 17 16 15 14 13 10 9 8 7 6 5 4 3 2 1

Contents

Preface

As soon as some ancient member in the great series
of the Primates came to be less arboreal, owing to
a change in its manner of procuring subsistence, or
to some change in the surrounding conditions, its
habitual manner of progression would have been
modified: and thus it would have been rendered more
strictly quadrupedal or bipedal.

—Charles Darwin, 1871

For the production of man a different apprenticeship
was needed to sharpen the wits and quicken the
higher manifestations of intellect—a more open
veldt country where competition was keener between
swiftness and stealth, and where adroitness of
thinking and movement played a preponderating role
in the preservation of the species.

—Raymond Dart, 1925

Our intellectual forebears explicitly grounded their notions of human origins in the ecological realm. Our understanding of human evolution has no doubt advanced since those times. We certainly know more about early hominin biology, diversity, and distributions through time and space. We also have a broadly informed sense of the plants and animals that populated the ancient landscapes on which our ancestors roamed, and a profoundly better notion of the environmental and climatic change they faced. We even have a steadily increasing understanding of the causal mechanisms behind such changes.

We have advanced less, however, in our understanding of how these disparate elements can be distilled to produce a coherent story of hominin evolution; and although this is at least partly a failure of theory, it is also the inevitable result of the often cloudy, controversial, and contradictory evidence that forms the pediment on which our evolutionary scenarios are built. If Hutchinson is correct in asserting that ecology is the theater in which the evolutionary play takes place, early hominin actors are shadowy, or worse, peripatetic figures. Major questions abound. Was *Ardipithecus ramidus* a forest creature or a denizen of wooded grasslands? Was *Australopithecus afarensis* an ecological generalist or a specialist limited to narrow microhabitats on otherwise diverse landscapes? Was *Paranthropus boisei* a frugivorous hard-object feeder or an amiable muncher of sedges and grasses? The answers to these questions lie to a large extent on which ecological data-sets we principally value, how we read them, and how they speak to one another. And while we have moved a long way in this regard, we still have a long way to go. In fact, for all of the recent developments in this field, it would probably be safe to say that we are less collectively confident of the ecological underpinnings of human evolution than we were twenty years ago, which is remarkable given the new tools at our disposal and the undiminished pace of fossil discovery. Who could have dreamed that this wealth of information would produce so little persistent light?

It was such concerns that led us to organize a workshop in October 2004 that brought together twelve researchers with varied interests including hominin paleontology, archaeology, primatology, paleoclimatology, sedimentology, and geochemistry to discuss the state of knowledge in their specialized subfields and in hominin paleoecology on the whole. After two days of convivial and stimulating discussion two things were clear. First, recent advances in the field and the lab were not only bettering our understanding of human evolution but potentially transforming it; and second, communicating the finer details of this research had become distressingly complex, given the increasing specialization of our individual fields of endeavor. And if communicating such knowledge proved difficult among a group of professionals, how much worse would it be to disseminate to a broader audience of graduate students, advanced undergraduates, and the interested public?

It was with the idea of confronting this burgeoning Babel that this volume was conceived. This is not a collection of specialized papers that recapitulate or expand upon presentations from the meeting. Rather, we tried to devise a book that would not only provide a good working knowledge of early hominin paleoecology, but would also provide a solid grounding in the sundry ways we construct such knowledge. We leave it to the reader to judge the degree to which this approach has been a success.

The book is divided into three sections: climate and environment (with a particular focus on the latter), adaptation and behavior, and modern analogs and models. The methods discussed in each represent only a portion of those employed by students of human evolution, although they are among those most frequently encountered in the literature. And because the story of human evolution is a long one, we made a decision to not cover material significantly postdating the origin of our genus *Homo*. Thus, the discussion of archaeology is relatively brief, as most of the story discussed herein precedes the appearance of an abundant archaeological record. We would argue, however, that this is an acceptable elision given the primacy often afforded stone tools and butchered animal bones in discussions of early hominin behavior, despite the fact that relatively few hominin taxa were indisputably stone toolmakers or -wielders.

Thus, we make no claim to have covered all of the relevant methodological bases herein, or even to have provided a comprehensive presentation of germane "facts." But despite these, and no doubt many other faults of the editors' making, we do believe that this book provides a generally accessible, yet scholarly, entrée into this fascinating and ever-evolving field. We look forward to the "chapters" that the next generation of scholars will write.

Acknowledgments

A GREAT MANY PEOPLE contributed to this book. We thank Stanley Ambrose, Marion Bamford, René Bobe, Chris Campisano, Daryl Codron, Sandi Copeland, Darryl de Ruiter, John Kingston, Brian Richmond, Mike Rogers, Mark Spencer, and Alan Walker for helping in one way or another. We also thank two anonymous reviewers for their helpful comments on the manuscript entire.

Jennifer Leichliter, Yasmin Rahman, and Sarah Taylor labored over the manuscript at various stages, and much of its coherence is due to their editorial ministrations. Darrin Pratt, Jessica d'Arbonne, Laura Furney, and the staff of the University Press of Colorado were unstintingly helpful and patient from the book's conception through the publication.

The editors also thank their colleagues, students, friends and family who have—in various ways—aided and abetted them in the production of this volume. MS, in particular, would like to thank Alia, Karim, and Yasmin for their inimitable support and forbearance, and Carmel Schrire for making this project possible.

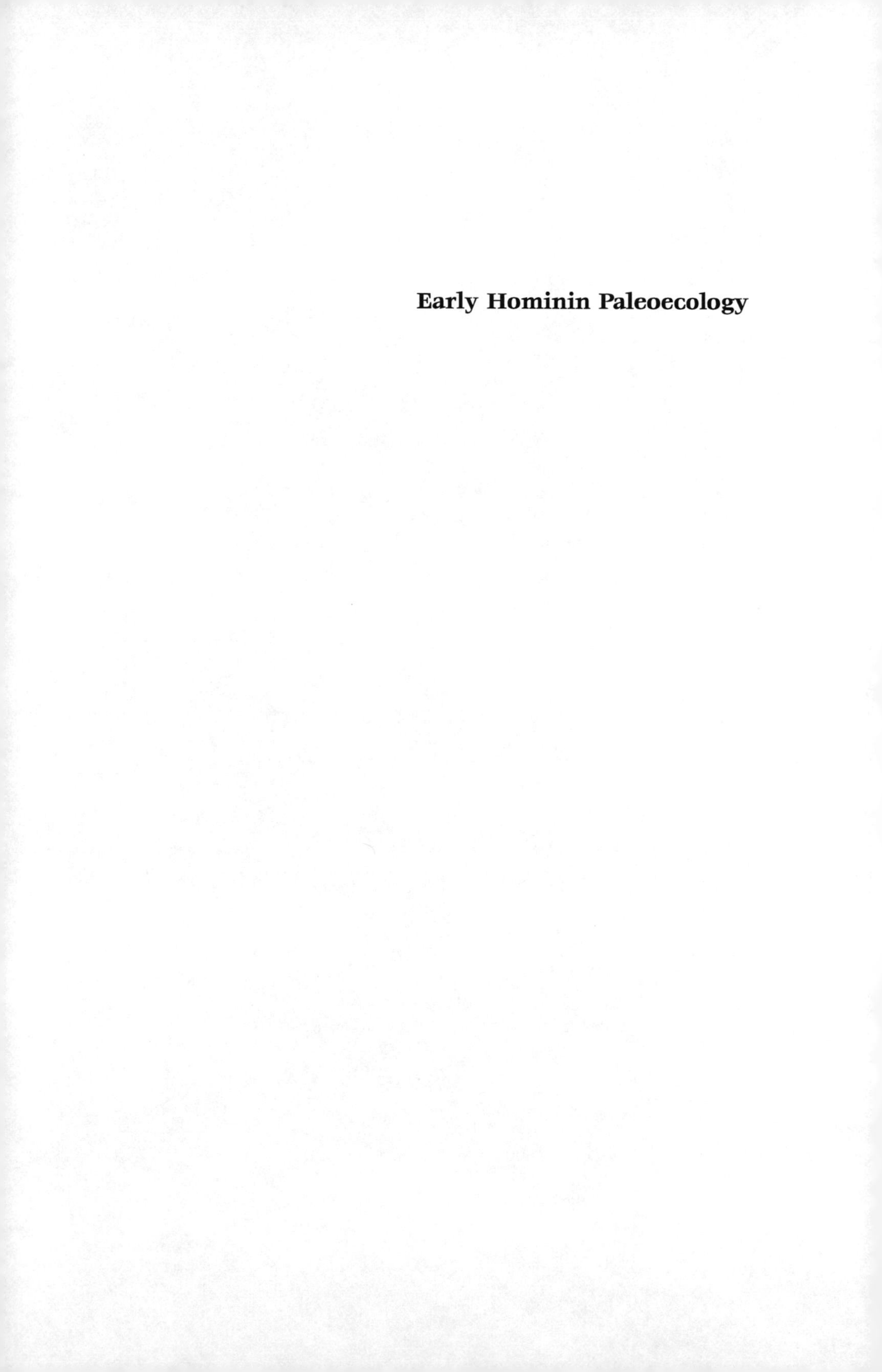

Early Hominin Paleoecology

PART I

Paleoclimate and Paleoenvironment

1

Faunal Approaches in Early Hominin Paleoecology

Kaye E. Reed, Lillian M. Spencer, and Amy L. Rector

The paleoecology of early hominin species is more than simply reconstructing the habitats in which they existed. Ultimately we would like to know the ecological context before and after speciation and extinction events, and about the interactions of hominins with their environment, including other species. A first step toward this goal is to discover as much information as possible regarding the climate, geomorphology, vegetation physiognomy (habitat structure), and the faunal community. These factors build on one another such that climate, soil properties, and geomorphology are responsible for the vegetation, which, in turn, plays a fundamental role in controlling what other life forms can be supported. An understanding of extant African habitats is necessary to reconstruct ancient vegetation physiognomy for early Pliocene hominins. An appreciation of living mammals is also important in interpreting Pliocene environments when using faunal techniques. The most common data recovered with early hominins are other mammalian fossils, and these are targeted here for explaining how reconstructions of habitat and community ecology can be approached. Faunal analyses can be compared with other types of research such as palynology, fossil botanical studies, and isotopic analyses of soils and teeth to arrive at a better understanding of hominin paleoecology.

Fossil mammals found within the same deposits as early hominins can be used to answer a variety of questions relating to evolutionary paleoecology. First, fossil

DOI: 10.5876/9781607322252:c01

mammals have been used as indicators of habitats since early paleontological studies (e.g., Ewer 1958; Brain 1967; Leakey and Harris 1987). More recent work on this topic has emphasized the importance of determining taphonomic histories before reconstructions are attempted (e.g., Behrensmeyer and Hill 1980; Brain 1981; Behrensmeyer 1991; Soligo and Andrews 2005; Andrews 2006), but this caveat is still only rarely addressed. The majority of African hominin paleoecological work falls into the category of using faunal analyses for reconstructing ancient habitats, and forms the bulk of the work reviewed here. Second, studies of contemporaneous fauna are critical for investigating aspects of community ecology, such as guild structure. This avenue of research can be also used to determine possible differences between ancient Plio-Pleistocene and extant communities (e.g., Janis et al. 2004). Third, faunal studies can give insights into how hominins might have interacted with specific members of their shared community. For example, study of the members of the carnivoran guild (Marean 1989; Lewis 1997) can lead to hypotheses about how hominins might have avoided predation or competed with predators for access to meat. Finally, faunal studies can be used to answer questions of patterns and processes in the evolution of both hominins and other mammalian lineages (e.g., Vrba 1988, 1995; Behrensmeyer et al. 1997; Potts 1998; Bonnefille et al. 2004).

Faunal approaches in hominin paleoecology can be assigned to two types of studies. The first is analyzing individual fossil species of mammals and other fauna found at particular localities. This information can be used to reconstruct habitats and to look at species interactions with hominins. It is also a critical precursor for community studies. The second type of study examines communities as a whole, which is necessary for studies of community ecology and also for investigating evolutionary patterns in hominin lineages.

A second dichotomy exists between the taxonomic and ecological/functional approaches to faunal research. In taxonomic analyses, phylogeny plays an important role. Taxonomic methodologies are used occasionally to reconstruct environments (e.g., Vrba 1980), but the usual focus using these methods is to examine biogeographic and species-turnover patterns (Behrensmeyer et al. 1997; Bobe and Eck 2001). The second approach is often referred to as taxon free because species diversity, ecological diversity, or the results of functional studies are ecological representations of each species. Damuth (1992) has argued that results derived from these types of taxon-free data transform species-specific fauna, and by extension assemblages or communities, into parameters to be incorporated into ecological patterns that can then be compared with any other faunal community in space and time since the parameters used are not taxon specific. For example, it might be difficult to compare an

Fossil Assemblages

Modern Assemblages

TAXONOMIC	**TAXON FREE**
• Environmental reconstruction based on single species • Environmental reconstruction based on groups of species (e.g., Bovidae) • Species-turnover patterns	• Ecomorphology of a single species to reconstruct diet, locomotion, or substrate use • Groups of species in communities or guilds to reconstruct habitats using ecological diversity patterns • Groups of species compared with living communities to find community structure differences

FIGURE 1.1. *Taxonomic and taxon-free methodologies are used to study modern and fossil assemblages.*

Australian *Macropoda* (kangaroo) to an African *Damaliscus* (topi) on a phylogenetic level, but to compare them as terrestrial grazers of similar body size is possible. Both phylogenetic and taxon-free approaches have been important in understanding hominin paleoecology as well as in developing evolutionary scenarios (Figure 1.1).

As mammal species are most often recovered in the greatest numbers from hominin fossil localities, much of our discussion is devoted to analyses of mammalian fauna. However, the results of any research using mammals should be compared with other types of analyses—such as the study of amphibian, bird, and reptile fossils; research on paleoclimate, pollen, and depositional environments; and isotopic analyses—depending on the ultimate goals of the research. In this chapter, we present a brief overview of existing African habitats and African mammal communities. We then discuss issues of taphonomy such as time-averaging, collection bias, and other factors that may bias faunal assemblages such that they obscure paleoecological reconstructions. Faunal analyses are only as good as the data derived from the fossil localities. Third, we provide an overview of the types of analyses mentioned above—those focusing on individual species and those focusing on the community from phylogenetic and taxon-free perspectives. We then survey research that has been used to investigate three areas in hominin paleoecology: reconstructing habitats, reconstructing community ecology, and investigating species interactions between hominins and other mammals.

AFRICAN HABITATS

Reconstructing past African habitats is usually based on comparisons to extant habitats. Today, ecologists often refer to existing habitats by the dominant plant species, such as miombo woodland. Usually the best that can be accomplished for ancient vegetation, however, is to reconstruct the habitat physiognomy (structure) in which fossil hominins have been recovered. Actual plant-species identification can only be done through palynological and paleobotanical studies when these types of remains are present. Habitat structure simply refers to the architecture of the floral species—for example, forest or bushland—rather than to the actual species. Within Africa the assumption is that the fundamental architecture of past and extant habitats is similar. Habitats from different continents may be inappropriate for comparison to fossil localities in Africa because vegetation structure can be labeled the same (e.g., forests) but exhibit significant differences (Archibold 1995). For this reason, habitats that are present in Africa today are probably the best analogs (but see Andrews et al. 1979; Andrews and Humphrey 1999; Mendoza et al. 2005). Mendoza et al. (2005) have shown that terrestrial ecosystems can be separated best if placed into three categories: arid habitats with no trees, humid evergreen forests, and wooded savannas. While this is undoubtedly true, almost all of the early hominin localities in Africa would likely fall into the wooded savanna category.

Extant African habitats range from primary rain forests to deserts. The amount of rainfall, temperature, sunlight, evapotranspiration, soil type, landscape, and weather patterns/seasonality are thus indicative of these habitats, and habitats can, in return, inform on these climatic conditions. In the tropical belt, however, the seasonal pattern and the amount of rainfall are the critical determining factors of the vegetation structure (Archibold 1995). Identifying ancient habitats will thus provide limited information on these aspects of the climate.

African forests consist of tall trees with multiple canopies (White 1983). Forests need mean annual rainfall of greater than 1500 mm and/or consistent groundwater, and long wet seasons if the moisture is derived from rain. These conditions create a closed architecture. Deserts usually have stunted trees, if any, and small, succulent plants and/or bushes. The mean annual rainfall is usually less than 200 mm. Deserts have extreme seasonality, that is, long periods without rainfall. The desert habitat is open.

Every other habitat in Africa today is savanna, covering 65 percent of the continent (Archibold 1995). This does not mean that all savanna habitats are the same or that they cannot be differentiated from one another: the nature of the physiognomy depends on the amount and seasonality of rainfall. Thus, when attempting to understand human (and broader mammalian) evolution, conceptually separating savannas from other terrestrial ecosystems is better for understanding the habitats in which hominins existed, and the community relationships within the environment. Many classifications of savannas have been made, but we follow White (1983) and utilize classifications that may be meaningful in reconstructing past habitats. Savannas characteristically have grasses as ground cover and other arid-adapted plants that can survive long dry seasons. Unlike on other continents, in Africa most of the woodland trees are deciduous (Archibold 1995). This means that leaf development will occur in the wet season and leaves will fall during the driest months of the year (Hopkins 1970). This contrasts with evergreen species in which leaf production occurs in the dry season. See Table 1.1 for subdivisions of savanna habitats.

The broad-based structural definitions of habitats are often the overall biomes of particular biogeographic regions, although there may be other habitats within them (White 1983), such as the Southern Savanna Grassland of South Africa (Rautenbach 1978). Various habitat structures often occur together in mosaic patterns within regions because of changes in soil types, subterranean water, and so on. River courses and lacustrine environments cause much of the mosaicism as they provide subterranean water that alters the general habitat close to the water. Thus it is possible to have riverine forests abutting almost desertlike habitats, such as along the Awash River that travels through extremely arid environs in Ethiopia. As most regions in Africa today possess a mosaic of habitats, it is reasonable to assume that ancient habitats were likely distributed in a similar manner. In eastern Africa, where the majority of hominin fossil localities occur within lacustrine and fluvial depositional environments, the vertical facies associations observed in the stratigraphy reflect ancient horizontal landscape associations (Miall 2000). That is, types of habitats across the landscape move horizontally through time due to common channel migration, change in fluvial regime (e.g., from meandering to braided), lake transgressions and regressions, and tectonic events.

ISSUES THAT CONFOUND THE RECONSTRUCTION OF HABITATS FOR FOSSIL LOCALITIES

The success of faunal analysis is related to the accurate estimation of taphonomic processes that contributed to the resultant fossil assemblages. If these

Table 1.1. Modern savanna habitat descriptions. While all of these are considered to be savannas as they have grass as a ground cover, finer descriptions better describe individual habitat structures.

Habitat Structure	*Mean Annual Rainfall (mm)*	*Description*	*Examples of Modern Localities*
Savanna	450–1100	Grass ground cover is ubiquitous; fires occasionally occur; main growth closely related to wet and dry seasons.[1]	
Closed Woodland	850–1100	Trees of between 8 and 20 m, crowns can contact but are not interlaced; less-developed grass cover; can have understory of bushes and shrubs.[2]	Guinea Woodland, Rwenzori National Park
Bushland	250–500	Bushes (multiple stems and ~3–7 m in height); cover at least 40% of ground surface; grasses secondary to bushes; can possess thickets of impassable bushes.[2]	Lake Mweru National Park, Rukwa Vally, Serengeti Bushland
Woodland	500–850	Trees (between 8 and 20 m in height); cover 30–40% of ground surface; some bushes, but these are often reduced by fire.[2]	Kapama Game Reserve, Sudan Woodland, Hluhluwe National Park, Kafue National Park
Wooded Grassland	450–500	Land mostly covered with grasses and occasional woody plants (10–40%), which may or may not include trees.[2]	Northern Senegal, parts of Kruger National Park
Shrubland	140–450	Shrubs of 10 cm to 2 m in height; plant structure caused by low rainfall, summer droughts.[2]	Kgalagadi Transfrontier Park; Modern Hadar
Scrub Woodland/ Transition Woodland	400–600	Transitional between woodland and bushland in which tree species are stunted due to poor soil, less rainfall, or both.[2]	Chobe National Park, Tarangire National Park, Amboseli National Park
Edaphic Grassland	600–800	Grasses associated with permanent or seasonally water-logged soils.[2]	Kafue Flats, Okavango Delta (grassland area)
Grassland	250–500	Grasses dominant with < 10% woody vegetation.[2]	Serengeti Plains, Southern Savanna Grassland
Ecotones	750–1200	Forests adjacent to grassland or heath.	Tongwe National Park, Aberdare National Park, Masai Mara

1. Bourliere and Haley (1983).
2. White (1983).

factors are not considered, the paleoecological reconstruction and subsequent evolutionary analyses derived from these comparisons may be inaccurate.

Taphonomy, strictly speaking, refers to the laws of burial (Efremov 1940). The term now usually refers to any alteration that may have occurred to a fossil at any time between the death of the animal and the fossil's placement in a museum. The taphonomic processes that have affected a fossil or a fossil assemblage may dictate that particular methods of analysis are inappropriate (Gifford 1981; Behrensmeyer 1991; Behrensmeyer et al. 2000). Therefore, it is important to discover the taphonomic information derived from both fossil fauna and other sources, such as the depositional environment, so that any incongruities resulting from these biases can be identified. Taphonomic processes that have influenced fossil localities can then be considered in the selection of methods to analyze the assemblage further (Behrensmeyer 1991). Taphonomy can be the focus of research and can therefore answer questions regarding modes of accumulation and pre- and postdepositional processes. However, we are more concerned in this chapter with briefly describing various confounding factors that may affect faunal analyses. When we explain methods of faunal analysis, we note which of these confounding aspects can be overcome as there are analytical methods that can minimize taphonomic overprint in faunal assemblages.

Time-averaging refers to the fact that most fossil deposits have accumulated over hundreds, if not thousands, of years. It is difficult to reconstruct a slice of time environment when the faunal accumulation used to predict the habitat is the result of some 10 kyr (thousand years) of deposition. We can never avoid this problem altogether, but analyses of particular species and their differences, if any, within deposits and also through time will help in this regard. That is, if species at the bottom of a section vary in dental dimensions from the same species at the top of the deposit, for example, then it is possible that what we have deemed as a single deposit is in fact more than one.

It is also necessary to determine if the fossil assemblage has been transported. An *autochthonous assemblage* is one in which no transport of specimens has occurred. *Allochthonous assemblages* refer to fossil deposits that have come from different habitats and yet appear to be a unified accumulation. These deposits would most likely be the result of fast-moving fluvial systems that wash many animals downstream during high-energy situations. These assemblages have a particular signature that can be interpreted before any ecological reconstruction is attempted. Lyman (1994) outlined taphonomic criteria with which to judge assemblages for identifying these biases, such as degree of abrasion and skeletal-part representation.

The *accumulating agent* refers to what or who was responsible for the fossil deposit—that is, animals dragging carcasses into caves, tar pits trapping animals, fluvial systems gathering up carcasses during flooding, hominin hunting or butchery practices, and so on. If one is interested in whether hominins were the hunters or the hunted (Brain 1975), then discerning the accumulating agent is the most important endeavor. Collection bias sometimes occurs when researchers are recovering fossils. It may be that there is no room in a museum for very large mammals, such as elephants, so they are left behind, or that the mode of collection (e.g., a walking survey) does not allow for the recovery of micromammals. Bone modification is the analysis of various alterations on the bone surface and can determine if the bones have been modified by carnivores, rodents, and/or hominins or if they have lain on the ground and weathered, or if they have been rolled or transported in fluvial settings (Behrensmeyer and Hill 1980; Behrensmeyer 1991; Lyman 1994).

Once the biases have been identified, researchers can select methods for reconstructing environments, identifying community structure, and examining species-turnover patterns that will minimize the effects of these biases. For example, if a fossil locality is depauperate in micromammals, it is likely that a comparison with extant faunal communities would be made only with macromammals. If hyenas have collected material of a certain size in the fossil record, then an extant database of animals should be created that takes this into consideration for subsequent comparisons with the fossil faunal assemblage. In other words, if one compares fossil communities with living communities to determine habitat, the comparison will work only if the same types of animals are being compared—that is, those of the same body size, such as micro-, mid-range, or macromammals, or those of the same mammalian orders, such as Artiodactyla or Primates (again of similar body sizes). Comparisons of recently deposited material—for example, assemblages acquired from hyena dens (Brain 1980) or fluvial flood remains, in which the originating habitat is known—will also minimize problems with reconstructing habitats when accumulating agents have modified the selection of fauna.

RECONSTRUCTING HABITATS

Taxonomic uniformitarianism

The taxonomic uniformitarian approach has been used most frequently in the reconstruction of environments (Dodd and Stanton 1990). Using this method, the ecology of a fossil species is reconstructed as similar to its closest living relatives. For example, a fossil bovid species is a member of a tribe that contains extant bovids, and the ranges of ecological parameters found in the living members of that tribe, such as tolerance to aridity, are attributed to the fossil species. While this taxonomic methodology can be accurate in reconstructing the paleobiology of some fossil taxa (Reed 1998), and therefore useful for subsequently reconstructing habitats, problems with this technique include an ecological bias that may result from the use of only one or two taxonomic groups (Cooke 1978), failure to consider morphological indicators of paleobiology in the fossil taxa (Spencer 1997), or failure to use the full range of extant behaviors for comparison of communities. Using taxonomic analogy, WoldeGabriel et al. (1994) suggested that the high abundances of an undescribed species of tragelaphin bovid and the numerous specimens of a colobine species at the *Ardipithecus* site of Aramis indicated the ancient habitat was wooded and closed. These suggestions are based on the fact that many extant tragelaphins, such as the bushbuck, *Tragelaphus scriptus*, and the bongo, *T. eurycerus*, are found in closed habitats, and most extant colobines are arboreal or at the least spend a great deal of time in trees (Fleagle 1999). However, extant tragelaphins also range into habitats of medium- and open-density woodlands and thus can be found in more arid environments (e.g., *T. imberbis*, lesser kudu, and *Taurotragus oryx*, eland). Several fossil colobine species were terrestrial and thus possibly lived in more open environments (Jablonski 2002; Frost and Delson 2002). While the habitat at Aramis may have been closed with many trees, taxonomic analogy does not definitively suggest this.

In an example of comparing species as if comparing actual habitats, Leakey et al. (2001) demonstrated that the Upper Lomekwi Member at West Turkana, from which the hominin *Kenyanthropus platyops* derives, differed in species composition from the Hadar site at a similar time period, and thus suggested that the Lomekwi habitat was more wet and closed than Hadar. The argument hinged on the fact that the extinct gelada baboon *Theropithecus darti* was recovered from Hadar, while *T. brumpti* was found at West Turkana. *Theropithecus brumpti* is frequently associated with more closed habitats (Krentz 1993). It is possible that the Upper Lomekwi Member is more closed and wet than Hadar, but an analysis of all of the fauna would provide more secure conclusions.

In the same vein as single-species taxonomic uniformitarianism, but using broader analogies, Vrba (1974) argued that members of the extant bovid tribes Alcelaphini and Antelopini (A & A) could be used to reconstruct habitats. Living members of these extant tribes are tolerant of arid conditions and are the majority of animals in open plains or grassland habitats. Vrba suggested that the presence of extinct members of these groups in similarly high proportions from Plio-Pleistocene fossil assemblages (either in high numbers of the same species or in high relative proportions of the tribe) would be indicative of similar habitats existing in the past.

Vrba (1980) and Greenacre and Vrba (1984) further used modern abundance data to calculate bovid tribal representation in order to determine a criterion for reconstructing habitats. In modern African game parks, when the percentage of antilopine and alcelaphin bovids contribute to more than 50 percent of the bovids on the landscape, inevitably these derive from an open woodland or grassland habitat. Percentages of bovids in these two tribes were then computed from the Plio-Pleistocene hominin sites of Sterkfontein, Swartkrans, and Kromdraai in South Africa. The results of the percentages of the A & A specimens suggested that Sterkfontein was the most closed of these localities with percentages of roughly 50 percent whereas the other sites had between 70 percent and 80 percent A & A bovids. Shipman and Harris (1988) extended this method to examine other bovid tribal representations at *Paranthropus* sites in East Africa. They suggested that high percentages of tragelaphin and aepycerotin bovids indicate closed, dry habitats, while high abundances of reduncin and bovin bovids signal closed, wet habitats. Their results suggested that robust australopithecines possibly preferred closed, wet habitats.

Taxonomic analogy is likely less effective the further back in time it is used because many artiodactyls, perissodactyls, and primates have undergone radiations in the last 2 myr (million years) (Vrba 1995). Nevertheless, as mentioned above this technique can be somewhat effective and is a baseline method for estimating the habitat of early hominins (WoldeGabriel et al. 1994; Leakey et al. 2001). Taphonomic considerations for all of the above examples would include understanding biases toward high abundances of any of the taxa (e.g., tragelaphins caught in a flood) or in the level of identification for collection of each species (e.g., collection of primate limb bones versus noncollection of bovid limb bones).

Taxon-free methodologies

Taxon-free methods depend on analyzing species or communities of fauna in such a way as to reconstruct ecological adaptations in individual species and ecological patterns in communities. The first method involves ecological or functional morphology in which individual species adaptations are analyzed from measurements of various functional systems and compared with modern taxa with similar adaptations (Kappelman 1988; Benefit and McCrossin 1990;

Plummer and Bishop 1994; Kappelman et al. 1997; Lewis 1997; Spencer 1997; Sponheimer et al. 1999; DeGusta and Vrba 2003, 2005a, 2005b). Habitats in which these species lived are thus derived from their adaptations. The second method is ecological-diversity analysis, in which distributions of the various adaptations in fossil communities are compared with extant communities of known habitats (Andrews et al. 1979; Reed 1997, 1998, 2008; Reed and Rector 2006; Rector and Reed 2010).

Ecological or functional morphology

Ecological morphology (also known as *ecomorphology*) links the fields of ecology and morphology (Wainwright and Reilly 1994). This discipline was formally defined in 1948 by Van der Klaauw as the study of the relationship between the morphology of an organism and its environments. Ecomorphological analyses operate under the assumption that the functional design of organisms can be related to their ecology (Wake 1992; Damuth 1992; Ricklefs and Miles 1994; Losos and Miles 1994).

A discussion of ecomorphology by Ricklefs and Miles (1994) articulated the limitations and advantages of this type of research. An important limitation is that morphology provides only a general indication of the possible range of behaviors available to an organism. This caution is especially relevant for bovids, as observations have documented a wide range of intraspecific variation in their diets despite similar morphology. Another limitation is that other aspects of an organism, such as physiology and behavior, are more responsive to short-term changes in the environment than is morphology. Therefore, it is important to combine morphological studies with analyses that are responsive to short-term changes such as microwear and isotope analysis of teeth. A final caveat is that morphologies can be difficult to compare between different classes of organisms: the morphology of a bird is not comparable with that of a mammal, for example.

In spite of these limitations, ecomorphology provides a number of advantages for addressing questions relevant to both ecologists and anatomists. Studies of ecomorphology may be used to address questions of convergence, evolution of function, community organization and evolution, and adaptive significance of morphological design (Losos and Miles 1994). The morphological characters used in an ecomorphological study are usually straightforward measurements that have high repeatability (Ricklefs and Miles 1994). Generally measurements are chosen to reflect biomechanical principles, which serve to strengthen the link between structure and function.

There is one major assumption on which ecomorphology is based: that the morphological phenotype provides information about the relationship between an organism and its environment (Wing and DiMichele 1992; Losos and Miles 1994; Ricklefs and Miles 1994). Though there have been criticisms of the automatic assumption that form is only related to function (Gould and Lewontin 1979), this premise is still prevalent in many studies of morphology. However, in the wake of Gould and Lewontin's criticism, this assumption has been used in a more rigorous and testable fashion (Wake 1992), and support for it has come from two different types of study: the correlation of ecological and morphological relationships, and the concordance of ecomorphological correlations between different species assemblages (Ricklefs and Miles 1994). It is widely accepted among biologists that a high degree of convergence between unrelated organisms indicates "a substantial role for natural selection in shaping or channeling functional attributes" (Wing and DiMichele 1992:140), and therefore the inference of function from form is justified.

The application of ecomorphology to paleontology is easily understood (Van Valkenburgh 1994). Most of paleontology is the study of morphology, because skeletal remains are often the only surviving evidence of extinct animals. The morphology of extinct animals is often used by paleontologists to address paleoecological questions about the relationships between particular morphologies and environments (Van Valkenburgh 1994).

Ecomorphological studies of extinct animals often consist of investigations of morphological adaptations within a single species or closely related groups of species. These studies are usually directed at reconstructing the lifeways (e.g., diet, locomotion, body size) of extinct animals, and can be used to address questions of convergence, morphological evolution through time, and the tempo and mode of evolution. Reconstructions at this level can be based on analogies with living relatives, but they are strengthened when based on physical laws that are equally applicable to past organisms as well as present, using the principles of uniformitarianism and analogy (Gould 1965; Janis 1994). For example, Sanson (1991) noted that inferring diet in extinct organisms is facilitated by the fact that convergence of the feeding apparatus is common across many animal groups, suggesting that there are constraints provided by the nature of the food type.

Faunal analysis at hominin sites has only relatively recently included the functional analysis of various taxa recovered with hominins to further understand their habitat and community ecology (Kappelman 1988; Benefit and McCrossin 1990; Plummer and Bishop 1994; Kappelman et al. 1997; Lewis 1997; Spencer 1997; Sponheimer et al. 1999; DeGusta and Vrba 2003, 2005a, 2005b). Functional morphology is useful for reconstructing the genetic potential of fossil taxa; that is, if the morphology of a particular bovid indicates that it was a mixed feeder (eating both leaves and grasses) then it had the *ability* to ingest both foods. Comparing these data with analyses of epigenetic data such as isotope analysis or micro- and mesowear of teeth can lead to better insights into actual and potential behavioral ecology.

The results of morphological analyses are often extended to infer the likely habitat in which the animals might have existed. For example, Benefit and McCrossin (1990) measured the molar shearing crests of both extant and extinct cercopithecine monkeys to examine trophic behavior. Using the resultant information, they argued that the proportion of foods eaten by each species is correlated with species habitat. They therefore suggested various habitats for hominid sites based on the presence of particular cercopithecines eating various percentages of fruits and leaves.

Spencer (1997) sought to determine morphological correlates in bovids to feeding in secondary grasslands to determine when this habitat became prevalent in Africa. Morphometric analyses of living bovids identified a number of cranial and mandibular traits that were correlated with diet (Figure 1.2). These results allowed reconstructions of diet in a variety of extinct bovids. Dietary reconstructions led to habitat reconstructions at a number of hominin sites. An important result from this study was the demonstration that diets and dietary morphology can differ within a tribe. An extinct reduncin, *Menelikia lyrocera,* did not resemble its close relatives. Also, one of the earliest members of the tribe Alcelaphini was reconstructed as a mixed feeder rather than a grass feeder (Spencer 1995). Finally, strong evidence for the presence of secondary grasslands does not appear until after 2 myr, coincident with the appearance of *Homo ergaster*. This result has implications for understanding the ecological transition from *Australopithecus* species to *Homo habilis sensu lato* and then to *Homo ergaster*.

Other analyses have examined postcranial elements in quadrupedal large mammals (Kappelman 1988, 1991; Kappelman et al. 1997; Plummer and Bishop 1994; Lewis 1997; DeGusta and Vrba 2003, 2005a, 2005b). Kappelman was the first to identify traits of the femur that could be related to locomotor behavior and thus to habitat preference. Plummer and Bishop (1994) extended this work to bovid metapodials, and showed that remains from Olduvai Bed I document a range of habitat types. Many of these studies used Discriminant Function Analysis (DFA) in order to assign fossil taxa to various categories of habitat cover. DFA is also useful for providing quantitative data regarding the ability of the method to classify the extant taxa correctly (DeGusta and Vrba 2003, 2005a, 2005b).

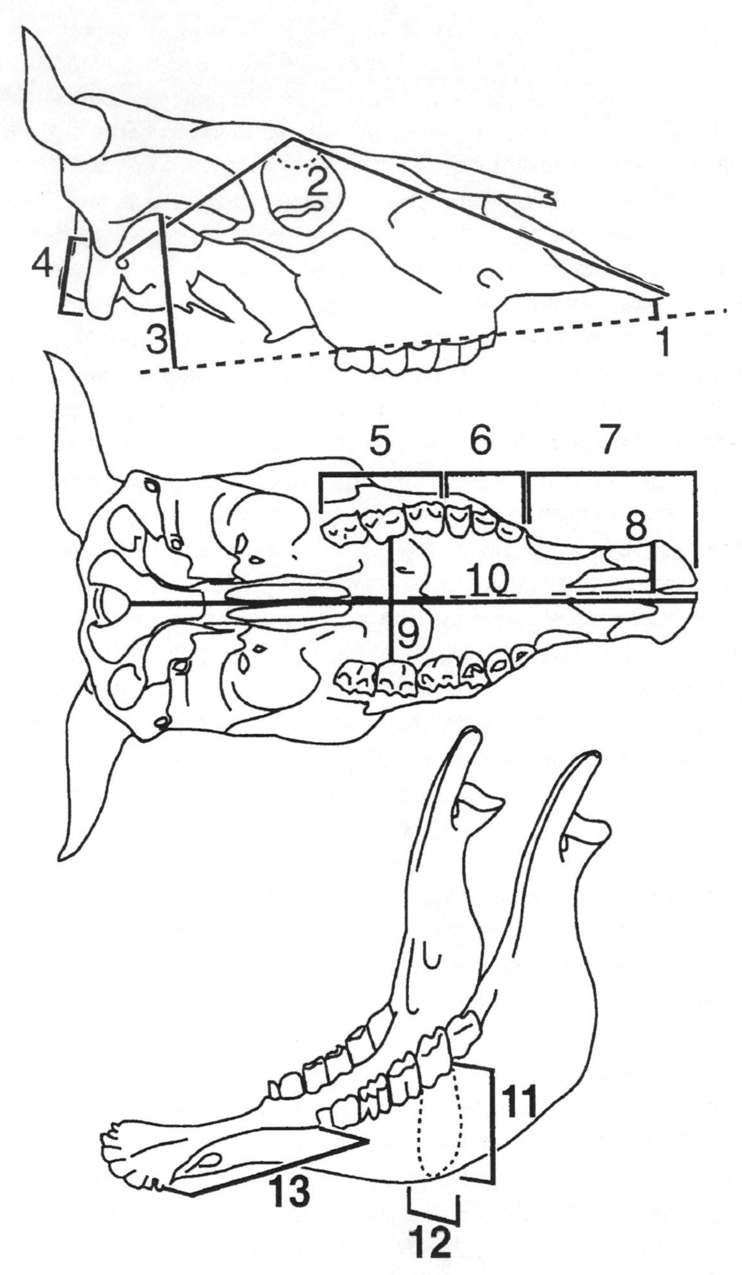

Figure 1.2. *Example of ecomorphological measurements made on bovid skulls to reconstruct dietary adaptations. Numbers indicate where measurements of the masticatory apparatus were taken. See Spencer (1997) for details. (After Spencer, 1995.)*

Some of the more interesting aspects of reconstructing the behavioral ecology of fossil taxa have been the results that show some taxa do not share a diet or substrate use with their extant relatives (e.g., Spencer 1997; Sponheimer et al. 1999; Frost and Delson 2002). Indeed, some taxa may have occupied a trophic niche for which there is no extant counterpart. These types of results enable the study of evolutionary patterns in particular lineages especially when compared with epigenetic data as available. They also show that taxonomic analogy may not be as reliable as one would like for reconstructing behavioral ecology in fossil taxa. Functional morphology and the use of the comparative method is an excellent means to infer the potential behavioral ecology of the organism being examined. However, an extension of this method to environmental reconstruction is possibly limited by the inclusion of only one taxonomic group, and, in the case of the cercopithecine trophic study mentioned above, the assumption that particular trophic resources represent definitive habitats. Taphonomic biases are limited when recreating the diet or locomotor behavior of particular species. However, reconstructing habitats by relying on the reconstructed behavior of a single species or a group of species merits review of collection and accumulating biases, differences in depositional environments, and so on.

Ecological structure or diversity analysis

This taxon-free faunal approach, usually used to reconstruct paleoenvironments, is concerned with the faunal community that existed with early hominins. Mammalian species exist cohesively in the various types of African habitats outlined above, and they partition resources such that the ecological adaptations exhibited by these mammals are somewhat predictable depending on the habitat structure (Andrews et al. 1979; Andrews 1989; Reed 1997, 1998; Mendoza et al. 2005). Each taxon is represented as an ecological entity and these behaviors are examined at the community level. For example, *Panthera pardus*, the leopard, is the ecological entity: "90 kg, terrestrial/arboreal, meat-eating animal." After all of the mammals from the community are assigned to various categories of trophic and substrate use, body-size categories, and so on, the numbers and/or percentages of each adaptation are calculated for the entire community, or in the case of fossil fauna, the relevant assemblage (i.e., single cave deposit, single stratigraphic level, or spatial location). Specific patterns of adaptations are equated with different African habitats. Of the many adaptations that mammals exhibit, six are significantly different among several types of habitats: aquatic, arboreal, and terrestrial locomotion; frugivory (combined with leaf or insect consumption); grazing; and fresh-grass grazing diets (Reed

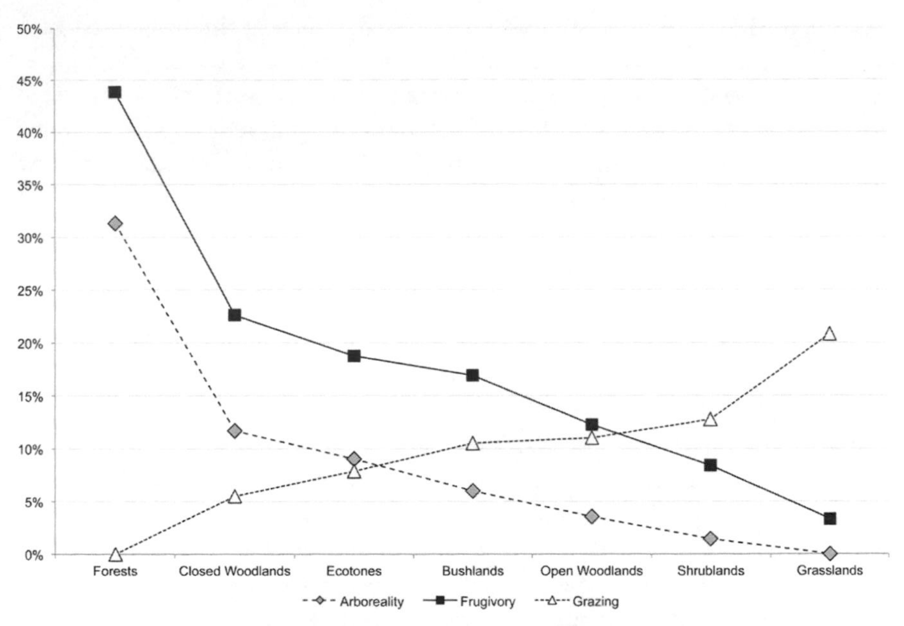

Figure 1.3. *Mean proportions of arboreal substrate use, frugivory, and grazing in groups of modern African habitats (forests through grasslands). Mammals included have a body mass greater than 500 grams. (Data from Reed, 1997)*

1998, 2008; Reed and Rector 2006; Rector and Reed 2010). The mammals that live together within a forest community, for example, exhibit higher proportions of arboreal substrate use than is found in any other habitat (Figures 1.3 and 1.4). Open grasslands have no arboreal animals but have high proportions of grazers. Wetland ecosystems also have high proportions of grazers and specialized fresh grass grazers—those mammals that focus on floodplain grasses. Sites that are near lakes and/or rivers have higher proportions of aquatic and fresh-grass grazing animals than sites that do not (Figure 1.5). Thus, the structure of the mammal communities as represented by trophic and substrate use is indicative of vegetation.

Analyzing fauna from a fossil assemblage using this approach requires several steps. First, fossils are assigned to categories using functional morphology if at all possible to discern dietary category, and in the case of carnivores, primates, and a few ungulates, to identify substrate use as well. Second, the fossil assemblage is contrasted with an extant comparative sample of communities from different habitats. Comparing fossil sites with extant communities has ranged from spectral analyses in which histograms of various adaptations have

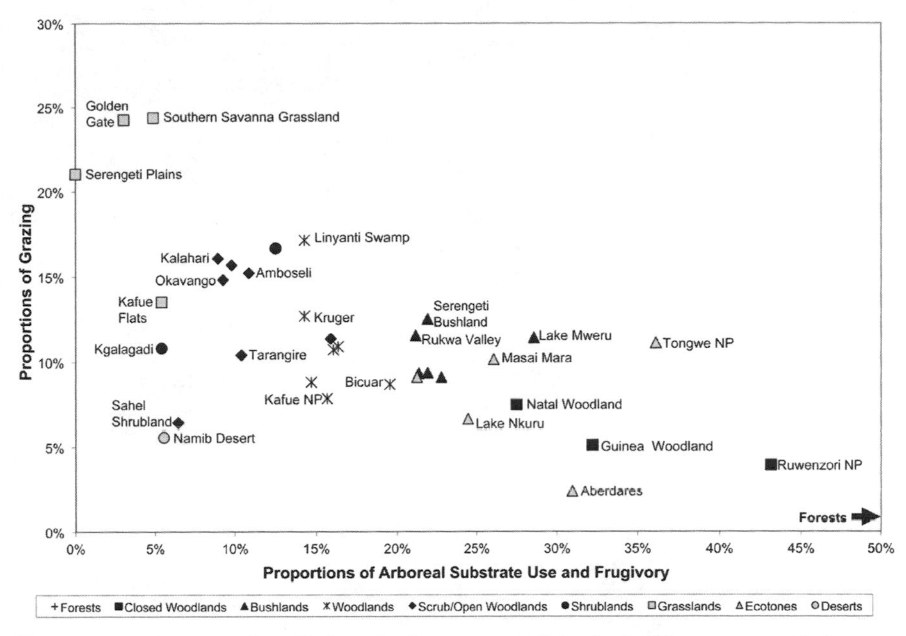

FIGURE 1.4. *Bivariate plot of arboreal substrate use and frugivory (percentages added together) vs. grazing mammals for modern sites of identifiable habitats. Wet, closed habitats are positioned toward the right, while more open, dry, and seasonal habitats are toward the left.*

particular shapes dependant on the habitat (Andrews et al. 1979; Andrews 1989; Gagnon 1997), bivariate plots of two adaptations (e.g., arboreality and frugivory; Reed 1997; Andrews and Humphrey 1999), to multivariate analyses, including principal components (PCA), discriminant function (DFA), and correspondence (CA) analyses of some or all adaptations in each community (Reed 1997, 1998, 2005, 2008; Sponheimer et al. 1999; Mendoza et al. 2005: Reed and Rector 2006; Rector and Reed 2010). Using ecological diversity analyses, habitats have been reconstructed for early hominin sites in eastern and southern Africa (Reed 1997).

RECONSTRUCTING COMMUNITY ECOLOGY

Guild structure

A *guild* is defined as "a group of species that exploit the same class of environmental resources in a similar way'" (Root 1967: 335). While being strictly defined in ecology, guilds and communities often represent the same group of species. For example, when discussing groups of primate species living in the same place,

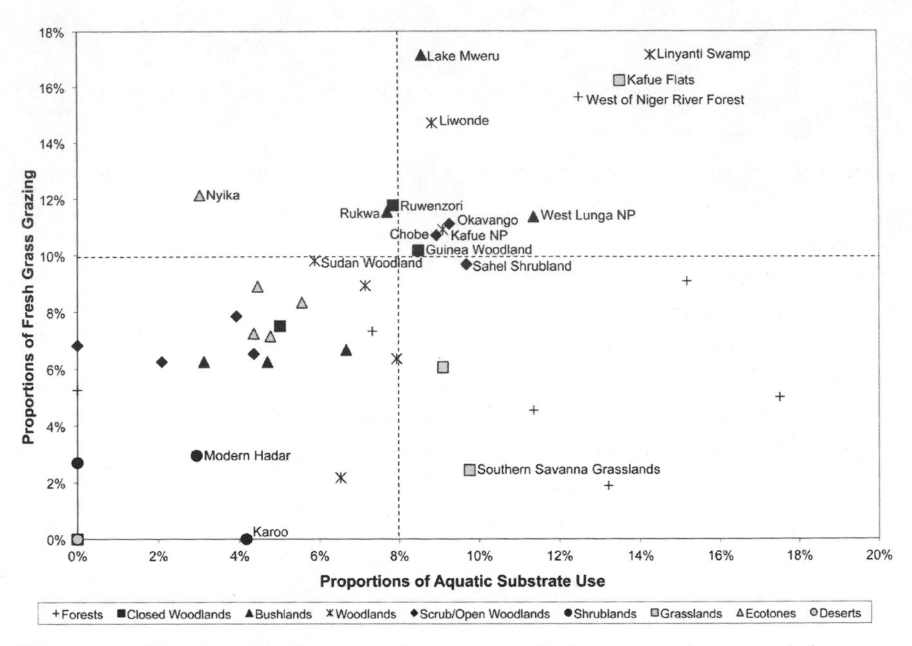

FIGURE 1.5. *Bivariate plot of aquatic substrate use vs. fresh-grass grazing animals for modern sites of identifiable habitats. Habitats in which there are wetlands, swamps, floodplains, and so on are located in the upper-right quadrant irrespective of overall habitat physiognomy and are labeled for reference.*

the term *primate community* is often used (Fleagle and Reed 1996). On the other hand, Lewis (1997) discussed *Homo* species as members of the carnivoran guild—that is, those mammals, irrespective of lineage, that consume meat. For our purposes, the concept of guild and community are defined by the focus of the researcher. For fossil assemblages, guilds usually refer to a subset of a community, unless community is preceded by a grouping adjective, such as *mammal* community or *carnivore* community. Fossil assemblages are usually considered to be samples of a living community or guild and as such need appropriate living analogs for comparisons. If a researcher is interested in a carnivoran guild, then extant samples are based on that concept.

Trophic, locomotor and body-size patterns in guilds and communities

All of the ecological information derived from the fossil assemblage is used to examine the ecological structure of the community. First, it is important to analyze the fossil assemblage taphonomically. Is there a particular body mass (trophic, substrate, etc.) class that is missing? Why is it missing? Are there both crania and postcrania available for study? Once the answers to these questions are determined, the second step is to develop an appropriate data set from modern communities so that the taphonomic biases are minimized (Soligo and Andrews 2005).

There are two primary questions to investigate regarding fossil communities. First, are the ecological patterns for a particular habitat (e.g., bushland) the same in the past? Second, if these patterns are not the same, what influences might have caused the differences and would those have affected hominins? The difference in the numbers of browsing mammals in modern and fossil communities provides an example. Browsers eat the leaves of dicot plants to the exclusion of other types of plants. Despite the fact that we might predict that there would be higher numbers of browsers in regions with more bushes and trees (forests, closed woodlands, and bushlands), modern browsers do not increase in numbers of species in these habitat types compared to others. We conclude that the distribution of browsing species across the modern African landscape is independent of habitat type. In fact, browsing is one of the few adaptations that is not significantly different between habitats (Reed 2008). Figure 1.6 illustrates an interesting phenomenon: in Pliocene fossil assemblages, browsers represent higher proportions of the faunal assemblages than is the case in extant habitats. This is also the case in the communities of the Miocene of North America (Janis et al. 2004). Is there something fundamentally different about past communities or habitats that allowed higher percentages of browsers? Janis et al. (2004) suggest higher plant productivity and/or higher levels of CO_2 in the atmosphere as possible reasons for the higher number of browsers in the Miocene. If this is the case in the African Pliocene, how would higher leafing productivity have affected early hominins? On the other hand, Soligo and Andrews (2005) suggested that taphonomic processes might inflate the numbers of large browsing species unless the correct modern comparative material is used in analyses. Further investigation into differences between modern and fossil communities will enable better understanding of the communities in which early hominins existed.

In another example, there are primate communities across the African continent today such that higher numbers of primates usually indicate more wet, forested habitats. These forest communities contain high numbers of cercopithecins and low numbers of colobines, papionins, and possibly great apes. In more open woodlands, there is usually just one papionin and cercopithecine species, with occasional colobines along river courses. In the more arid

Faunal resemblance coefficients

Analyses of this type do not provide a reconstruction of the habitat of a particular site but quantitatively show the similarity of one site to another. The use of similarity coefficients or faunal resemblance indices (FRIs) facilitates the comparison of the fauna between two sites by examining a resultant index that ranges from 0 (no similarity) to 1 (total similarity). Originally developed by Simpson, this type of index is often used to detect taxonomically distinct biogeographic areas (Flynn 1986). High similarity between assemblages has also been used to suggest that communities were environmentally analogous, such as one study of Miocene hominoid sites (Van Couvering and Van Couvering 1976). While this may be true with geographically penecontemporaneous sites, these comparisons may not be accurate when examining sites over time and space, because the indices may reflect differences in chronology or geography rather than ecology. In fact, Flynn (1986) argued that the Simpson Similarity Coefficient, in particular, was designed to minimize ecological differences and therefore it is erroneous to suggest environmental similarity using this type of index. Similarity coefficients can be used to examine mammalian communities of changing taxa over time (i.e., relative chronology; Flynn 1986) and across space (i.e., biogeographical differences).

Indices can be used solely between two sites giving a percentage of similarity, or indices can be calculated between many pairs of habitats and examined through cluster analyses and other multivariate analyses (Reed and Lockwood 2001). Taxonomic similarity measures among members (strata) of the same site will either reflect change in species composition or show that faunal assemblages are similar through time. Similar faunal groups in a sedimentary sequence may mean that there was little habitat change through time and mammals remained the same over long periods. As can be seen in Figure 1.7, the early part of the Hadar sequence from which *Australopithecus afarensis* has been recovered is fairly similar with the same species of animals uniting the Basal, Sidi Hakoma, Denen Dora, and basal Kada Hadar Members. The differences among these units could be the result of taphonomic or depositional differences during each particular period or indicate slight habitat changes back and forth through time. The Kada Hadar 2 Member (KH2), however, is in a fairly isolated cluster with a large jump in dissimilarity from the other members and represents a species turnover (Reed 2008). Similarly, the Makaamitalu (MAKA) area from which *Homo* has been recovered shows a major species turnover from the rest of the deposit. When sites are compared across geographic regions using similarity coefficients, possible contemporaneous localities can be identified, although

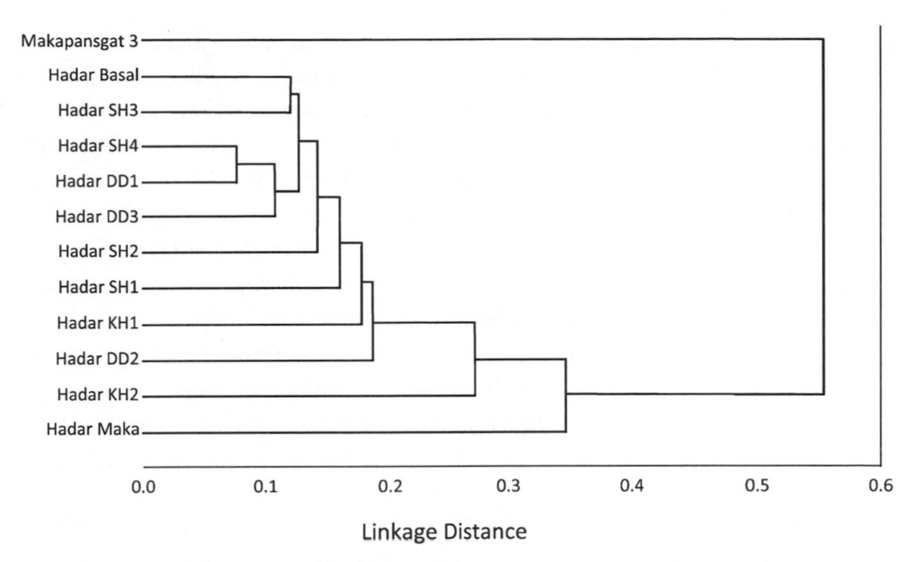

Figure 1.7. *Results of a cluster analysis of the submembers of Hadar (Ethiopia) and Makapansgat (South Africa), demonstrating faunal-turnover patterns.*

this can be complicated by biogeography. This can also be seen in Figure 1.7 as Makapansgat Member 3, despite sharing several taxa with the Hadar assemblages, is very distinct, likely indicating a different biogeographic realm and/or a possible difference in age.

The advantage to examining species turnover using more than one taxonomic group is that minor discrepancies that may result from using single higher taxa, such as Cercopithecoidea (e.g., Delson 1984), are minimized. Sites that are found to be chronologically similar through radiometric techniques and relative dating methods can be further compared both ecologically and biogeographically. Ecological reconstructions of communities that are based on other methods of faunal analysis can then be interpreted with reference to the chronology of fossil assemblages. This chronology, along with ecological and biogeographical analyses, will allow ecological patterns to be studied through time and across space.

Bobe and Eck (2001) used this methodology on the American collection of the Shungura Formation of Omo in Southern Ethiopia and discovered a rapid turnover in bovid taxa that occurred between 2.9 and 2.7 mya, which likely indicated a change from wet to drier environments. These researchers proposed that changes of relative abundances in various taxa across time are more likely to reflect environmental alterations than speciation and extinction events. It is important in light of this statement to have research that considers changes in species abundances as well as species representations across the landscape in order to understand their associations with depositional environments with respect to the principle that vertical facies associations observed in the stratigraphy reflect horizontal landscape associations (Miall 2000). Alemseged (2003) used abundance data from roughly the same region collected by the French paleontological team and found a species-abundance turnover at approximately 2.3 mya, possibly coinciding with the appearance of *P. boisei*. These studies used both the analysis of abundances of bovids over time as well as correspondence analyses to examine the relationships of various tribes of bovid taxa with extant and extinct habitats.

SPECIES INTERACTIONS BETWEEN HOMININS AND OTHER MAMMALS

As discussed above, an important facet of the paleoecology of early hominin species is elucidating the interaction between hominins and contemporary large mammals. This information is important for understanding the ecological context of hominin evolution. Unfortunately, this is one of the most difficult aspects of paleoecology for reconstructing the fossil record. Previous efforts in this arena have focused on the possible interactions between hominins and large carnivores. Archaeological studies have concentrated on hominin carcass acquisition and processing abilities, since these data are readily available in the fossil record. Therefore, it has been pointed out that it is necessary to understand how carnivores would have influenced hominin dietary strategies, both as competitors for prey and as providers of carcasses for scavenging (Blumenschine 1987; Marean 1989; Lewis 1997; Domínguez-Rodrigo 2001).

Studies of modern carnivore behaviors in the context of the East African savanna habitats have demonstrated riparian woodlands are the habitat in which carcasses are likely to survive the longest (Blumenschine 1987; Domínguez-Rodrigo 2001). Using these modern communities as an analog for the past, it has been suggested that early hominins would have found a niche in the carnivore guild that consisted of inhabiting riparian woodlands to exploit both food and the relative safety provided by trees. Efforts to strengthen this analogy by analysis of the paleobiology of the extinct carnivores, especially the sabertoothed felids, indicated that there was a niche available for exploitation of hominins that consisted of scavenging sabertooth kills in closed woodlands (Marean 1989). However, more detailed analyses of the entire carnivoran guild of the east African Plio-Pleistocene by Lewis (1997) suggested that there were fewer scavenging opportunities available to hominins than previously thought because of the large number of carnivores filling the ecomorphospace. Lewis indicated that scavenging opportunities would have been greater in East Africa relative to South Africa, because of the lack of a large bone-cracking carnivore in the East African carnivoran paleoguild.

The basic interactions of carnivores with the earliest hominins, however, are probably those of predator and prey. Brain (1981) refined the art of taphonomic analyses by showing that australopithecines were the hunted (or at least scavenged) rather than the hunters. Behrensmeyer (2008) noted that the AL–333 assemblage from Hadar was likely the result of carnivore attack on a group of early hominins. These studies do not focus on competition, but rather indicate the difficulty of early hominin survival in savanna mosaic habitats.

CONCLUSION

Faunal analyses that explore paleoecological patterns are critical to understanding hominin evolution. Faunal analyses are used as secondary indicators of habitat because it has been shown that mammalian adaptations correlate with various types of vegetation (i.e., habitats). Ecomorphological studies allow predictions of adaptations, based on taxonomic analyses, to be tested against modern comparative samples for better understanding of life in the past. Species-turnover patterns, combined with climatic information, assist in understanding how mammals of particular groups and early hominins may have been affected by climate-induced habitat change. No faunal analyses are complete without a consideration of taphonomy, because alterations to fossil assemblages are the norm. Without understanding taphonomic processes, any faunal analysis used to reconstruct environments or examine species turnovers is likely flawed. We hope that future faunal analyses will be able to build on the past and offer more refinements to reconstructing diets, habitats, and patterns of eurytopic and stenotopic species through time, and to provide valuable insight into hominin evolution.

REFERENCES

Alemseged, Z. 2003. "An Integrated Approach to Taphonomy and Faunal Change in the Shungura Formation (Ethiopia) and Its Implication for Hominid Evolution." *Journal of Human Evolution* 44 (4): 451–78. http://dx.doi.org/10.1016/S0047-2484(03)00012-5. Medline:12727463

Andrews, P. 1989. "Paleoecology of Laetoli." *Journal of Human Evolution* 18 (2): 173–81. http://dx.doi.org/10.1016/0047-2484(89)90071-7.

Andrews, P. 2006. "Taphonomic Effects of Faunal Impoverishment and Faunal Mixing." *Palaeogeography, Palaeoclimatology, Palaeoecology* 241 (3-4): 572–89. http://dx.doi.org/10.1016/j.palaeo.2006.04.012.

Andrews, P., and L. Humphrey. 1999. "African Miocene Environments and the Transition to Early Hominins." In *Paleoclimate and Evolution with Emphasis on Human Origins*, ed. E. S. Vrba, G. H. Denton, T. C. Partridge, and L. C. Burckle, 282–300. New Haven, CT: Yale University Press.

Andrews, P., J. M. Lord, and E. M. Nesbitt-Evans. 1979. "Patterns of Ecological Diversity in Fossil and Modern Mammalian Faunas." *Biological Journal of the Linnean Society* 11:177–205.

Archibold, O. W. 1995. *Ecology of World Vegetation*. New York: Chapman and Hall. http://dx.doi.org/10.1007/978-94-011-0009-0

Behrensmeyer, A. K. 1991. "Terrestrial Vertebrate Accumulations." In *Taphonomy: Releasing the Data Locked in the Fossil Record*, ed. P. A. Allison and D.E.G. Briggs, 291–335. New York: Plenum Publishers.

Behrensmeyer, A. K. 2008. "Paleoenvironmental Context of the Pliocene A.L. 333 "First Family" Hominin Locality, Hadar Formation, Ethiopia." *GSA Special Papers* 446:203–14.

Behrensmeyer, A. K., and A. P. Hill, eds. 1980. *Fossils in the Making: Vertebrate Taphonomy and Paleoecology*. Chicago: University of Chicago Press.

Behrensmeyer, A. K., S. M. Kidwell, and R. A. Gastaldo. 2000. "Taphonomy and Paleobiology." *Paleobiology* 26 (Suppl. 4): 103–47. http://dx.doi.org/10.1666/0094-8373(2000)26[103:TAP]2.0.CO;2.

Behrensmeyer, A. K., N. E. Todd, R. Potts, and G. E. McBrinn. 1997. "Late Pliocene Faunal Turnover in the Turkana Basin, Kenya and Ethiopia." *Science* 278 (5343): 1589–94. http://dx.doi.org/10.1126/science.278.5343.1589. Medline:9374451

Benefit, B. R., and M. L. McCrossin. 1990. "Diet, Species Diversity, and Distribution of African Fossil Baboons." *Kroeber Anthropological Society Papers* 71–72:79–93.

Blumenschine, R. J. 1987. "Characteristics of an Early Hominid Scavenging Niche." *Current Anthropology* 28 (4): 383–407. http://dx.doi.org/10.1086/203544.

Bobe, R., A. K. Behrensmeyer, and R. E. Chapman. 2002. "Faunal Change, Environmental Variability and Late Pliocene Hominin Evolution." *Journal of Human Evolution* 42 (4): 475–97. http://dx.doi.org/10.1006/jhev.2001.0535. Medline:11908957

Bobe, R., and G. G. Eck. 2001. "Responses of African Bovids to Pliocene Climatic Change." *Paleobiology* 27 (Suppl. 2): 1–48. http://dx.doi.org/10.1666/0094-8373(2001)027<0001:ROABTP>2.0.CO;2.

Bonnefille, R., R. Potts, F. Chalié, D. Jolly, and O. Peyron. 2004. "High-Resolution Vegetation and Climate Change Associated with Pliocene *Australopithecus afarensis*." *Proceedings of the National Academy of Sciences of the United States of America* 101 (33): 12125–9. http://dx.doi.org/10.1073/pnas.0401709101. Medline:15304655

Brain, C. K. 1967. "Procedures and Some Results in the Study of Quaternary Cave Fillings." In *Background to Evolution in Africa*, ed. W. W. Bishop and J. D. Clark, 285–301. Chicago: University of Chicago Press.

Brain, C. K. 1975. "An Interpretation of the Bone Accumulation from the Kromdraai Australopithecine Site, South Africa." In *Paleoanthropology, Morphology, and Paleoecology*, ed. R. H. Tuttle, 225–244. The Hague: Mouton. http://dx.doi.org/10.1515/9783110810691.225

Brain, C. K. 1980. "Some Criteria for the Recognition of Bone-Collecting Agencies in African Caves." In *Fossils in the Making: Vertebrate Taphonomy and Paleoecology*, ed. A. K. Behrensmeyer and A. P. Hill, 108–130. Chicago: University of Chicago Press.

Brain, C. K. 1981. *The Hunters or the Hunted? An Introduction to African Cave Taphonomy*. Chicago: University of Chicago Press.

Bourliere, F., and M. Haley. 1983. "Present Day Savannas: An Overview." In *Ecosystems of the World: Tropical Savannas*, ed. F. Bourliere, 1–17. New York: Elsevier.

Cooke, H.S.B. 1978. "Faunal Evidence for the Biotic Setting of Early African Hominids." In *Early Hominids of Africa*, ed. C. J. Jolly, 267–284. New York: St. Martin's Press.

Cousins, S. H. 1991. "Species Diversity Measurement: Choosing the Right Index." *Trends in Ecology & Evolution* 6 (6): 190–2. http://dx.doi.org/10.1016/0169-5347(91)90212-G. Medline:21232454

Cruz-Uribe, K. 1988. "The Use and Meaning of Species Diversity and Richness in Archaeological Faunas." *Journal of Archaeological Science* 15 (2): 179–96. http://dx.doi.org/10.1016/0305-4403(88)90006-4.

Damuth, J. D. 1992. "Taxon-Free Characterization of Animal Communities." In *Terrestrial Ecosystems through Time*, ed. A. K. Behrensmeyer, J. D. Damuth, W. A. DiMichele, R. Potts, H.-D. Sues, and S .L. Wing, 183–204. Chicago: University of Chicago Press.

DeGusta, D., and E. Vrba. 2003. "A Method for Inferring Paleohabitats from the Functional Morphology of Bovid Astragali." *Journal of Archaeological Science* 30 (8): 1009–22. http://dx.doi.org/10.1016/S0305-4403(02)00286-8.

DeGusta, D., and E. Vrba. 2005a. "Methods for Inferring Paleohabitats from Discrete Traits of the Bovid Postcranial Skeleton." *Journal of Archaeological Science* 32 (7): 1115–23. http://dx.doi.org/10.1016/j.jas.2005.02.011.

DeGusta, D., and E. Vrba. 2005b. "Methods for Inferring Paleohabitats from the Functional Morphology of Bovid Phalanges." *Journal of Archaeological Science* 32 (7): 1099–113. http://dx.doi.org/10.1016/j.jas.2005.02.010.

Delson, E. 1984. "Cercopithecid Biochronology of the African Plio-Pleistocene: Correlation among Eastern and Southern Hominid-Bearing Localities." *Courier Forschungsinstitut Senckenberg* 69:199–218.

deMenocal, P. B. 2004. "African Climate Change and Faunal Evolution during the Pliocene-Pleistocene." *Earth and Planetary Science Letters* 220 (1-2): 3–24. http://dx.doi.org/10.1016/S0012-821X(04)00003-2.

Dodd, J. R., and R. J. Stanton. 1990. *Paleoecology: Concepts and Applications*. New York: John Wiley and Sons.

Domínguez-Rodrigo, M. 2001. "A Study of Carnivore Competition in Riparian and Open Habitats of Modern Savannas and Its Implications for Hominid Behavioral Modelling." *Journal of Human Evolution* 40 (2): 77–98. http://dx.doi.org/10.1006/jhev.2000.0441. Medline:11161955

Efremov, J. A. 1940. "Taphonomy: A New Branch of Paleontology." *Pan American Geologist* 74:81–93.

Ewer, R. F. 1958. "The Fossil Suidae of Makapansgat." *Proceedings of the Zoological Society of London* 1303:329–72.

Feibel, C. S. 1999. "Basin Evolution, Sedimentary Dynamics and Hominid Habitats in East Africa: An Ecosystem Approach." In *African Biogeography, Climate Change, and Human Evolution*, ed. T. Bromage and F. Schrenk, 276–281. Oxford: Oxford University Press.

Fleagle, J. G. 1999. *Primate Adaptation and Evolution*. 2nd ed. New York: Academic Press.

Fleagle, J. G., and K. E. Reed. 1996. "Comparing Primate Communities: A Multivariate Approach." *Journal of Human Evolution* 30 (6): 489–510. http://dx.doi.org/10.1006/jhev.1996.0039.

Flynn, J. J. 1986. "Faunal Provinces and the Simpson Coefficient." *University of Wyoming Special Paper 3: Contributions to Geology*, 317–38.

Frost, S. R., and E. Delson. 2002. "Fossil Cercopithecidae from the Hadar Formation and Surrounding Areas of the Afar Depression, Ethiopia." *Journal of Human Evolution* 43 (5): 687–748. http://dx.doi.org/10.1006/jhev.2002.0603. Medline:12457855

Gagnon, M. Feb-Mar 1997. "Ecological Diversity and Community Ecology in the Fayum Sequence (Egypt)." *Journal of Human Evolution* 32 (2-3): 133–60. http://dx.doi.org/10.1006/jhev.1996.0107. Medline:9061555

Gifford, D. P. 1981. "Taphonomy and Paleoecology: A Critical Review of Archaeology's Sister Disciplines." In *Advances in Archeological Method and Theory*, vol. 4, ed. M. B. Schiffer, 365–438. New York: Academic Press.

Gould, S. J. 1965. "Is Uniformitarianism Necessary?" *American Journal of Science* 263 (3): 223–8. http://dx.doi.org/10.2475/ajs.263.3.223.

Gould, S. J., and R. C. Lewontin. 1979. "The Spandrels of San Marco and the Panglossian Paradigm: A Critique of the Adaptationist Programme." *Proceedings of the Royal Society of London. Series B. Biological Sciences* 205 (1161): 581–98. http://dx.doi.org/10.1098/rspb.1979.0086. Medline:42062

Greenacre, M., and E. Vrba. 1984. "Graphical Display and Interpretation of Antelope Census Data in African Wildlife Areas, Using Correspondance Analysis." *Ecology* 65 (3): 984–97. http://dx.doi.org/10.2307/1938070.

Hopkins, B. 1970. "Vegetation of the Olokemeji Forest Reserve, Nigeria, VII: The Plants on the Savanna Site with Special Reference to Their Seasonal Growth." *Journal of Ecology* 58 (3): 795–825. http://dx.doi.org/10.2307/2258535.

Jablonski, N. G. 2002. "Fossil Old World Monkeys: The Late Neogene Radiation." In *The Primate Fossil Record*, ed. W. C. Hartwig, 255–299. Cambridge: Cambridge University Press.

Janis, C. M. 1994. "The Sabertooth's Repeat Performances." *Natural History* 103: 78–83.

Janis, C. M., J. Damuth, and J. Theodor. 2004. "The Species Richness of Miocene Browsers, and Implications for Habitat Type and Primary Productivity in the North American Grassland Biome." *Palaeogeography, Palaeoclimatology, Palaeoecology* 207 (3-4): 371–98. http://dx.doi.org/10.1016/j.palaeo.2003.09.032.

Kappelman, J. 1988. "Morphology and Locomotor Adaptations of the Bovid Femur in Relation to Habitat." *Journal of Morphology* 198 (1): 119–30. http://dx.doi.org/10.1002/jmor.1051980111. Medline:3199446

Kappelman, J. 1991. "The Paleoenvironment of *Kenyapithecus* at Fort Ternan." *Journal of Human Evolution* 20 (2): 95–129. http://dx.doi.org/10.1016/0047-2484(91)90053-X.

Kappelman, J., T. Plummer, L. Bishop, A. Duncan, and S. Appleton. Feb-Mar 1997. "Bovids as Indicators of Plio-Pleistocene Paleoenvironments in East Africa." *Journal of Human Evolution* 32 (2-3): 229–56. http://dx.doi.org/10.1006/jhev.1996.0105. Medline:9061558

Krentz, H.B. 1993. "Postcranial Anatomy of Extant and Extinct Species of *Theropithecus*." In *Theropithecus: The Rise and Fall of a Primate Genus*, ed. N. G. Jablonski, 383–422. Cambridge: Cambridge University Press. http://dx.doi.org/10.1017/CBO9780511565540.015

Leakey, M. G., and J. Harris. 1987. *Laetoli*. Oxford: Clarendon Press.

Leakey, M. G., F. Spoor, F. H. Brown, P. N. Gathogo, C. Kiarie, L. N. Leakey, and I. McDougall. 2001. "New Hominin Genus from Eastern Africa Shows Diverse Middle Pliocene Lineages." *Nature* 410 (6827): 433–40. http://dx.doi.org/10.1038/35068500. Medline:11260704

Lewis, M. E. Feb-Mar 1997. "Carnivoran Paleoguilds of Africa: Implications for Hominid Food Procurement Strategies." *Journal of Human Evolution* 32 (2-3): 257–88. http://dx.doi.org/10.1006/jhev.1996.0103. Medline:9061559

Losos, J. B., and D. B. Miles. 1994. "Adaptation, Constraint, and Comparative Method: Phylogenetic Issues and Methods." In *Ecological Morphology: Integrative Organismal Biology*, ed. P. C. Wainwright and S. Reilly, 60–98. Chicago: University of Chicago Press.

Lyman, R. L. 1994. *Vertebrate Taphonomy*. Cambridge: Cambridge University Press.

Magurran, A. E. 1988. *Ecological Diversity and Its Measurement*. Princeton, NJ: Princeton University Press.

Marean, C. W. 1989. "Sabertooth Cats and Their Relevance for Early Hominid Diet and Evolution." *Journal of Human Evolution* 18 (6): 559–82. http://dx.doi.org/10.1016/0047-2484(89)90018-3.

Marean, C. W., N. Mudida, and K. E. Reed. 1994. "Holocene Paleoenvironmental Change in the Kenyan Central Rift as Indicated by Micromammals from Enkapune-Ya-Muto Rockshelter." *Quaternary Research* 41 (3): 376–89. http://dx.doi.org/10.1006/qres.1994.1042.

Mendoza, M., C. M. Janis, and P. Palmqvist. 2005. "Ecological Patterns in the Trophic-Size Structure of Large Mammal Communities: A 'Taxon-Free' Characterization." *Evolutionary Ecology Research* 7:1–26.

Miall, A. D. 2000. *Principles of Sedimentary Basin Analysis*. 3rd ed. New York: Springer-Academic Press.

Peet, R. K. 1974. "The Measurement of Species Diversity." *Annual Review of Ecology and Systematics* 5 (1): 285–307. http://dx.doi.org/10.1146/annurev.es.05.110174.001441.

Plummer, T. W., and L. Bishop. 1994. "Hominid Paleoecology at Olduvai Gorge, Tanzania, as Indicated by Antelope Remains." *Journal of Human Evolution* 27 (1-3): 47–75. http://dx.doi.org/10.1006/jhev.1994.1035.

Potts, R. 1998. "Environmental Hypotheses of Hominin Evolution." *Yearbook of Physical Anthropology* 41 (Suppl. 27): 93–136. http://dx.doi.org/10.1002/(SICI)1096-8644(1998)107:27+<93::AID-AJPA5>3.0.CO;2-X. Medline:9881524

Rautenbach, I. L. 1978. "A Numerical Re-Appraisal of Southern African Biotic Zones." *Bulletin of Carnegie Museum of Natural History* 6:175–87.

Rector, A. L., and K. E. Reed. Sep-Oct 2010. "Middle and Late Pleistocene Faunas of Pinnacle Point and Their Paleoecological Implications." *Journal of Human Evolution* 59 (3-4): 340–57. http://dx.doi.org/10.1016/j.jhevol.2010.07.002. Medline:20934090

Reed, K. E. Feb-Mar 1997. "Early Hominid Evolution and Ecological Change through the African Plio-Pleistocene." *Journal of Human Evolution* 32 (2-3): 289–322. http://dx.doi.org/10.1006/jhev.1996.0106. Medline:9061560

Reed, K. E. 1998. "Using Large Mammal Communities to Examine Ecological and Taxonomic Organization and Predict Vegetation in Extant and Extinct Assemblages." *Paleobiology* 24:384–408.

Reed, K. E. 2002. "The Use of Paleocommunity and Taphonomic Studies in Reconstructing Primate Behavior." In *Reconstructing Primate Fossil Behavior in the Fossil Record*, ed. M. Plavcan, R. Kay, C. van Schaik, and W. L. Jungers, 217–259. New York: Kluwer Academic/Plenum Press.

Reed, K. E. 2005. "African Plio-Pleistocene Mammal Communities: Do Unique Compositions Indicate Distinct Vegetation?" *Journal of Vertebrate Paleontology* Meetings Suppl.

Reed, K. E. 2008. "Paleoecological Patterns at the Hadar hominin Site, Afar Regional State, Ethiopia." *Journal of Human Evolution* 54 (6): 743–68. http://dx.doi.org/10.1016/j.jhevol.2007.08.013. Medline:18191177

Reed, K. E., and C. A. Lockwood. 2001. "Identifying Patterns of Migration and Endemism in African Mammal Localities." *American Journal of Physical Anthropology* 32 (*Suppl.*): 123–4.

Reed, K. E., and A. L. Rector. 2006. "African Pliocene Paleoecology: Hominin Habitats, Resources, and Diets." In *Evolution of the Human Diet: The Known, the Unknown, and the Unknowable*, ed. P. S. Ungar, 262–288. Oxford: Oxford University Press.

Ricklefs, R. E., and D. B. Miles. 1994. *Ecological and Evolutionary Inferences from Morphology: An Ecological Perspective*. Chicago: University of Chicago Press.

Root, R. B. 1967. "The Niche Exploitation Pattern of the Blue-Grey Gnatcatcher." *Ecological Monographs* 37 (4): 317–50. http://dx.doi.org/10.2307/1942327.

Sanson, G. D. 1991. "Predicting the Diet of Fossil Mammals." In *Vertebrate Palaeontology of Australasia*, ed. P. Vickers-Rich, J. M. Monaghan, R. F. Baird, and T. H. Rich, 201–228. Lilydale: Pioneer Design Studio.

Shipman, P., and J. M. Harris. 1988. "Habitat Preference and Paleoecology of *Australopithecus boisei* in Eastern Africa." In *Evolutionary History of the "Robust" Australopithecines*, ed. F. E. Grine, 343–384. New York: Aldine de Gruyter.

Soligo, C., and P. J. Andrews. 2005. "Taphonomic Bias, Taxonomic Bias and Historical Non-Equivalence of Faunal Structure in Early Hominin Localities." *Journal of*

Human Evolution 49 (2): 206–29. http://dx.doi.org/10.1016/j.jhevol.2005.03.006. Medline:15975630
Spencer, L. M. 1995. Antelopes and Grasslands: Reconstructing African Hominid Environments. PhD dissertation, SUNY Stony Brook.
Spencer, L. M. Feb-Mar 1997. "Dietary Adaptations of Plio-Pleistocene Bovidae: Implications for Hominid Habitat Use." *Journal of Human Evolution* 32 (2-3): 201–28. http://dx.doi.org/10.1006/jhev.1996.0102. Medline:9061557
Sponheimer, M., K. E. Reed, and J. A. Lee-Thorp. 1999. "Combining Isotopic and Ecomorphological Data to Refine Bovid Paleodietary Reconstruction: A Case Study from the Makapansgat Limeworks Hominin Locality." *Journal of Human Evolution* 36 (6): 705–18. http://dx.doi.org/10.1006/jhev.1999.0300. Medline:10330334
Van Couvering, J.A.H., and J. A. Van Couvering. 1976. "Early Miocene Mammal Fossils from East Africa: Aspects of Geology, Faunistics, and Paleoecology." In *Human Origins: Louis Leakey and the East African Evidence*, ed. G. L. Isaac and E. R. McCown, 155–207. Menlo Park, CA: Staples Press.
Van der Klaauw, C. J. 1948. "Ecological Studies and Reviews, IV: Ecological Morphology." *Biobliotheca Biotheoretica* 4:27–111.
Van Valkenburgh, B. 1994. "Ecomorphological Analysis of Fossil Vertebrates and Their Paleocommunities." In *Ecological Morphology: Integrative Organismal Biology*, ed. P. C. Wainwright and S. Reilly, 140–168. Chicago: University of Chicago Press.
Vrba, E. S. 1974. "Chronological and Ecological Implications of the Fossil Bovidae at the Sterkfontein Australopithecine Site." *Nature* 250 (5461): 19–23. http://dx.doi.org/10.1038/250019a0.
Vrba, E. S. 1980. "The Significance of Bovid Remains as Indicators of Environment and Prediction Patterns." In *Fossils in the Making: Vertebrate Taphonomy and Paleoecology*, ed. A. K. Behrensmeyer and A. P. Hill, 247–271. Chicago: University of Chicago Press.
Vrba, E. S. 1988. "Late Pliocene Climatic Events and Hominid Evolution." In *Evolutionary History of the "Robust" Australopithecines*, ed. F. E. Grine, 405–426. New York: Aldine de Gruyter.
Vrba, E. S. 1995. "The Fossil Record of African Antelopes (Mammalia, Bovidae) in Relation to Human Evolution and Paleoclimate." In *Paleoclimate and Evolution with Emphasis on Human Origins*, ed. E. S. Vrba, G. H. Denton, T. C. Partridge, and L. C. Burckle, 385–424. New Haven, CT: Yale University Press.
Wainwright, P. C., and S. M. Reilly, eds. 1994. *Ecological Morphology: Integrative Organismal Biology*. Chicago: University of Chicago Press.
Wake, M. H. 1992. "Morphology, the Study of Form and Function in Modern Evolutionary Biology." In *Oxford Surveys in Evolutionary Biology*, vol. 8, ed. D. Futuyma and J. Antonovics, 289–340. New York: Oxford University Press.
Wing, S. L., and W. A. DiMichele. 1992. "Ecological Characterization of Fossil Plants." In *Terrestrial Ecosystems through Time*, ed. A. K. Behrensmeyer, J. D. Damuth, W. A. DiMichele, R. Potts, H.-D. Sues, and S. L. Wing, 139–180. Chicago: University of Chicago Press.

White, F. 1983. *The Vegetation of Africa: A Descriptive Memoir to Accompany UNESCO/AETFAT/UNSO Vegetation Maps of Africa*. Paris, France: UNESCO.
WoldeGabriel, G., T. D. White, G. Suwa, P. Renne, J. de Heinzelin, W. K. Hart, and G. Heiken. 1994. "Ecological and Temporal Placement of Early Pliocene Hominids at Aramis, Ethiopia." *Nature* 371 (6495): 330–3. http://dx.doi.org/10.1038/371330a0. Medline:8090201

2

Facies Analysis and Plio-Pleistocene Paleoecology

CRAIG S. FEIBEL

SEDIMENTARY ROCKS ARE THE RESULT OF chains of complex genetic processes, including particle formation, transport, and accumulation, along with subsequent histories of in situ modification following deposition. Some processes may produce a variety of sedimentary products with only slight variation in the available components or environmental conditions. And some sedimentary products may result from one of a variety of processes or conditions. Typically, however, the suite of characteristics recognizable in sedimentary rocks provides detailed and often unambiguous clues to the processes and environmental conditions during their formation. These suites of characteristics are what enable us to recognize sedimentary facies. Analysis of associations of facies, along with their vertical and lateral relationships, allows the interpretation of sedimentary environments and the reconstruction of ancient landscapes. The ability to recreate ancient landscapes and the sequence of changes in environmental character through time is a valuable tool in ecological reconstruction from ancient records. Placing fossil or archaeological assemblages into this framework can greatly enhance our understanding of the conditions of site formation, the processes active before, during, and after assemblage, and the broader environmental setting of a particular locality.

A *facies* refers to "the sum of lithologic and paleontologic characteristics of a sedimentary rock from which its origin and the environment of its formation may be

DOI: 10.5876/9781607322252:c02

inferred" (Teichert 1958). The concept is an invaluable tool for the description and interpretation of sedimentary rocks, as it allows the easy integration of lithological characteristics, primary and secondary sedimentary structures, and pedogenic (soil) overprints. A facies is a distinctive rock type, recognizable on the basis of one or more defining characteristics, and it carries with it an interpretation of the processes known to result in that particular assemblage of characteristics. From a practical point of view we can consider facies strictly in terms of rock characters, or we can interpret them in terms of the related environmental conditions. In application, it is often more convenient to discuss the environmental interpretations that result from facies recognition, but it is essential to bear in mind that the interpretation rests upon recognition of characteristics, and that in some cases multiple processes can result in similar or identical characters. The basic units characterized as facies have fundamental significance in terms of environmental dynamics, and when considered sequentially within sedimentary packages, reflect discrete components of ancient landscapes. Using Walther's Law, that sequential facies packages reflect lateral relationships upon the landscape, the broader character of ancient landscapes can be reconstructed in considerable detail (Reading 1978).

SEDIMENTARY ROCK CHARACTERISTICS

The formation of sedimentary rocks results from the assemblage of components in a sedimentary basin or trap, or on an available surface. The components of sedimentary rocks are typically distinguished in three groups: particles (clasts or grains and their coatings), voids (empty space), and plasma (fluids filling void space or coating particles). The particles themselves are of primary interest, but in certain cases the voids (determining porosity and permeability) and plasma (often mediating cementation and coherence) can be significant. A common sedimentary analysis involves *granulometry*, the investigation of particle size (*texture*), shape (*sphericity*), and population distribution (*sorting*). Perhaps more significant, however, is the rock *fabric*, the arrangement or organization of those particles. The sorting, packing, and arrangement of particles, along with the accumulation of secondary cements, strongly controls rock properties and is highly reflective of processes involved in the history of that rock. It is also important to consider that some sedimentary particles are what may be termed *superparticles*, formed by the amalgamation of individual particulate components. This is a common process in soil development and in the behavior of very small particles (clays) or biogenic materials.

The genesis of individual sedimentary particles may proceed along a variety of pathways, and is often as complex as the rocks they assemble to form. The major groups of sedimentary particles are *detrital*, comprising weathered or fragmented precursor rocks; *chemical*, precipitated directly from a solution; *biogenic*, produced by biological activity; and *volcaniclastic*, resulting from the explosive fragmentation in volcanic settings. There are some special cases and gray areas in sedimentary particle characterization as well. Most detrital sediment is the product of weathering of silicate rocks, and is commonly termed *siliciclastic*. But carbonate rocks may also be subjected to weathering and erosion, producing detritus that is not siliciclastic in nature. The boundary between biological and chemical processes is not always distinct, and biochemical particles result. Biological processes may also produce components of naturally occurring rocks that are not mineral based, such as coals and peats, which are grouped as organic sediments. The significance of characterization at this level is that the individual components of a sedimentary rock often present major clues to the processes involved in rock formation. These primary lithologic characterizations will also be a central key in recognizing and naming sedimentary facies.

In Earth surface environments, sedimentary particles are available for removal (*erosion*), transport, and deposition. These processes may be dominated by simple influence of gravity (e.g., a rockfall), but most commonly they are mediated by fluid flow. In most sedimentary environments, water is the predominant fluid, but air can also be significant. Ice may also be important in certain settings. Two aspects of fluid flow are central to the processes by which sediments are eroded, transported, and deposited. These are the intensity of flow and its duration. At the deposition stage, these may impart to the sediment a suite of *primary depositional characteristics*, usually based on the way in which fluid dynamics interacts with granulometry to produce sediment fabric. Flow intensity is directly related to the size of particles that may be eroded and entrained, thus there is a good relationship between particle size and flow strength. Duration of flow determines how long the process has to interact with particles, and thus the degree to which it can affect certain properties. Sorting is the prime example, as a short duration flow has limited potential to separate particles based on size, shape, and density, and usually results in a poorly sorted deposit. A long-duration flow or repetitive-flow processes (e.g., oscillating waves) have greater capability to sort sediment into more uniform populations of particles. Without significant sorting, heterogeneous deposits may result with a *massive* or featureless fabric. Sorting may result in a wide variety of primary depositional structures such as graded bedding or cross-bedding. These more complex arrangements of particles are commonly informative as to flow characteristics and *bedforms*, the three-dimensional components of a depositional landscape.

Although most of the characteristics of a sedimentary rock are determined through the processes of particle genesis and deposition, postdepositional processes may impart significant additional characteristics to a sediment body, or overprint primary features. These processes are broadly classed as *diagenetic* or transformational processes. For sedimentary rocks deposited in terrestrial environments, an important stage in diagenesis is the group of early postdepositional processes, most significantly *pedogenesis* or soil formation. Postdepositional transformative processes may impart new characters to a sediment body, may transform components to different phases, or may alter the fabric of the sediment. Again, many of the processes involved at this stage have definitive characteristics through which process may be inferred.

The net result of these formative processes, both genetic and diagenetic, is a rock body (a *lithosome*) with a suite of characteristics. Some dominant characters are very common, such as being composed of sand-sized particles, and are thus not particularly definitive to process and environment of formation. A combination of characteristics, however, often limits the range of processes and environmental settings, and these are what generally define a facies. Thus a trough cross-bedded sand represents deposition by a migrating series of dunes. That characterization alone may not be sufficient to distinguish the environmental setting between an aeolian dune field, a river channel, or a channelized flow on the seafloor. The association of facies, however, is nearly always sufficient to place the individual facies within its broader sedimentary context.

Two additional points in approaching facies analysis must be emphasized here: the significance of scale and the importance of surfaces. A central aspect of facies characterization and understanding relates to the scale of observation and analysis. Some features of sedimentary rocks can be observed at a fine scale, and indeed some facies analysis is done at the microscopic level. The level of observation, however, must be scaled to the magnitude of the features and the processes involved. Large-scale trough cross-beds, produced by migrating dunes, are commonly tens of centimeters in thickness and meters across. Thus they are features observable primarily at the outcrop scale and could not be recognized through a microscope. Most facies studies, and particularly the approaches highlighted here, depend heavily upon outcrop-scale analysis. A second consideration is the impact of surfaces, the breaks between sedimentary units. Since facies analysis revolves around the tangible components of the sedimentary record, the rocks themselves, it is easy to miss the importance of the surfaces that bound individual sediment bodies. In a temporal sense, these surfaces typically reflect more time than the actual accumulation of the rock record. In terms of process, the recognition of characteristic surfaces can provide crucial additional data for analysis. Erosional surfaces, easily recognized by scalloped scours and downcutting, reflect transitions from an accumulational regime to an erosional phase. Soil surfaces reflect often long periods of time when accumulation is minimal, but pedogenesis of subjacent materials may record important environmental variables. Every surface in the sedimentary record reflects a shift or transition in ambient process, and thus a major portion of the sedimentary history of a sequence is written in these intangible boundaries.

SEDIMENTARY FACIES ANALYSIS

The characterization of sedimentary rocks by facies has a long history, dating back to introduction of the term by Gressley in 1838 (Teichert 1958). Methodologies and particularly the terminology used to undertake this type of analysis have changed considerably through time. A minor revolution in facies analysis took place in the late 1970s and early 1980s, and thus two superficially different approaches may be encountered in literature from the not-too-distant past.

Early analyses grouped "facies" as genetically related packages of sediment, with numerous characteristic components as dominant or minor ingredients in the mix. Thus a package composed of lenticular sediment bodies, in an upward-fining sequence of gravel and sand, with a progression of primary sedimentary structures from trough cross-bedded sands, through planar-bedded sands, to ripple-marked sands would be classed as a facies, and interpreted to represent a meandering river channel or point-bar sequence. In this approach, the individual facies were often named in a descriptive fashion, such as the "lenticular conglomerate and sandstone facies." This approach successfully recognized packages of sediment, with characteristic components in a common but not invariable progression, which reflected a suite of processes in a particular depositional environment, such as a meandering river channel. Continued development of this approach, however, focused attention on the range of variability in the ways in which such packages could be compiled, and led to the recognition that the best fundamental units of analysis were the individual components of these packages. The "revolution" of the 1980s thus moved the appellation *facies* to the fundamental rock unit (in better keeping with its original intent) and viewed the larger packages as *facies associations*. This is the approach adopted by most sedimentologists today, and it is used in the examples set forth here.

A typical facies classification scheme groups sedimentary rocks by primary lithological characteristics, usually particle genesis and/or grain size, and further distinguishes individual facies with primary or secondary features. Thus a study may involve a series of facies comprised of gravels, and distinguished by bedding characteristics, or a series of biogenic deposits, further distinguished by fossil content. Recurring packages of these facies are recognized as facies associations, typically characterized by the dominant facies and their common progression, along with minor facies that may or may not fill out the association. In this approach, the individual facies reflect discrete combinations of components and processes. The associations reflect how these facies come together within natural depositional systems (or in postdepositional landscapes). The degree of potential variability in component facies and their sequence within a facies association is a direct reflection of the real complexity of Earth surface systems.

A number of excellent volumes focus on methodology and examples of this facies approach (Walker 1979; Walker and James 1992; Posamentier and Walker 2006). It is also important to note that much of our understanding of sedimentary facies is based on an interplay between the rock record and modern landscapes. Thus many investigations link ancient facies characters directly to modern analogs (e.g., Miall 1978). To illustrate the facies classification approach, and to explore its utility in ecological reconstructions, an example from the Koobi Fora Formation in northern Kenya is presented here.

FACIES CLASSIFICATION IN THE KOOBI FORA FORMATION

Geologists investigating sedimentary strata of the Koobi Fora region in northern Kenya developed a facies scheme to characterize the deposits and interpret depositional environments beginning in the 1970s (Bowen 1974; Burggraf and Vondra 1982). The initial facies characterization (using the early descriptive approach) recognized four major depositional environments in this nonmarine setting: alluvial fan, fluvial, deltaic, and lacustrine. Continued study and an increased focus on details in localized portions of the sequence resulted in a further elaboration and subdivision of the scheme, recognizing variants on the fluvial and lacustrine facies (Burggraf et al. 1981; White et al. 1981; summarized in Feibel 1983). This development reflected both the natural evolution of thought with more focused study as well as trends in sedimentological philosophy at the time. A comprehensive synthesis of this facies scheme as a model for rift-valley sedimentation was presented by Burggraf and Vondra (1982).

The mid-1980s brought an extensive reevaluation of Koobi Fora stratigraphy (Brown and Feibel 1986) as well as the application of refined facies approaches to the Koobi Fora deposits. Much of the following analysis was first set out in an early report (Feibel and Brown 1986) and subsequently refined and expanded

(Feibel 1988; Feibel et al. 1991). Gathogo (2003; Gathogo and Brown 2006) adopted this scheme in his study of the Ileret region.

The Koobi Fora Formation (Figure 2.1) attains a composite thickness of some 560 meters (Brown and Feibel 1986), exposed discontinuously over about 1200 square kilometers to the northeast of Lake Turkana. These strata reflect deposition in a variety of fluvial, lacustrine and deltaic environments within a series of half-grabens over the interval from circa 4.2 to 0.6 mya (million years ago) (Feibel 1994). Within this setting, the strata express considerable lateral variation reflecting a complex ancient landscape, as well as significant shifts through time. The latter reflect both the dynamics of an evolving landscape as well as extrinsic factors such as climate change, which is an important control on sedimentary dynamics. The Koobi Fora region provides an excellent laboratory in which to explore facies relationships and their significance, because the extensive area of outcrop coupled with superb temporal and stratigraphic control (Brown and Feibel 1991; McDougall and Brown 2006; Brown et al. 2006) allows for an investigation of both temporal and spatial variability over a dynamic landscape.

Facies at Koobi Fora

In the following treatment, individual facies are delineated following the pattern set out by Miall (1977). Dominant lithology along with primary and secondary structures, biogenic character, and bounding surfaces were the primary criteria in identifying the facies. The initial facies characterization developed in this way (Feibel and Brown 1986; Feibel 1988) has required only minor modifications to accommodate continued analysis of the deposits at Koobi Fora, as well as other strata of Miocene to Recent age in the Turkana Basin (e.g., Gathogo 2003; Gathogo and Brown 2006). The only significant extension of the scheme relates to the expanded recognition of characteristic fossil soils, or *pedotypes* (Wynn and Feibel 1995; Wynn 1998), that has further elaborated environments of postdepositional modification.

In this analysis, twenty-one sedimentary facies are described to characterize the Plio-Pleistocene deposits of Koobi Fora (Table 2.1). The major lithological components of significance there are gravels (G), sands (S), fine clastics (muds, M), bioclastic sediments (B), and pedogenically modified sediments (P). Because the relationship between process and product is central both to recognizing sedimentary facies and to their interpretation, a brief discussion of the individual facies and their characteristics and interpretation is presented here.

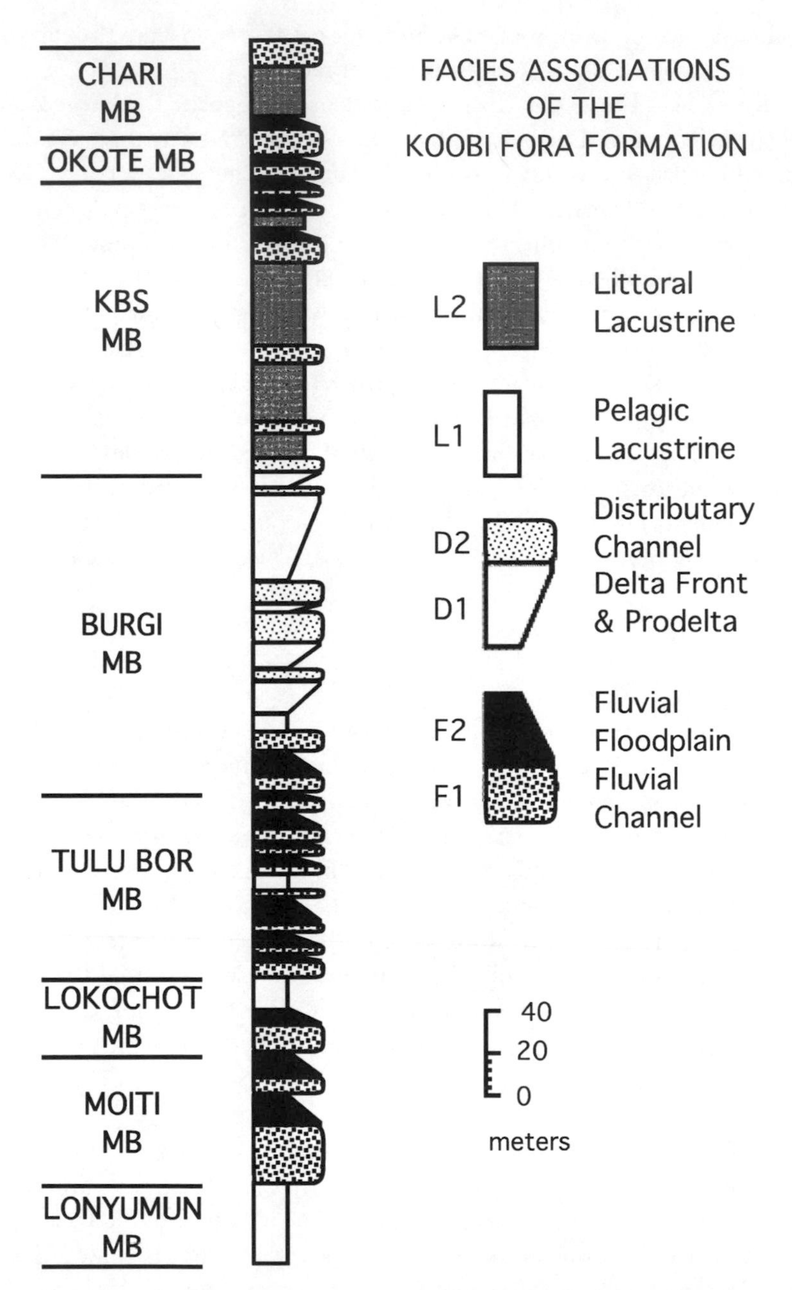

Figure 2.1. *Composite stratigraphic column of the Koobi Fora Formation (after Brown and Feibel, 1986). Dominant facies associations and stratigraphic subdivisions are indicated. Note that occurrences of the crevasse association are too small to represent at this scale.*

Table 2.1. Facies classifications applied to the Koobi Fora deposits.

Code	*Facies*	*Features*	*Interpretation*
Gms	gravel, matrix-supported, massive	massive	debris flow
Gm	gravel, massive or weakly-bedded	horizontal bedding, clast imbrication	longitudinal bar, lag deposits
Gp	gravel, bedded	planar cross-beds	linguoid bars and bar accretion
Sm	sand, massive	massive	bioturbated?
Se	sand, bedded	epsilon cross-strata	point-bar surfaces
St	sand, bedded	trough cross-beds	dunes
Sp	sand, bedded	planar cross-beds	linguoid bars, transverse bars, sand waves
Sr	sand, laminated	ripple marks	ripples
Sh	sand, laminated	horizontal lamination	planar bed flow
Fm	sand, silt, clay; massive	massive, bioturbation	suspension deposits (overbank or lacustrine), biologically reworked
Fl	sand, silt, clay; laminated	planar laminated (to thinly bedded)	suspension deposits
Fb	sand, silt, clay; bedded	thin to medium bedded	flow reduction deposits; levee
Ba	cryptalgal carbonate	mats, oncolites and stromatolites	biogenic deposit, cryptalgal
Bd	diatomite	massive to weakly laminated	biogenic deposit, diatomaceous
Bo	ostracod carbonate	massive	biogenic deposit, ostracods
Bm	molluscan carbonate	massive or weakly bedded	biogenic deposit, mollucs
Be	*Etheria* carbonate	agglutinated	oyster bioherm, reef
Pa	mud with sand veins	anastomosing veins	desiccation-cracked vertic epipedon with sand fill
Pv	clay	slickensided dish fractures	compressional vertic subsoil horizon
Pp	clay	prismatic structure	structural subsoil horizon
Pk	pedogenic carbonate	massive or laminated, rhizoliths	Bk horizon

Gravel (G) Facies. The massive gravel (Gm) and planar cross-bedded gravel (Gp) facies are typical of gravel bars in a variety of fluvial systems. The large particle size and shape characteristics of cobbles and pebbles often render stratification weak or absent. In this case, then, the massive nature of the deposit is a primary characteristic. Thin horizontal beds and planar cross-beds typically represent accumulation on the upper surfaces and fronts of bars respectively. Discontinuous thin occurrences of the Gm facies are common at the base of many fluvial cycles, representing the gravel lag accumulated along a channel bottom. At Koobi Fora, these gravels are typically composed of granule- to pebble-sized quartzofeldspathic material. In exposures close to the basin margin, or in the upper half of the formation, these lags may be dominated by well-rounded volcanic clasts. The active reworking of floodplain muds by cutbank erosion leads in some cases to the accumulation of soil carbonate nodules in a Gm facies. These are locally significant, and have been observed to reach up to 0.5 meters in thickness. In rare cases, such as Locality 261–1 (Coffing et al. 1994), the Gm lag may be extensive, and dominated by fossil bone and teeth to the point of forming a "bone bed."

The matrix-supported gravel (Gms) facies is a rare occurrence, in which very poorly sorted material is chaotically mixed, and as a result the gravel component is matrix supported, that is the individual large clasts are often not in contact, but appear to "float" in the matrix of finer sediment. This is a characteristic feature of high-viscosity flows such as debris flows. In the Koobi Fora examples these likely result from short-term fluidization of channel debris, and travel distances are likely short.

Sand (S) Facies. Sands are a prominent component of the sedimentary section of the Koobi Fora Formation (Feibel 1988). Diagenetic cementation is fairly limited in these deposits, and it is likely that many examples of the massive sand (Sm) facies represent disruption of primary fabrics in the near-surface environment. Two levels of organization of primary sedimentary structures are apparent in the sand facies: the structures produced by migration of individual bedforms and the structures produced by channel-form evolution (Miall 1985). Migrating dunes of sand produce trough cross-stratification (St), ranging in size from small-scale (centimeters) to giant (meters) depending on the magnitude of the bedform. Planar cross-bedding in sands (Sp) results from the migration of sand waves. Horizontally laminated sands (Sh) and ripple-laminated sands (Sr) are produced by the movement of thin layers of sand as sheets and ripples. Within a channel, all of these bedforms may be organized along the gently inclined surface of a point bar. This compound-bar form may leave in the sedimentary record a larger-scale structure, referred to as epsilon cross-stratification (Se; Allen 1963a), the signature product of a laterally accreting point bar.

Mud (M) Facies. Fine-grained accumulations ranging from pure clays through silty clays, silts, and sandy clay/silts are all considered collectively here, in part as they form a closely related continuum and in part because they are subject to very similar accumulation and modification processes. Muds typically accumulate through the slow settling of suspended particles in a low-energy environment. Most commonly this reflects lacustrine or floodplain water bodies, but it is likely that aeolian contributions are pervasive but difficult to recognize.

Massive muds (Fm) may result from settling processes in which accumulational discontinuities are too subtle to be reflected in lamination or bedding. Massive muds may also result from pervasive disruption of a primary fabric through burrowing of organisms or rooting, without leaving diagnostic evidence of the biological effects.

Pulsed accumulation of fines, such as that which results from annual floodwater influx may result in laminated or thinly bedded muds (Fl). Larger-scale influx or source-proximal settings (e.g., a levee) may be reflected in bedded muds (Fb).

Mudstones with obvious soil overprint are considered separately in the pedogenic (P) category, but it should be noted that those facies exist in a continuum with the unmodified muds considered here.

Bioclastic (B) Facies. A wide range of organisms produce skeletal material that may dominate a sedimentary facies. Biomineralization in the form of carbonate minerals (calcite and aragonite) are most common, but siliceous forms are also well represented. In most cases body fossils supply ample evidence for classification, but certain biological groups (e.g., algae and bacteria) produce biosedimentary structures in which body fossils may be rare or absent. The algal biofacies (Ba) is most prominently reflected in various stromatolite forms but also in mat forms and more problematic biosedimentary crusts. Diatomites (Bd) are typically massive, low-density siliceous accumulations of diatom frustules, but may also be preserved as more dense recrystallized rocks when diagenetically altered.

Ostracods (calcareous bivalved crustaceans) are nearly ubiquitous in aquatic habitats, but less commonly accumulate to produce discrete biogenic deposits (ostracodites, Bo). Molluscs may be diverse and abundant in aquatic settings, and may be concentrated in both life and death accumulations as distinctive facies (Bm). A distinctive variant of molluscan biofacies is the bioherm (reef) produced by the Nile oyster *Etheria* (Be).

Pedogenic (P) Facies. Early diagenesis of typically fine-grained lithofacies in the near-surface environment results in a broad suite of distinctive features that characterize fossil soils (e.g., Retallack 1990).

The Pa facies was first recognized by Feibel (1983) along the Koobi Fora Ridge. The marginal lacustrine strata there have mudstones that are characterized by sand-filled veins in an anastomosing network. In rare examples where a plan-view of the facies can be seen, the veins have a polygonal pattern. That the veins represented sand-filled crack networks was obvious, but their genesis was not understood until recently. While it was apparent that the drying of smectitic (shrink-swell) clays would open the crack networks, it was unclear how the infill occurred. An aeolian source was possible, but it seemed that introduction of sand by water would cause the clays to swell and close the cracks before they could fill. Recent observations on the Koobi Fora landscape have shown, however, that the crack networks may open up *beneath* an overlying sand cover, and that the fill is more likely a passive process than an active one. In any case, these sand-filled crack networks are a common feature of vertisols worldwide (Wilding and Puentes 1988). In the Koobi Fora Ridge examples, the Pa facies often occurs alone, as an isolated soil horizon. Elsewhere, however, the Pa facies often occurs stratigraphically above the Pv facies, which is a genetically related subsoil.

The same shrink-swell clays responsible for the Pa facies also dominate the Pv facies. This, however, is a subsoil horizon. It is usually observed beneath a Pa facies unit, or alone in an erosionally truncated profile. The process that dominates the Pv is not the contraction (shrink) phase of clay dynamics, but the expansion (swell) phase. Because the Pv is a subsoil horizon, the expansion of swelling clays must displace surrounding material. Significant stress is built up in the process, and the clay fails along arcuate ("dish-shaped") fractures. Movement along these fractures generates the characteristic slickensides in the surfaces of the clay peds.

The Pk facies, commonly referred to as the calcic or "Bk" horizon in soil parlance, is perhaps the most commonly recognized characteristic facies of ancient soils. The utility of pedogenic carbonate in isotopic analysis for paleovegetation reconstructions (e.g., Quade and Levin, Chapter 3, this volume) and paleoprecipitation estimates (Retallack 1994) have been particularly important.

Facies associations

The packaging of individual facies into characteristic associations establishes the critical transition from discrete sedimentary elements to genetically sig-

TABLE 2.2. Facies associations applied to the Koobi Fora deposits.

Association Code	*Environment*	*Dominant Facies*	*Minor Facies*
F1a	Meandering Fluvial Channel	Se, St, Sp, Sh, Sr	Gm
F1b	Braided Fluvial Channel	Sp	Fm
F2	Fluvial Floodplain	Fm, Pk, Pss, Pa	Pp
L1	Pelagic Lacustrine	Fm, Fl, Bd	Bo
L2	Marginal Lacustrine	Fm, Fl, Bm, Ba, Bo, Pa	St, Sr
D1	Prodeltaic and Delta Front	Fl, Fm	St, Sr
D2	Distributary Channel	Ss, St, Sp, Sr	
C1	Crevasse Channel	St, Sr	Sm
C2	Crevasse Splay	Sm, Fb, Pv	Sr

nificant sequences that reflect components of ancient landscapes. The facies associations described here reflect the specific case of the Koobi Fora Plio-Pleistocene record, but amply demonstrate the methodology of linking facies to establish a context for ancient communities. Much of the description of Koobi Fora facies associations below was first presented by Feibel (1988) and has been expanded in light of subsequent research (Table 2.2).

Fluvial Facies Associations. Three fluvial facies associations are expressed in the deposits of the Koobi Fora Formation (Figure 2.2). These reflect two distinct types of fluvial channel form (F1a and F1b), and a highly variable development of associated floodplain environments (F2).

The Fla sequence typically begins with a basal erosional scour having 10–40 centimeters of local relief, and several meters of relief where channel margins can be seen. The surface of the erosional scour is marked by thin lenticular accumulations of a gravel lag, consisting of quartzo-feldspathic granules and pebbles, along with clay pebbles. These gravel lags represent the Gm facies. This is overlain by a fining-upward progression of sands displaying trough cross-bedding (St), horizontal lamination (Sh), and ripple lamination (Sr), and capped by the fine-grained floodplain deposits. Commonly, however, this ideal sequence has been truncated by a succeeding channel, or primary stratification features have been destroyed by bioturbation. In the Fla sequences of the Lokochot Member and the Tulu Bor Member, channel accretion surfaces (Se facies) are preserved by cementation differences. This type of sequence has been well documented from both modern and ancient examples, and represents deposits of a meandering river channel (Allen 1965, 1971). One extremely well-preserved Fla sequence of the Tulu Bor Tuff (Feibel 1988) allows a calculation of paleochannel width at that point in time. Using the relationship determined by Leeder (1973), a bankfull depth of 9.4 meters measured in one section suggests a bankfull width of about 214 meters. The actual magnitude may have been slightly larger, as the section is truncated at the top. The F1b channel deposits have a less well-developed basal erosional surface, and commonly lack the lenticular Gm accumulations. They are dominated by planar crossbedded sands (Sp) in cosets of 10 to 30 centimeters, with interbeds of horizontally laminated sands (Sh). Capping floodplain deposits are either thin (5–10 cm) or lacking entirely. F1b channels are restricted to the lower Lokochot Member and the lower KBS Member.

The F1b sequence matches modern and ancient examples of braided-stream channels (Smith 1972; Miall 1977). In particular, the features of this association are very similar to the Platte type braided-river profile, a sand-dominated system where sedimentation occurs primarily along the transverse and linguoid bars of a shallow, perennial braided stream. Fluvial floodplain deposits, termed the F2 association (Feibel and Brown 1986), occur superposed on an F1 sequence. They are characterized by their fine-grained nature (dominantly silts and clays), lack of aquatic invertebrate fossils, abundant pedogenic alteration (facies Pv, Pp, Pk; see section on pedogenesis for details), and bioturbation.

Deposits of the F2 association are common throughout the formation. They are generally unremarkable in weathered section except for their darker colors and locally abundant surface lags of white pedogenic carbonate concretions (a Pk lag). Where modern streams have excavated natural exposures, however, the F2 deposits exhibit a wide array of distinctive structures, primarily related to pedogenesis.

Another aspect of significance within the fluvial facies association is the character of the transition from the F1, or channel component, to the F2, or floodplain interval. This transition ranges in character from very gradual, through stepwise in character, to a very abrupt transition. The gradual transition reflects the classical upward-fining sequence seen in the point-bar environment, where a gradual migration of the channel leads to a lowering of energy conditions, and is seen in fining of sedimentary particles and a shift in primary sedimentary structures (Allen 1963b). The F1–F2 transition in this case is marked by a continuous gradation from fine sands, through silts, to clays. It is not uncommon, however, for the transition to be recorded in a more stepwise transition, with sharp stratigraphic breaks and steps in particle-size change. In some such cases, reversals occur where the general upward fining trend is interrupted by a shift to coarser particles. In some cases this reflects minor complexity on a point-bar

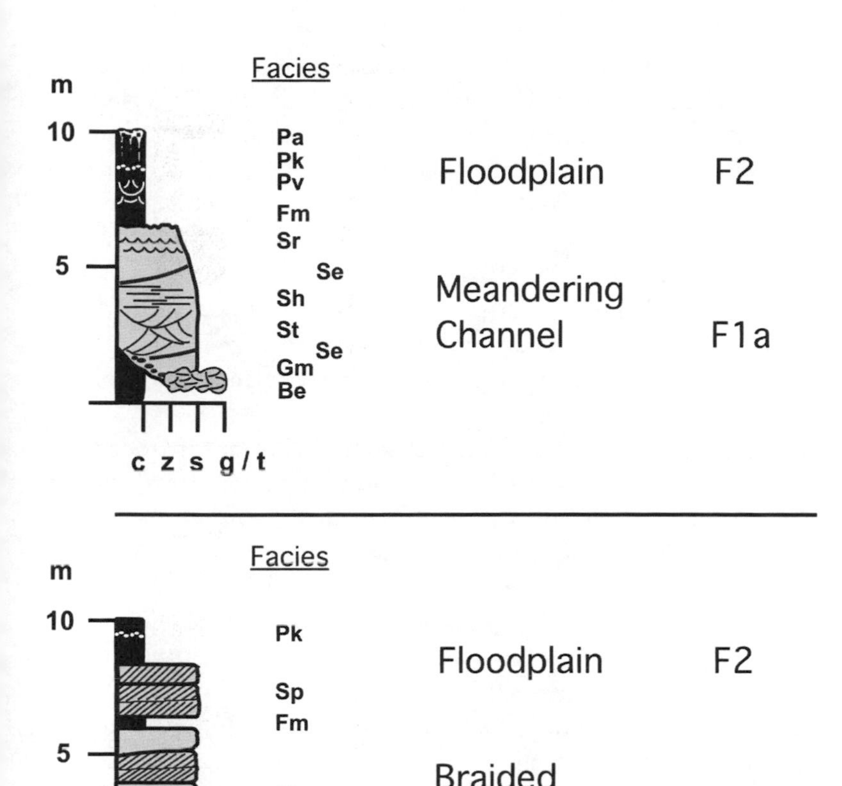

FIGURE 2.2. *Characteristic features and sequences of the fluvial facies associations. Idealized stratigraphic columns are depicted, showing thickness and dominant lithology (c ,clay; z, silt; s, sand; g, gravel; t, tuff). Facies codes are described in Table 2.1.*

surface, such as chute cutoffs. In other examples the abrupt shifts reflect episodic accumulation. An excellent example of this is seen in a complete point-bar sequence preserved in the Tulu Bor Tuff (Feibel 1988; 1999). In this case, the upward-fining progression in the tuff is broken by three abrupt shifts to much finer-grade material. These are interpreted to record three flood–slack couplets during the accumulation history of the tuff, and may actually record three annual events of flood-season and waning-stage flow. The extreme example of abrupt F1–F2 transition is documented in fluvial associations in which a fluvial couplet is seen. In this case, a relatively coarse sand F1 association is overlain abruptly by a clay F2, with no transitional particle-size gradation. These fluvial couplets are interpreted to reflect avulsion events (Smith et al. 1989), in which the position of the active channel has abruptly shifted on the floodplain, and a channel F1 association is left to be overlain by distal floodplain vertical accretion.

Lacustrine Facies Associations. Deposits of lacustrine facies comprise about one-third of the Koobi Fora sediments. Characteristic sequences of the lacustrine facies allow them to be grouped into two major associations (Figure 2.3), one representing an open-water or pelagic lacustrine environment (Ll) and one the marginal lacustrine setting (L2).

The features that characterize the Ll facies association are sedimentation from suspension, with a low clastic input. Claystones predominate, either in massive (Fm) or finely laminated (Fl) facies. Biogenic accumulations may be associated, including diatomites (Bd) and scattered or abundant ostracods (Bo). Molluscs also occur scattered throughout these facies in some localities. This association is not well represented in the Koobi Fora sequence, suggesting that either dominantly marginal conditions or proximity to a source of clastic detritus (resulting from deltaic encroachment) were more typical of the lacustrine phases. Pelagic lacustrine deposits are known from the lower Lonyumun Member, Lokochot Member, mid–Tulu Bor Member, and upper Burgi Member.

The littoral lacustrine (L2) association is marked by a heterogeneous mixture of facies, often closely and repetitively interbedded. The term *littoral* is used here in its lacustrine sense, meaning "the shore region where the water is shallow enough for continuous mixing and for photosynthesis to the bottom" (Beadle 1981). In practice this implies that most of the littoral zone is affected to some degree by wave action. The facies common in this environment include molluscan carbonates (Bm) that are dominantly life assemblages but that can be seen to grade laterally into more arenaceous, reworked shoreline deposits representing death assemblages. Another common facies in this assemblage is the cryptalgal biolithite (Ba), which includes carbonates in the form of oncolites, stromatolites, and cryptalgal mats. Some ostracodites (Bo) occur in this association, although more commonly the ostracods are relatively dispersed. The association is dominated volumetrically by fine-grained clastics, commonly massive (Fm) but locally with well-developed pedogenic features (Pv). Thin (1–2 m) beds of laminated siltstones and fine-grained sandstones (Fl) also occur in this association. These are actually a very thin variant of the prodeltaic association discussed below. The characteristic facies and sequences of the littoral lacustrine association are best developed in the uppermost upper Burgi and the

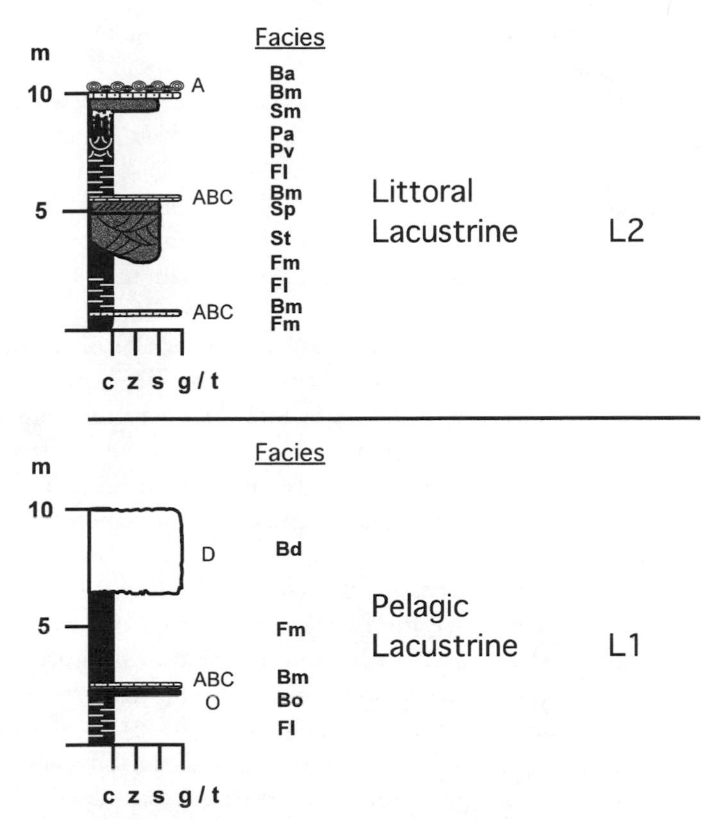

Figure 2.3. *Characteristic features and sequences of the lacustrine facies associations. Idealized stratigraphic columns are depicted, showing thickness and dominant lithology (c, clay; z, silt; s, sand; g, gravel; t, tuff). Facies codes are described in Table 2.1.*

KBS Members. The association is also well represented in the upper Okote and Chari Members at Ileret.

A major distinction between the two lacustrine associations is the frequency of subaqueous/subaerial cyclicity and the degree of development of littoral biotic accumulations. The facies of the pelagic lacustrine association imply conditions minimally affected by nearshore processes (wave activity, high clastic influx). The facies of the littoral lacustrine association are indicative of a closer proximity to the shoreline and a relatively frequent lateral migration across the shore transition. There is no indication that the pelagic lacustrine association developed in a deepwater (profundal) environment. There are rare examples of desiccation and minor soils development in the pelagic clays of the Lonyumun Member. Analysis of the deltaic facies associations (following section) suggests that maximum lake depths were on the order of 40 meters. One of the most distinctive features of the littoral lacustrine environment at Koobi Fora is the well-developed molluscan carbonate (Bm) facies. Feibel (1983) suggested that one of the best available analogs for this facies was modern Lake Chad. There, relatively shallow (2–3 m) waters host extremely abundant mollusc populations that are compositionally almost identical to those of the Koobi Fora Formation (Lévèque 1972). Recent history has also shown that this sort of environment is quite unstable, and changes in lake level result in large fluctuations in lake area.

Deltaic Facies Associations. Facies associations attributable to deltaic environments comprise some 20 percent of the Koobi Fora deposits. Temporally, however, they are overrepresented, as the characteristic high-sedimentation rates of the deltaic environment produces an extremely thick section over a relatively short period of time. The deltaic environment can be characterized by two facies associations, one representing the prodelta and delta-front environments (D1) and the other the product of distributary channels (D2). The characteristic features of the deltaic facies associations are illustrated in Figure 2.4.

The prodelta and delta-front association (D1) is dominated by laminated fine-grained clastics (Fl), including clays, silts, and fine sands, which occur in thick (10–20 m) packages. The laminated sediments are commonly capped by a massive (commonly bioturbated) fine-grained sequence (Fm) or by delta-front sands (Sm). As mentioned earlier, the distributary channel association (D2) is very similar to the F1 associations, and is most easily distinguished by its sequential position atop a D1–D2 couplet. The medium- to coarse-grained sands of this association occur as narrow lenticular bodies within the D1 deposits, or as broader lenticular deposits overlying them. Trough (St) and planar (Sp) cross-bedding and ripple lamination (Sr) is generally weakly developed, or has been destroyed by bioturbation. The D1 and D2 associations are well developed in the upper Lokochot Member and in the upper Burgi Member. They are not represented in association with the lakes of the Lonyumun Member or Tulu Bor Member. A very thin variant of these associations is represented in the KBS through Chari Members. The D1 interval is only 1–2 meters thick, and the D2 association is characterized by a deeply erosional base. These thin intervals of D1–D2 sedimentation are characteristically associated with the littoral lacustrine deposits of the upper Koobi Fora Formation.

The well-developed D1–D2 sequences of the Lokochot and upper Burgi Members allow for determination of water depths (deVries Klein 1974). For the Lokochot Member lake, the deltaic infilling suggests water depths of 10–15 meters. The upper Burgi Member lake records several intervals of deltaic pro-

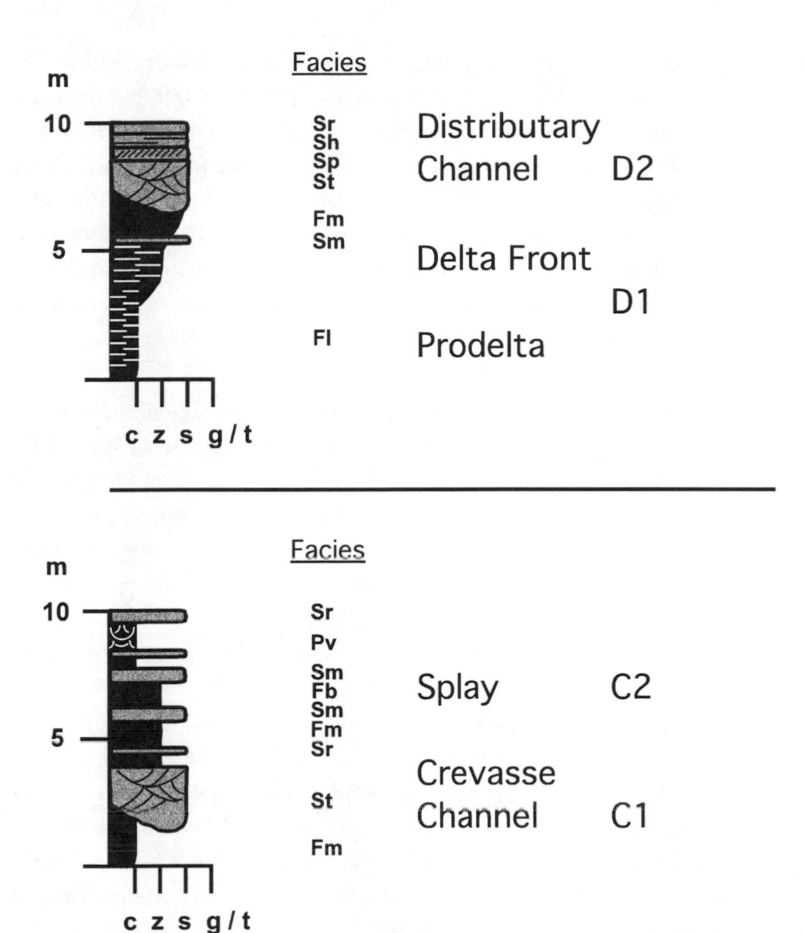

FIGURE 2.4. *Characteristic features and sequences of the deltaic (*top*) and crevasse (*bottom*) facies associations. Idealized stratigraphic columns are depicted, showing thickness and dominant lithology (c, clay; z, silt; s, sand; g, gravel; t, tuff). Facies codes are described in Table 2.1.*

gradation, with water depths of different phases ranging from 20 to 40 meters (Feibel and Brown 1986). It is also possible to use this method to characterize the water bodies present at Koobi Fora in the KBS, Okote, and Chari Members. These intervals of upward-coarsening cycles are only 1–2 meters in thickness, and suggest that the water bodies present had depths of that order. Whether this represents the total depth of a lake or only a marginal zone cannot be determined from the data available.

The extremely well-developed lamination of silts and very fine-grained sands in the D1 association has important implications. The lamination generally occurs on a 0.5-centimeter scale and is interpreted to represent a strongly seasonal influx of detrital clastic material. A similar sort of lamination is seen in modern Lake Turkana, where it results from the seasonal nature of the annual Omo River floods (Yuretich 1979). A similar origin is inferred for the Fl deposits of the Dl association. This implies that the monsoonal seasonality of rainfall characteristic of the Ethiopian Highlands today was a feature of the Plio-Pleistocene as well.

Crevasse Facies Association. A small-scale variant sharing some characteristics of the deltaic facies association is the crevasse facies association (Figure 2.4). These packages occur throughout the Koobi Fora Formation, but are particularly well developed in intervals of high sediment accumulation, such as the Okote Tuff Complex (Brown and Feibel 1985) and comparable strata at Ileret (Quinn and Lepre 2005). Crevasse systems develop as breakouts of major channels, with the crevasse itself cutting through channel bank or levee deposits, and a splay forming as a minidelta on the adjacent floodplain or interdistributary bay (Elliott 1974; Bristow et al. 1999).

The crevasse channel (C1) component is typically dominated by trough cross-bedded sands (St) and climbing ripple-laminated sands (Sr) within an erosionally scoured channel-form body. In some examples, the basal portion of this channel is lined with coarse tuff, which may be pumiceous. The C1 component merges laterally with the splay complex (C2), which is dominated by thin massive (Sm) or rippled (Sr) sands, bedded silts (Fb), and clays with pedogenic overprint (Pv). Crevasse splays result in rapid localized accumulation on the floodplain or distributary bay, and may be significant in the burial and preservation of fossil or artifact accumulations.

FROM FACIES ASSOCIATIONS TO LANDSCAPES AND PALEOECOLOGY

The sedimentary records that accompany fossil and archaeological assemblages are a rich archive of environmental indicators upon which paleoecological reconstructions can be based. Facies analysis provides a useful methodology for recognizing fundamental rock units, genetically significant associations, and for their use in reconstructing both depositional and postdepositional sedimentary environments. The direct connection between fossils or artifacts and their sedimentary context provides an immediate link between the sedimentary proxies and the associated records, but in addition the facies provide a means of extending the analysis in two dimensions: across spatial landscapes and through time.

A facies analysis is a first step in reconstructing the dynamic landscape that supported ancient communities, and for understanding the interplay of processes that preserved fossil and artifact assemblages. It provides some information on crucial variations of the hydrologic system in the region, central to ecological relationships, as well as a basis for reconstructing patterns of vegetation structure based on soil and moisture relationships (Tinley 1982). Although the geological substrate represents a nonbiological actor in ecological relationships, it is nonetheless a crucial element in paleoecological reconstructions. Upon this reconstructed landscape of sediments, landforms, and associated moisture constraints, it is possible to array the biological and cultural evidence, the sum of the paleontological and archaeological records, and investigate their spatial and temporal relationships and interactions, which is the central goal of paleoecology.

ACKNOWLEDGMENTS

The evolution of this facies approach to the Koobi Fora sequence has benefited from the contributions of many geologists through the years. The geologists of Koobi Fora, particularly Dan Burggraf and Howard White in Carl Vondra's lab, as well as Ian Findlater, built the foundations of this study. Frank Brown was a valuable sounding board for ideas over many years. Jim Aronson illuminated the importance of crevasse splays, and Chris Lepre and Rhonda Quinn recognized their importance at Ileret. Research at Koobi Fora was supported by Richard and Meave Leakey and the National Museums of Kenya, to whom I owe a tremendous debt of gratitude. Financial support for much of this work was provided by the National Science Foundation and the Leakey Foundation. Matt Sponheimer's inspiration in developing this workshop and volume, and his endless patience in seeing it through, are greatly appreciated.

REFERENCES

Allen, J.R.L. 1963a. "Henry Clifton Sorby and the Sedimentary Structures of Sands and Sandstones in Relation to Flow Conditions." *Geologie & Mijnbouw* 4:223–8.

Allen, J.R.L. 1963b. "The Classification of Cross-Stratified Units with Notes on Their Origin." *Sedimentology* 2 (2): 93–114. http://dx.doi.org/10.1111/j.1365-3091.1963.tb01204.x.

Allen, J.R.L. 1965. "A Review of the Origin and Characteristics of Recent Alluvial Sediments." *Sedimentology* 5 (2): 89–191. http://dx.doi.org/10.1111/j.1365-3091.1965.tb01561.x.

Allen, J.R.L. 1971. "Rivers and Their Deposits." *Science Progress, Oxford* 59:109–22.
Beadle, L. C. 1981. *The Inland Waters of Tropical Africa*. 2nd ed. London: Longman. 475 pp.
Bowen, B. E. 1974. The Geology of the Upper Cenozoic Sediments in the East Rudolf Embayment of the Lake Rudolf Basin, Kenya. PhD dissertation. Iowa State University, Ames. 164 pp.
Bristow, C. S., R. L. Skelly, and F .G. Ethridge. 1999. "Crevasse Splays from the Rapidly Aggrading, Sand-Bed, Braided Niobrara River, Nebraska: Effect of Base-Level Rise." *Sedimentology* 46 (6): 1029–47. http://dx.doi.org/10.1046/j.1365-3091.1999.00263.x.
Brown, F. H., and C. S. Feibel. 1985. "Stratigraphical Notes on the Okote Tuff Complex at Koobi Fora, Kenya." *Nature* 316 (6031): 794–7. http://dx.doi.org/10.1038/316794a0.
Brown, F. H., and C. S. Feibel. 1986. "Revision of Lithostratigraphic Nomenclature in the Koobi Fora Region, Kenya." *Journal of the Geological Society* 143 (2): 297–310. http://dx.doi .org/10.1144/gsjgs.143.2.0297.
Brown, F. H., and C. S. Feibel. 1991. "Stratigraphy, Depositional Environments and Palaeogeography of the Koobi Fora Formation." In *Stratigraphy, Artiodactyls and Palaeoenvironments*, ed. J. M. Harris, 1–30. Koobi Fora Research Project, Volume 3. Oxford: Clarendon Press.
Brown, F. H., B. Haileab, and I. McDougall. 2006. "Sequence of Tuffs between the KBS Tuff and the Chari Tuff in the Turkana Basin, Kenya and Ethiopia." *Journal of the Geological Society* 163 (1): 185–204. http://dx.doi.org/10.1144/0016-764904-165.
Burggraf, D. R., Jr., and C. F. Vondra. 1982. "Rift Valley Facies and Paleoenvironments: An Example from the East African Rift System of Kenya and Southern Ethiopia." *Zeitschrift fur Geomorphologie* 42:43–73.
Burggraf, D. R., Jr., H. J. White, H. J. Frank, and C. F. Vondra. 1981. Hominid Habitats in the Rift Valley, Part 2. In *Hominid Sites: Their Geologic Settings*, ed. G. Rapp Jr. and C. F. Vondra, 115–147. American Association for the Advancement of Science Selected Symposium 63.
Coffing, K., C. Feibel, M. Leakey, and A. Walker. 1994. "Four-Million-Year-Old Hominids from East Lake Turkana, Kenya." *American Journal of Physical Anthropology* 93 (1): 55–65. http://dx.doi.org/10.1002/ajpa.1330930104. Medline:8141242
Elliott, T. 1974. "Interdistributary Bay Sequences and Their Genesis." *Sedimentology* 21 (4): 611–22. http://dx.doi.org/10.1111/j.1365-3091.1974.tb01793.x.
Feibel, C. S. 1983. "Stratigraphy and Paleoenvironments of the Koobi Fora Formation along the Western Koobi Fora Ridge, East Turkana, Kenya." MS thesis, Iowa State University, Ames. 104 pp.
Feibel, C. S. 1988. "Paleoenvironments from the Koobi Fora Formation, Turkana Basin, Northern Kenya." PhD dissertation, University of Utah, Salt Lake City. 330 pp.
Feibel, C. S. 1994. "Controls on Sedimentation in a Plio-Pleistocene, Fluvial-Dominated Rift Basin, the Turkana Basin of East Africa." AAPG Annual Meeting, Denver. *Abstracts* 3: 148.
Feibel, C. S. 1999. "Tephrostratigraphy and Geological Context in Paleoanthropology." *Evolutionary Anthropology* 8 (3): 87–100. http://dx.doi.org/10.1002/(SICI)1520-6505 (1999)8:3<87::AID-EVAN4>3.0.CO;2-W.

Feibel, C. S., and F. H. Brown. 1986. "Depositional History of the Koobi Fora Formation, Northern Kenya." *Proceedings of the Second Conference on the Geology of Kenya.*

Feibel, C. S., J. M. Harris, and F. H. Brown. 1991. "Palaeoenvironmental Context for the Late Neogene of the Turkana Basin." In *Stratigraphy, Artiodactyls and Paleoenvironments,* ed. J. M. Harris, 321–70. Koobi Fora Research Project, Volume 3. Oxford: Clarendon Press.

Gathogo, P. N. 2003. "Stratigraphy and Paleoenvironments of the Koobi Fora Formation of the Ileret Area, Northern Kenya." 160. MS thesis, University of Utah, Salt Lake City.

Gathogo, P. N., and F. H. Brown. 2006. "Stratigraphy of the Koobi Fora Formation (Pliocene and Pleistocene) in the Ileret Region of Northern Kenya." *Journal of African Earth Sciences* 45 (4-5): 369–90. http://dx.doi.org/10.1016/j.jafrearsci.2006.03.006.

deVries Klein, G. 1974. "Estimating Water Depths from Analysis of Barrier Island and Deltaic Sedimentary Sequences." *Geology* 2 (8): 409–12. http://dx.doi.org/10.1130/0091-7613(1974)2<409:EWDFAO>2.0.CO;2.

Leeder, M. R. 1973. "Fluviatile Fining-Upward Cycles and the Magnitude of Paleochannels." *Geological Magazine* 110 (03): 265–76. http://dx.doi.org/10.1017/S0016756800036098.

Lévèque, C. 1972. "Mollusques Benthiques du Lac Tchad: Ècologie, Ètude des Peuplements et Estimation des Biomasses." *Cah. Orstom Hydrobiol* 6:3–45.

McDougall, I., and F. H. Brown. 2006. "Precise 40Ar/39Ar Geochronology for the Upper Koobi Fora Formation, Turkana Basin, Northern Kenya." *Journal of the Geological Society* 163 (1): 205–20. http://dx.doi.org/10.1144/0016-764904-166.

Miall, A. D. 1977. "A Review of the Braided River Depositional Environment." *Earth-Science Reviews* 13 (1): 1–62. http://dx.doi.org/10.1016/0012-8252(77)90055-1.

Miall, A. D., ed. 1978. *Fluvial Sedimentology.* Memoir 5. Calgary: Canadian Society of Petroleum Geologists.

Miall, A. D. 1985. "Architectural-Element Analysis: A New Method of Facies Analysis Applied to Fluvial Deposits." *Earth-Science Reviews* 22 (4): 261–308. http://dx.doi.org/10.1016/0012-8252(85)90001-7.

Posamentier, H. W., and R. G. Walker. 2006. *Facies Models Revisited.* SEPM Special Publication 84. Tulsa: SEPM. 532 pp.

Quinn, R. L., and C. J. Lepre. 2005. "Environmental Context of Early Pleistocene Hominins from the Ileret Subregion (Area 1A) of Koobi Fora, Kenya." *American Journal of Physical Anthropology* 126 (S40): 169.

Reading, H. G. 1978. "Facies." In *Sedimentary Environments and Facies,* ed. H. G. Reading, 4–14. New York: Elsevier.

Retallack, G. J. 1990. *Soils of the Past.* London: Unwin Hyman. 520 pp. http://dx.doi.org/10.1007/978-94-011-7902-7

Retallack, G. J. 1994. "The Environmental Factor Approach to the Interpretation of Paleosols." In *Factors of Soil Formation: A Fiftieth Anniversary Perspective,* ed. R. Amundson, J. Harden, and M. Singer, 31–64. SSSA Special Publication 33. Madison, WI: Soil Science Society of America.

Smith, N. D. 1972. "Some Sedimentological Aspects of Planar Cross-Stratification in a Sandy Braided River." *Journal of Sedimentary Petrology* 42:624–34.

Smith, N. D., T. A. Cross, J. P. Dufficy, and S. R. Clough. 1989. "Anatomy of an Avulsion." *Sedimentology* 36 (1): 1–23. http://dx.doi.org/10.1111/j.1365-3091.1989.tb00817.x.
Teichert, C. 1958. "Concepts of Facies." *Bulletin of the American Association of Petroleum Geologists* 42:2718–44.
Tinley, K. L. 1982. "The Influence of Soil Moisture Balance on Ecosystem Patterns in Southern Africa." In *Ecology of Tropical Savannas*, ed. B. J. Huntley and B. H. Walker, 175–192. Berlin: Springer-Verlag. http://dx.doi.org/10.1007/978-3-642-68786-0_9
Walker, R. G., ed. 1979. *Facies Models*. Geoscience Canada, Reprint Series 1. Toronto: Geological Association of Canada. 211 pp.
Walker, R. G., and N. P. James, eds. 1992. *Facies Models: Response to Sea Level Change*. Toronto: Geological Association of Canada. 454 pp.
White, H. J., D. R. Burggraf, Jr., R. B. Bainbridge, Jr., and C. F. Vondra. 1981. "Hominid Habitats in the Rift Valley: Part 1." In *Hominid Sites: Their Geologic Settings*, ed. G. Rapp Jr. and C. F. Vondra, 57–113. American Association for the Advancement of Science Selected Symposium 63.
Wilding, L. P., and R. Puentes, eds. 1988. *Vertisols: Their Distribution, Properties, Classification and Management*. College Station: Texas A&M University Printing Center.
Wynn, J. G. 1998. "Paleopedological Characteristics Associated with Intervals of Environmental Change from the Neogene Turkana Basin, Northern Kenya." MS thesis, University of Utah, Salt Lake City. 103 pp.
Wynn, J. G., and C. S. Feibel. 1995. "Paleoclimatic Implications of Vertisols within the Koobi Fora Formation, Turkana Basin, Northern Kenya." *Journal of Undergraduate Research* 6 (1): 32–42.
Yuretich, R. F. 1979. "Modern Sediments and Sedimentary Processes in Lake Rudolf (Lake Turkana) Eastern Rift Valley, Kenya." *Sedimentology* 26 (3): 313–31. http://dx.doi.org/10.1111/j.1365-3091.1979.tb00912.x.

3

East African Hominin Paleoecology:
Isotopic Evidence from Paleosols

Jay Quade and
Naomi E. Levin

Pedogenic carbonate ($CaCO_3$) occurs in most arid to semiarid settings globally and is widely recognized in the geological record back at least to the Silurian. Because of its abundance, there is considerable interest in the carbon ($\delta^{13}C$) and oxygen ($\delta^{18}O$)[1] isotopic composition of pedogenic carbonate in paleoenvironmental reconstruction. Early (pre-1984) models of the soil-isotopic system were strongly influenced by studies of carbonate (speleothem) formation in caves, particularly as developed by Hendy (1971) and Hendy et al. (1972). Versions of the cave-based model presented in Salomons et al. (1978) or simpler views found in Magaritz and Amiel (1980) were widely used into the 1980s (e.g., Talma and Netterberg 1983; Magaritz et al. 1981; Dever et al. 1982; Rabenhorst et al. 1984) to interpret the paleoenvironmental record from carbonates in soils.

All these studies recognized the potentially important influence of plants on the carbon isotopic composition of pedogenic carbonate, and of local rainwater, soil temperature, and soil evaporation on its oxygen isotopic composition. However, the soil system, particularly for carbon, was then viewed to be a complex function of other, mainly nonplant variables, such as kinetic outgassing of CO_2, and no model recognized

1. Isotopic data are reported using the δ notation, expressed in ‰, where $\delta = ((R_{sample}/R_{standard})-1) \times 1000$, and $R = {}^{13}C/{}^{12}C$ or ${}^{18}O/{}^{16}O$. The $\delta^{13}C$ and $\delta^{18}O$ values of carbonates are commonly calculated in reference to the Vienna Pee Dee Belemnite (VPDB) standard.

DOI: 10.5876/9781607322252:c03

the fundamental contribution of gaseous diffusion to the carbon isotopic composition of pedogenic carbonate. These pre-1984 models of the carbon isotopic system in pedogenic carbonate were superseded by the Cerling (1984) soil-diffusion model, which with some modest revisions is widely embraced by the geologic community. The Cerling model has been used to interpret the isotopic composition of pedogenic carbonate in paleoenvironmental terms on all the continents except Antarctica. In the last fifteen years it has been widely applied to reconstructing paleoenvironments associated with hominins in Africa.

We begin our chapter with a review of basic criteria for pedogenic carbonate recognition in the field, since interpretation of the isotopic results depends critically on the analysis of pedogenic carbonate only, and not other types of secondary carbonate common in the geologic record. The carbon isotopic composition of pedogenic carbonate is closely linked to the proportion of C_3 to C_4 plants growing on the soil surface. As in all equatorial latitudes, East Africa hosts both plant types, the distribution of which we review. We then briefly present the soil-diffusion model developed by Cerling (1984) and as modified by subsequent studies.

In comparison to the carbon isotopic system, oxygen isotopes in pedogenic carbonates in East Africa have received much less attention. However, when approached systematically, pedogenic carbonate $\delta^{18}O$ records can yield valuable information on local water-balance and climate. In East Africa, it is clear that pedogenic carbonates with low $\delta^{18}O$ values formed at times in the past when rainfall patterns in East Africa were different from today.

We illustrate our presentation with examples from our own recent research on the record preserved at Gona, Ethiopia (Figure 3.1). Gona archives a long (~6 to < 0.2 mya) record of human evolution, and has been the focus of intensive paleontologic, geologic, and isotopic scrutiny for almost a decade (Semaw et al. 2003; Quade et al. 2004; Levin et al. 2004; Semaw et al. 2005; Kleinsasser et al. 2008; Levin et al. 2008; Quade et al. 2008). Gona provides excellent examples of the opportunities and limitations of the use of carbon and oxygen isotopes from soils in paleoenvironmental reconstruction. We finish by synthesizing the large body of new isotopic data from Gona with other pedogenic carbonate data sets from East Africa.

PEDOGENIC CARBONATE, CLIMATE, AND VEGETATION IN EAST AFRICA

Pedogenic carbonate today occurs globally in soils where annual rainfall is less than ~1 m. In wetter climates, soil solutions do not attain calcite saturation

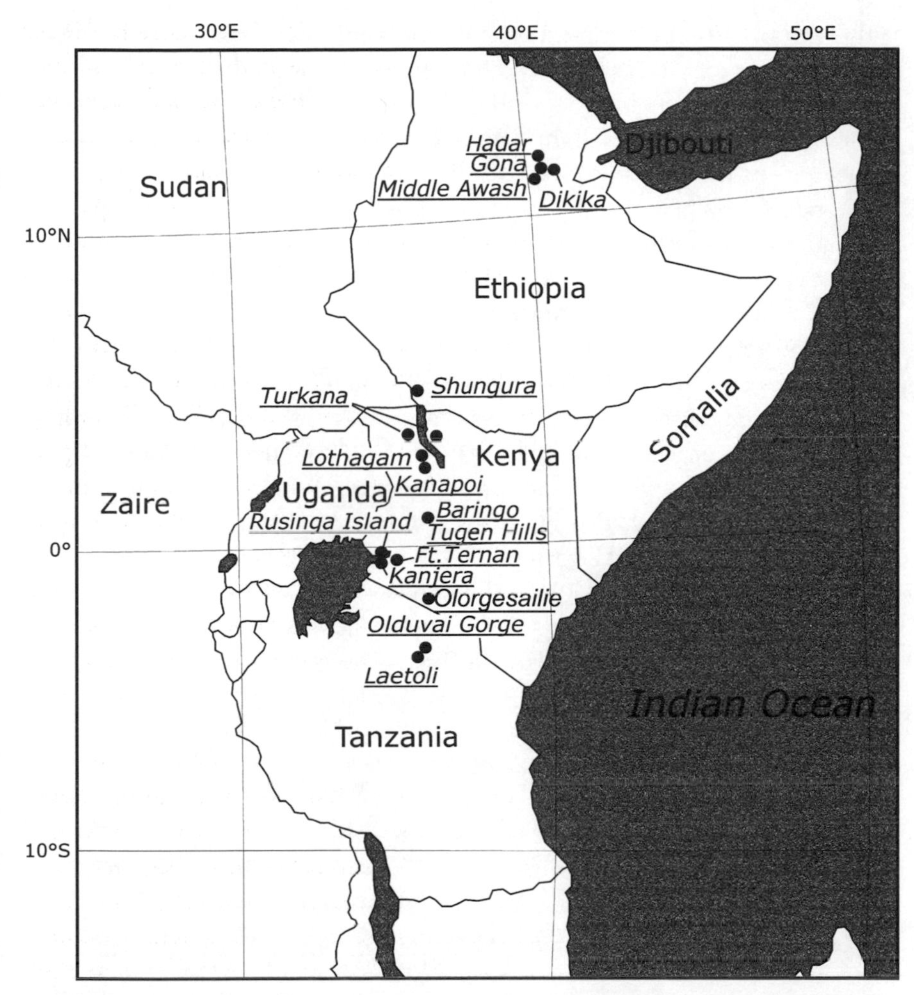

Figure 3.1. *Study areas with stable isotope results from East Africa, as discussed in the text.*

and precipitate pedogenic carbonate prior to soil flushing by the next rainstorm event. Climate regimes with rainfall less than 1 meter/year encompass most of East Africa today at lower elevations, explaining the abundance of pedogenic carbonate under much of the modern and ancient East African landscape. Rainfall in Kenya only exceeds 1 meter/year along the narrow coastal strip and in highland areas above 1500 meters in south and west-central Kenya. Pedogenic carbonate is ubiquitous at lower elevations of the Ethiopian Rift, where it persists up to about 1700 meters in elevation.

The carbon isotope composition of pedogenic carbonate is determined largely by the proportions of C_3 and C_4 plants growing in the immediate vicinity. Globally, C_3 plants include virtually all trees, almost all shrubs, and those grasses growing under low-light and cool conditions. C_4 plants occur in only eighteen plant families, and are mostly grasses and sedges. Twenty-one percent of all grasses are C_4, and compared to C_3 grasses they tend to favor high-light and higher temperature conditions (i.e., warm growing seasons) (Sage 1999). Cacti, succulents, and other CAM plants can be found in some ecosystems in East and South Africa, but they are not abundant in most pedogenic carbonate-forming settings.

C_4 grasses are favored over C_3 grasses in climate regimes where the mean monthly temperature of the warmest month exceeds about 22°C and where the rainy season occurs during these warm months (Collatz et al. 1998). Hence, in East Africa today, virtually all grasses growing below 2000 meters are C_4 grasses, except where shaded by a closed-forest canopy (Tieszen et al. 1979). The temperature threshold favoring C_4 grasses may have been higher in the past if pCO_2 (concentration of CO_2 in the atmosphere) was higher (Collatz et al. 1998), confining C_4 grasses to slightly lower elevations. Lower pCO_2 would have the opposite effect. Other C_4 plants in East Africa include a few herbaceous plants (e.g., *Blepharis* sp., *Tribulus* sp.) that are relatively uncommon except in disturbed ground, and sedges (Hesla et al. 1982), which tend to grow in swampy settings where pedogenic carbonate does not form.

Biochemical variants of C_4 grasses include NAD malic enzyme (NAD-ME), NADP malic enzyme (NADP-ME), and phosphenoenolpyruvate (PCK) subtypes. All these variants occur in East Africa and have slightly different $\delta^{13}C$ values. The proportion of NAD-ME and PCK variant grasses increases with decreasing rainfall (Taub 2000); consequently, these grasses dominate areas where most pedogenic carbonate forms. $\delta^{13}C$ values of NAD-ME and PCK variant grasses average $-13.3 \pm 1.4‰$ today, based on our compilation of data ($n = 105$) presented in Cerling et al. (2003a) for East Africa. Settings for these grasses include xeric bushland and savanna, and open-canopy forest. From a floristic point of view, conditions are dry and warm enough for C_4 grasses to be present in some proportion in about 90 percent of Kenya and in the rifted areas of Ethiopia.

Irrespective of setting, C_3 plants dominate anywhere that forest or scrub dominates over grass, and C_3 grasses dominate over C_4 grasses at elevations above about 2000 meters in East Africa (Tieszen et al. 1979). Continuous, closed-canopy forest is relatively uncommon in East Africa and is confined to a few coastal areas and the high mountains above about 1000 meters flanking the rift (Pratt et al. 1966). High available-moisture indices and rainfall associated with soils in high-elevation settings suggest that they should be largely noncalcareous. Hence, upland forests and moorlands are not likely to be archived in the pedogenic carbonate isotopic record. However, lowland closed-canopy forest (hence pure C_3)—gallery or groundwater forests—should in theory be preserved in some pedogenic carbonate isotopic records, a point to which we will return when examining the Gona record. Based on data from open-canopy forest and bushland in Kenya today, C_3 plants average −27.0 ± 1.3‰ (Cerling et al. 2003a).

When using isotopic averages from modern plants to reconstruct paleovegetation patterns, it is important to keep in mind that the $\delta^{13}C$ value of atmospheric CO_2 ($\delta^{13}C_{atm}$) has decreased in the last 150 years due to fossil-fuel burning (the Suess Effect). We must correct for this in order to use the $\delta^{13}C$ values of modern plants as a basis for interpreting the $\delta^{13}C$ value of ancient pedogenic carbonates. The shift in $\delta^{13}C_{atm}$ values due to the Suess Effect is estimated to be −1.5 ± 0.1‰ since 1850 (Friedli et al. 1986), and we adopt this estimate for the deeper geologic past. As a result, prior to 1850, pure C_3 biomass in the drier areas of East Africa is assumed to have averaged −25.5 ± 1.3‰ (= −27.0 + 1.5‰) in $\delta^{13}C$ (VPDB) prior to 1850, and −11.8 ± 1.4‰ for pure C_4 plants.

PALEOSOL CARBONATE RECOGNITION

The use of the soil-diffusion model as an interpretive framework for carbon isotope values in soils assumes that the material being analyzed is well-preserved pedogenic carbonate. *Pedogenic carbonate* refers to carbonate precipitated in soils due to weathering processes at the land surface, as opposed to other forms of carbonate, such as detrital limestone or secondary cements formed after burial. Identification of pedogenic carbonate in the geologic record can be difficult. Many sections that we have studied turned out not to contain pedogenic carbonate, although other secondary, nonsoil forms were abundant. Mistaken sampling of nonpedogenic carbonate likely contributes to the large scatter of $\delta^{13}C$ values in some "pedogenic carbonate" sample sets from Africa and elsewhere.

There is no single distinguishing characteristic of pedogenic carbonate, as it can show considerable overlap in appearance and texture with nonpedogenic carbonates. The latter commonly include lacustrine, spring, and groundwater carbonates. Experience has taught us to rely on a suite of features for accurate identification of pedogenic carbonate over other secondary forms of calcite cementation.

Association with other soil features

Pedogenic carbonate forms a diagnostic horizon (termed a *Bk* horizon) in many modern African soils, generally in the lower part of weathering profiles. In undisturbed modern profiles, Bk horizons are overlain by bioturbated, reddened, clay-enriched horizons (designated *Bw* or *Bt*) and an organic-rich A horizon. Both Bw and Bt horizons tend to be leached of carbonate. Once buried, the organic-rich A horizons rarely persist, but the clay-rich horizons often survive. In paleosols (fossil soils), carbonates are likely pedogenic if found within a distinct Bk horizon and associated with other pedogenic features typical of B horizons, such as clear ped structures, Mn- or Fe-oxides, and clay cutans (clay skins) on ped surfaces. Another common soil feature is deep cracking of the soil due to shrinking and swelling of soils and clays, producing dish-shaped structures in outcrop and leaving vertical marks from the shrink–swell movement (slickensides) on the ped surfaces. Such "vertic" features are characteristic of modern Vertisols (e.g., Ivanov et al. 1985; Debele 1985) and buried paleo-Vertisols across the region, including at Gona (Quade et al. 2004).

Shape and texture of nodules

Pedogenic carbonate in modern fine-grained soils tends to form as filaments or as coatings on peds and clasts in the early stages, maturing into coarse, 1–3-centimeter-diameter "popcorn" nodules with time. With further exposure these nodular zones can coalesce into continuously cemented ledges. The nodules themselves are composed of fine-grained micrite to microspar; coarser-grained sparry carbonate tends to be associated with other forms of postpedogenic cementation (Chadwick et al. 1988; Deutz et al. 2002). A representative sample of nodules should be examined in thin-section to verify their texture.

It is important to stress that micritic but nonpedogenic nodules are extremely common in the geological record. In such cases, nodules preserving sedimentary bedding or fossils such as aquatic snails and ostracodes are clearly nonpedogenic, whereas in many other cases the distinction is not clear based on textural criteria alone.

Reworking of nodules into contemporaneous channel lags

Secondary carbonate is likely pedogenic if it can be found reworked into contemporaneous channels. The presence of reworked carbonate in channel lags demonstrates that cementation occurred very early, prior to deep burial.

The reworked nodules should be identical in texture, shape, and isotopic composition to *in situ* paleosol carbonates. The clast load of shallowly entrenched, minor channels provides a better test of contemporaneity than those in major channels, since minor channels tend to rework material only from the local area.

Detrital contamination

One concern in the analysis of any pedogenic carbonate is the possible presence of detrital carbonate, which can be physically incorporated into growing pedogenic carbonate nodule, contaminating the primary pedogenic cements. Detrital carbonate is abundant in many basin deposits, often eroded from carbonate rocks (e.g., limestone, dolostone, lacustrine marl) exposed in the paleowatersheds of the basin. In many mature soils this carbonate is leached out of much of the profile in the early stages of weathering, but it can persist deeper in many soil profiles.

The best approach to detecting detrital contamination of soil nodules is to check for carbonate in unweathered parent material above and below the paleosol. If present, then one should sample nodules from portions of the soil profile where the matrix is leached of its original carbonate. In many East African soils, carbonate is uncommon in parent material since volcanic rocks tend to dominate the bedrock geology of rift basins.

CARBONATE FORMATION AND ISOTOPIC EQUILIBRIUM IN SOILS

The widely cited Equation 1 governs carbonate dissolution (right to left) and precipitation (left to right):

$$Ca^{2+}\,(aq) + 2HCO_3^-\,(aq) \leftrightarrow CaCO_3\,(s) + CO_2\,(g) + H_2O\,(aq) \qquad (1)$$

As is clear from the reaction, CO_2 degassing and soil dewatering can both drive the reaction to the right, leading to carbonate formation. Soil dewatering, partly by plants (evapotranspiration) and partly by evaporation, is probably the dominant mechanism. Some early models of pedogenic carbonate formation invoked the simple stoichiometry of Equation 1 to suggest that half of the C in solution as HCO_3^- derives from plant-derived CO_2 and the other half derives from dissolution of antecedent $CaCO_3$ such as marine limestone. We now know that this view is misleading for soils because it fails to take into account open-system, isotopic equilibrium considerations in addition to bulk chemical ones. The pore space of soils, even in drier regions, can be filled with high concentrations (up to 10,000 ppm) of CO_2, the result of root respiration

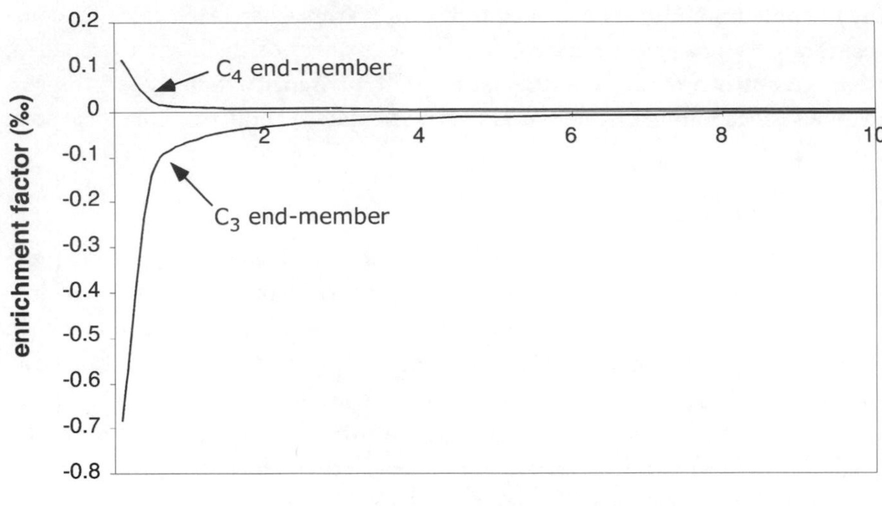

Figure 3.2. *Depletion or enrichment due to contamination by detrital carbonate ($\delta^{13}C$ (VPDB) = 0‰) of end-member C_3 ($\delta^{13}C$ = −12‰) or C_4 ($\delta^{13}C$ = +2‰) pedogenic carbonate values as a function of soil respiration rate, assuming carbonate accumulation rates of 0.5 g/cm²/1000 yrs. Contamination by detrital carbonates such as limestone has a negligible effect on the $\delta^{13}C$ value of pedogenic carbonate where respiration rates exceed ~0.5 mmoles/m²/hr.*

and tissue decay of plants. The atmosphere contains much lower levels (270 ppm prior to 1850) of CO_2 than typical soil pore space. The partial pressure differential of CO_2 between the atmosphere and soil sets up a strong diffusion gradient upwards from the soil to the atmosphere. The resulting flux out of a typical arid to subhumid soil is on the order of 1–10 mmoles/m²/hr, in contrast with the very slow rate of pedogenic carbonate formation in most soils, around 0.005 mmoles/m²/hr (Machette 1985). The latter provides an upper limit on the amount of Ca^{2+}, and hence nonplant carbon, that limestone dissolution can contribute, since the ratio of Ca/C from limestone dissolution is 1:1 (Equation 1). Assuming these conditions, it means that the ratio of carbon from plant CO_2 versus limestone dissolution in dissolved soil HCO_3^- is ≥ 175:1. Hence dissolution of limestone would have a negligible effect on the carbon isotopic value of the resultant pedogenic carbonate at all but extremely low soil-respiration rates (< 0.5 mmoles/m²/hr; Figure 3.2). Such low rates are only encountered in hyper-arid deserts such as the Atacama and do not occur in East Africa. Therefore, from the point of view of mass balance, plant-derived CO_2 dominates the carbon isotopic system of most soils, at least at deeper soil levels.

Another critical feature of soils, particularly drier soils, is that they tend to dewater slowly in the days to weeks after a rainfall event, and soil water remains under tension. The thin water films produce high surface-to-volume ratios in soil water, promoting rapid isotopic exchange between soil air and soil moisture. The long dewatering times (≥ 1 day) of most soils greatly exceed the rate constants of carbonate reactions (seconds to minutes) between carbon species (Fantidis and Ehhalt 1970; Dulinski and Rozanski 1990; Szaran 1997). These considerations make it likely that carbon isotopic equilibrium is maintained between solid and dissolved carbon species during pedogenic carbonate formation.

These two features—abundant CO_2 produced by plants compared to the small amount of C in pedogenic carbonate, and slow dewatering rates in soils—ensure that pedogenic carbonate undergoes continuous, open-system exchange with a large, isotopically stable plant CO_2 reservoir.

Plant-derived CO_2—and hence the local proportions of C_3 to C_4 plants growing at the soil surface—therefore can be predicted to determine the $\delta^{13}C$ value ($\delta^{13}C_{sc}$) of pedogenic carbonate found deep (> 50 cm) in soils where plant-derived soil CO_2 is more abundant. In shallow soils, atmospheric CO_2 comprises an important fraction of total soil CO_2, and the proportion of C_3 to C_4 plants cannot be calculated from $\delta^{13}C_{sc}$ values. For this reason, it is essential to obtain pedogenic carbonate at depth, generally greater than 50 centimeters, as discussed below. (Text continues on p. 72.)

In-depth Analysis

SOIL-DIFFUSION MODEL

The one-dimensional soil-diffusion model developed in Cerling (1984) and modified in Davidson (1995) and Quade et al. (2007) provides the quantitative framework for interpretation of pedogenic carbonate isotopic results. A number of key assumptions and boundary conditions underpinning the original soil-diffusion model are worth restating here.

(1) Exponentially decreasing soil CO_2 production with soil depth. A variety of evidence points to an exponential decrease in soil CO_2 production (ϕ_s) with soil depth (z) (Richter, 1987). The general form of this exponential decrease is:

$$\phi_s = \phi_s^0 e^{-z/k} \qquad (2)$$

where k is a constant that mainly varies according to the depth distribution of plant roots and soil organic matter, and ϕ_s^0 is the production rate of CO_2 at the

In-depth Analysis—*continued*

soil surface (z = 0). Diffusive transfer of soil CO_2 from the soil to the atmosphere should follow Fick's Second Law, which states that:

$$\frac{\partial^2 C}{\partial t^2} = D^s \frac{\partial_2 C}{\partial z^2} + \phi^s$$

where: D_s = the diffusion coefficient of CO_2 in soils (in cm^2/sec)
C = soil CO_2 concentration (in $mmoles/cm^3$)
t = time (in seconds)
z = depth (in cm)
ϕ_s = soil CO_2 production (in $mmoles/cm^2/sec$)

Soil CO_2 concentrations should be relatively stable (i.e., at steady-state) over the hours to weeks that pedogenic carbonate forms after a soil-wetting event, such that

$$\frac{\partial^2 C}{\partial t^2} = 0 ,$$

hence:

$$\frac{d^2 C}{dz^2} = -\frac{\phi_s}{D_s} \qquad (3)$$

and the partial differential becomes a simple differential with only two variables remaining. Substitution of Equation (2) into Equation (3) yields

$$\frac{d^2 C}{dz^2} = -\frac{\phi_s^0 e^{-z/k}}{D_s}$$

The solution to this second order differential equation is:

$$C^s = -\frac{k^2 \phi_s^0}{D_s}(1 - e^{-z/k}) + C^a \qquad (4)$$

using several boundary conditions, including $C_s = C_a$ at z = 0, where C_s and C_a are the concentrations of soil and atmospheric CO_2, respectively (see Quade et al., 2007).

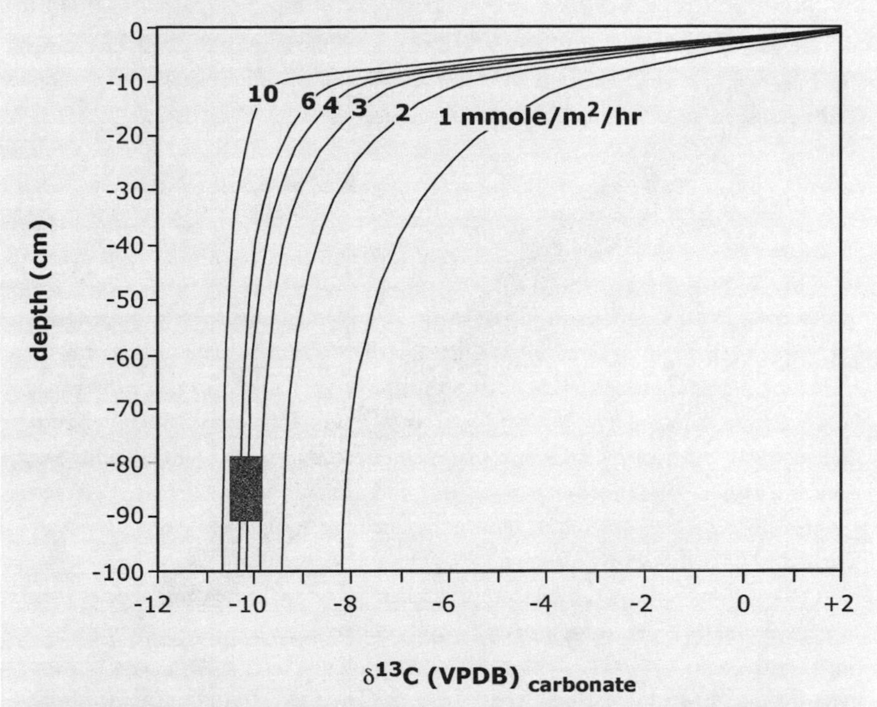

FIGURE 3.3. *The $\delta^{13}C$ (VPDB) value of pedogenic carbonate as a function of soil depth at the varying soil respiration rates indicated. Model conditions are a uniform soil CO_2 production function and conditions for Gona, Ethiopia ($T = 23.5°C$, $\delta^{13}C_a = -6.5$ ‰, $\delta^{13}C_{resp.} = -24.3$‰, $pCO_{2atm} = 270$ ppm, porosity = 0.4). Most East African pedogenic carbonate probably forms at respiration rates between 3 and 9 mmoles/m²/hr, introducing a ~0.8‰ uncertainty in the resultant $\delta^{13}C$ value of pedogenic carbonate at > 50 cm depth in the soil.*

Two key unknowns in Equation 4 are soil CO_2 production rates at the surface (ϕ_s^0), and the characteristic CO_2 production depth, or k, for the exponential CO_2 production function (Equation 2). We estimate the production rates to vary between 3 and 9 mmoles/m²/hr in most East African soils, producing a 0.8‰ uncertainty in the $\delta^{13}C_{sc}$ value produced by C_3 plants (Figure 3.3). We estimate k to be about 20–30 centimeters in most East African ecosystems; that is, most CO_2-producing tissue, both live and dead, is in the upper 30 centimeters of the soil, and not uniformly distributed over 1 meter of soil depth, as modeled by the original solutions to the one-dimensional soil diffusion models (e.g., Cerling 1984).

Determination of C_s contributes to estimation of the average $\delta^{13}C$ value of some mixture of C_3 and C_4 plants growing at a soil site ($\delta^{13}C_{plant}$), using the approximation presented in Davidson (1995) but solved in terms of $\delta^{13}C_{plant}$:

$$\delta^{13}C^{plant} = \frac{\delta^{13}C_s - 4.4 - \delta^{13}C_a R(z) + 4.4R(z)}{1.0044(1-R(z))} \quad (5)$$

where $\delta^{13}C_s$ and $\delta^{13}C_a$ are the $\delta^{13}C$ values of soil CO_2 and air, respectively, and $R(z) = C_a / C_s$ at the depth of sampling *z*. This equation subsumes the effects of kinetic enrichment in ^{13}C of soil CO_2 during diffusion. This arises from the fact that $^{13}CO_2$ diffuses more slowly than $^{12}CO_2$ from the soil. This sets up a temperature-independent enrichment of CO_2 in the soil compared to the CO_2 flux leaving the soil of 4.2 to 4.4‰ (Davidson, 1995). This enrichment factor has been observed empirically (Cerling et al. 1991a) and can be calculated using Stephen–Stokes Law applied to CO_2 diffusing through air (Cerling 1984).

(2) Slow, open-system carbon isotopic exchange of dissolved carbon species with plant-derived CO_2. At isotopic equilibrium, pedogenic carbonate will be enriched in ^{13}C with respect to soil CO_2 by 9–11‰, depending on temperature (25–5°C). The $\delta^{13}C_s$ values required to solve for the $\delta^{13}C_{plant}$ in Equation (5) can be obtained from the $\delta^{13}C$ value of pedogenic carbonate ($\delta^{13}C_{sc}$),

$$\delta^{13}C = \frac{(\delta^{13}C_{sc} + 1000)}{\left(\frac{11.98(\pm 0.13) - 0.12(\pm 0.1)T}{1000} + 1\right)} - 1000 \quad (6)$$

where $11.98(\pm 0.13) - 0.12(\pm 0.1)T$ expresses the temperature-dependent (T = temperature in °C) enrichment factor between calcite and gaseous CO_2 (Romanek et al. 1992). Inspection of this equation shows that the enrichment factor is not strongly sensitive to temperature, with a slope of ~0.12‰/°C.

Substitution of Equation (6) into Equation (5) allows $\delta^{13}C_{plant}$ to be calculated from $\delta^{13}C_{sc}$. The fraction C_4 biomass that once grew on the landscape can in turn be calculated from:

$$\text{fraction } C_4 \text{ biomass} = (\delta^{13}C_{plant} - \delta^{13}C_{C3})/(\delta^{13}C_{C4} - \delta^{13}C_{C3}) \quad (7)$$

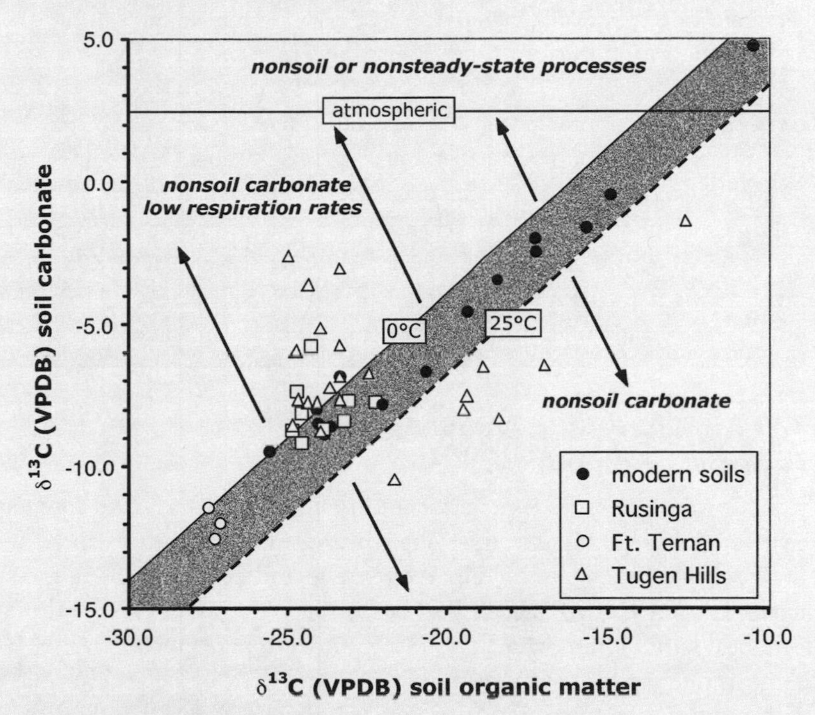

FIGURE 3.4. *The* $\delta^{13}C$ *(VPDB) value of soil organic matter versus that of co-existing pedogenic carbonate. The shaded zone denotes* $\Delta_{sc\text{-}om}$ $(= \delta^{13}C_{sc} - \delta^{13}C_{om})$ *values of 13.4–16.4‰ predicted by the soil diffusion model of* Cerling (1984), *and observed in most modern soils. The zone depicted also assumes uniform soil* CO_2 *production with depth. Exponential soil* CO_2 *production with a k value of ~30 cm would shift the shaded zone to the right by ≤ 0.4‰.* $\Delta_{sc\text{-}om}$ *values from Ft. Ternan and some Rusinga Island paleosols fall within this range, supporting a pedogenic origin. Most values from the Tugen Hills reported by* Kingston (1992) *fall outside the shaded zone (*$\Delta_{sc\text{-}om} \neq$ *13–16‰), a situation generally produced by low soil respiration rates and/or a non-soil origin for carbonates.*

where $\delta^{13}C_{plant}$ = the estimated carbon isotopic value for bulk plant cover, from Equations (5) and (6),

$\delta^{13}C_{C3}$ = the average carbon isotopic value of xerophytic East African C_3 vegetation = −25.5 ± 1.3‰ (for preindustrial conditions), and

$\delta^{13}C_{C4}$ = the average carbon isotopic value of xerophytic East African C_4 vegetation (for NAD-ME and PCK metabolisms) = −11.8 ± 1.4‰.

In-depth Analysis—*continued*

In summary, where soil respiration rates are high (> 3–5 mmoles/m^2/hr), the isotopic composition of pedogenic carbonate should differ (Δ_{sc-om}, where the subscript "om" denotes soil organic matter) from that of coexisting plant cover by about 13–16‰ for the low (< 500 ppmV) atmospheric CO_2 case. This difference ($\Delta_{sc-om} = \delta^{13}C_{sc} - \delta^{13}C_{om}$) involves both equilibrium fractionation between $CaCO_3$ and CO_2 of 9–11‰, and kinetic diffusion effects of ~4.4‰. This difference is observed in modern soils, verifying the general validity of the model (Figure 3.4). In practice, however, there is considerable uncertainty in the exact $\delta^{13}C_{sc}$ value of the C_3 end-member (± 17%, see error analysis discussed in Quade et al., 2004), due to ranges in possible $\delta^{13}C_{plant}$ values for C_3 and C_4 plants, and to uncertainties in soil temperature and in soil respiration rates.

THE OXYGEN ISOTOPIC COMPOSITION OF PEDOGENIC CARBONATE

The $\delta^{18}O$ values of pedogenic carbonate ($\delta^{18}O_{sc}$) are determined by factors largely related to local climate, in strong contrast to $\delta^{13}C_{sc}$ values, which, as we already discussed, are determined by the makeup and density of plant cover on a soil site. The key climate factors that determine $\delta^{18}O_{sc}$ values are (1) the $\delta^{18}O$ value of local rainfall from which the pedogenic carbonate precipitates, (2) local soil temperature, and (3) evaporative modification of $\delta^{18}O$ values of local rainfall within the soil, as the soil is dewatered. Of these factors, (1) and (3) are the most important in hot, evaporitic settings like East Africa. As regards (1), $\delta^{18}O$ values in local rainfall are determined by several factors such as storm source, local elevation, and the extent of rainout from local cloud masses. In the next section we examine these factors in more detail, and in the following section on the isotopic record from Gona, show how the apparently complex determinants of $\delta^{18}O_{sc}$ values can be separated and useful paleoclimate information obtained. (Text continues on p. 75.)

In-depth Analysis

BACKGROUND

The oxygen isotopic composition of pedogenic carbonate ($\delta^{18}O_{sc}$) is determined by several variables summarized in the following general Equation 8:

$$\Delta\delta^{18}O_{sc} = \frac{\partial\delta^{18}O_{sc}}{\partial T}(\Delta T) + \frac{\partial\delta^{18}O_{p}}{\partial T}(\Delta T) + \frac{\partial\delta^{18}O_{p}}{\partial A}(\Delta A) + \frac{\partial\delta^{18}O_{sea}}{\partial V}(\Delta V) + \frac{\partial\delta^{18}O_{sw}}{\partial RH}(\Delta RH) + \frac{\partial\delta^{18}O_{p}}{\partial E}(\Delta E)$$

This array of factors can conveniently be grouped for discussion purposes under three headings:

1. TEMPERATURE EFFECTS

Temperature can potentially influence $\delta^{18}O_{sc}$ values in two ways, through the influence of (1) soil temperature on fractionation between water and calcite during pedogenic carbonate formation, and (2) of air temperature on $\delta^{18}O$ values of precipitation ($\delta^{18}O_p$).

In Equation (8) above,

$$\frac{\Delta\delta^{18}O_{sc}}{\partial T}$$

expresses the dependence of the $\delta^{18}O_{sc}$ value on soil temperature, as calculated from the fractionation factor ($\alpha_{c\text{-}w}$) between calcite and water

$$1000 \ln \alpha^{c\text{-}w} = \frac{18030}{T} - 32.42,$$

with temperature (T) in K (Kim and O'Neil, 1997). The temperature dependence of calcite-water fractionation equates to ~0.22–0.24‰/1°C, meaning that (1) the large uncertainties in the $\delta^{18}O$ value of soil water ($\delta^{18}O_{soilw}$) produce major uncertainties in reconstructed paleotemperature, but conversely, (2) large uncertainties in paleotemperature produce relatively small uncertainties in reconstructed $\delta^{18}O_{soilw}$. This is why the $\delta^{18}O$ value of pedogenic carbonate is a poor paleothermometer but can be useful for constraining paleo-$\delta^{18}O_{soilw}$ values (but see Passey et al. 2010 and Quade et al. 2013 for recent work on clumped isotope carbonate paleothermometry). Moreover, soil temperature approaches mean annual temperature (MAT) at depth in soils (Hillel 1982), further reducing potentially large seasonal temperature extremes. So, in the example from Gona, our estimate of long-term MAT of 25 ± 5°C only introduces a ~2‰ uncertainty in our estimates of paleo-$\delta^{18}O_{soilw}$ values.

In Equation (8),

$$\frac{\partial\delta^{18}O_p}{\partial T}$$

expresses the dependence of the oxygen isotope value of precipitation ($\delta^{18}O_p$) on

IN-DEPTH ANALYSIS—*continued*

temperature. At higher latitudes there is strong correlation between these factors (the "Dansgaard Relationship") (Dansgaard 1964), but at low latitudes there is not, hence we drop this term from further consideration.

2. SOIL WATER EVAPORATION

In Equation (8),

$$\frac{\partial \delta^{18}O_{soilw}}{\partial RH}$$

expresses the dependence of the oxygen isotope value of soil water ($\delta^{18}O_{soilw}$) on relative humidity. Evaporation can strongly increase $\delta^{18}O_{soilw}$ values with respect to parent $\delta^{18}O_p$ values (Hsieh et al. 1998; Quade et al. 2007). When relative humidity is ~100%, $\delta^{18}O_{soilw} = \delta^{18}O_p$. Evaporative enrichment in ^{18}O of water increases linearly with decreased relative humidity, as described by Craig and Gordon (1965). In semiarid settings akin to much of East Africa, enrichment in ^{18}O of soil water has been widely demonstrated (Hsieh et al. 1998; Quade et al. 2007). In such settings, $\delta^{18}O_p$ values reconstructed from $\delta^{18}O_{sc}$ values are maximum estimates.

3. δ^{18}O COMPOSITION OF PRECIPITATION ($\Delta^{18}O_p$)

At low latitudes $\delta^{18}O_p$ is influenced by several factors, including rainfall amount and changes in the $\delta^{18}O$ value of sea water (the source of moisture).

In Equation (8): The dependence of $\delta^{18}O_p$ on rainfall amount is expressed by

$$\frac{\partial \delta^{18}O_p}{\partial A},$$

referred to as the Amount Effect. It is the dominant control on $\delta^{18}O_p$ at low latitudes above 20°C (Straight et al. 2004), and is ~1.5‰/10 cm rainfall globally (Rozanski et al. 1993). In most of East Africa, mean annual temperature only drops below 20°C above ~1700–1800 m.

The dependence of the oxygen isotope value of sea water ($\delta^{18}O_{sea}$) on ice volume is expressed by

$$\frac{\partial \delta^{18}O_{sea}}{\partial V}.$$

In-depth Analysis—*continued*

The ocean is the source of most atmospheric moisture, hence changes in $\delta^{18}O_{sea}$ should translate directly into changes in $\delta^{18}O_p$. Well-documented changes in global ice volume have changed the isotopic composition of the oceans < 1.0–1.5‰ in the past 50 myr (million years) (Lear et al., 2000). Since the mid-Miocene these changes have been small, < 0.5‰.

The dependence of $\delta^{18}O_p$ on elevation is expressed by

$$\frac{\partial \delta^{18} O_p}{\partial E}.$$

Elevation may have changed substantially at some sites in East Africa in the Neogene. The global average relationship is about −2.8‰/km (Poage and Chamberlain 2001).

In addition, changes in the location of moisture source areas could affect local $\delta^{18}O_p$ values, although any accompanying changes in rainfall amount should already be subsumed under the

$$\frac{\partial \delta^{18} O_p}{\partial A}$$

term.

GONA: A CASE STUDY

Over the past ten years we have conducted intensive isotopic studies at Gona in north-central Ethiopia (Figure 3.1). In this chapter we essentially double the number of published analyses (Levin et al. 2004) and refine the age of some published samples based on new age information.

Gona archives over 6 myr (~6 to < 0.2 mya [million years ago]) of basin sedimentation and human evolution (Kleinsasser et al. 2008; Levin et al. 2008; Quade et al. 2008). Numerous hominin remains have been discovered, including those of *Homo erectus* in the younger Busidima Formation (2.7 to < 0.2 mya), *Australopithecus afarensis* in the Hadar Formation (3.4 to 2.9 mya), and *Ardipithecus ramidus* at 4.5 to 4.3 mya (Semaw et al. 2005) from the Sagantole Formation. The Busidima Formation is entirely fluvial, having been deposited by the ancestors to the modern Awash River and its tributaries that flow through and dissect the area today (Quade et al. 2004). The Hadar Formation at Gona is fluviolacustrine, and was deposited by a through-flowing (fluvial) or impounded (lacustrine) Awash River. The Sagantole Formation is also fluviolacustrine, and was deposited closer to the flanks of the paleo–Awash Valley formed by major basalt flows. The only major gap in the record occurs between the top (4.2 mya) of the Sagantole Formation and bottom (≥ 3.5 mya) of the Hadar Formation at Gona. Fossil soils and teeth abound in the geologic record, providing a rich history of landscape and paleodietary change through time (Levin et al. 2008). Age control on the deposits is provided by a combination of published $^{40}Ar/^{39}Ar$ dates on felsic tuffs, magnetostratigraphy, and tephrostratigraphy (Quade et al. 2004, 2008).

Paleosols are readily recognizable in both the Busidima and Sagantole Formations. They are characterized by dark brown color (5 YR 4/2 dry), deep vertical cracking, and slickensides along fractures, all features of the soil class known as Vertisols common across East Africa today. Pedogenic carbonate occurs at all depths in soils but tends to be more concentrated in the lower half of paleosol profiles. Carbonate assumes several different forms, from isolated 1–5-centimeter nodules and rhizoliths (calcareous root castes) to plates found in vertic fractures. All of these carbonate types are found reworked into contemporaneous secondary channels common throughout the Busidima and Sagantole Formations, demonstrating that the nodules must have been present as the paleosols were being buried.

Detrital carbonate is not present in the parent sediments of Gona: bedded siltstones in the sections are noncalcareous. Modal analyses of sandstones show that they are free of detrital carbonate and are composed of quartz, feldspar, and volcanic lithic fragments. The relatively soft, reworked pedogenic carbonate nodules in the coarse bedload of paleochannels do not contribute appreciably to the sand- and silt-size fractions that ultimately serve as parent material to the paleosols.

$\delta^{13}C_{sc}$ RECORD AT GONA

About 285 analyses of pedogenic carbonate at Gona reveal a pattern of gradually increasing $\delta^{13}C_{sc}$ values upsection, from values as low as –11.9‰ in the Sagantole Formation to values commonly as high as –0.3‰ in the upper Busidima Formation. We also see significant scatter in $\delta^{13}C_{sc}$ values from level to level. For example, in the copiously sampled Sagantole Formation and portions of the middle Busidima Formation, the spread in values over short time intervals can be up to 6.5‰.

How do we interpret the large scatter in $\delta^{13}C_{sc}$ values within single stratigraphic levels at Gona? To answer this we must first consider how much time each of our analyses represents. Recent U-series dating of desert pedogenic carbonates suggest sequential accretion of carbonate with little isotopic disturbance of individual layers (Sharp et al. 2003). This suggests that carbonate

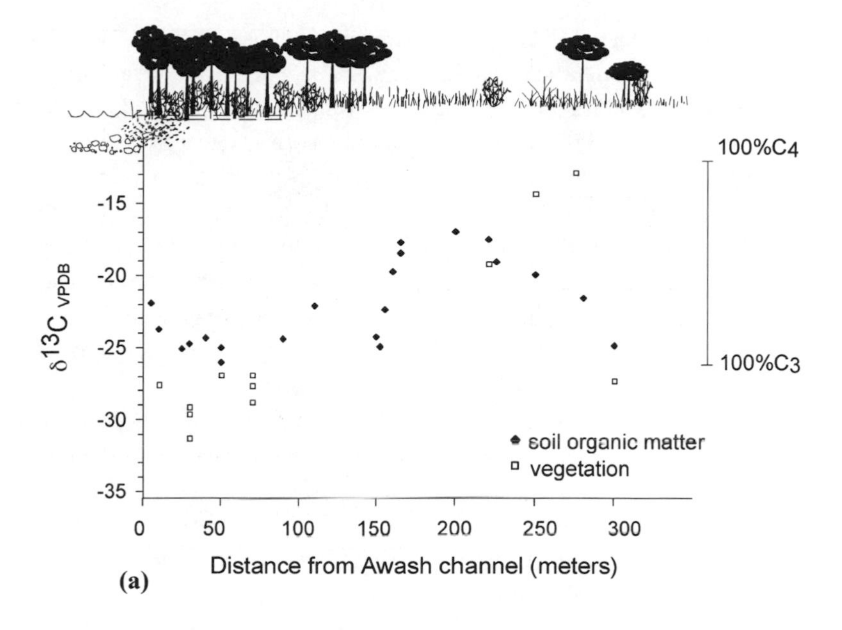

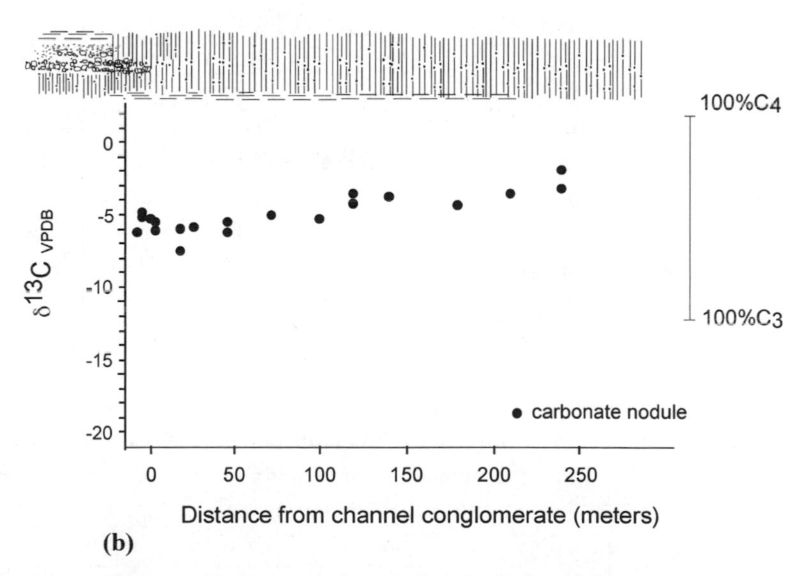

Figure 3.5. *The $\delta^{13}C$ value of (a) vegetation and soil organic matter with distance from the modern Awash River channel, and (b) paleosol carbonate with distance from a paleochannel of the ancestral Awash River at Gona. The systematic increase in $\delta^{13}C$ values away from the channel—both modern and ancient—reflects an increase in C_4 plants.*

cements preserve a series of isotopic "snapshots" of local vegetation. The duration of the snapshot depends on sample size and rate of carbonate accumulation. Our samples of carbonate are typically 0.1–1.0 milligrams, ground and homogenized, then subsampled for analysis. Pedogenic carbonate in the southwestern United States accumulates at a rate of 0.3 to 0.6 g/cm^2-1000 years. If representative of rates in East Africa, this suggests that our typical sample averages about 100–1,000 years of time. This would represent a few to many average life spans of plants, which for trees is hundreds of years, and for perennial grasses, a few years. Therefore, $\delta^{13}C_{sc}$ values are recording short temporal (and hence small spatial) snapshots of what was apparently an ecologically highly heterogeneous (C_3 and C_4) landscape at Gona, not a well-mixed one.

An illustration of the brevity of this landscape snapshot of our samples is visible in the $\delta^{13}C_{sc}$ values from pedogenic carbonate in a single paleosol lateral to a paleochannel at the 2.7-mya level at Gona (Figure 3.5). The $\delta^{13}C_{sc}$ values increase systematically with distance from the channel. This is exactly the pattern seen today with distance from the Awash River at Gona: a narrow strip of gallery forest (all C_3) gives way to edaphic grassland (largely C_4). The modern Awash is a perennial meandering system that migrates laterally over time. Gallery forest should closely follow the lateral migration of the river. Apparently, the migration time of the paleo–Awash River represented in the Busidima Formation was slower than the average duration of paleosol carbonate-nodule formation (Figure 3.5).

Against this background, we can use Equation (7) to reconstruct variation in the fraction C_3/C_4 biomass through time. The overall range in $\delta^{13}C_{sc}$ values at Gona shows that the percentage of C_4 grasses ranged from 0 percent (–12.5‰) to 85 percent (–0.3‰). By the time of the oldest soil samples at Gona, 4.5 mya, C_4 grasses are already well established on the landscape, although subordinate in their overall contribution to soil CO_2 compared to C_3 trees and shrubs. This suggests that the habitats available to *Ardipithecus ramidus* found at these stratigraphic levels ranged from open woodlands at one extreme to evenly mixed grasses and trees/shrubs on the other. This picture of mixed habitat at 4.5 mya is strongly supported by the dominant contribution of C_4 grasses to the diets of large herbivores found in direct association with *Ar. ramidus* (Semaw et al. 2005; Levin et al. 2008). From the Hadar Formation at Gona, and nearby Hadar and Dikika, $\delta^{13}C_{sc}$ values display the same wide range as those in the Sagantole Formation (Figure 3.6a), suggesting a similar range of habitats available to *Australopithecus afarensis*.

The C_4 grasses increase at some Gona sites at 2.7 mya (Figure 3.6a). This increase lies stratigraphically just below the first appearance of stone tools at 2.58

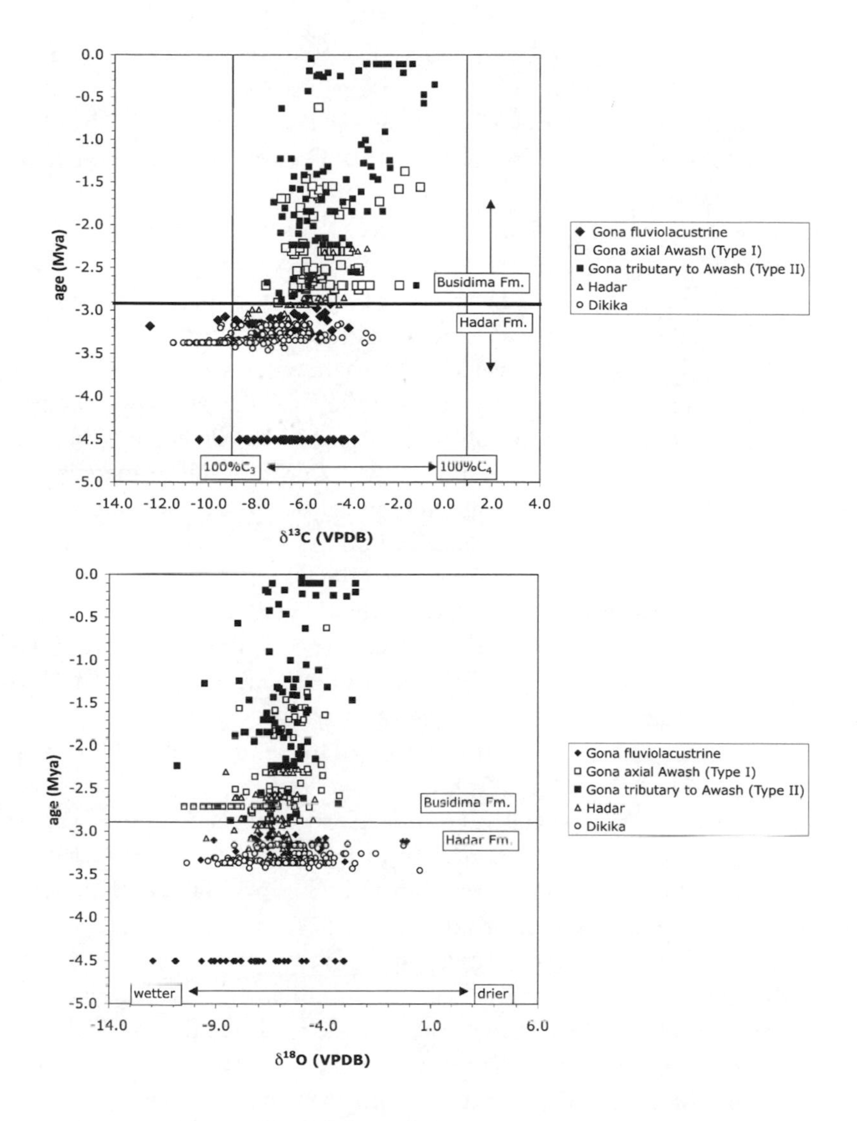

FIGURE 3.6. *(a) The $\delta^{13}C$ (VPDB) value and (b) the $\delta^{18}O$ (VPDB) value of pedogenic carbonate from the northern Awash River region, Ethiopia. Paleosols found in association with Type I (open squares) channels (Quade et al. 2004) formed on floodplains of the ancestral Awash River, whereas paleosols found with Type II channels (solid squares) formed along large and small tributaries to the Awash River on distal alluvial fans fringing the valley axis. These channel types are not distinguishable prior to 2.9 mya, where paleosols occur in a general fluvio-lacustrine association (solid diamonds). Isotope data are from Levin et al. (2004) and unpublished. Hadar data are from Aronson et al. (2008), and Dikika data from Wynn et al. (2006).*

mya (Quade et al. 2004; Semaw et al. 2003) and astride the Hadar–Busidima Formation boundary, a point to which we return below. Some $\delta^{13}C$ values are close to 0‰, representing the earliest evidence of continuous savannalike grasslands in the Gona record. A wide spectrum of $\delta^{13}C$ values between –10 and 0‰ persists through the top of the record.

One critical consideration in long sections such as at Gona is the potential bias produced by the changing position of paleosols on the paleolandscape. At Gona there were significant changes in this respect. In the older (4.5–4.3 mya) Sagantole Formation, paleosols developed on fluvial deposits deposited by small, perennial river systems that flowed into a shallow lake. Carbon isotopes in paleosols only archive the vegetation growing in this alluvial corridor between the lake and the extensive basalt tablelands to the west. The basalt fields—possible habitat for the ardipithecines found in the alluvial and lake-margin facies of the Sagantole Formation—remain unsampled due to the lack of pedogenic carbonate development. During Hadar (≥ 3.5 to 2.9 mya) and lower Busidima Formation time (2.7 to 1.6 mya), paleosols developed on floodplains on the margin of the ancestral Awash River or lakes formed by impoundment of the river. For most of upper Busidima Formation time (< 1.6 mya), the ancestral Awash River shifted eastwards out of the project area, and the distal portions of alluvial fans transgressed across much of the project area.

We originally speculated that the shift toward higher $\delta^{13}C$ values in the upper part of the Busidima Formation at Gona was related to this local facies shift, and not to changes in regional vegetation patterns toward more C_4 grasses (Quade et al. 2004; Levin et al. 2004). With a much larger data-set in hand, we no longer favor this view because the shift to higher $\delta^{13}C$ values starting at 2.7 mya preceded the facies shift after 1.6 mya.

We examined the potential relationship between facies and $\delta^{13}C$ values by subdividing all the $\delta^{13}C$ values by their local facies association. The facies can be readily distinguished by channel type (Quade et al. 2004). Type I channels are characteristic of the axial drainage system occupied by the ancestral Awash River. The paleosols in this association formed on the floodplain of the axial Awash River, under the gallery forest (*Tamarix–Ficus*) and edaphic grasslands visible on the floodplain today. Type II channels were deposited by large and small tributaries to the paleo-Awash that flowed across basin-margin alluvial fans. Today, these areas are covered by *Acacia–Commiphora* woodland and xeric grasslands. Our plot (Figure 3.6a) of $\delta^{13}C$ values shows that the expansion of C_4 grasses is not just associated with Type II channels. Paleosols on the ancestral Awash floodplain (with Type I channels) also witnessed the expansion of C_4 grasses.

Recent studies by Aronson et al. (2008) in nearby Hadar, Wynn et al. (2006) in Dikika, and by Quinn et al. (2007) from Koobi Fora east of Lake Turkana, all make strong cases that major depositional shifts explain the accompanying increases in $\delta^{13}C_{sc}$ values. In these areas the shift in environments is from closed or semiclosed basins filled with lakes to open fluvial systems drained by large rivers. Rift basins often undergo this transition as isolated basins gradually expand and fill with lake sediments, to be eventually interlinked by large externally drained rivers (Gawthorpe and Leeder 2000; Quade et al. 2008). The transition from the Hadar Formation to the Busidima Formation 2.9 to 2.7 mya is one classic example of this change in the northern Awash (Quade et al. 2008), and the Koobi Fora Formation at Turkana between 2.0 and 1.5 mya is another (Quinn et al. 2007).

In this process, hydrologic conditions change from poorly drained within closed basins to well drained within hydrographically open fluvial systems. In general, ecosystems will respond with the expansion of grass at the expense of forest as drainage increases. Both Aronson et al. (2008) and Wynn et al. (2006) attributed the modest increases in δ^{13}Csc values across this boundary to the facies shift that marks the Hadar–Busidima Formation boundary in the northern Awash (Figure 3.6a). Quinn et al. (2007) make the same case in the Koobi Fora Formation at Turkana. These modest environmental shifts clearly relate to the gradual tectonic evolution that most East African rift basins probably experienced, and not to regional climate change.

A related observation for the record from Gona is the large spread in carbon isotopic values at any one stratigraphic level. This scatter is real and reflects the diverse nature of vegetation cover across the landscape at any one time. This ecologic diversity is still readily apparent today. The closed gallery forest along the banks of the Awash today is entirely C_3, as apparent above ground and in the carbon isotopic composition of modern-soil organic matter (Levin et al. 2004). Laterally, gallery forest can give way abruptly—over less than 10 meters—to nearly pure edaphic (groundwater supported) C_4 grassland. Edaphic grassland is locally interrupted by nongallery forest trees. Beyond the margins of the floodplain, distal alluvial-fan areas are vegetated by locally wooded savanna (Quade et al. 2004). The carbon isotopic record suggests that this same heterogeneity in vegetation characterized the setting for all of the Busidima Formation (Levin et al. 2004).

In many ways our efforts at landscape reconstruction at Gona mirror the studies of Sikes et al. (1999) at Olorgesailie and more recently Behrensmeyer et al. (2007) from Pakistan. Both studies undertook major sampling campaigns along single paleosol horizons in order to try to reconstruct paleoenvironments in the spatial rather than the temporal dimension. As at Gona, both studies detected lateral variations in $\delta^{13}C$ values that likely reflect heterogeneity in the proportion of C_3 and C_4 plants in the landscape. The challenge for all such studies is to place fossil hominins on this reconstructed landscape. At Olorgesailie, the environment studied at 0.99 mya turned out to be largely grassland (see further discussion below), and hence this was the setting for *Homo erectus*. A key advantage of the Olorgesailie study was that the paleosols studied were relatively immature, meaning the isotopic "snapshot" was relatively short and more ecologically specific. This kind of reconstruction may not always be possible elsewhere. At Gona, for example, many of the Oldowan archaeological sites occur in mature Vertisols (Quade et al. 2004). With so much time represented in such soils, it is unclear where the hominin occupation of the site fits into the life of the paleosol. Indeed, pedogenesis seems to have overprinted the original parent floodplain sediments containing the artifacts, and hence pedogenesis (and the nodules that we analyzed) postdates the occupation. In these cases, $\delta^{13}C_{sc}$ values only provide a general and time-averaged picture of the range of environments available to the hominins.

The $\delta^{18}O_{sc}$ record at Gona

The existing Gona record has been presented and discussed in detail by Levin et al. (2004); here we contribute new $\delta^{18}O_{sc}$ results to the discussion. We sampled paleosol nodules at depths of greater than 30 centimeters in soils to reduce the potential for evaporative enrichment (see Quade, Cerling, and Bowman 1989a). In addition, we measured the $\delta^{18}O$ value of modern local waters in order to place our $\delta^{18}O_{sc}$ results in the context of modern climate conditions. In future we intend to compare $\delta^{18}O_{sc}$ results to other coeval carbonates (aragonitic shell, fossil teeth) to estimate degree of evaporative enrichment (aridity) of primary $\delta^{18}O_p$ values.

The $\delta^{18}O_{sc}$ record at Gona exhibits two main trends: (1) an increase in minimum $\delta^{18}O_{sc}$ values from the early Pliocene through the Pleistocene, and (2) a large range in $\delta^{18}O_{sc}$ values at all well-sampled intervals. Averaged $\delta^{18}O_{sc}$ values binned at 100-kyr (thousand year) increments exhibit a 2‰ increase from the early Pliocene to the late Pleistocene (Figure 3.7).

Following from the introduction to oxygen isotopes above, we consider changes in (1) soil-water evaporation, (2) the $\delta^{18}O$ value of rainfall, and (3) paleoelevation as the first-order determinants of $\delta^{18}O_{sc}$ values at Gona. Although it is difficult to quantify the effects of soil-water evaporation on $\delta^{18}O_{sc}$ values, minimum $\delta^{18}O_{sc}$ values from well-sampled time intervals can be used to estimate temporal changes in $\delta^{18}O_p$ values because they represent soil-water compositions least affected by evaporation. Comparison of the Gona $\delta^{18}O_{sc}$ minima to modern water $\delta^{18}O_p$ values allows us to evaluate if, and by how much, $\delta^{18}O_p$values have changed through time.

Unevaporated groundwater from shallow local aquifers at Gona today has a $\delta^{18}O$ (VSMOW)[2] value of −2.9‰. This value likely closely approximates modern mean-annual precipitation (Levin et al. 2004). The $\delta^{18}O_{sc}$ (VPDB) values formed in equilibrium with this water at 26°C (modern MAT for Gona) would equal −5.5‰, but could range from −4.3 to −6.3‰ if formed at temperatures between 20 and 30°C (Kim and O'Neil 1997). Pedogenic carbonates formed on late Pleistocene gravels capping the whole Gona sequence dated to less than 0.2 mya have a minimum $\delta^{18}O_{sc}$ value of −6.4‰, whereas in the early Pliocene, minimum $\delta^{18}O_{sc}$ values reach as low as −11.9‰. These values are much less than the lowest possible $\delta^{18}O_{sc}$ values for pedogenic carbonates formed in equilibrium with modern waters, and suggest that rainfall in the early Pliocene had $\delta^{18}O_p$ values ~6.5‰ lower than today. This closely matches the estimates made by Hailemichael et al. (2002) of decreases in $\delta^{18}O_p$ values of 6–7‰ over the last 3.4 myr.

Having separated out the maximum contribution of evaporation to $\delta^{18}O_{sc}$ increases, how do we evaluate the contribution of the remaining terms in Equation 8 to this ~6.5‰ increase in $\delta^{18}O_p$ (and hence $\delta^{18}O_{sc}$ values) values through time? First, we note that an approximately 5°C decrease in average T since the mid-Miocene would increase $\delta^{18}O_{sc}$ values by only ~1‰ (since

$$\frac{\partial(\varepsilon_{c\text{-}w})}{\partial T}(\Delta T) = 0.22‰/°C \times 5°C \approx 1‰).$$

Gradual increase in ice volume

$$(\frac{\partial\delta^{18}O_{sea}}{\partial V})$$

would have further increased $\delta^{18}O_p$ (and therefore $\delta^{18}O_{sc}$) values by ~0.5‰. We consider the effects of local elevation change to have been small

$$(\frac{\partial\delta^{18}O_p}{\partial E} = (\Delta E) = 0),$$

2. VSMOW is an abbreviation for Vienna Standard Mean Ocean Water which is a standard used to calculate $\delta^{18}O$ values of waters. The $\delta^{18}O$ values calculated from the VSMOW and VPDB standards can be compared using the following relationship: $\delta^{18}O_{VSMOW} = 1.03091 * \delta^{18}O_{VPDB} + 30.91$.

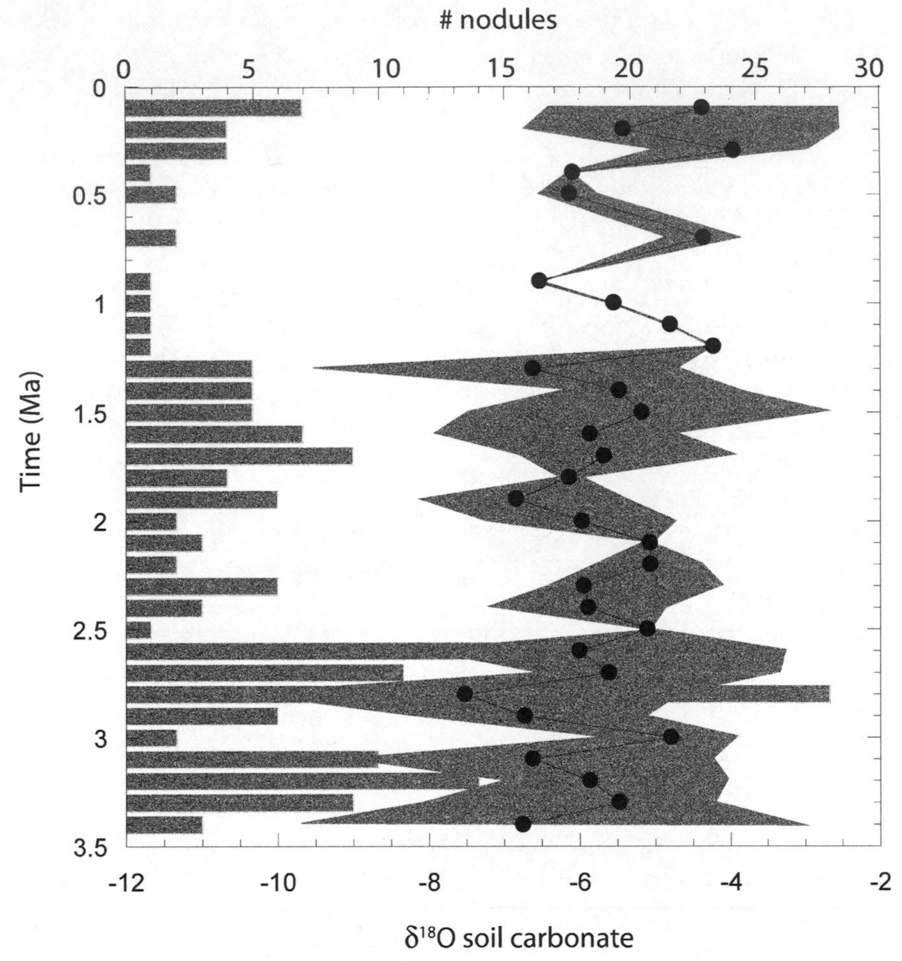

FIGURE 3.7. *Averaged $\delta^{18}O_{sc}$ values from Gona binned, in 100-kyr increments, with range of $\delta^{18}O_{sc}$ values at each interval marked by gray shading. Number of nodules (analyses) included in each 100 kyr are plotted by underlying histogram. Early Pliocene data are not included in this plot due to uncertainties of the chronologic relationships among these samples.*

since the large increases in $\delta^{18}O_p$ implied by the $\delta^{18}O_{sc}$ record are seen throughout East Africa, as discussed in the next section of this chapter. That leaves two factors, changes in moisture source and rainfall amount, to account for the remaining 5.0‰ (6.5‰–1.5‰) postulated change in $\delta^{18}O_p$ values.

The maximum reduction in rainfall over the past 5 myr can be constrained as follows. Pedogenic carbonate is present throughout the section, requiring rainfall amounts under about 1 meter/year, based on the rainfall limits of carbonates in modern soils. Current mean annual rainfall at Gona is 50 centimeters/year. Therefore, a change of up to about 50 centimeters/year is permitted at Gona by the continuous presence of pedogenic carbonate in the record. From Equation 9,

$$\frac{\partial \delta^{18}O_p}{\partial A}(\Delta A)=(1.5‰/10\ \text{cm} \times 30\ \text{cm}) = 4.5‰.$$

Thus, a near 30-percent reduction in rainfall over the past 5 myr would account for the 4.5‰ increase in $\delta^{18}O_p$ values left over after accounting for the other factors. We regard this estimate as an upper limit, since the sensitivity of $\delta^{18}O_p$ to rainfall amount in East Africa today appears to be less than the global average value of 1.5‰/10 cm (see Levin et al. 2004; Levin et al. 2009). This in turn makes it possible that changes in air mass sources made some contribution to the 6.5‰ increase in $\delta^{18}O_p$ through time.

Amid changes in $\delta^{18}O_p$ values discernible from $\delta^{18}O_{sc}$ minima, there is a large range in $\delta^{18}O_{sc}$ values within time periods that represent spatial and temporal environmental variation (Figure 3.7). This range in values is captured in the Gona record where carbonates were intensively sampled, within nodules, throughout single paleosols, and across paleolandscapes. Sampling density correlates to variability at single stratigraphic levels (Figure 3.7). Of the fifty-five nodules from Gona that have been microsampled, the maximum range of $\delta^{18}O_{sc}$ values within a single nodule is 2.9‰ (minimum 0.1‰) (Levin, 2002). This degree of intranodule variation occurs within the same set of nodules that exhibited a 5‰ range in $\delta^{18}O_{sc}$ values within 250 meters of one paleosol. Within 100-kyr intervals, $\delta^{18}O_{sc}$ values of different nodules range up to 6.9‰ (Figure 3.7). These ranges likely represent a combination of changing soil-water evaporation conditions (changes in local geography or environmental aridity) and temporal (glacial–interglacial) changes in $\delta^{18}O_p$ values.

The variation in the $\delta^{18}O_{sc}$ record at Gona demonstrates significant environmental variability within time periods and across landscapes. Records of terrestrial variability for East Africa are important in light of proposals that African climate became more variable as glaciations intensified from the mid-Pliocene through the Pleistocene (deMenocal 2004; Bobe and Behrensmeyer 2004). The amount of variability preserved in the $\delta^{18}O_{sc}$ record depends on the amount of variability that actually existed and the frequency with which that variability is recorded. The latter depends on depositional rate, duration of carbonate formation, and sampling density. Figure 3.7 shows that uneven sampling density throughout the Gona $\delta^{18}O_{sc}$ record limits our ability to assess true variability in

$\delta^{18}O_{sc}$ through time. The variation exhibited in the Gona $\delta^{18}O_{sc}$ record suggests that $\delta^{18}O_{sc}$ values could be an effective tool for documenting terrestrial variability if carbonates were uniformly and intensively sampled throughout the entire record.

The isotopic record from paleosols in East Africa and globally

Current thinking suggests that C_4 metabolism first emerged independently, in a classic example of parallel evolution, among many grass families sometime in the mid-Tertiary, and later among the dicots. Molecular clock analyses indicate that C_4 grasses emerged 20–30 mya (Kellog 1999), somewhat older than the oldest securely identified macroscopic remains of C_4 plants at 12.5 mya (Tidwell and Nambudiri 1989). This was a time of major temperature decline (Zachos, Stott, and Lohmann 1994), aridification, and possibly pCO_2 decline (Pearson and Palmer 2000). A key question is how extensive C_4 biomass was during this early period. So far, C_4 biomass apparently does show up at this time in several isotopic records in Africa (Kingston et al. 1994; Morgan et al. 1994) and North America (Fox and Koch 2004), but C_4 biomass never dominated many tropical to subtropical ecosystems as it does today.

From the global perspective, carbon isotopic results from pedogenic carbonates and from fossil teeth combine to suggest that C_4 grasses experienced a dramatic expansion beginning perhaps about 8 mya nearer to the equator and 2–3 myr later at higher latitudes. This time-transgressive pattern of expansion led Cerling et al. (1997a) to suggest that a drop in pCO_2 in the late Miocene may have been the cause. Although C_3 grasses function more efficiently (as measured by quantum yield) than C_4 plants at low temperatures, the productivity of the latter is higher at lower pCO_2. Cerling et al. (1997a) suggested that as pCO_2 declined, C_4 grasses expanded, first at low latitudes. However, more recent studies have failed to confirm a decline in pCO_2 in the late Miocene (Pagani et al. 1999), and so the cause for the expansion remains unknown.

It is important to keep in mind that in many regions, C_4 grasses displaced C_3 shrubs and trees, not grasses. This suggests that the causes for C_4 grass expansion also involved a shift in growth forms (forest to grassland) as well as in plant metabolism (C_3 to C_4). A well-documented example of this comes from Nepal, where pollen evidence shows that C_4 grasses expanded between 8 and 6 mya at the expense of temperate to subtropical forest, confirming carbon isotopic results from the same sections (Quade et al. 1989b). Grass pollen composed a small fraction (8–33%) of total pollen prior to 8 mya, and over 80 percent after 6.5 mya (Hoorn et al. 2000). For East Africa, it remains to be seen whether the more gradual shift toward greater C_4 biomass starting in the late Miocene involved just a shift in photosynthetic pathway (C_3 grasses to C_4 grasses) or both photosynthetic pathway and growth form (C_3 trees and shrubs to C_4 grasses). The distinction may be important, since the conditions that would favor grassland over forest—such as increased aridity, fire, and disturbance (Sarmiento 1984; Bond and van Wilgen 1996; Sankaran et al. 2005)—differ from those that might favor C_4 grasses over C_3 grasses (warm rather than cool growing seasons, aridity, low pCO_2).

The oxygen isotope record from Gona (and East Africa; see next section) clearly displays a long-term increase, which we attribute to aridification and changes in moisture source. Recent studies in East Africa specifically link percentage of woody cover in modern savannas to mean annual rainfall (Sankaran et al. 2005). This implies that the late Miocene expansion of grasses and aridification may be causally linked.

Whatever the cause, the similarity in the timing of the low- to midlatitude C_4 grass expansion 8–5 mya is tantalizingly similar to the timing of the ape–human split also thought to occur in Africa during this period (Kumar and Hedges 1998; Chen and Li 2001; Salem et al. 2003). The once suggested linkage between early hominin development and grassland expansion (e.g., Dart 1925; Washburn 1978; Coppens 1994; Pilbeam 1996) has been challenged, particularly recently (Clarke 2003; Pickford and Senut 2001; Pickford et al. 2004). However, at least from a carbon isotope perspective, the link remains viable in several study areas (Chad: Zazzo et al. 2000; Lothagam: Cerling et al. 2003b; Gona: Levin et al. 2004 and Semaw et al. 2005). At Gona, as already noted, carbon isotopes from both pedogenic carbonate and fossil teeth suggest a substantial presence of C_4 grasses in the same area where numerous remains of *Ardipithecus ramidus* are found (Levin et al. 2004; Semaw et al. 2005; Levin et al. 2008). But what does the rest of the African record show?

A large body of isotopic data on pedogenic carbonates, starting with the pioneering study of Cerling and Hay (1986) at Olduvai Gorge, has been collected in East Africa, from sections spanning the last 18 myr. It is very difficult to evaluate the quality of the isotopic data from all these studies. Of the more systematic uncertainties, identification of pedogenic carbonate is a key challenge. Difficulties with identification are understandable, and the criteria for identification in many studies are often not very clear. Examples of studies that make a strong case for analysis of only pedogenic carbonate—employing some or all of the criteria we described at the beginning of this chapter—include those from Rusinga Island (Bestland and Retallack 1993; Bestland and Krull 1999), Lothagam (Cerling, Harris, and Leakey 2003b), Kanapoi (Wynn

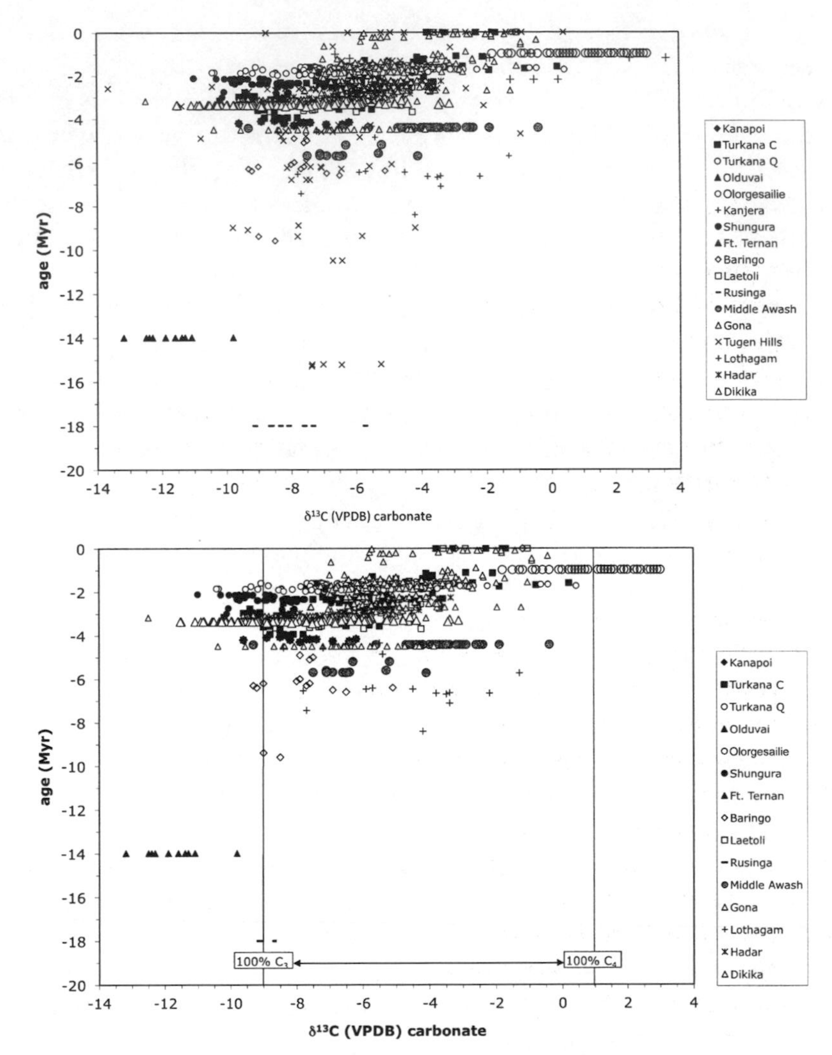

Figure 3.8. *(a) The* $\delta^{13}C$ *(VPDB) value of pedogenic carbonate compiled from studies in East Africa: Kanapoi (*Wynn 2000*); Turkana C (*Cerling et al. 1988; Wynn 2004*) and Turkana Q (*Quinn et al. 2007*); Olduvai (*Cerling and Hay 1986; *Olorgesailie (*Sikes et al. 1999*); Kanjera (*Plummer et al. 1999*); Shungura (*Levin et al. 2011*); Ft. Ternan (*Cerling et al. 1991b*); Baringo (*Cerling 1992*); Laetoli (*Cerling 1992*); Rusinga (*Bestland and Krull 1999*); Middle Awash (*WoldeGabriel et al. 2001*); Hadar (*Aronson et al. 2008*); Gona (*Quade et al. 2004; Levin et al. 2004*); and Tugen Hills (*Kingston 1992*). (b) All of the above data excluding studies where the pedogenic associations are either not consistent or not clearly documented. Data used in these plots can be found in the compilation by Levin (2013).*

2000), the Middle Awash (WoldeGabriel et al. 2001; 2009), Shungura (Levin et al. 2011), Laetoli and Baringo (Cerling 1992), and Olorgesalie (Sikes et al. 1999). In other studies, the published data had to be carefully culled to exclude all nonpedogenic carbonate analyses (Turkana [Cerling et al. 1988]; Olduvai Gorge [Cerling and Hay 1986]; Gona [Levin et al. 2004; Quade et al. 2004]). In some studies, the distinction between soil and nonsoil analyses was unclear (e.g., Plummer et al. 1999; Kingston et al. 1994). In other studies, such as Liutkus et al. (2005), phreatic carbonates were studied, which are useful for reconstruction of $\delta^{18}O_p$ values but not of $\delta^{13}C_{plant}$ values.

Another way to evaluate the East African data sets is to quantify the isotopic difference ($\Delta_{sc\text{-}om} = \delta^{13}C_{sc} - \delta^{13}C_{om}$) between coexisting pedogenic carbonate ($\delta^{13}C_{sc}$) and organic matter ($\delta^{13}C_{om}$). In modern soils at moderate to high respiration rates and free of detrital contamination, the observed difference (Figure 3.4) is 13.4–16.4‰. This difference can be quantified for ancient soils as a test of their pedogenic origin and postburial resistance to alteration. Unfortunately, soil organic matter is not widely preserved in the geologic record and only a fraction of the East African data can be so tested. We find that the paleosols from Fort Ternan pass the test, whereas those from Rusinga Island and the Tugen Hills are inconsistent. In the Rusinga Island case we can reject the $\delta^{13}C_{sc}$ values falling outside the "soil envelope" (Figure 3.4). For the Tugen Hills, most $\Delta_{sc\text{-}om}$ pairs fall outside the soil envelope (Kingston 1992, 958; 1999, 77), suggesting that nonpedogenic carbonates are included in the data set.

We present the carbon isotopic data set for carbonates from all East African sites (Figure 3.8a), and the same data set excluding analyses not clearly shown to be pedogenic (Figure 3.8b). Based only on the latter results (Figure 3.8b), the carbon isotopic record can be subdivided into four periods:

18 TO 8.4 MYA: A MAINLY C_3 WORLD

Sampling density is very sparse and is confined to three sites in west-central Kenya: Rusinga Island, Fort Ternan, and a few samples from Baringo. Nearly all $\delta^{13}C_{sc}$ values are −13.5 to −8.0‰. Given the uncertainties of the method described previously, this range would be consistent with anywhere from 80 to 100% C_3 biomass.

Inclusion of the data from Tugen Hills (Kingston et al. 1994) and from Rusinga Island (Bestland and Krull 1999) with $\Delta_{sc\text{-}om}$ > 17‰ would raise the estimate of the proportion of C_4 biomass during this period to 50 percent at some sites (Figure 3.8a).

8.4 TO 7 MYA: C_4 GRASSES APPEAR

Only three soil data-points from one area, Lothagam, are available from this period (Cerling et al. 2003b), a critical interval when C_4 grasses were expanding at low latitude sites globally (Quade et al. 1989b; Latorre et al. 1997; Cerling et al. 1997a). Two of the analyses from Lothagam show that C_4 grasses were clearly on the landscape by 8.4 and 7.1 mya, and analysis of fossil-tooth enamel from Lothagam confirms that some mammals had a C_4-dominated diet before 7 mya (Cerling et al. 2003b).

7 TO 2.7 MYA: EXPANSION OF C_4 GRASSES

Sampling density and the number of sites improves greatly after 5 mya. A picture of mixed C_3 and C_4 habitats emerges from a number of sites, including Baringo, Gona, Kanapoi, Koobi Fora, and Lothagam. Most sampled habitats during this time range from open woodlands to wooded grasslands with an even mix of C_3 and C_4 biomass. Another important feature includes a very large spread in $\delta^{13}C_{sc}$ values as sampling density increases. Both the Middle Awash and Lothagam areas show evidence of C_4 grass savannas (> 80% C_4 grass) during this period. From a strictly isotopic perspective, many of the remains of *Ardipithecus ramidus* from the Middle Awash (4.42 mya; WoldeGabriel et al. 2009) can be placed in an open, C_4 grass-dominated setting (Cerling et al. 2010).

2.7 MYA TO PRESENT: DEVELOPMENT OF MODERN SAVANNAS IN MOST OF EAST AFRICA

Starting between 2.7 and 2.0 mya, C_4 grasses undergo a major expansion seen first at Gona, and later, by 2 mya, in other areas such as Turkana. This is before the mid-Pleistocene facies shift at Gona discussed previously; we view it as a fundamental habitat shift between Hadar Formation (3.4 to 2.9 mya) and Busidima Formation time (< 2.7 mya). At Gona this is also coincident with the *minimum* age for the first appearance of stone toolmaking at 2.58 mya (Quade et al. 2004; Semaw et al. 2003).

This general picture of gradual and somewhat uneven expansion, starting in the late Miocene, of C_4 biomass is strongly supported by carbon isotopic evidence from fossil teeth. The ecologic picture painted by paleodietary isotope data is unlike soils in that animals are selective in their feeding preferences and move around on the landscape. Carbon isotopic evidence from fossil teeth therefore provides a more qualitative than quantitative estimate of C_4

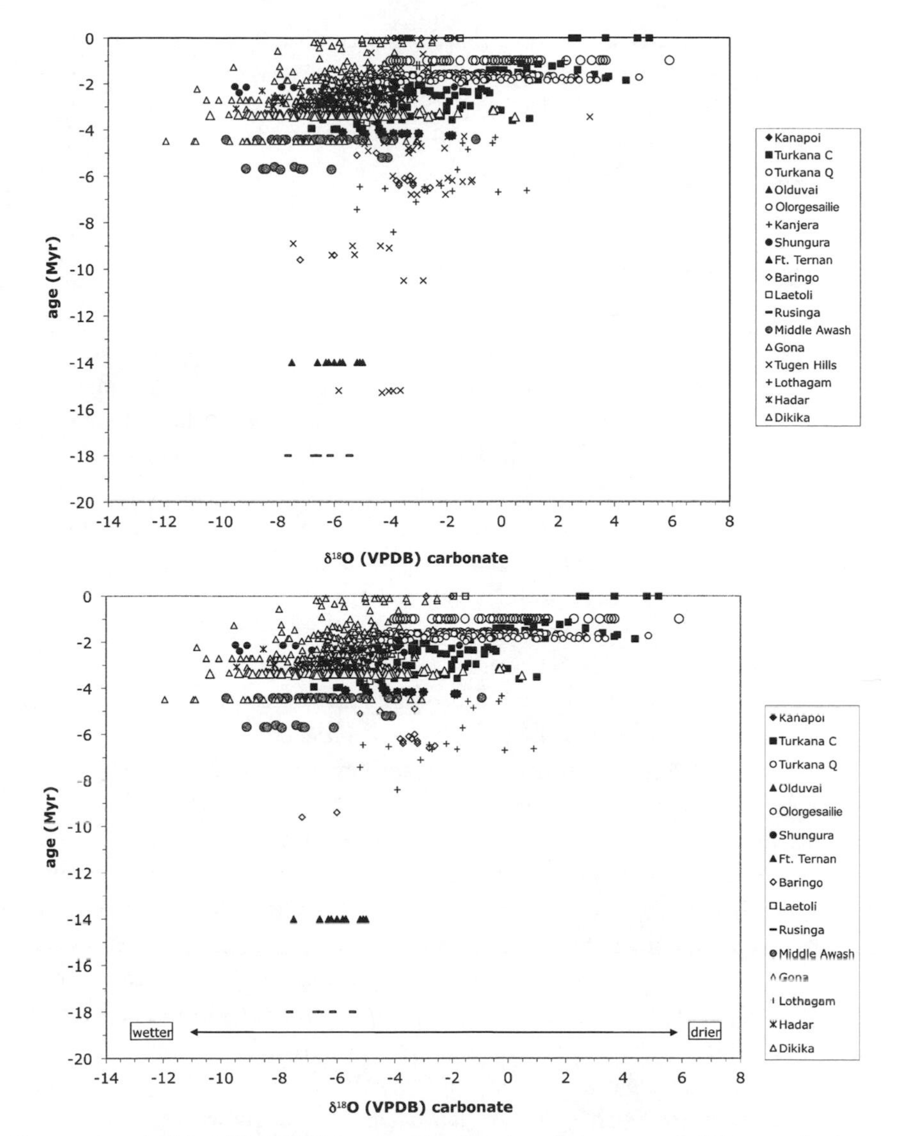

FIGURE 3.9. *(a) The $\delta^{18}O$ (VPDB) value of pedogenic carbonate compiled from studies in East Africa. See* Figure 3.8 *caption for sources of data. (b) All of the above data excluding studies where the pedogenic associations are either not consistent or not clearly documented.*

biomass than soils, and a spatially broader snapshot of the paleoecology. For the pre-7-mya period, fossils from Fort Ternan (Cerling et al. 1997b) and mixed taxa reported by Morgan et al. (1994) from the Tugen Hills both show that C_3

biomass dominated paleodiets, but that in the case of the Tugen Hills, some C_4 plants were likely consumed. For the 6 to 4.5 mya period, data sets from Lothagam (Cerling et al. 2003b), Gona (Semaw et al. 2005; Levin et al. 2008), and Chad (Zazzo et al. 2000) point to mixed C_3/C_4 habitats and the dominance of C_4 grass in the diet of most large herbivores.

East African $^{18}O_{sc}$ record

The $\delta^{18}O_{sc}$ records from Kenya and Tanzania show an increase in $\delta^{18}O_{sc}$ values (minimum values and ranges) that began in the late Miocene (Figure 3.9). The only long-term $\delta^{18}O_{sc}$ record from Ethiopia is preserved at Gona, where minimum $\delta^{18}O_{sc}$ values have increased since the early Pliocene, but where upper limits (about −3‰) for $\delta^{18}O_{sc}$ values have remained relatively constant for all well-sampled intervals. The long-term increases in $\delta^{18}O_{sc}$ values across East Africa argue against local elevation increases (Bonnefille et al. 1987, but see also Bonnefille et al. 2004) as the cause, and favor a regionwide climatic explanation.

The $\delta^{18}O_{sc}$ values from Ethiopian paleosols are systematically lower than those from Tanzanian and Kenyan paleosols. At all time intervals since the late Miocene, there is only a 2–3‰ overlap between the most positive Ethiopian $\delta^{18}O_{sc}$ values and the most negative Tanzanian and Kenyan $\delta^{18}O_{sc}$ values. These compiled East African $\delta^{18}O_{sc}$ data (Figure 3.9) reflect the complexity of local and regional East African climate. The offset between the Ethiopian and the Kenyan and Tanzanian records demonstrates that these areas were influenced by different climatic regimes. But what does the offset mean? Do the lower $\delta^{18}O_{sc}$ values at Gona, Shungura and the Middle Awash imply less soil-water evaporation, more rainfall, higher elevation, or a moisture source with lower $\delta^{18}O_p$ values? Contemporaneous records from the mid-Pliocene to the late Pleistocene at Gona and in Turkana are a good place to start to explore this offset.

Between 3.4 and 2.0 mya, $\delta^{18}O_{sc}$ values from Gona and Shungura overlap (combined range, −9.5‰ to −1.0‰) and are lower than $\delta^{18}O_{sc}$ values from the Nachukui and Koobi Fora Formations (combined range, −6.8‰ and +1‰). There are two scenarios that may explain this: (1) pedogenic carbonates from the paleo-Awash and paleo-Omo River floodplains formed from rainfall with low $\delta^{18}O_p$ values, perhaps sourced in the Ethiopian highlands, whereas pedogenic carbonates from the Koobi Fora and Nachukui Formations formed from waters with higher initial $\delta^{18}O_p$ values; or (2) higher $\delta^{18}O_{sc}$ values in the Koobi Fora and Nachukui Formations reflect more soil-water evaporation and more arid conditions than pedogenic carbonates from the Shungura or Busidima (Gona) Formations. We do not know which explanation is correct. More sampling throughout the Turkana Basin is needed to understand the relationship between $\delta^{18}O_p$ values and local topography, environment, and climate.

The well-developed tephrostratigraphic record in East Africa affords geologists a unique opportunity to generate multiple synchronous $\delta^{18}O_{sc}$ records that can be interlinked regionally and to the marine record. The spatial variation evident in the existing $\delta^{18}O_{sc}$ record demonstrates that terrestrial East African environments did not respond uniformly to global climate change throughout the Pliocene and Pleistocene. We need more data to resolve terrestrial environmental variability, filling in the gaps for the Pliocene and Pleistocene $\delta^{18}O_{sc}$ record in East Africa, and generating a more complete record for the Miocene.

CONCLUDING REMARKS

Rather than summarize our results, this is what we think needs to be done to further capitalize on the already powerful contribution of carbon and oxygen isotopes from pedogenic carbonates to paleoecologic reconstruction in East Africa:

1. Confine the use of the soil-diffusion model of Cerling (1984) to paleosol carbonates alone. Avoid other types of secondary carbonate common in the geologic record. And hold all future publications on isotopes from soils to a high standard for paleosol carbonate identification. We suggested a list of identification criteria toward the beginning of our chapter.
2. Conduct much more work on the pre-7-mya record, both spatially and temporally. There are very few studies from this time period, and the period between 7 and 14 mya in particular. The spatial distribution of studies is highly restricted to a small area of western Kenya. A key question is whether the higher proportion of C_4 plants suggested by the study of Kingston et al. (1994) is sustained by other records. Palynologic and paleobotanic studies would assist in understanding if the late Miocene expansion in C_4 grasses at the expense of C_3 plants was strictly metabolic or also involved a life-form shift. More extensive paleodietary analysis of large herbivores will greatly augment the pedogenic carbonate isotope record.
3. Strong controversy surrounds the paleoenvironmental context of *Ardipithecus ramidus* in Ethiopia. One view is that this hominin lived mainly in a forest or woodland and was adapted mainly to tree-dwelling (White et al. 2009; WoldeGabriel et al. 2009). Our own perspective is

that *Ar. ramidus* lived in a range of environments that included open, C_4 grass-dominated settings (Cerling et al. 2010). The isotopic evidence has and will continue to make a key contribution to our understanding of the paleoenvironmental context of all hominins, including *Ar. ramidus*.

ACKNOWLEDGMENTS

Thure Cerling has been a dedicated mentor and supportive colleague over many years, and we are both greatly in his debt. We would like to thank the ARCCH of the Ministry of Youth, Sports, and Culture in Ethiopia, and all the members of the Gona project for their help in the field, especially Sileshi Semaw and Scott Simpson, and Melanie Everett for her assistance in the laboratory. The Gona research was supported by grants from the Wenner-Gren Foundation, National Geographic, the Leakey Foundation, and the National Science Foundation (SBR-9910974 and RHOI).

REFERENCES

Aronson, J. L., M. Hailemichael, and S. M. Savin. 2008. "Hominid Environments at Hadar from Paleosol Studies in a Framework of Ethiopian Climate Change." *Journal of Human Evolution* 55 (4): 532–50. http://dx.doi.org/10.1016/j.jhevol.2008.04.004. Medline:18571216

Behrensmeyer, A. K., J. Quade, T. Cerling, J. Kappelman, I. A. Khan, P. Copeland, L. Roe, J. Hicks, P. Stubblefield, B. J. Willis, et al. 2007. "The Structure and Rate of Late Miocene Expansion of C4 Plants: Evidence from Lateral Variation in Stable Isotopes in Paleosols of the Siwalik Group, Northern Pakistan." *Geological Society of America Bulletin* 119 (11–12): 1486–505. http://dx.doi.org/10.1130/B26064.1.

Bestland, E. A., and E. S. Krull. 1999. "Palaeoenvironments of Early Miocene Kisingiri Volcano Proconsul Sites: Evidence from Carbon Isotopes, Palaeosols and Hydromagmatic Deposits." *Journal of the Geological Society* 156 (5): 965–76. http://dx.doi.org/10.1144/gsjgs.156.5.0965.

Bestland, E. A., and G. Retallack. 1993. "Volcanically Influenced Calcareous Palaeosols from the Miocene Kiahera Formation, Rusinga Island, Kenya." *Journal of the Geological Society* 150 (2):293–310. http://dx.doi.org/10.1144/gsjgs.150.2.0293.

Bobe, R., and A. K. Behrensmeyer. 2004. "The Expansion of Grassland Ecosystems in Africa in Relation to Mammalian Evolution and the Origin of the Genus Homo." *Palaeogeography, Palaeoclimatology, Palaeoecology* 207 (3-4): 399–420. http://dx.doi.org/10.1016/j.palaeo.2003.09.033.

Bond, W. J., and B. van Wilgen. 2006 [1996]. Plants and Fire. New York: Chapman.

Bonnefille, R., A. Vincens, and G. Buchet. 1987. "Palynology, Stratigraphy and Palaeoenvironment of a Pliocene Hominid Site (2.9–3.3 M.Y.) at Hadar, Ethiopia." *Palaeogeography, Palaeoclimatology, Palaeoecology* 60:249–81. http://dx.doi.org/10.1016/0031-0182(87)90035-6.

Bonnefille, R., R. Potts, F. Chalié, D. Jolly, and O. Peyron. 2004. "High-Resolution Vegetation and Climate Change Associated with Pliocene *Australopithecus afarensis*." *Proceedings of the National Academy of Sciences of the United States of America* 101 (33): 12125–9. http://dx.doi.org/10.1073/pnas.0401709101. Medline:15304655

Cerling, T. E. 1984. "The Stable Isotopic Composition of Modern Soil Carbonate and Its Relation to Climate." *Earth and Planetary Science Letters* 71 (2): 229–40. http://dx.doi.org/10.1016/0012-821X(84)90089-X.

Cerling, T. E. 1992. "Development of Grasslands and Savannas in East Africa during the Neogene." *Palaeogeography, Palaeoclimatology, Palaeoecology* 97 (3): 241–7. http://dx.doi.org/10.1016/0031-0182(92)90211-M.

Cerling, T. E., J. R. Bowman, and J. R. O'Neil. 1988. "An Isotopic Study of a Fluvial-Lacustrine Sequence: The Plio-Pleistocene Koobi Fora Sequence; East Africa." *Palaeogeography, Palaeoclimatology, Palaeoecology* 63 (4): 335–56. http://dx.doi.org/10.1016/0031-0182(88)90104-6.

Cerling, T. E., J. M. Harris, S. H. Ambrose, M. G. Leakey, and N. Solounias. 1997b. "Dietary and Environmental Reconstruction with Stable Isotope Analyses of Herbivore Tooth Enamel from the Miocene Locality of Fort Ternan, Kenya." *Journal of Human Evolution* 33 (6): 635–50. http://dx.doi.org/10.1006/jhev.1997.0151. Medline:9467773

Cerling, T. E., J. M. Harris, and M. G. Leakey. 2003b. "Isotope Paleoecology of the Nawata and Nachukui Formations at Lothagam, Turkana Basin, Kenya." In *Lothagam: The Dawn of Humanity in Eastern Africa*, ed. M. G. Leakey and J. M. Harris, 605–24. New York: Columbia University Press.

Cerling, T. E., J. M. Harris, B. J. MacFadden, J. R. Ehleringer, M. G. Leakey, J. Quade, and V. Eisenman. 1997a. "Global Vegetation Change through the Miocene/Pliocene Boundary." *Nature* 389 (6647): 153–58. http://dx.doi.org/10.1038/38229.

Cerling, T. E., J. M. Harris, and B. H. Passey. 2003a. "Diets of East African Bovidae Based on Stable Isotope Analysis." *Journal of Mammalogy* 84 (2): 456–70. http://dx.doi.org/10.1644/1545-1542(2003)084<0456:DOEABB>2.0.CO;2.

Cerling, T. E., and R. L. Hay. 1986. "An Isotopic Study of Paleosol Carbonates from Olduvai Gorge." *Quaternary Research* 25 (1): 63–78. http://dx.doi.org/10.1016/0033-5894(86)90044-X.

Cerling, T. E., N. E. Levin, J. Quade, J. G. Wynn, D. L. Fox, J. D. Kingston, R. G. Klein, and F. H. Brown. 2010. "Comment on the Paleoenvironment of *Ardipithecus ramidus*." *Science* 328 (5982): 1105, author reply 1105. http://dx.doi.org/10.1126/science.1185274. Medline:20508112

Cerling, T. E., J. Quade, S. H. Ambrose, and N. E. Sikes. 1991b. "Fossil Soils, Grasses, and Carbon Isotopes from Fort Ternan, Kenya: Grassland or Woodland?" *Journal of Human Evolution* 21 (4): 295–306. http://dx.doi.org/10.1016/0047-2484(91)90110-H.

Cerling, T. E., D. K. Solomon, J. Quade, and J. R. Bowman. 1991a. "On the Isotopic Composition of Carbon in Soil Carbon Dioxide." *Geochimica et Cosmochimica Acta* 55 (11): 3403–5. http://dx.doi.org/10.1016/0016-7037(91)90498-T.

Chadwick, O. A., J. M. Sowers, and R. A. Amundson. 1988. "Morphology of Calcite Crystals in Clast Coatings from Four Soils in the Mojave Desert Region." *Soil Science Society of America Journal* 53 (1):211–19. http://dx.doi.org/10.2136/sssaj1989.036159950053000010038x.

Chen, F. C., and W. H. Li. 2001. "Genomic Divergences between Humans and Other Hominoids and the Effective Population Size of the Common Ancestor of Humans and Chimpanzees." *American Journal of Human Genetics* 68 (2): 444–56. http://dx.doi.org/10.1086/318206. Medline:11170892

Clarke, R. J. 2003. "Newly Revealed Information on the Sterkfontein Member 2 *Australopithecus* Skeleton." *South African Journal of Science* 98:523–6.

Collatz, G., J. Berry, and J. Clark. 1998. "Effects of Climate and Atmospheric CO_2 Partial Pressure on the Global Distribution of C_4 Grasses: Present, Past, and Future." *Oecologia* 114 (4): 441–54. http://dx.doi.org/10.1007/s004420050468.

Coppens, Y. 1994. "East Side Story: The Origin of Mankind." *Scientific American* (May 1994): 62–69.

Craig, H., and L. Gordon. 1965. "Deuterium and Oxygen-18 Variation in the Ocean and the Marine Atmosphere." In *Stable Isotopes in Oceanographic Studies and Paleotemperatures*, ed. E. Tongiorgi, 9–130. Spoleto.

Dansgaard, W. 1964. "Stable Isotopes in Precipitation." *Tellus* 16 (4): 436–68. http://dx.doi.org/10.1111/j.2153-3490.1964.tb00181.x.

Dart, R. A. 1925. "*Australopithecus africanus*: The Man Ape of South Africa." *Nature* 115 (2884): 195–9. http://dx.doi.org/10.1038/115195a0.

Davidson, G. R. 1995. "The Stable Isotopic Composition and Measurement of Carbon in Soil CO_2." *Geochimica et Cosmochimica Acta* 59 (12): 2485–9. http://dx.doi.org/10.1016/0016-7037(95)00143-3.

Debele, B. 1985. "The Vertisols of Ethiopia: Their Properties and Management." *World Soils Reports* (UN, FAO, ROME) 56: 31–54.

DeMenocal, P. B. 2004. "African Climate Change and Faunal Evolution during the Plio-Pleistocene." *Earth and Planetary Science Letters* 220 (1-2): 3–24. http://dx.doi.org/10.1016/S0012-821X(04)00003-2.

Deutz, P., I. P. Montañez, and H. C. Monger. 2002. "Morphology and Stable and Radiogenic Isotope Composition of Pedogenic Carbonates in Late Quaternary Relict Soils, New Mexico, USA: An Integrated Record of Pedogenic Overprinting." *Journal of Sedimentary Research* 72 (6): 809–22. http://dx.doi.org/10.1306/040102720809.

Dever, L., R. Durand, J.C.H. Fontes, and P. Vachier. 1982. "Geochemie et Teneurs Isotopiques des Systemes Saisonniers de Dissolution de la Calcite dans un Sol sur Craie." *Geochimica et Cosmochimica Acta* 46 (10): 1947–56. http://dx.doi.org/10.1016/0016-7037(82)90132-6.

Dulinski, M., and K. Rozanski. 1990. "Formation of $^{13}C/^{12}C$ Ratios in Speleothems: A Semi-Dynamic Model." *Radiocarbon* 32:7–16.

Fantidis, J., and D. H. Ehhalt. 1970. "Variations of the Carbon and Oxygen Isotopic Composition in Stalagmites and Stalactites: Evidence of Non-Equilibrium Isotopic

Fractionation." *Earth and Planetary Science Letters* 10 (1): 136–44. http://dx.doi.org/10.1016/0012-821X(70)90075-0.
Fox, D., and P. L. Koch. 2004. "Carbon and Oxygen Isotopic Variability in Neogene Paleosol Carbonates: Constraints on the Evolution of C_4 Grasslands." *Palaeogeography, Palaeoclimatology, Palaeoecology* 207 (3-4): 305–29. http://dx.doi.org/10.1016/j.palaeo.2003.09.030.
Friedli, H., H. Lötscher, H. Oeshger, U. Siegenthaler, and B. Stauffer. 1986. "Ice Core Record of $^{13}C/^{12}C$ Ratio of Atmospheric CO_2 in the Past Two Centuries." *Nature* 324 (6094): 237–8. http://dx.doi.org/10.1038/324237a0.
Gawthorpe, R. L., and M. R. Leeder. 2000. "Tectono-Sedimentary Evolution of Active Extensional Basins." *Basin Research* 12 (3-4): 195–218. http://dx.doi.org/10.1046/j.1365-2117.2000.00121.x.
Hailemichael, M., J. L. Aronson, S. Savin, M.J.S. Tevesz, and J. G. Carter. 2002. "$\delta^{18}O$ in Mollusk Shells from Pliocene Lake Hadar and Modern Ethiopian Lakes: Implications for the History of the Ethiopian Monsoon." *Palaeogeography, Palaeoclimatology, Palaeoecology* 186 (1-2): 81–99. http://dx.doi.org/10.1016/S0031-0182(02)00445-5.
Hendy, C. H. 1971. "The Isotopic Geochemistry of Speleothems, I: The Calculation of the Effects of Different Modes of Formation on the Isotopic Composition of Speleothems and Their Applicability as Paleoclimatic Indicators." *Geochimica et Cosmochimica Acta* 35 (8): 801–24. http://dx.doi.org/10.1016/0016-7037(71)90127-X.
Hendy, C. H., T. A. Rafter, and T. G. MacIntosh. 1972. "The Formation of Carbonate Nodules in Soils of the Darling Downs, Queensland, Australia, and the Date of the Talgai Cranium." In *Proceedings of the 8th International Radiocarbon Dating Conference, Lower Hutt, New Zealand*, ed. T. A. Rafter and T. G. Taylor, 417–437. Wellington: Royal Society of New Zealand.
Hesla, A.B.I., L .L. Tieszen, and S. K. Imbamba. 1982. "A Systematic Survey of C_3 and C_4 Photosynthesis in the Cyperaceae of Kenya, East Africa." *Photosynthetica* 16:196–205.
Hillel, D. 1982. *Introduction to Soil Physics*. London: Academic Press.
Hoorn, C., T. Ohja, and J. Quade. 2000. "Palynological Evidence of Vegetation Development and Climatic Change in the Sub-Himalayan Zone (Neogene, Central Nepal)." *Palaeogeography, Palaeoclimatology, Palaeoecology* 163 (3-4): 133–61. http://dx.doi.org/10.1016/S0031-0182(00)00149-8.
Hsieh, J. C., O. Chadwick, E. Kelly, and S. Savin. 1998. "Oxygen Isotopic Composition of Soil Water: Quantifying Evaporation and Transpiration." *Geoderma* 82 (1-3): 269–93. http://dx.doi.org/10.1016/S0016-7061(97)00105-5.
Ivanov, V. V., D. Lemma, B. G. Rozanov, and T. A. Soklova. 1985. "Composition of Ethiopian Upland Soils." *Pochvovendeniye* 3:29–39.
Kellog, E. A. 1999. "Phylogenetic Aspects of the Evolution of C_4 Photosynthesis." In *C_4 Plant Biology*, ed. R. F. Sage and R. K. Monson, 411–444. San Diego, CA: Academic Press.
Kim, S., and J. R. O'Neil. 1997. "Equlilibrium and Nonequilibrium Oxygen Isotope Effects in Synthetic Carbonates." *Geochimica et Cosmochimica Acta* 61 (16): 3461–75. http://dx.doi.org/10.1016/S0016-7037(97)00169-5.

Kingston, J. D. 1992. "Stable Isotopic Evidence for Hominid Paleoenvironments in East Africa." PhD dissertation. Harvard University, Cambridge, MA. 162 pp.

Kingston, J. D. 1999. "Environmental Determinants in Early Hominid Evolution: Issues and Evidence from the Tugen Hills, Kenya." In *Late Cenozoic Environments and Hominid Avolution: A Tribute to Bill Bishop*, ed. P. Andrews and P. Banham, 69–84. London: Geological Society.

Kingston, J. D., A. Hill, and B. D. Marino. 1994. "Isotopic Evidence for Neogene Hominid Paleoenvironments in the Kenya Rift Valley." *Science* 264 (5161): 955–9. http://dx.doi.org/10.1126/science.264.5161.955. Medline:17830084

Kleinsasser, L., J. Quade, W. C. McIntosh, N. E. Levin, S. W. Simpson, and S. Semaw. 2008. "Stratigraphy and Geochronology of the Late Miocene Adu-Asa Formation at Gona, Ethiopia." In *The Geology of Early Humans in the Horn of Africa*, ed. J. Quade and J. G. Wynn, 33–65. Geological Society of America Special Paper. Boulder, CO: Geological Society of America.

Kumar, S., and S. B. Hedges. 1998. "A Molecular Timescale for Vertebrate Evolution." *Nature* 392 (6679): 917–20. http://dx.doi.org/10.1038/31927. Medline:9582070

Latorre, C., J. Quade, and W. C. McIntosh. 1997. "The Expansion of C_4 Grasses and Global Change in the Late Miocene: Stable Isotope Evidence from the Americas." *Earth and Planetary Science Letters* 146 (1-2): 83–96. http://dx.doi.org/10.1016/S0012-821X(96)00231-2.

Lear, C. H., H. Elderfield, and P. A. Wilson. 2000. "Cenozoic Deep-Sea Temperatures and Global Ice Volumes from Mg/Ca in Benthic Foraminiferal Calcite." *Science* 287 (5451): 269–72. http://dx.doi.org/10.1126/science.287.5451.269. Medline:10634774

Levin, N. 2002. "Isotopic Evidence for Plio-Pleistocene Environmental Change at Gona, Ethiopia." MS thesis, University of Arizona, Tucson. 47 pp.

Levin, N. E. 2013. "Compilation of East Africa Soil Carbonate Stable Isotope Data." *Integrated Earth Data Applications*. doi: 10.1594/IEDA/100231

Levin, N. E., F. H. Brown, A. K. Behrensmeyer, R. Bobe, and T. E. Cerling. 2011. "Paleosol Carbonates from the Omo Group: Isotopic Records of Local and Regional Environmental Change in East Africa." *Palaeogeography, Palaeoclimatology, Palaeoecology* 307 (1-4): 75–89. http://dx.doi.org/10.1016/j.palaeo.2011.04.026.

Levin, N. E., J. Quade, S. Simpson, S. Semaw, and M. Rogers. 2004. "Isotopic Evidence for Plio-Pleistocene Environmental Change at Gona, Ethiopia." *Earth and Planetary Science Letters* 219 (1-2): 93–110. http://dx.doi.org/10.1016/S0012-821X(03)00707-6.

Levin, N. E., S. W. Simpson, J. Quade, T. E. Cerling, and S. R. Frost. 2008. "Herbivore Enamel Carbon Isotopic Composition and the Environmental Context of *Ardipithecus* at Gona, Ethiopia." In *The Geology of Early Humans in the Horn of Africa*, ed. J. Quade and J. G. Wynn, 215–234. Geological Society of America Special Paper. Boulder, CO: Geological Society of America.

Levin, N. E., E. J. Zipser, and T. E. Cerling. 2009. "Isotopic Composition of Waters from Ethiopia and Kenya: Insights into Moisture Sources for Eastern Africa." *Journal of Geophysical Research* 114 (D23): D23306. http://dx.doi.org/10.1029/2009JD012166.

Liutkus, C., J. D. Wright, G. M. Ashley, and Nancy E. Sikes. 2005. "Paleoenvironmental Interpretation of Lake-Margin Deposits Using $\delta^{13}C$ and $\delta^{18}O$ Results from Early Pleistocene Carbonate Rhizoliths: Olduvai Gorge, Tanzania." *Geology* 33 (5): 377–80. http://dx.doi.org/10.1130/G21132.1.

Machette, M. N. 1985. "Calcic Soils in the Southwestern United States." In *Soils and Quaternary Geology of the Southwestern United States*, ed. D. L. Weide. Geological Society of America. Special (SAUS) 203 pp.

Magaritz, M., and A. J. Amiel. 1980. "Calcium Carbonate in a Calcareous Soil from the Jordan Valley, Israel; Its Origin, as Revealed by Stable Carbon Isotope Method." *Soil Science Society of America Journal* 44 (5): 1059–62. http://dx.doi.org/10.2136/sssaj1980.03615995004400050037x.

Magaritz, M., A. Kaufman, and D. H. Yaalon. 1981. "Calcium Carbonate Nodules in Soils: $^{18}O/^{16}O$ and $^{13}C/^{12}C$ Ratios and ^{14}C Contents." *Geoderma* 25 (3-4): 157–72. http://dx.doi.org/10.1016/0016-7061(81)90033-1.

Morgan, M. E., J. D. Kingston, and B. D. Marino. 1994. "Carbon Isotope Evidence for the Emergence of C_4 Plants in the Neogene from Pakistan and Kenya." *Nature* 367 (6459): 162–5. http://dx.doi.org/10.1038/367162a0.

Pagani, M., K. H. Freeman, and M. A. Arthur. 1999. "Late Miocene Atmospheric CO_2 Concentrations and the Expansion of C4 Grasses." *Science* 285 (5429): 876–9. http://dx.doi.org/10.1126/science.285.5429.876. Medline:10436153

Passey, B. H., N. E. Levin, T. E. Cerling, F. H. Brown, and J. M. Eiler. 2010. "High-Temperature Environments of Human Evolution in East Africa Based on Bond Ordering in Paleosol Carbonates." *Proceedings of the National Academy of Sciences of the United States of America* 107 (25): 11245–9. http://dx.doi.org/10.1073/pnas.1001824107. Medline:20534500

Pearson, P. N., and M. R. Palmer. 2000. "Atmospheric Carbon Dioxide Concentrations over the Past 60 Million Years." *Nature* 406 (6797): 695–9. http://dx.doi.org/10.1038/35021000. Medline:10963587

Pickford, M., and B. Senut. 2001. "The Geological and Faunal Context of Late Miocene Hominid Remains from Lukeino, Kenya." *Comptes rendus de l'Académie des Sciences Paris* 332:145–52.

Pickford, M., B. Senut, and C. Mourer-Chauvire. 2004. "Early Pliocene Tragulidae and Peafowls in the Rift Valley, Kenya: Evidence for Rainforest in East Africa." *Comptes Rendus. Palévol* 3 (3): 179–89. http://dx.doi.org/10.1016/j.crpv.2004.01.004.

Pilbeam, D. 1996. "Genetic and Morphological Records of the Hominoidea and Hominid Origins: A Synthesis." *Molecular Phylogenetics and Evolution* 5 (1): 155–68. http://dx.doi.org/10.1006/mpev.1996.0010. Medline:8673283

Plummer, T., L. C. Bishop, P. Ditchfield, and J. Hicks. 1999. "Research on Late Pliocene Oldowan Sites at Kanjera South, Kenya." *Journal of Human Evolution* 36 (2): 151–70. http://dx.doi.org/10.1006/jhev.1998.0256. Medline:10068064

Poage, M. A., and C. P. Chamberlain. 2001. "Empirical Relationships between Elevation and the Stable Isotope Composition of Precipitation and Surface Waters: Consideration for Studies of Paleoelevation Change." *American Journal of Science* 301 (1): 1–15. http://dx.doi.org/10.2475/ajs.301.1.1.

Pratt, D. J., P. J. Greenway, and M. D. Gwynne. 1966. "A Classification of East African Rangeland, with an Appendix on Terminology." *Journal of Applied Ecology* 3 (2): 369–82. http://dx.doi.org/10.2307/2401259.

Quade, J., T. E. Cerling, and J. R. Bowman. 1989a. "Systematic Variations in the Carbon and Oxygen Isotopic Composition of Pedogenic Carbonate along Elevation Transects in the Southern Great Basin, United States." *Geological Society of America Bulletin* 101 (4): 464–75. http://dx.doi.org/10.1130/0016-7606(1989)101<0464:SVITCA>2.3.CO;2.

Quade, J., T. E. Cerling, and J. R. Bowman. 1989b. "Development of Asian Monsoon Revealed by Marked Ecological Shift during the Latest Miocene in Northern Pakistan." *Nature* 342 (6246): 163–6. http://dx.doi.org/10.1038/342163a0.

Quade, J., J. Eiler, M. Daëron, and H. Achythan, 2013. "The clumped isotope paleothermometer in soils and paleosol carbonate." *Geochimica et Cosmochimica Acta* 105, 92–107.

Quade, J., N. Levin, S. Semaw, S. Simpson, M. Rogers, and D. Stout. 2004. "Paleoenvironments of the Earliest Stone Toolmakers, Gona, Ethiopia." *Geological Society of America Bulletin* 116 (11):1529–44. http://dx.doi.org/10.1130/B25358.1.

Quade, J., N. L. Levin, S. W. Simpson, R. Butler, W. McIntosh, S. Semaw, L. Kleinsasser, G. Dupont-Nivet, P. Renne, and N. Dunbar. 2008. "The Geology of Gona, Ethiopia." In *The Geology of Early Humans in the Horn of Africa*, ed. J. Quade and J. G. Wynn, 1–31. Geological Society of America Special Paper. Boulder, CO: Geological Society of America.

Quade, J., J. A. Rech, C. H. Latorre, J. Betancourt, E. Gleason, and M. Kalin-Arroyo. 2007. "Soils at the Hyperarid Margin: The Isotopic Composition of Soil Carbonate from the Atacama Desert." *Geochimica et Cosmochimica Acta* 71 (15): 3772–95. http://dx.doi.org/10.1016/j.gca.2007.02.016.

Quinn, R. L., C. J. Lepre, J. D. Wright, and C. S. Feibel. 2007. "Paleogeographic Variations of Pedogenic Carbonate $\delta^{13}C$ Values from Koobi Fora, Kenya: Implications for Floral Compositions of Plio-Pleistocene Hominin Environments." *Journal of Human Evolution* 53 (5): 560–73. http://dx.doi.org/10.1016/j.jhevol.2007.01.013. Medline:17905411.

Rabenhorst, M. C., L. P. Wilding, and L. T. West. 1984. "Identification of Pedogenic Carbonates Using Stable Carbon Isotope and Microfabric Analyses." *Soil Science Society of America Journal* 48 (1): 125–32. http://dx.doi.org/10.2136/sssaj1984.03615995004800010023x.

Richter, J. 1987. *The Soil as a Reactor*. Catena Paperback. Cremlingen, Germany: Catena-Verlag Publisher.

Romanek, C. S., E. T. Grossman, and J. W. Morse. 1992. "Carbon Isotopic Fractionation in Synthetic Aragonite and Calcite: Effects of Temperature and Precipitation Rate." *Geochimica et Cosmochimica Acta* 56 (1): 419–30. http://dx.doi.org/10.1016/0016-7037(92)90142-6.

Rozanski, K., L. Araguas-Araguas, and R. Gonfiantini. 1993. "Isotopic Patterns in Modern Global Precipitation." In *Continental Indicators of Climate, Proceedings of Chapman Conference*, ed. P. Swart, J. A. McKenzie, and K. C. Lohman, 1–36. Jackson Hole, Wyoming. American Geophysical Union Monograph 78.

Sage, R. F. 1999. "Why C_4 Photosynthesis?" In *C_4 Plant Biology*, ed. R. F. Sage and R. K. Monson, 3–16. San Diego: Academic Press. http://dx.doi.org/10.1016/B978-012614440-6/50002-1
Salem, A. H., D. A. Ray, J. Xing, P. A. Callinan, J. S. Myers, D. J. Hedges, R. K. Garber, D. J. Witherspoon, L. B. Jorde, and M. A. Batzer. 2003. "Alu Elements and Hominid Phylogenetics." *Proceedings of the National Academy of Sciences of the United States of America* 100 (22): 12787–91. http://dx.doi.org/10.1073/pnas.2133766100. Medline:14561894
Salomons, W., A. Goudie, and W. G. Mook. 1978. "Isotopic Composition of Calcrete Deposits from Europe, Africa and India." *Earth Surface Processes* 3 (1): 43–57. http://dx.doi.org/10.1002/esp.3290030105.
Sankaran, M., N. P. Hanan, R. J. Scholes, J. Ratnam, D. J. Augustine, B. S. Cade, J. Gignoux, S. I. Higgins, X. Le Roux, F. Ludwig, et al. 2005. "Determinants of Woody Cover in African Savannas." *Nature* 438 (7069): 846–9. http://dx.doi.org/10.1038/nature04070. Medline:16341012
Sarmiento, G. 1984. *The Ecology of Neotropical Savannas*. Cambridge, MA: Harvard University Press.
Semaw, S., M. J. Rogers, J. Quade, P. R. Renne, R. F. Butler, M. Dominguez-Rodrigo, D. Stout, W. S. Hart, T. Pickering, and S. W. Simpson. 2003. "2.6-Million-Year-Old Stone Tools and Associated Bones from OGS-6 and OGS-7, Gona, Afar, Ethiopia." *Journal of Human Evolution* 45 (2): 169–77. http://dx.doi.org/10.1016/S0047-2484(03)00093-9. Medline:14529651
Semaw, S., S. W. Simpson, J. Quade, P. R. Renne, R. F. Butler, W. C. McIntosh, N. Levin, M. Dominguez-Rodrigo, and M. J. Rogers. 2005. "Early Pliocene Hominids from Gona, Ethiopia." *Nature* 433 (7023): 301–5. http://dx.doi.org/10.1038/nature03177. Medline:15662421
Sharp, W. D., K. R. Ludwig, O. A. Chadwick, R. Amundson, and L. L. Glaser. 2003. "Dating of Fluvial Terraces by ^{230}Th/U on Pedogenic Carbonate, Wind River Basin, Wyoming." *Quaternary Research* 59 (2): 139–50. http://dx.doi.org/10.1016/S0033-5894(03)00003-6.
Sikes, N. E., R. Potts, and A. K. Behrensmeyer. 1999. "Early Pleistocene Habitat in Member 1 Olorgesailie Based on Paleosol Stable Isotopes." *Journal of Human Evolution* 37 (5): 721–46. http://dx.doi.org/10.1006/jhev.1999.0343. Medline:10536089
Straight, W. H., R. Barrick, and D. A. Eberth. 2004. "Reflections of Surface Water, Seasonality and Climate in Stable Oxygen Isotopes from Tyrannosaurid Tooth Enamel." *Palaeogeography, Palaeoclimatology, Palaeoecology* 206 (3-4): 239–56. http://dx.doi.org/10.1016/j.palaeo.2004.01.006.
Szaran, J. 1997. "Achievement of Carbon Isotope Equilibrium in the System HCO_3 (solution)—CO_2 (gas)." *Chemical Geology* 142 (1-2): 79–86. http://dx.doi.org/10.1016/S0009-2541(97)00077-6.
Talma, A. S., and F. Netterberg. 1983. "Stable Isotope Abundances in Calcretes." In *Residual Deposits: Surface Related Weathering Processes and Materials*, ed. R.C.L. Wilson, 221–233. Oxford: Blackwell Scientific Publications.

Taub, D. R. 2000. "Climate and the U.S. Distribution of C4 grass Subfamilies and Decarboxylation Variants of C4 Photosynthesis." *American Journal of Botany* 87 (8): 1211–5. http://dx.doi.org/10.2307/2656659. Medline:10948007

Tidwell, W. D., and E.M.V. Nambudiri. 1989. "*Tomlisonia thomassonii*, a Permineralized Grass from the Upper Miocene Ricardo Formation, California." *Review of Palaeobotany and Palynology* 60 (1-2): 165–77. http://dx.doi.org/10.1016/0034-6667(89)90075-4.

Tieszen, L. L., M. M. Senyimba, S. K. Imbamba, and J. H. Troughton. 1979. "The Distribution of C_3 and C_4 Grasses and Carbon Isotope Discrimination along an Altitudinal and Moisture Gradient in Kenya." *Oecologia* 37:337–50.

Washburn, S. L. 1978. "The Evolution of Man." *Scientific American* 239 (3): 194–8, 201–2, 204 passim. http://dx.doi.org/10.1038/scientificamerican0978-194. Medline:100881

White, T. D., S. H. Ambrose, G. Suwa, D. F. Su, D. DeGusta, R. L. Bernor, J. R. Boisserie, M. Brunet, E. Delson, S. Frost, et al. 2009. "Macrovertebrate Paleontology and the Pliocene Habitat of *Ardipithecus ramidus*." *Science* 326 (5949): 87–93. http://dx.doi.org/10.1126/science.1175822. Medline:19810193

WoldeGabriel, G., S. H. Ambrose, D. Barboni, R. Bonnefille, L. Bremond, B. Currie, D. DeGusta, W. K. Hart, A. M. Murray, P. R. Renne, et al. 2009. "The Geological, Isotopic, Botanical, Invertebrate, and Lower Vertebrate Surroundings of *Ardipithecus ramidus*." *Science* 326 (5949): 65 (summary), 65e1–65e5 (text). http://dx.doi.org/10.1126/science.1175817. Medline:19810191

WoldeGabriel, G., Y. Haile-Selassie, P. R. Renne, W. K. Hart, S. H. Ambrose, B. Asfaw, G. Heiken, and T. White. 2001. "Geology and Palaeontology of the Late Miocene Middle Awash Valley, Afar Rift, Ethiopia." *Nature* 412 (6843): 175–8. http://dx.doi.org/10.1038/35084058. Medline:11449271

Wynn, J. G. 2000. "Paleosols, Stable Carbon Isotopes, and Paleoenvironmental Interpretation of Kanapoi, Northern Kenya." *Journal of Human Evolution* 39 (4): 411–32. http://dx.doi.org/10.1006/jhev.2000.0431. Medline:11006049

Wynn, J. G. 2004. "Influence of Plio-Pleistocene Aridification on Human Evolution: Evidence from Paleosols of the Turkana Basin, Kenya." *American Journal of Physical Anthropology* 123 (2): 106–18. http://dx.doi.org/10.1002/ajpa.10317. Medline:14730645

Wynn, J. G., Z. Alemseged, R. Bobe, D. Geraads, D. Reed, and D. C. Roman. 2006. "Geological and Palaeontological Context of a Pliocene Juvenile Hominin at Dikika, Ethiopia." *Nature* 443 (7109): 332–6. http://dx.doi.org/10.1038/nature05048. Medline:16988711

Zachos, Z. C., L. D. Stott, and K. C. Lohmann. 1994. "Evolution of Early Cenozoic Marine Temperatures." *Paleoceanography* 9 (2): 353–87. http://dx.doi.org/10.1029/93PA03266.

Zazzo, A., H. Bocherens, M. Brunet, A. Beauvilain, D. Billiou, H. T. Mackaye, P. Vignaud, and A. Mariotti. 2000. "Herbivore Paleodiet and Paleoenvironmental Changes in Chad during the Pliocene Using Stable Isotope Ratios of Tooth Enamel Carbonate." *Paleobiol* 26 (2): 294–309. http://dx.doi.org/10.1666/0094-8373(2000)026<0294:HPAPCI>2.0.CO;2.

4

Tectonics, Orbital Forcing, Global Climate Change, and Human Evolution in Africa

Mark A. Maslin,
Beth Christensen,
and Katy E. Wilson

Africa is home to many significant hominin sites (Figure 4.1). If we are to understand the driving forces behind human evolution, it is essential to reconstruct changes in past regional environments in Africa. There are, however, significant difficulties in making correlations between hominin speciation events, local tectonic and environmental changes, and regional and global climate records. Many hypotheses about climate-driven hominin evolution and dispersions in Africa have been based on either short and discontinuous records from on-land formations where the fossils are recovered, or continuous low-resolution records from offshore marine sediment cores; however, neither provides the opportunity for direct, in situ measurement of climate change at the sites of hominin evolution. And indeed, as many of the chapters in this volume attest, these records allow measurements of climate proxies, rather than direct measures of climate itself.

This chapter illustrates attempts to relate geologic and climatic records with hominin evolution. We identify three main forcing factors for regional and global climate changes: local tectonics, regional orbital forcing, and global climate changes. We also seek to tie published data to possible explanations for evolutionary change. Previous research that focused on the link between the geologic record and climate change (e.g., Vrba 1995) has provided key insights into the role global climate change played in hominin evolution. Over a decade later, we have a more developed understanding

DOI: 10.5876/9781607322252:c04

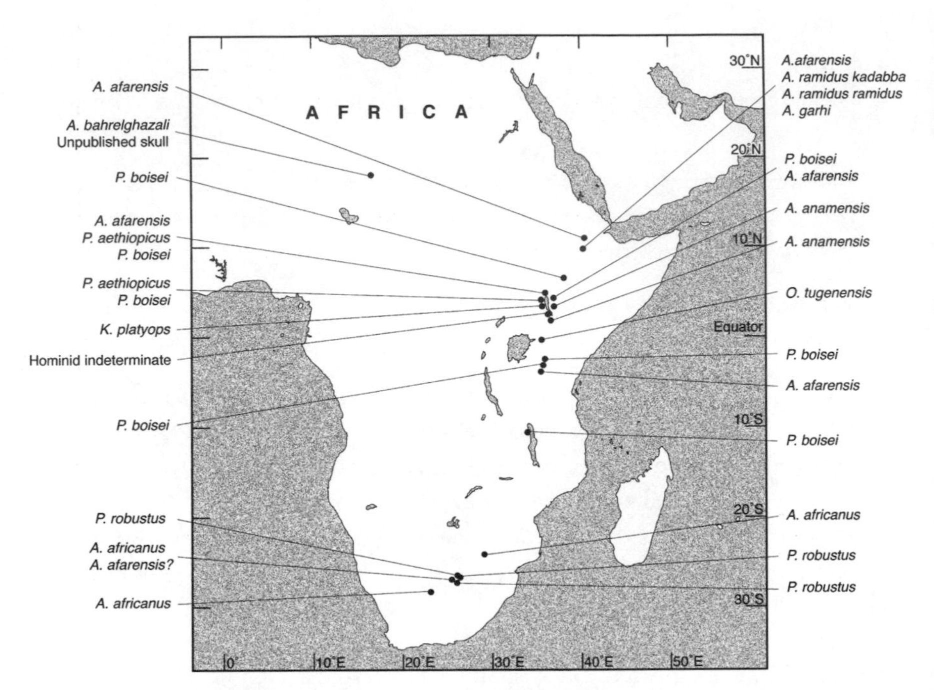

FIGURE 4.1. *Locations of key hominin species in Africa.*

of the global climate system and of Plio-Pleistocene changes in African climate. It is our hope that this chapter will highlight not only recent advances but also the next steps that might be taken toward understanding the role of African climate on hominin evolution.

PATTERNS OF ENVIRONMENTAL CHANGE

Environmental responses

To understand how early human and mammalian evolution and migration may have been influenced by environmental factors, we need to understand how these changes can occur. Environmental changes at any scale are a response to external and/or internal forcing mechanisms acting on the global and local climate. External forcing mechanisms include (1) changing orbital parameters, which alter the net radiation budget of the Earth, and (2) tectonic uplift, which changes the altitude and topography of a region. Examples of internal forcing mechanisms are the carbon dioxide (CO_2) content of the atmosphere, which modulates the greenhouse effect, and ocean circulation, which drives heat and

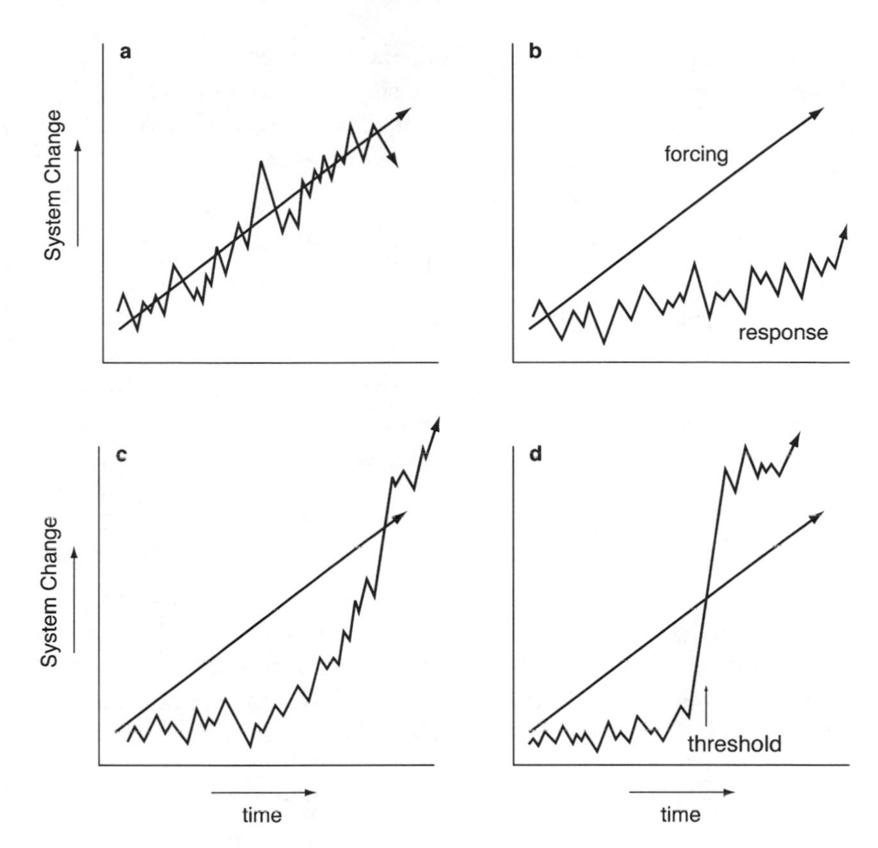

FIGURE 4.2. *Four alternative responses of the global climate system to internal or external forcing. (See text.)*

moisture transport between regions. We can explore the ways in which regional environments may respond to internal or external forcing agents by examining the following scenarios:

- Linear and synchronous response (Figure 4.2a): the forcing produces a direct response in the climate system, the magnitude of which is in proportion to the forcing.
- Muted or limited response (Figure 4.2b): the forcing may be extremely strong, but the climate system is buffered and therefore exhibits little response.
- Delayed or nonlinear response (Figure 4.2c): the climate system may have a slow response to the forcing or is in some way buffered at first. After an initial period the climate system responds to the forcing but in a nonlinear way.

- Threshold response (Figure 4.2d): initially there is no or very little response in the climate system to the forcing; all the response takes place in a very short period of time in one large step at some threshold. In many cases the response may be much larger than one would expect from the size of the forcing. This can be referred to as a response overshoot.

Though these are purely theoretical models of how the environment can respond, they provide useful guidelines for understanding how climate may respond to a perturbation. Moreover, these scenarios can be applied at a range of spatial and temporal scales. East African tectonics and associated systems provide good examples for illustrating the power of these models. If we examine tectonic uplift in East Africa in terms of uplift rate and altitude, we could speculate that it was a linear response (4.2a), which would tail off at higher altitudes when erosion became significant (4.2b). In terms of altitude and rainfall, we would then expect a threshold response (4.2d) due to a rain-shadow effect that was created when the uplift passed a critical altitude. However, the vegetation response (i.e., species composition) to rainfall changes may be delayed as forests have the ability to recycle water and thus buffer the system (4.2c). This chapter illustrates how these scenarios interact with forcing factors at both regional and global scales, ultimately producing environmental changes that may have precipitated the evolution of our ancestors.

Bifurcations

An added complication when assessing the causes of environmental change is the possibility that climate thresholds contain bifurcations: that is, the forcing required to go in one direction through the threshold is different from the forcing required to go in the other direction. This implies that once a climate threshold has occurred in a particular region it is more difficult to reverse it. A bifurcated climate system has been inferred from ocean models that mimic the impact of fresh water on deep-water formation in the North Atlantic (e.g., Rahmstorf et al. 1996). However, this is not a unique example. Such an asymmetrical relationship can be applied to many forcing mechanisms and corresponding responses of the local environment. As Figure 4.3 demonstrates, a bifurcated climate system reveals the possibility of different relationships between climate and the forcing mechanism. This is common in natural systems, such as in cases in which inertia or the shift between different states of matter needs to be overcome. This may be reversible (Figure 4.3a, b), but not always (Figure 4.3c). In some instances, the control variable must exceed the value of

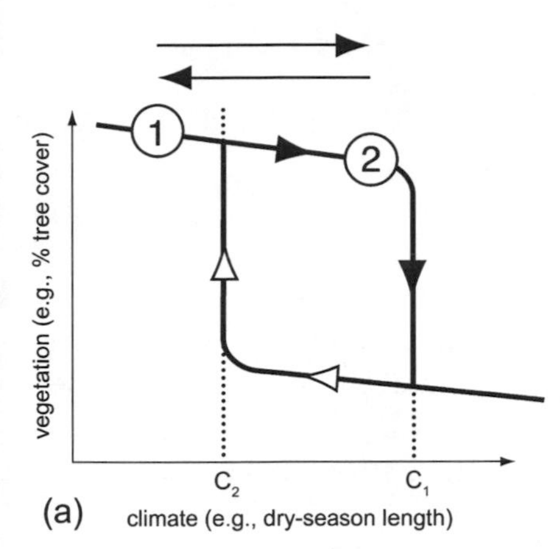

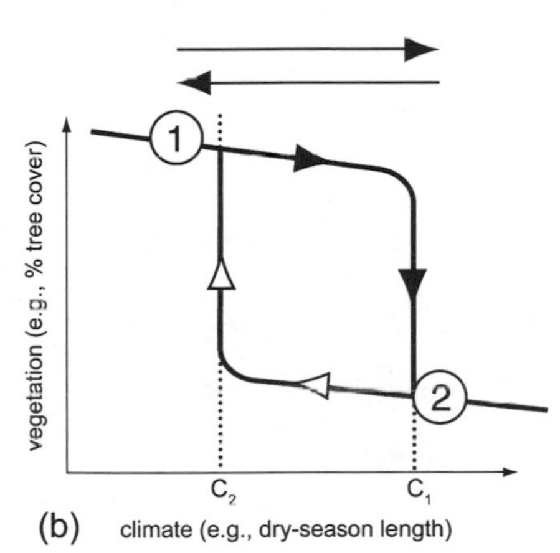

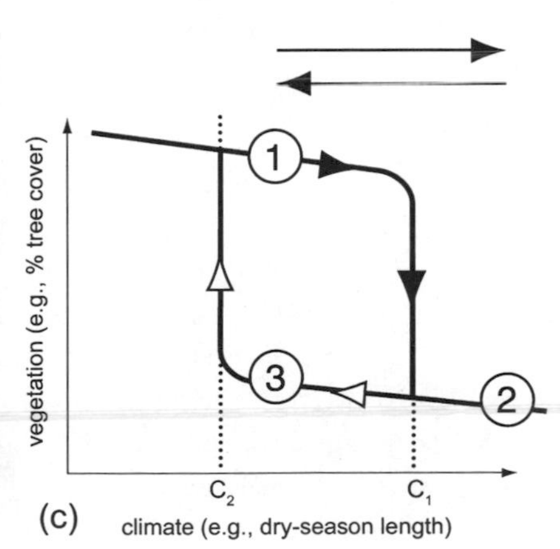

FIGURE 4.3. *Bifurcation of the vegetation-climate relationship. In this example, the control variable is dry-season length, and responding variable is the percentage land cover by trees. Point 1 represents potential starting positions with respect to the climate-vegetation bifurcation. Point 2 or 3 represent the final position depending on whether the bifurcation system allows for reversibility. Arrows represent the pathway that must be taken when the dry-season length alters. The difference is due to the recycling of moisture within the forest, which provides a buffer to the increased dry-season length.*

(a) An insensitive system in which the percentage tree cover does not vary greatly with large changes in the dry season as it does not pass the threshold due to efficient recycling of moisture within the forest.

(b) The dry-season length increases beyond a critical threshold point C_1 (Point 2 in figure) and thus there is a major change in percentage of tree cover. However, by returning the dry-season length to its original state, the system is reversible and the percentage of tree cover can recover (Point 1 in figure). But note the return does not follow the same pathway because "buffering" of the system via recycling of water within the forest is absent.

(c) The dry-season length increases beyond a critical threshold point C_1 (Point 2). However, returning the dry-season length to its original state does not reverse the change in percentage of tree cover and the system is stuck at Point 3. An additional change to the dry-season length is required to overcome the bifurcation and the critical threshold point C_2 and return the percentage of tree cover back to the original Point 1. If this additional change is not possible within the system, the threshold becomes an irreversible one.

its previous equilibrium state to get over the threshold and return the system to its prethreshold state.

Let us consider these examples in terms of dry-season length in the tropics and the percentage of rainforest cover. We know that there is a strong correlation between dry-season length and the amount of rainforest in an area; we also know that this is a bifurcated relationship as shown in Figure 4.3 (Maslin 2004). In the case represented by Figure 4.3a, increasing the dry-season length has no effect on the amount of forest cover as the forest is resilient enough to resist these drought conditions due to the recycling of water. In the case represented in Figure 4.3b, increasing the dry-season length reduces the amount of forest cover, though a significant increase in dry-season length is required as the forest recycles moisture and thus can resist quite extreme drought conditions. If the dry-season length returns to its previous, prethreshold level, then so does the forest cover, but it only comes back after a significant reduction in dry-season length since there is no buffering of the system as there is no forest to recycle water. In the case of Figure 4.3c, increasing the dry-season length reduces the amount of forest cover. However, simply returning the dry-season length to the normal level does not return forest cover to its original state. Because of the bifurcation, a much shorter dry season is required to get forest cover back to its previous level. Analysis of both past and future climate change is complicated by bifurcations, and it should be determined whether a bifurcation has occurred and whether evolution of the system was reversible (Figure 4.3). This is important when trying to relate reconstructions of climate change to patterns in human evolution. For example, a proxy for rainfall will only tell you how much rain there was, not what its effect was on vegetation, as there is no clear relationship between these variables.

Problems of time, scale, and biological response

The response time of different parts of the climate system is an important factor in climate forcing. Figure 4.4 shows the response times of different internal systems that vary from hundreds of millions of years to days. It is complex because these processes are constantly changing but at different rates, and they have different responses to possible external or internal forcing. Investigation of past environmental change in Africa requires integration of the different scale processes and the associated forcing mechanisms. Thus, this problem of time greatly complicates analysis. For example, there are long-term processes, such as tectonic uplift and global climate change; medium-scale changes, such as orbital forcing; and short-term variations, such as the appearance and dis-

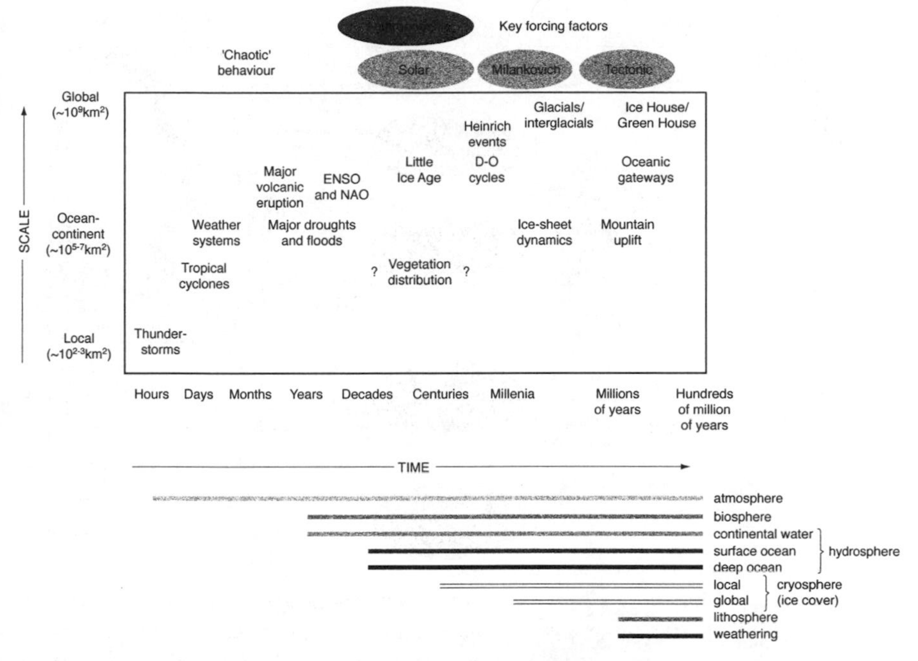

FIGURE 4.4. *The spatial and temporal dimensions of the Earth's climate system plotted on logarithmic scales. The key forcing functions and response time of different sections of the global climate system are also shown. D-O cycles = Dansgaard–Oeschger cycles.*

appearance of lakes and changes in vegetation. In terms of spatial scale there are processes that affect the whole Earth, such as global climate changes (e.g., glacial–interglacial cycles and changes in atmospheric CO_2); regional-scale changes, such as tectonic uplift and basin formation; and local changes, which may involve changes in vegetation or the drainage network. Great care must be taken when relating evolution to reconstructed environmental changes that the appropriate temporal and spatial scale is used. For example, changes in one lake may not have affected evolution if it was a local effect and the population could migrate to another nearby lake. However, if all the lakes in a region dried out at the same time, it may have significantly impacted the local population and stimulated evolution (Trauth et al. 2005, 2010).

The broad changes, which can be reconstructed from the geological record, do not, however, provide details of extreme events. In terms of individuals, extreme events—such as a 1-in-10-year drought, or an El Niño Southern Oscillation (ENSO) that forces a 3-to-7-year massive flood event—selectively control survival and are very important to the evolutionary process. With any

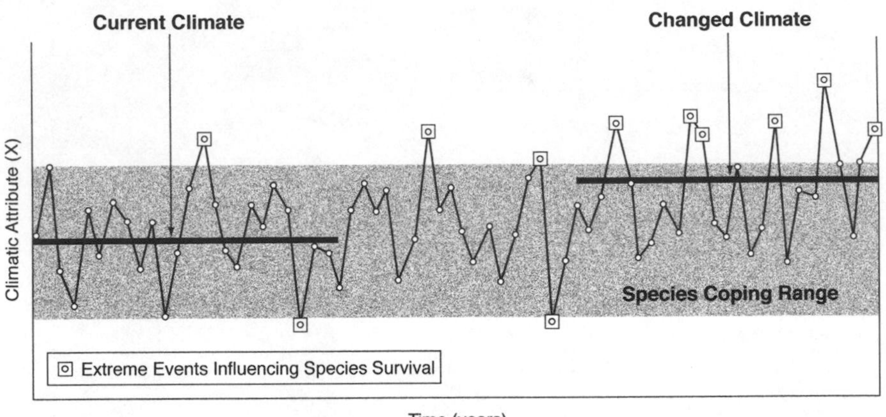

FIGURE 4.5. *Hypothetical "species environmental coping range" compared to a changing climate. Note the increased number of extreme events that may provide a selective survival pressure on the species.*

regional climate change the chance that these extreme events will occur may either increase or decrease. This could be due to the changing frequency of extreme events or the changing sensitivity of the region to the climate variations. This can be modeled assuming each species has an environmental coping range—that is, a range of weather conditions that the species can tolerate. Figure 4.5 shows the theoretical effect of combining the species environmental coping range with gradual climate change. In the initial climate, the coping range encompasses nearly all the variation in weather with only one or two extreme events. As the climate moves gradually toward its new average, many more extreme events occur as the species coping range is fixed. The increased occurrence of these "extreme events" will therefore apply selective pressures on individuals to survive extreme events.

ENVIRONMENTAL FORCING FACTORS

Tectonics

Global. Long-term climate change seems to be primarily modulated by tectonic changes both at the global and local scale. Figure 4.6 shows the stacked (composite) benthic foraminiferal oxygen- and carbon isotope records for the last 75 myr (million years) compared to the history of the polar ice sheets and key tectonic and biotic events. The first major continental ice sheets began on Antarctica about 35 mya (million years ago) with the opening of the Tasmania–

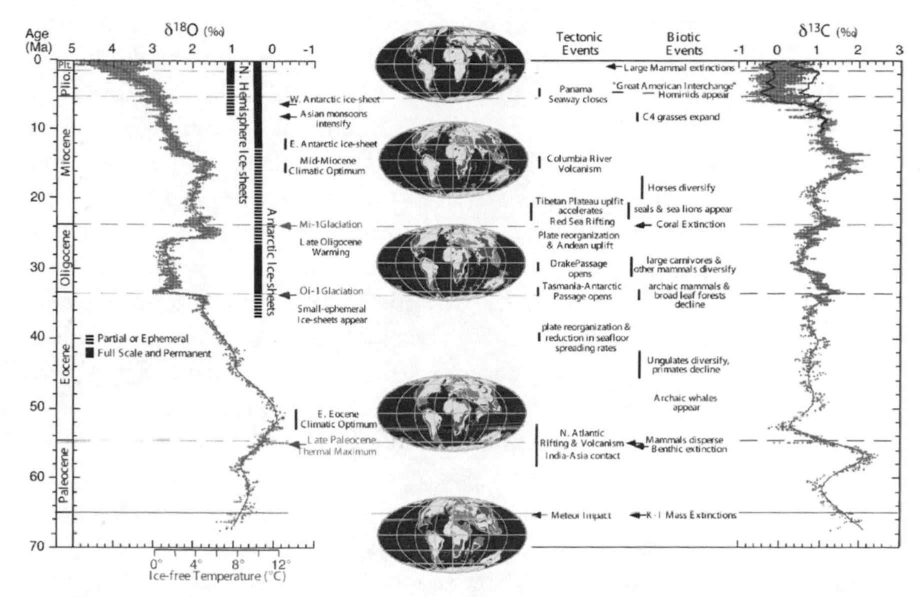

FIGURE 4.6. *Summary of the major climatic, tectonic, and biological events over the last 70 myr (Zachos et al. 2001). The benthic foraminiferal oxygen isotope record on the left side provides an indication of global temperature trends and also indicates the expansion of both the Antarctic and Northern Hemisphere ice sheets.*

Antarctic and Drake passages (Kennett 1996; Huber et al. 2004; Stickley et al. 2004), which was coupled with reduced global atmospheric CO_2 levels (DeConto and Pollard 2003). The resultant "Ice House" climate mode reached a zenith with late Cenozoic global cooling and Northern Hemisphere glaciation. The glaciation of the Northern Hemisphere has been ascribed to the uplift of Tibet (Ruddiman et al. 1988; Ruddiman and Kutzbach 1990; Qiang et al. 2001), to the restriction of the Indonesian seaway (Cane and Molnar 2001), and to the closure of the Panama isthmus (Haug and Tiedemann 1998; Bartoli et al. 2005), though the exact role of atmospheric CO_2 still needs to be elucidated (Sundquist and Visser 2004). The major climate changes during this long-term global cooling are described in detail later, in the "Late Cenozoic Global Climate Transitions" section, as they seem to have had a major influence on African climate.

East Africa. On a regional scale, tectonics can also cause significant changes in climate. In East Africa, long-term climatic change is also controlled by tectonics, with the progressive formation of the East African Rift Valley leading to increased aridity and an increase in the development of fault-graben basins as catchments for lakes. Trauth et al. (2005, 2007) suggest that volcanism in East Africa may have started as early as 45–33 mya in the Ethiopian Rift. By 33 mya, volcanism occurred in northern Kenya. Magmatic activity of the central and southern segments of the rift in Kenya and Tanzania did not start until between 15 and 8 mya (e.g., Bagdasaryan et al. 1973; Crossley and Knight 1981; McDougall and Watkins 1988; Ebinger et al. 2000; George et al. 1998). Rifting begins with updoming at the site of future separation, and downwarping away from the site, and is followed by the actual rifting and separation as half-grabens (land that has subsided with a fault on one side) are formed on either side of the rift. While the early (33 mya) stages of rifting were characterized by general updoming and downwarping, in the later stages faulting progressed from north to south (Figure 4.7). Major faulting in Ethiopia between 20 to 14 mya was followed by east-dipping fault development in northern Kenya between 12 and 7 mya, which was then superseded by normal faulting on the western side of the Central and Southern Kenya Rift between 9 and 6 mya (Baker et al. 1988 Blisniuk and Strecker 1990; Ebinger et al. 2000). These early half-grabens were subsequently faulted antithetically between about 5.5 and 3.7 mya, which generated a full-graben (a block of subsided land with faults on either side) morphology (Baker et al. 1988; Strecker et al.1990). Prior to the full-graben stage, the large Aberdare volcanic complex (elevation in excess of 4000 m), an important Kenyan orographic barrier, was established (Williams et al. 1983). By 2.6 mya, the graben was further segmented in the Central Kenya Rift by western-dipping faults, creating the 30-kilometer-wide intrarift Kinangop Plateau and the tectonically active 40-kilometer-wide inner rift (Baker et al. 1988; Strecker et al. 1990). After 2 mya many of the Kenya lake basins continued to fragment due to volcanic activity, including the Barrier volcano complex separating Lake Turkana and the Suguta Valley (~1.4 mya eastern side and ~0.7 mya western side), and the Emuruangogolak (~1.3 mya), and Namarunu volcanos (0.8 mya), which separate Lake Baringo and the Suguta Valley (Dunkley et al. 1993; McDougall et al. 2012). Hence lakes before 1.4 mya may have extended from the Omo National Park in the north to Lake Baringo in the south, covering over 16,000 square kilometers. In the Tanzanian sector of the rift, sedimentation in isolated basins began at about 5 mya. A major phase of rift faulting occurred at 1.2 mya and produced the present-day rift escarpments (Foster et al. 1997).

These tectonic events are associated with a variety of biotic changes. There is evidence from both soil carbonate (Levin et al. 2004; Wynn 2004) and *n*-alkane carbon isotopes (Feakins et al. 2005, 2007) that there was a progressive vegetation shift from domination by C_3 plants to domination by C_4 plants during the

Plio-Pleistocene. This vegetation shift has been ascribed to increased aridity due to the progressive rifting of East Africa (deMenocal 2004). Rainfall modeling by Sepulchre et al. (2006) demonstrates the huge influence of East African regional uplift. They predict a marked decrease in rainfall as the wind patterns became less zonal as uplift increased. In their model, as elevation increased, a rain-shadow effect occurred that reduced the moisture available for rain on the eastern sides of the mountains and valleys and produced the strong aridification seen in paleoenvironmental records (Sepulchre et al. 2006).

In addition to contributing to the aridification of Africa, the tectonic events described above also produced numerous basins suitable for the formation of lakes. Figure 4.7 illustrates that tectonics was essential for the production of isolated basins in the East African Rift Valley within which lakes can form. The southward propagation of rifting, including the formation of faults and magmatic activity, is also reflected in the earliest formation of lake basins in the northern parts of the rift. Many lakes formed long after the basins were produced. For example, the Middle and Upper Miocene saw the beginning of lakes in the Afar, Omo–Turkana, and Baringo–Bogoria basins, but the oldest lacustrine sequences in the central and southern segments of the rift in Kenya and Tanzania are of early Pliocene age (Tiercelin and Lezzar 2002). Paleolakes in the northern part of the East African Rift Valley thus formed earlier than in the south (see Figure 4.7). However, if tectonics was the sole control over the appearance and disappearance of lakes, then either a north–south or a northwest–southeast temporal pattern would be expected. In contrast, what is observed is the synchronous appearance of large, deep lakes across a large geographical area at specific times (Trauth et al. 2005), possibly suggesting some other regional climatic control.

Southern Africa. While the late Cenozoic tectonic and moisture history of East Africa is well constrained temporally (Levin et al. 2004; Wynn 2004; Feakins et al. 2005, 2007; Trauth et al. 2005, 2007; Sepulchre et al. 2006), this is not the case for southern Africa (Dupont et al. 2005; Maslin and Christensen 2007). Far from the major rifting centers, the region lacks the volcanic (basalt and ash) deposits that are characteristic of East Africa and that permit the development of radiometric-based chronologies. Instead, most of the continental evidence for past climate change is derived from estimations of denudation rate and landscape development using fission-track thermochronology and/or cosmogenic isotope analysis, or by reconstructing regional geomorphological features associated with specific tectonic events. Estimations of denudation rate and landscape development for Southwest Africa (Brown et al. 1994, 2000;

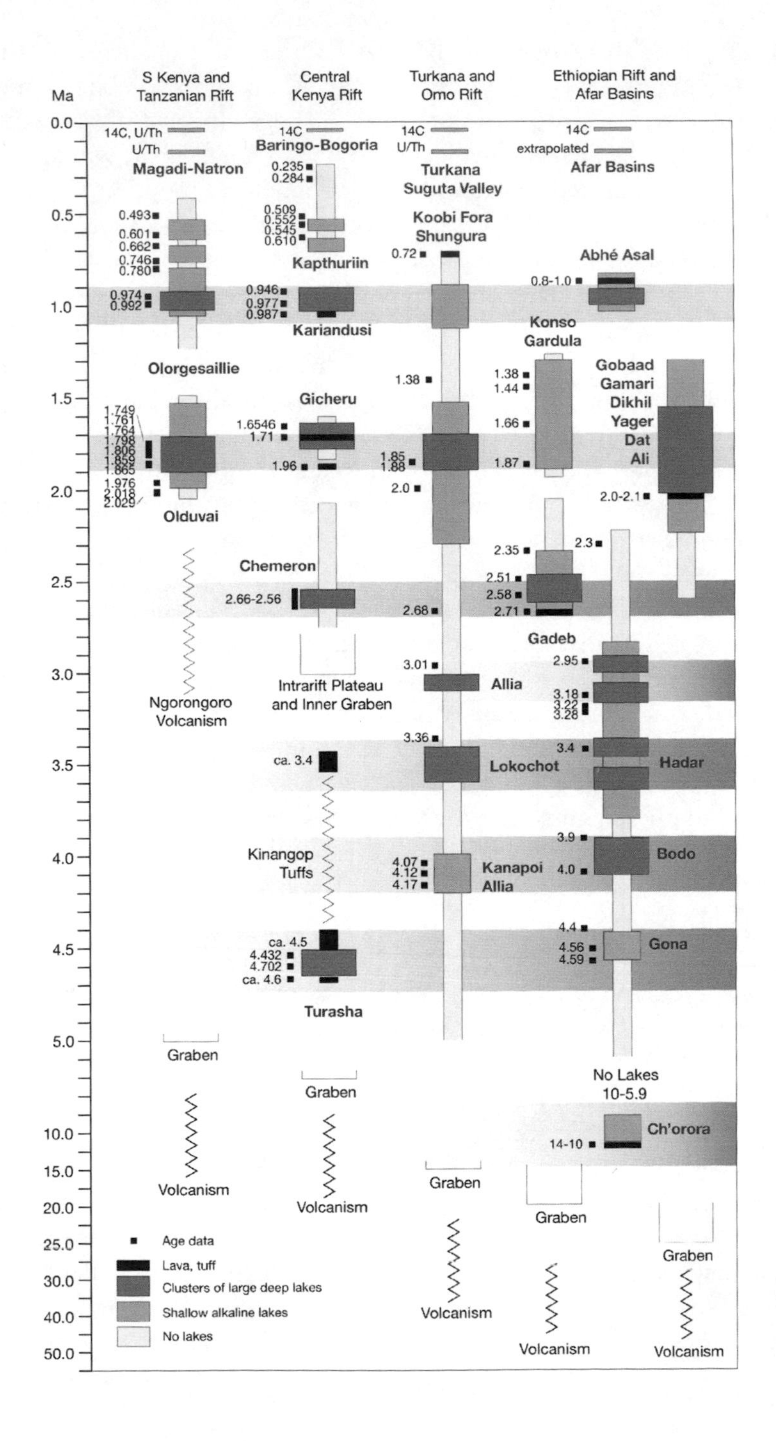

Ma
S Kenya and Tanzanian Rift
Central Kenya Rift
Turkana and Omo Rift
Ethiopian Rift and Afar Basins
14C, U/Th
U/Th
Magadi-Natron
14C
Baringo-Bogoria
14C
U/Th
Turkana
Suguta Valley
14C
extrapolated
Afar Basins
Koobi Fora
Shungura
Kapthuriin
Abhé Asal
Kariandusi
Konso
Gardula
Olorgesaillie
Gicheru
Gobaad
Gamari
Dikhil
Yager
Dat
Ali
Olduvai
Chemeron
Gadeb
Allia
Ngorongoro
Volcanism
Intrarift Plateau
and Inner Graben
Lokochot
Hadar
Kinangop
Tuffs
Kanapoi
Allia
Bodo
Gona
Turasha
Graben
Graben
No Lakes
10-5.9
Ch'orora
Volcanism
Volcanism
Graben
Graben
Graben
Volcanism
Volcanism
Volcanism
Age data
Lava, tuff
Clusters of large deep lakes
Shallow alkaline lakes
No lakes

FIGURE 4.7. *Compilation of tectonic features and prominent lake periods of the eastern branch of the East African Rift System. Tectonic features and events compiled from Williams et al. 1983; Baker et al. 1988; Strecker et al. 1990; Foster et al. 1997; Ebinger et al. 2000. Paleoenvironmental and radiometric age data in millions of years for Olduvai Basin (Walter et al. 1991; Ashley and Hay 2002); Magadi–Natron and Olorgesailie basins (Potts 1998; Potts et al. 1999; Behrensmeyer et al. 2002); Natron has one persistent lacustrine interval (a member of the Moinik Formation called the Moinik Clays) dated to 1.1–1.0 mya (Deino, pers. communication); Gicheru Basin (Baker et al. 1988; Strecker 1991; Boven 1992); Naivasha Basin (Richardson and Richardson 1972; Richardson and Dussinger 1986; Strecker et al. 1990; Trauth et al. 2003, 2005); Nakuru–Elmenteita Basin (Evernden and Curtis 1965; Washbourn-Kamau 1970, 1975, 1977; Richardson and Richardson 1972; Richardson and Dussinger 1986; Strecker 1991; Boven 1992; Trauth et al. 2005); Baringo–Bogoria Basin (Owen 2002; Deino et al. 2006; Kingston et al. 2007); Suguta Basin (Butzer et al. 1969; Hillaire-Marcel et al. 1986; Sturchio et al. 1993); Omo–Turkana Basin (McDougall and Watkins 1988; Brown and Feibel 1991; Leakey et al. 1995, 1998); Ethiopian Rift (Williams et al. 1979; Gasse 1990; WoldeGabriel et al. 2000); and Afar Basin (Gasse 1990).*

Cockburn et al. 1999, 2000; Gallagher and Brown 1999; Raab et al. 2002, 2005) and Southeast Africa (Fleming et al. 1999; van der Beek et al. 2002; Brown et al. 2002) indicate denudation rates peaking at between 100 and 200 meters/myr during the late Cretaceous (80 to 60 mya). Subsequent denudation rates for Namibia are 15 meters/myr or lower, and are typical for a passive margin setting (Raab et al. 2005); similar results have also been found for the Drakensberg Escarpment (Brown et al. 2002). This suggests that since 60 mya there has been little or no uplift in Southwest Africa and that tectonic activity has been quiescent.

Continental paleoclimate records are also limited in this region due to a lack of unambiguous chronological control (e.g., Vaal River terraces [de Wit et al. 2000], Florisbad Pan deposits [de Wit et al. 2000], Tswaing Crater [Partridge et al. 1997], or lake deposits [e.g., Scott 2000]). Further complicating the attempt to reconstruct continental climate in this region is the fact that there are no long, continuous Plio-Pleistocene records of sedimentation. The Tswaing Crater pan (an impact crater in the northeast South African interior) is the only long-lived basin with approximately continuous accumulation of sediment; however, this only spans a relatively short interval from about 200 kya (thousand years ago) to the present (Partridge et al. 1997). Nonetheless, it has provided some insight to Southeast African rainfall showing that it has varied at a 23-kyr (thousand year) periodicity (following the solar insolation for 30°S.).

Unlike East Africa, where lakes dominate, the major repositories for Plio-Pleistocene sediments and fossils in southern Africa are cave sites, which have a complex stratigraphy that has often been further disturbed by quarrying. They also often lack unambiguous chronostratigraphic control (Hopley et al. 2007). Thus, the record of southern African continental climate is inferred primarily from age associations between faunas in East Africa and similar forms in South Africa, as well as from associations between geomorphological events and global climate changes. Although the data are patchy and often equivocal, the need for such a record for the contextualization of mammal evolutionary data has led to a general acceptance of its veracity (Vrba 2000; Maslin and Christensen 2007).

It is believed that a major change in southern African climate occurred during the late Cenozoic, resulting from the development of a strong cross-continental temperature gradient associated with the onset of the Benguela Current (BC) in the South Atlantic Ocean (Tyson and Partridge 2000; Christensen et al. 2002). The upwelling associated with the BC initiated arid conditions in Southwest Africa, which were ameliorated in the Pliocene when streams were rejuvenated and grasslands expanded (Tyson and Partridge 2000). Uplift of the southeastern and eastern hinterlands, including East Africa and Zimbabwe, created rain-shadow areas to the west of these regions (Tyson and Preston-Whyte 2000) and is thought to have enhanced aridification. Tyson and Preston-Whyte (2000) suggest the timing of this uplift was not well constrained and that it occurred in association with the late Pliocene intensification of Northern Hemisphere Glaciation (Li et al. 1998; Maslin et al. 1998). Subsequently, Trauth et al. (2005, 2007) have shown East African uplift started at about 12 mya in northern Ethiopia and progressively shifted southward and occurred between about 4.5 mya to 3.5 mya in central Kenya and between 3.2 mya to 2.3 mya in southern Kenya and Tanzania. This may coincide with a long-term drying trend between 3.5 mya and 1.7 mya in Southwest Africa that has been inferred from pollen records from ODP (Ocean Drilling Program) Site 1082 off the coast of the Namib Desert. This record also suggests a possible rapid increase in local aridity at about 2.2 mya (Dupont et al. 2005). Arguments to the contrary suggest that the late Miocene was more arid and that there is no indication of major change during the Pliocene (Diester-Haass et al. 2002; Giraudeau et al. 2002). Moreover a recently published marine record of *n*-alkane carbon isotopes off the coast of Southwest Africa shows virtually no change in water stress over the last 3.5 myr (Maslin et al. 2012) suggesting the climate of Southwest Africa was very stable over this period.

What We Do Not Know. There are still two main elements of regional tectonics in Africa that are not completely understood. First is the exact timing and altitude of the uplift. Much work has been done on the timing of key tectonic features, but uplift rates and maximum altitude are still unconstrained, especially for southern Africa. These factors control local rainfall patterns and thus are important for understanding the evolution of African climate. So although we know that progressive uplift and rifting has caused East Africa to dry, we do not know precisely when, or at what pace, these changes occurred.

The second factor is what effect tectonics had on vegetation. This is crucial for understanding hominin evolution. For example, 10 myr before the occurrence of significant doming of East Africa, were there rainforests in East Africa? If so, when did the forest fragment? When did grasslands become important, and more specifically, dominant? Was there a vegetative corridor between southern Africa and eastern Africa that may have facilitated dispersal and exchange between populations? At the moment we have detailed knowledge of vegetation and environmental conditions at sites containing hominin remains for East Africa and South Africa. However, these provide information only on the niche that our ancestors inhabited and not on the wider environment or regional climatic pattern. Moreover many of these reconstructions are based on large sequences covering a wide range of time and may be affected by time-averaging (Hopley and Maslin 2010), which means short-term and large-scale variations in environment are combined. Sepulchre et al. (2006) uniquely approached these problems by providing a reconstruction of climate and vegetation with greatly reduced relief over eastern Africa. The next step will be to produce detailed time-slices through the last 8 myr with both relief and global climate applied to a regional climate and vegetation model. Another novel approach to link local environmental niches with wider regional climate patterns is employed by Lee-Thorp et al. (2007), who analyzed $^{13}C/^{12}C$ ratios of fossil tooth-enamel to provide a semiquantitative measure of "open" versus "closed" habitats. However, this technique is limited to the occurrence of mammal fossils. Two further ways to try to reconstruct past vegetation are (1) examination of long, continuous ocean and lake cores, and (2) the use of coupled climate–vegetation models.

Orbital forcing

The oscillation between glacial and interglacial climates, which is the most fundamental environmental characteristic of the Quaternary period, is likely primarily forced by changes in the Earth's orbital parameters (Hays et al. 1976). However, there is not necessarily a direct cause-and-effect relationship between climate cycles and the Earth's orbital parameters due to the dominant effect of feedback mechanisms internal to the Earth's climate system. An illustration of this is that the insolation (solar radiation energy) received at the critical latitude of 65°N was very similar 18 kya during the Last Glacial Maximum (LGM) to what it is today (Laskar 1990; Berger and Loutre 1991). The three main orbital parameters—eccentricity, obliquity, and precession—are described below.

Eccentricity. The shape of the Earth's orbit changes from near circular to an ellipse. This varies over a period of about 100 kyr with an additional long cycle of about 400 kyr (in detail there are two distinct spectral peaks near 100 kyr and another at 413 kyr). Described another way, the long axis of the ellipse varies in length over time. In recent times, the Earth is closest (146 million km) to the sun on January 3; this position is known as perihelion. On July 4 the Earth is most distant from the sun (156 million km) at the aphelion (Figure 4.8). Changes in eccentricity cause only very minor variations in insolation, but can have significant seasonal effects. If the orbit of the Earth were perfectly circular there would be no seasonal variation in solar insolation. Today, the average amount of radiation received by the Earth at perihelion is ~351 W/m^2 and is 329 W/m^2 aphelion. This represents a difference of about 6 percent, but at times of maximum eccentricity (ellipse length) over the last 5 myr, the difference could have been as large as 30 percent. Milankovitch (1949) suggested that northern ice sheets are more likely to form when the sun is more distant in summer, so that each year some of the previous winter's snow does not melt. The intensity of solar radiation reaching the Earth diminishes in proportion to the square of the planet's distance, so modern global insolation is reduced by nearly 7 percent between January and July. As Milankovitch suggested, this produces a situation that is favorable for snow accumulation, but in the Northern rather than the Southern Hemisphere. The more elliptical the shape of the orbit becomes, the more the seasons will be exaggerated in one hemisphere and moderated in the other. The other effect of eccentricity is to modulate the effects of precession (see below). However, it is essential to note that eccentricity is by far the weakest of all three orbital parameters.

Obliquity. The tilt of the Earth's axis of rotation with respect to the plane of its orbit (the plane of the ecliptic—i.e., the path of the sun) varies between 21.8° and 24.4° over a period of 41 kyr. It is the tilt of the axis of rotation that causes seasonality. In summer, the hemisphere that is tilted toward the sun is warmer because it receives more than 12 hours of sunlight, and the sun is higher in the sky. At the same time in the opposite hemisphere, the axis of rotation is tilted away from the sun; it is colder because it receives less than 12 hours of sunlight and the sun is lower in the sky. The larger the obliquity, the larger the difference between summer and winter.

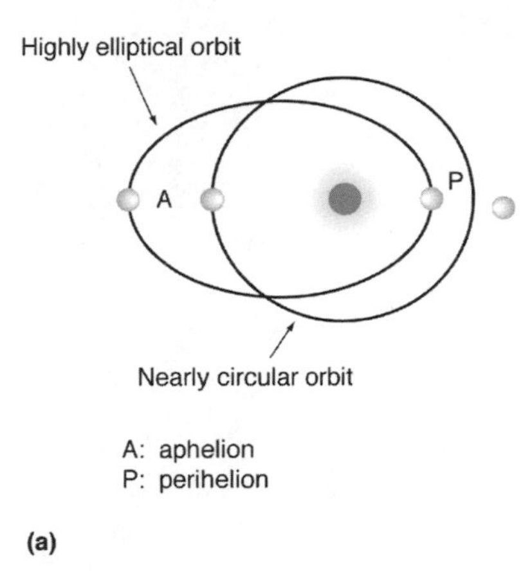

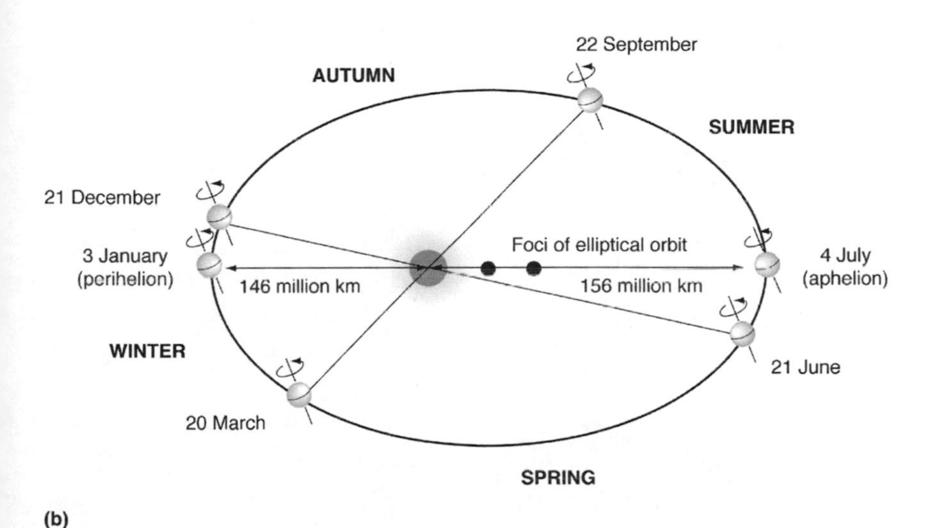

FIGURE 4.8. *Changes in the shape of the Earth's orbit around the Sun. (a) The shape of the orbit changes from near circular to elliptical in shape. The position along the orbit when the Earth is closest to the Sun is termed the perihelion and the position when it is farthest from the Sun is the aphelion. (b) The present-day orbit and its relationship to the seasons, solstices, and equinoxes. (After Wilson 2000.)*

As Milankovitch (1949) suggested, the colder the Northern Hemisphere summers, the more likely it is that snowfall accumulation results in the gradual buildup of glaciers and ice sheets. This is why there seems to be a straightfor-

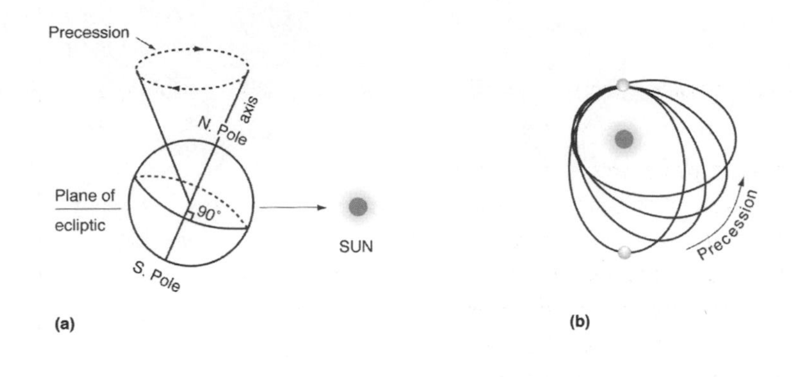

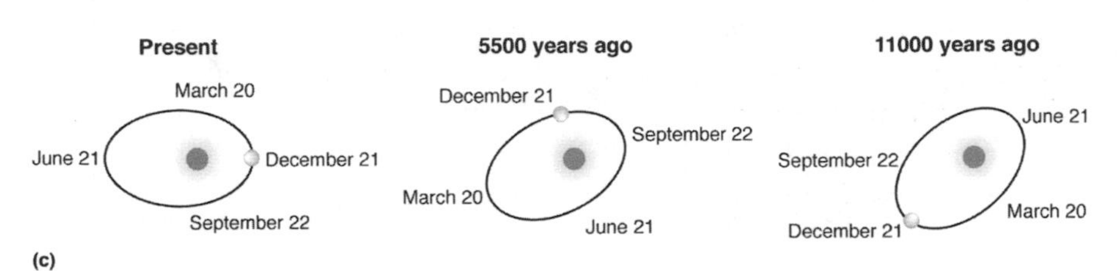

FIGURE 4.9. *The components of the precession of the equinoxes. The precession of (a) the Earth's axis of rotation, (b) the Earth's orbit, and (c) the equinoxes. (After Wilson 2000.)*

ward explanation for the glacial–interglacial cycles that occurred every 41 kyr prior to the Mid-Pleistocene Revolution (see below). Raymo and Nisancioglu (2003) have shown that obliquity controls the differential heating between high and low latitudes (i.e., the atmospheric meridional flux of heat, moisture, and latent heat) that exerts a dominant control on global climate. This is because the majority of heat transport between 30° and 70°N is by the atmosphere: thus a linear relationship between obliquity, northward heat transport, and glacial–interglacial cycles can be envisioned.

Precession. There are two components of precession that relate to the elliptical orbit of the Earth and its axis of rotation. The Earth's rotational axis moves around a full circle, or precesses, every 27 kyr (Figure 4.9). This is similar to the gyrations of the rotational axis of a spinning toy top. Precession causes a change in the Earth–Sun distance for any particular date, such as the beginning of the Northern Hemisphere summer (Figure 4.9). The combined effects of the precessional components are illustrated in Figure 4.10.

It is the combination of the different orbital parameters that results in the classically cited precessional periodicities of 23 kyr and 19 kyr. Combining the

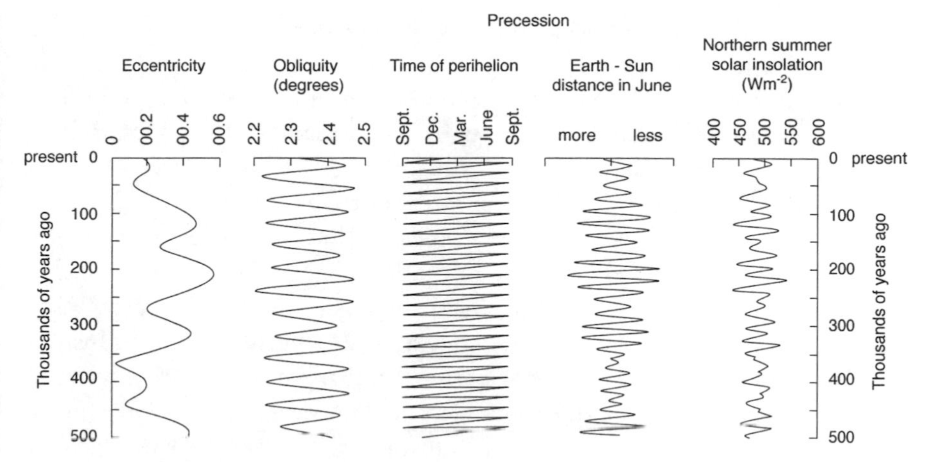

Figure 4.10. *Variations in the Earth's orbital parameters: eccentricity, obliquity, and precession, and the resultant Northern Hemisphere 65°N insolation for the last half million years. (After Wilson 2000.)*

precession of the axis of rotation plus the precessional changes in orbit produces a period of 23 kyr. However, the combination of eccentricity and precession of the axis of rotation also results in a period of 19 kyr. These two periodicities combine so that perihelion coincides with the summer season in each hemisphere on average every 21.7 kyr. Precession has the most significant impact in the tropics (in contrast to the impact of obliquity at the equator, which is zero). So although obliquity clearly influences high-latitude climate change, which may ultimately influence the tropics, the direct effects of insolation in the tropics are due to eccentricity-modulated precession alone.

Combining the Effects of Eccentricity, Obliquity, and Precession. Combining the effects of eccentricity, obliquity, and precession provides the means for calculating insolation for any latitude through time (e.g., Milankovitch 1949; Berger and Loutre 1991). Changes in insolation may not seem significant until one considers that the maximum difference in solar radiation in the last 600 kyr (see Figure 4.10) is equivalent to the difference between the amount of summer radiation received today at 65°N and that received now at 77°N, over 550 km to the north. Although this is an oversimplification, this would bring the current glacial limit from mid-Norway down to the latitude of Scotland. The key factor is that each of the orbital parameters has a different effect depending on latitude. For example, obliquity has greater influence the higher the latitude, while precession has its greatest influence in the tropics.

Glacial–Interglacial Cycles. Milankovitch (1949) initially suggested that the critical factor in determining the development of glacial cycles was total summer insolation at about 65°N; in order for an ice sheet to grow, some additional ice must survive each successive summer. In contrast, the Southern Hemisphere is limited in its response because the expansion of ice sheets is curtailed by the Southern Ocean around Antarctica. The conventional view of glaciation is that low summer insolation in the temperate Northern Hemisphere allows ice to survive summer and build up on the northern continents (see detailed references in Maslin et al. 2001). Orbital forcing in itself is, however, insufficient to drive the observed glacial–interglacial variability in climate. Instead, the Earth system amplifies and transforms the changes in insolation received at the Earth's surface through various feedback mechanisms. As snow and ice accumulate due to initial changes in insolation, the ambient environment is modified. This is primarily through an increase in albedo (reflected sunlight), and a reduction in the absorption efficiency of incident solar radiation that results in a suppression of local temperatures. This promotes the accumulation of more snow and ice and a further modification of the ambient environment (e.g., the classic "ice albedo" feedback; Figure 4.11).

Another feedback is triggered when the ice sheets, particularly the North American Laurentide Ice Sheet, become large enough to affect atmospheric circulation. This alters the storm path across the North Atlantic Ocean and prevents the warm-water Gulf Stream and North Atlantic Drift from penetrating as far north as they do today. This change in surface ocean currents, combined with the general increase in meltwater in the Nordic Seas and Atlantic Ocean due to the presence of large continental ice sheets, ultimately leads to a reduction in the production of deep water. The sinking of water to form what is called the North Atlantic Deep Water (NADW) is the start of the thermohaline deep-ocean system that is a critical part of the global climate system as it transfers heat and salinity from the North Atlantic Ocean into the Southern Hemisphere. Any reduction in this NADW formation reduces the amount of warm water pulled northwards toward the Nordic and Labrador seas, which leads to increased cooling in the Northern Hemisphere and, ultimately, the expansion of the northern ice sheets. It also reduces the amount of heat transported southward to the Southern Hemisphere, contributing to global cooling (see detailed references in Maslin et al. 2001).

However, the action of these and similar "physical climate" feedbacks on variations in the amount of solar insolation received at the Earth's surface are not sufficient to allow for a complete accounting of the timing and magnitude of glacial–interglacial cycles. There is mounting evidence that other feedback

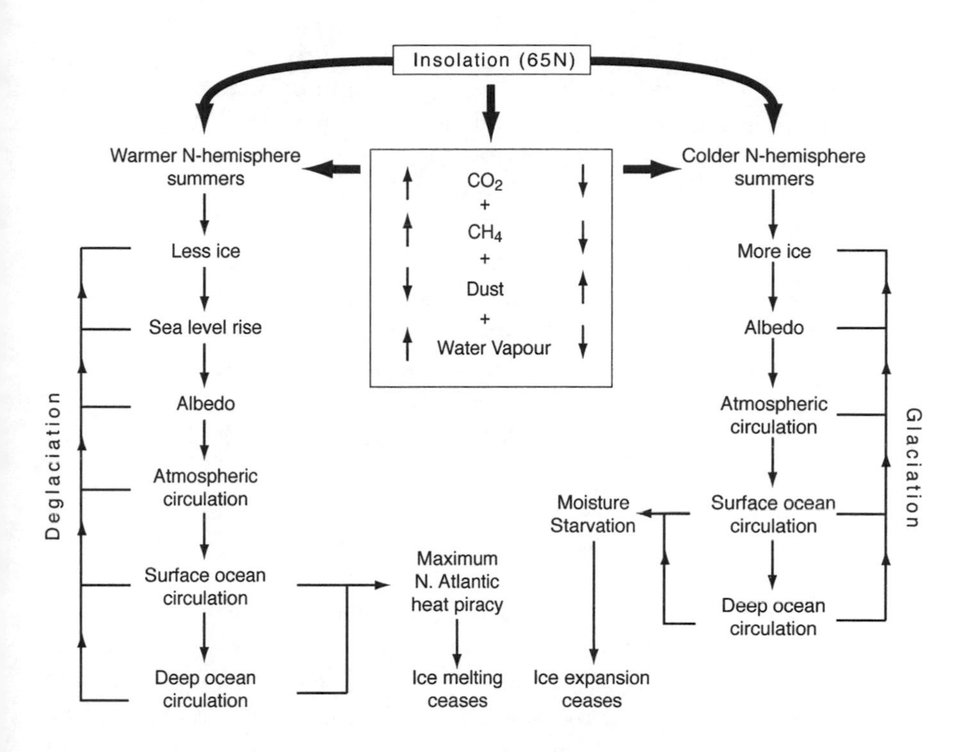

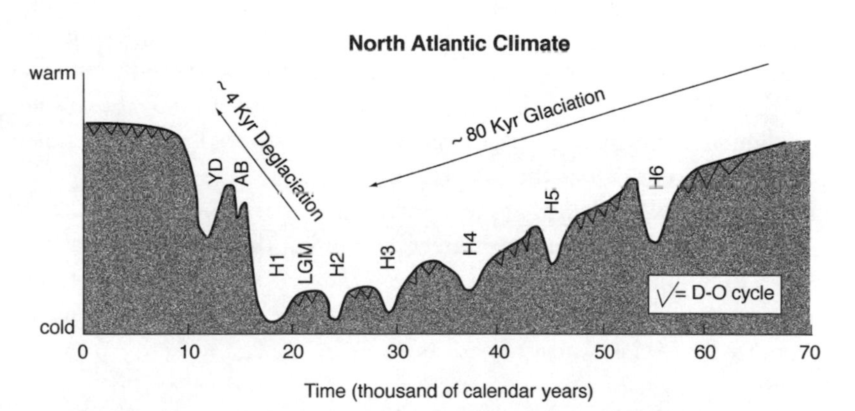

Figure 4.11. *Summary of the conventional view of the feedback mechanism forced by insolation at 65°N which drives glaciation and deglaciation. AB = Allerød–Bølling Interstadial, D-O = Dansgaard–Oeschger cycles, H = Heinrich events, LGM = Last Glacial Maximum, and YD = Younger Dryas. (Adapted from Maslin et al. 2001.)*

mechanisms may be equally if not more important in driving the long-term climate system. The role of "greenhouse" gases in the atmosphere that absorb outgoing infrared radiation is critical (e.g., Berger 1988; Saltzman et al. 1993; Berger and Loutre 1991; Li et al. 1998). Any reduction in the atmospheric concentration of atmospheric constituents such as CO_2, CH_4, and water vapor will drive a general global cooling (Figure 4.11), which in turn furthers glaciation. We already know that these properties varied considerably over glacial–interglacial cycles. For CO_2 and CH_4, this variability is recorded in air bubbles trapped in polar ice (Petit et al. 1997) and is coeval with changes in global ice volume (Shackleton 2000; Ruddiman and Raymo 2003). Glacial periods are by their very nature drier, which reduces atmospheric water vapor, and they provide clear evidence that water vapor production of the equatorial Pacific zone was greatly curtailed during the last five glacial periods (Lea et al. 2000).

Feedbacks are also central to the climate system's exit from glaciation. Deglaciation occurs much quicker than glaciation because of the natural instability of ice sheets (Maslin et al. 2001). In the case of the last deglaciation, or Termination I, the transition from glacial to interglacial lasted only 4 kyr, even including a brief return to glacial conditions known as the Younger Dryas period (see detailed references in Maslin et al. 2001). Two controls were operating during this time: (1) an increase in summer insolation at 65°N (e.g., Imbrie et al. 1993), which drives melting of the Northern Hemisphere ice sheets and allows them to retreat, and (2) the rise of atmospheric CO_2 and CH_4, which promoted global warming and encouraged further melting of large continental ice sheets (Shackleton 2000; Ruddiman and Raymo 2003; Ridgwell et al. 2003). With an initial rise in sea level, large ice sheets adjacent to the oceans are undercut, which again increases sea level. This sea-level feedback mechanism can be extremely rapid. Once the ice sheets are in retreat, the feedback mechanisms that promoted glaciation are reversed (Figure 4.11). These feedbacks are prevented from producing a runaway effect by limiting the amount of heat that moves from the South Atlantic to the North Atlantic to maintain the interglacial deep-water overturning rate and by a limit on the amount of carbon that can be exported into the atmosphere.

Direct Influence of Orbital Forcing on Tropical Africa. Orbital forcing has an obvious impact on high-latitude climates and influenced late Cenozoic global-climate transitions, but it also has a huge influence on the tropics, particularly through precession and its effect on seasonality, and thus on rainfall. There is a growing body of evidence for precessional forcing of moisture availability in the tropics, both in East Africa during the Pliocene (deMenocal 1995, 2004;

Denison et al. 2005; Deino et al. 2006; Kingston et al. 2007; Hopley et al. 2007; Lepre et al. 2007) and elsewhere in the tropics during the Pleistocene (Bush et al. 2002; Clemens and Prell 2003, 2007; Trauth et al. 2003; Cruz et al. 2005; Wang et al. 2004; Tierney et al. 2008; Verschuren et al. 2009; Ziegler et al. 2010). The precessional control on tropical moisture has also been clearly illustrated by climate modeling (Clement et al. 2004), which showed that a 180° shift in precession could change annual precipitation in the tropics by at least 180 millimeters/year and cause a significant shift in seasonality. This is on the same order of magnitude as the effect of a glacial–interglacial cycle in terms of the hydrological cycle. In contrast, precession has almost no influence on global or regional temperatures. Support for increased seasonality during these extreme periods of climate variability also comes from mammalian community structures (Reed 1997; Bobe and Eck 2001; Reed and Fish 2005) and hominin paleodiet reconstructions (Teaford and Ungar 2000).

In northern and eastern Africa there are excellent records of precessional forcing of climate, including (1) East Mediterranean marine-dust abundance (Larrasoaña et al. 2003), which reflects the aridity of the eastern Algerian, Libyan, and western Egyptian lowlands located north of the central Saharan watershed; (2) sapropel formation in the Mediterranean Sea, which is thought to be caused by increased Nile River discharge (Lourens et al. 2004); and (3) dust records from ocean-sediment cores adjacent to West Africa and Arabia (Clemens and Prell 2003, 2007; deMenocal 1995, 2004; Ziegler et al. 2010). There is also a growing body of evidence for precessional forcing of East African lakes. Deino et al. (2006) and Kingston et al. (2007) have found that the major lacustrine episodes of the Baringo–Bogoria Basin in the Central Kenyan Rift between 2.7 and 2.55 mya actually consisted of five paleolake phases separated by a precessional cyclicity of about 23 kyr (Figure 4.12). The occurrences of the lakes are in phase with increased freshwater discharge and thus sapropel formation in the Mediterranean Sea (Lourens et al. 2004), and also coincide with dust-transport minima recorded in sediments from the Arabian Sea (deMenocal 1995, 2004; Clemens et al. 1996). Hence, the lake records from East Africa and the dust records from the Arabian Sea document extreme climate variability with precessionally forced wet and dry phases. Precessional forcing of vegetation change also occurred at this time in Southwest Africa, independent of glacial–interglacial cycles (Denison et al. 2005). There is also evidence for precessional forcing of another lake phase (between 1.9 and 1.7 mya) in the KBS Member of the Koobi Fora Formation in the northeast Turkana Basin in Kenya (Brown and Feibel 1991; Lepre et al. 2007). During the same period an oxygen isotope record from the Buffalo Cave flowstone (Makapansgat Valley, Limpopo Province,

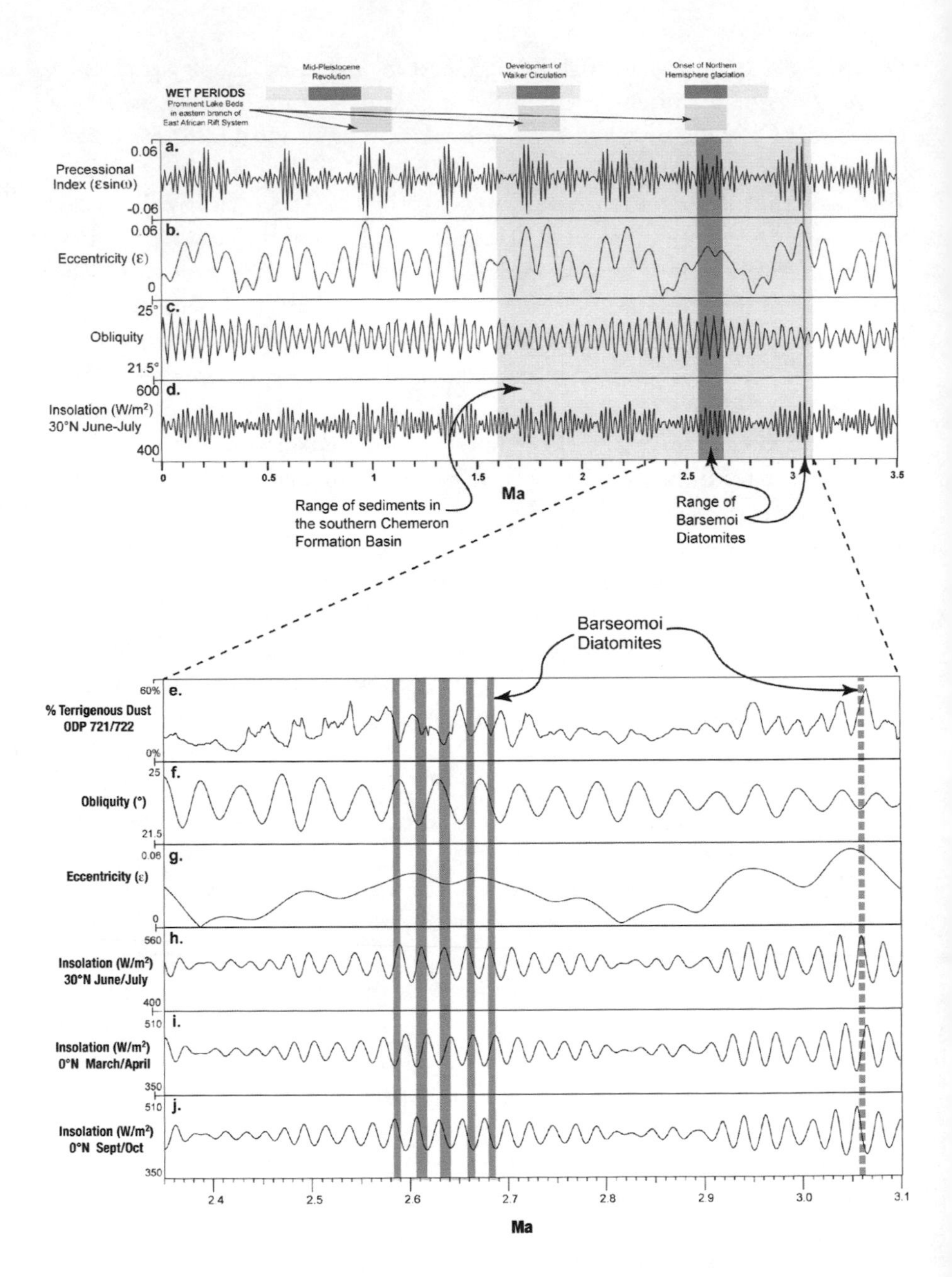

Mid-Pleistocene Revolution
Development of Walker Circulation
Onset of Northern Hemisphere glaciation
WET PERIODS
Prominent Lake Beds in eastern branch of East African Rift System
Precessional Index (εsinω)
Eccentricity (ε)
Obliquity
Insolation (W/m²) 30°N June-July
Ma
Range of sediments in the southern Chemeron Formation Basin
Range of Barsemoi Diatomites
Barseomoi Diatomites
% Terrigenous Dust ODP 721/722
Obliquity (°)
Eccentricity (ε)
Insolation (W/m²) 30°N June/July
Insolation (W/m²) 0°N March/April
Insolation (W/m²) 0°N Sept/Oct
Ma

Figure 4.12. *(a–d) Temporal range of the Barsemoi diatomite package and general range of local Chemeron Formation sediments situated within shifting patterns of calculated astronomical parameters (INSOLA program) (Laskar et al. 2004) spanning the last 3.5 myr. Proposed humid phases are based on known lacustrine sequences documented in the East African Rift Valley (Trauth et al. 2005). (e–j) Absolute ages of Barsemoi diatomite horizons (depicted as vertical shaded bands) juxtaposed against: (e) ODP 721/722 aeolian dust curves (deMenocal and Bloemendal 1995), (f, g) calculated obliquity and eccentricity curves (Paillard et al. 1996) (http://www.agu.org/eos_elec/96097e.html), and (h–j) relevant insolation curves (Laskar et al. 2004) for the 2.35–3.1 mya interval. Lake highstands, characterized by diatomite occurrence, correlate most closely with insolation curves for 30°N during June/July (h), indicating intensified precipitation maxima corresponding to an intensified low-level monsoonal flow. Pleistocene sedimentary records at Lake Naivasha (120 km south of the Baringo Basin in the Rift Valley) indicate that lake highstands follow maximum equatorial solar radiation in March and September, supporting the concept of a direct relationship between increased insolation heating and an intensification of the intertropical convergence and convective rainfall over East Africa (i, j) (Trauth et al. 2003). There is no obvious correlation between wet intervals in the Baringo Basin as reflected by diatomites and spring/fall equatorial insolation peaks. Diatomites correlate well with low input of terrestrial (aeolian) dust that, if representative of source-area aridity, indicate more humid conditions (deMenocal and Bloemendal 1995). The oldest Barsemoi diatomite (at ca. 3.06 mya) is depicted as a dashed line to indicate that the age of the lake represented is not as precisely constrained as the other diatomites. (Adapted from Kingston et al. 2007.)*

South Africa) shows clear evidence of precessionally forced changes in rainfall (Hopley et al. 2007). Pronounced cyclical deposition or evidence of short-term climatic change has also been suggested for the Shungura Formation, Olduvai Gorge, the Hadar Formation, and Olorgesailie; these short-term changes have been tentatively linked to orbitally forced climate change (see Figure 4.7 caption for detailed references).

Although the direct influence of orbital forcing on African climate seems straightforward, isolating the driving forces is extremely complex. First, high-latitude orbital forcing influences glacial–interglacial cycles, which in turn influence African climate through changes in (1) pole–equator temperature gradients; (2) sea-surface temperatures (SST); (3) wind strength and direction; and (4) atmospheric carbon dioxide, methane, and water-vapor content. Studies of Quaternary paleovegetation records indicate that equatorial African ecosystems are highly sensitive to glacial–interglacial cycles, and that these are associated with atmospheric CO_2 shifts and vegetation–soil feedbacks, resulting in rapid shifts in pollen assemblage indices (Lezine 1991; Bonnefille and Mohammed 1994; Elenga et al. 1994), charcoal fluxes (Verardo and Ruddiman 1996), and the relative proportions of C_3 (trees, shrubs, cold-season grasses and sedges) and C_4

(warm-season grasses and sedges) biomarkers (Huang et al. 1999; Ficken et al. 2002; Schefuß et al. 2003). Prior to the Mid-Pleistocene Revolution (MPR) at about 1 mya, these changes varied every 41 kyr, whereas after the MPR there is the quasiperiodic 100-kyr cycle. Second, indirect effects are coincident with the direct orbital effects on African climate. These are mainly precessional changes in seasonality, and in the strength and north–south migration of the monsoons. After the MPR, we can consider four "climate phases": (1) glacial with positive precession, (2) glacial with negative precession, (3) interglacial with positive precession, and (4) interglacial with negative precession. All of these phases or periods have a distinct effect on the moisture, temperature, and greenhouse-gas content of the atmosphere. An additional complication has recently been proposed by Berger et al. (2006). During any year in the tropics there are two insolation maxima—when the sun is over the equator (spring and autumn equinoxes)—and two insolation minima—when the sun is over the Tropics of Cancer and Capricorn (the summer and winter solstices). The magnitude of the maxima and minima, and thus of insolation at equinox and solstice, are controlled by precession. Berger et al. (2006) calculated the maximum insolation cycle and showed that it peaks about every 11.5 kyr. This is because as the spring equinox's insolation maxima are reduced, the autumnal equinox's insolation maxima are increased. Berger et al. (2006) also calculated the maximum seasonality, defined as the difference between the maximum and minimum solar insolation in any one year, and found that this produced a cyclic seasonality of 5 kyr. Theoretically, paleoclimate records in tropical Africa could respond to orbital forcing of seasonality at both 11.5-kyr and 5-kyr intervals.

What We Do Not Know. We are only just starting to understand the complex relationship between orbital forcing and African climate. This complexity is in part due to the fact that much of Africa is influenced by high-latitude orbital forcing via glacial–interglacial cycles and local direct orbital forcing, which is dominated by precession. Preliminary evidence supports a strong precessional (23 kyr) control on moisture availability in East Africa (Deino et al. 2006; Kingston et al. 2007) and South Africa (Hopley et al. 2007) and on vegetation in Southwest Africa (Denison et al. 2005). The work of Berger et al. (2006) also provides new insight into the complex relationship between direct tropical orbital forcing and seasonality. However, what is now required are high-resolution paleoclimate records from key periods of time to identify whether moisture availability is influenced by glacial–interglacial cycles (~23 Ka, ~11.5 Ka, and/or ~5 Ka) in different parts of Africa. At the moment we have only just started to understand the influence of local orbital forcing on Africa and the possibility that this may be more important than more global changes in climate such as glacial–interglacial cycles. This is a revolutionary concept that opens a new field of research and is already starting to provide a new view of African paleoclimate and its influence on human evolution.

LATE CENOZOIC GLOBAL CLIMATE TRANSITIONS

It is not always easy to isolate "global" transitions, as many climate changes are highly regional. However, during the period of early human evolution in Africa there are approximately five major transitions that can be considered global. First is the emergence and expansion of C_4 grass-dominated biomes that took place during the Middle to Upper Miocene (Morgan et al. 1994; Kingston et al. 1994; Edwards et al. 2010; Brown et al. 2011) and may have been driven by lower atmospheric CO_2 levels. This is a global climate event as C_4 grass-dominated biomes had long-lasting impacts on continental biotas, including major shifts in vegetation structure, characterized in Africa by shrinking forests and the emergence of more open landscapes accompanied by large-scale evolutionary shifts in faunal communities. Ségalen et al. (2007) show that the exact timing of the emergence of C_4 grasses is disputed, leading to contrasting views of the patterns of environmental change and their links to faunal shifts, including those of early hominins. They have evaluated existing isotopic evidence available for central, eastern, and southern Africa and review previous interpretations in light of these data. They demonstrate that pedogenic and biogenic carbonate $\delta^{13}C$ data suggest that evidence for C_4 plants before about 8 mya is weak, and that clear evidence for the emergence of C_4 biomass exists only from 8 to 7 mya. In mid-latitude sites, C_4 plants appeared later.

The second global event is the Messinian Salinity Crisis. The tectonic closure of the Strait of Gibraltar led to the transient isolation of the Mediterranean Sea from the Atlantic Ocean. During this isolation the Mediterranean Sea desiccated several times, resulting in the formation of vast evaporite deposits. The Messinian Salinity Crisis was a global climate event because nearly 6 percent of all dissolved salts in the oceans were removed, changing the alkalinity of the seawater. The onset of the Messinian Salinity Crisis was 5.96 ± 0.02 mya and full isolation occurred 5.59 mya (Krijgsman et al. 1999; Roveri et al. 2008). Normal marine conditions were reestablished with the Terminal Messinian Flood at 5.33 mya (Bickert et al. 2004) and a significant amount of salt was returned to the world's oceans via the Mediterranean–Atlantic gateway. At present, very little is known concerning the effect of the Messinian Salinity Crisis on North and East African climate. This would make an ideal study for a coupled climate–vegetation model. The last three global climate transitions are the intensification of Northern Hemisphere Glaciation, the development of the Walker Circulation, and the Mid-Pleistocene Revolution, which are discussed in greater detail below.

INTENSIFICATION OF NORTHERN HEMISPHERE GLACIATION (INHG)

The earliest recorded onset of significant global glaciation during the last 100 myr was the widespread continental glaciation of Antarctica at about 35 mya (e.g., Zachos et al. 2001). In contrast, the earliest recorded glaciation in the Northern Hemisphere occurred between 10 and 6 mya (e.g., Wolf-Welling et al. 1995). Marked expansion of continental ice sheets in the Northern Hemisphere was the culmination of a longer term, high-latitude cooling, which began with late Miocene glaciation of Greenland and the Arctic and continued through the major increases in global ice volume at around 2.5 mya despite a prolonged warm period during mid-Pliocene (Tiedemann et al. 1994; see Figure 4.13). The terminology for this event is confusing as some authors use the Onset of Northern Hemisphere Glaciation (oNHG) to mean the first significant ice sheets, which could have been as early as the Oligocene, while others use it to refer to the onset of large-scale glacial–interglacial cycles at about 2.5 mya. To avoid confusion we refer to the 2.5-mya event as the iNHG.

Evidence from Ocean Drilling Program marine records suggests that long-term cooling in the late Cenozoic led to four key steps in the iNHG (Maslin et al. 1998): (1) global cooling, which started as early as 3.2 mya, (2) glaciation of the Eurasian Arctic and Northeast Asia regions at approximately 2.74 mya, (3) glaciation of Alaska at 2.70 mya, and (4) the significant glaciation of the Northeast American continent at 2.54 mya. There are many hypothesized causal mechanisms to explain the iNHG that have focused on major tectonic events and their modification of both atmospheric and ocean circulation (Hay 1992; Raymo 1994a; Maslin et al. 1998). These include the uplift and erosion of the Tibetan–Himalayan plateau (Ruddiman et al. 1988; Raymo 1991, 1994b), the deepening of the Bering Straits (Einarsson et al. 1967) and/or the Greenland–Scotland ridge (Wright and Miller 1996), the restriction of the Indonesian seaway (Cane and Molnar 2001), and the emergence of the Panama isthmus (Keigwin 1978, 1982; Keller et al. 1989; Mann and Corrigan 1990; Haug and Tiedemann 1998).

Ruddiman et al. (1988), Ruddiman et al. (1989), and Ruddiman and Kutzbach (1991) suggested that the iNHG was caused by progressive uplift of the Tibetan–

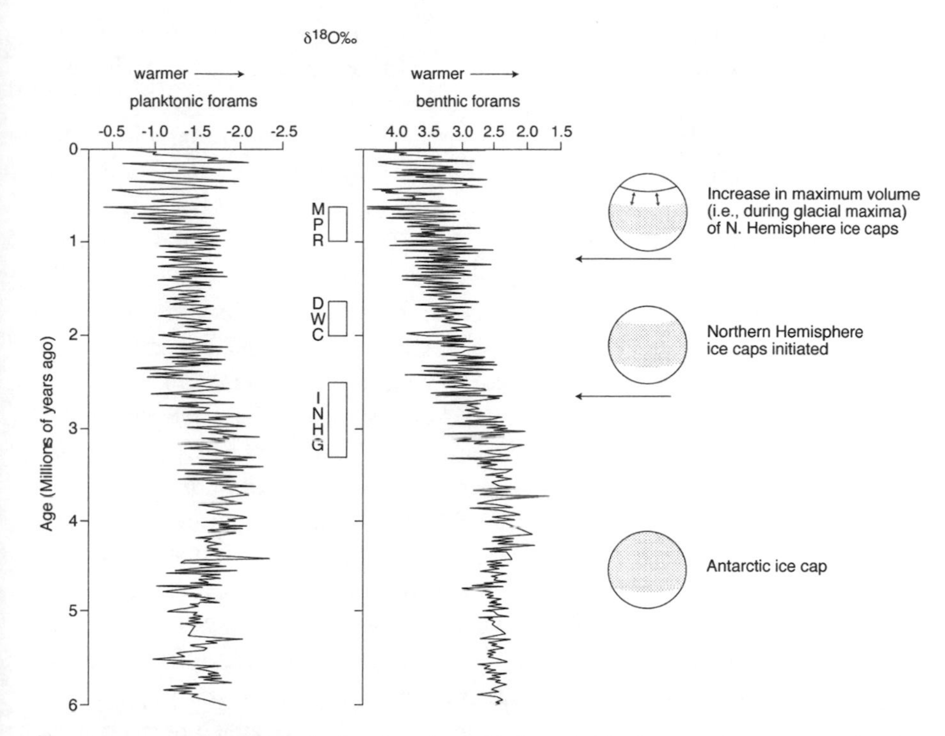

FIGURE 4.13. *Generalized planktonic and benthic foraminiferal oxygen isotope curves for the last 6 myr, compared with the intensification of Northern Hemisphere glaciation (iNHG), development of the Walker Circulation (DWC), and the Mid-Pleistocene Revolution (MPR).*

Himalayan and Sierran–Coloradan regions. This may have altered the circulation of atmospheric planetary waves such that summer ablation (melting of ice accumulations) was decreased, allowing accumulation of snow and ice in the Northern Hemisphere (Boos and Kuang 2010). However, most of the Himalayan uplift occurred much earlier, between 20 and 17 mya (Copeland et al. 1987; Molnar and England 1990), and the Tibetan Plateau reached its maximum elevation during the late Miocene (Harrison et al. 1992; Quade et al. 1989). Raymo et al. (1988), Raymo (1991, 1994b). Raymo and Ruddiman (1992) also suggested that the uplift caused a massive increase in tectonically driven chemical weathering, which removed CO_2 from the atmosphere, promoting global cooling. This idea, however, suffers from a number of major drawbacks (Maslin et al. 2001), including evidence that there was no decrease in atmospheric CO_2 during the Miocene (Pagani et al. 1999; Pearson and Palmer 2000).

The second key tectonic control invoked as a trigger for the iNHG is the closure of the Pacific–Caribbean gateway by the Panama isthmus. Haug and Tiedemann (1998) suggest the landmass began to emerge at 4.6 mya and that the gateway finally closed at 1.8 mya. However, there is still considerable debate on the exact timing of the closure (Burton et al. 1997; Vermeij 1997; Frank et al. 1999). The closure of the Panama gateway as an explanation has been described as paradoxical (Berger and Wefer 1996) since there is considerable debate over whether it would have helped or hindered the iNHG.

Tectonic forcing alone cannot explain the rapid changes in both the intensity of glacial–interglacial cycles and mean global-ice volume. It has therefore been suggested that changes in orbital forcing may have been an important mechanism contributing to the gradual global cooling and the subsequent rapid iNHG (Lourens 1994; Maslin et al. 1995, 1998; Haug and Tiedemann 1998). This theory extends the ideas of Berger et al. (1993) by recognizing distinct phases during the Pleistocene and late Pliocene, characterized by the relative strength of the different orbital parameters during each interval. Maslin et al. (1998) and Haug and Tiedemann (1998) have suggested that the observed increase in the amplitude of orbital obliquity cycles from 3.2 mya onwards may have increased the seasonality of the Northern Hemisphere, thus initiating the long-term global-cooling trend. The subsequent sharp rise in the amplitude of precession and, consequently, in insolation at 60°N between 2.8 and 2.55 mya may have forced the rapid glaciation of the Northern Hemisphere.

We still do not know exactly what caused the Northern Hemisphere to glaciate at about 2.5 mya. A plausible hypothesis involves aspects of all the proposed mechanisms. Tibetan uplift could have caused long-term cooling during the late Cenozoic. The closure of the Panama isthmus then may have delayed the intensification of Northern Hemisphere Glaciation but ultimately provided the moisture that allowed intensive glaciation to develop at high latitudes. The global climate system seems to have reached a threshold at about 3 mya, when orbital configuration may have pushed global climate across this threshold, resulting in the buildup of all the major Northern Hemisphere ice sheets in a little over 200 kyr. It also seems that the iNHG was just one stage in the progressive reorganization of the global climate system and was followed by the development of the Walker Circulation (DWC) at about 1.9 mya and the Mid-Pleistocene Revolution (MPR) at about 1 mya.

In terms of the tropics and particularly Africa, there is evidence for more extreme climate from 2.7 mya onwards. DeMenocal (1995, 2004) documents a significant increase in the amount of dust coming off the Sahara and Arabia, indicating aridity in that region in response to the iNHG, while there is also evidence for the growth and decline of large lakes between 2.7 and 2.5 mya in the Baringo–Bogoria Basin (Deino et al. 2006; Kingston et al. 2007) and on the eastern shoulder of the Ethiopian Rift and in the Afar Basin (Williams et al. 1979; Bonnefille 1983). With regard to the relationship between African climate and hominin evolution, Vrba (1995) first attempted to link iNHG to radiations in bovid species at about 2.5 mya. Vrba (1995) put forward the turnover-pulse hypothesis (TPH), which suggested that increased aridity associated with the iNHG led to larger, more juvenilized descendants in the bovid lineages and, by inference, among hominins as well. Rodent (e.g., Wesselman 1985) and other data initially appeared to support the TPH conclusions. However, the relationship between major climate transitions and mammalian evolution may not be as simple as the TPH suggests.

Development of the Walker circulation (DWC)

Until recently the iNHG and the MPR were the only two major climate changes recognized in the last 4 myr. This is because in terms of global-ice volume very little happens between 2.5 and 1 mya: glacial–interglacial cycles occur about every 41 kyr and are of a similar magnitude. However, a clear shift in long-term records of sea-surface temperature from the Pacific Ocean is evident at 1.9–1.6 mya (Ravelo et al. 2004; McClymont and Rosell-Melé 2005), when a strong east–west temperature gradient develops across the tropical Pacific Ocean. This change is also matched by a significant increase in seasonal upwelling off the coast of California. Ravelo et al. (2004) suggest this is evidence for the development of a stronger Walker circulation, as strong easterly trade winds are required to initiate the enhanced east–west SST gradients. They suggest this switch was part of the gradual global cooling and at about 2 mya the tropics and subtropics switched to the modern mode of circulation with relatively strong Walker circulation and cool subtropical temperatures. Change in the Walker circulation at about 1.9 mya coincides with numerous changes in the tropics. For example, Lee-Thorp et al. (2007) use $^{13}C/^{12}C$ ratios from fossil mammals to suggest that although there was a general trend toward more open environments after 3 mya, the most significant environmental shift toward open, grassy landscapes occurred after 2 mya rather than 2.6–2.4 mya as earlier suggested. There is also evidence for the occurrence of large deep lakes in East Africa at about 2 mya (Trauth et al. 2005, 2007; Figure 4.7). The DWC provided another interesting twist on African climate: only once a strong east–west temperature gradient is established in the Pacific Ocean can El Niño South Oscillation (ENSO) operate. There is strong documentary evidence that ENSO has a large influence on the climate of modern East Africa. Extreme annual climate events driven by ENSO may have had a major impact on human evolution. If so, we need to understand when ENSO first started to have a significant impact on East Africa and how it has changed through the Plio-Pleistocene.

Mid-pleistocene revolution (MPR)

The MPR, which is also known as the Mid-Pleistocene Transition, is the marked prolongation and intensification of glacial–interglacial climate cycles initiated sometime between 900 and 650 kya (Mudelsee and Stattegger 1997). From the iNHG to the MPR, global climate conditions appear to have responded primarily to the obliquity orbital periodicity (Imbrie et al. 1992; see Figure 4.13). The consequences of this are glacial–interglacial cycles with a mean period of 41 kyr. After about 800 kya, glacial–interglacial cycles occur with a much longer mean period (~100-kyr periodicity), with a marked increase in the amplitude of global-ice volume variations. The ice-volume increase may in part be attributed to the prolonging of glacial periods and thus of ice accumulation (Pisias and Moore 1981; Prell 1984; Shackleton et al. 1988; Berger and Jansen 1994; Tiedemann et al. 1994; Mudelsee and Stattegger 1997; Raymo 1997). The amplitude of ice-volume variation may also have been impacted by the extreme warmth of many of the post-MPR interglacial periods; similar interglacial conditions can only be found at about 1.1 mya, about 1.3 mya, and before about 2.2 mya (Figure 4.13). The MPR, in addition to marking a change in periodicity, also marks a dramatic sharpening of the contrast between warm and cold periods. Mudelsee and Stattegger (1997) used time-series analysis to review deep-sea evidence spanning the MPR and summarized the salient features. They suggest that the MPR was actually a two-step process, with the first transition between 940 and 890 kya, when there is a significant increase in global-ice volume, and the dominance of a 41-kyr climate response. This situation persists until the second step, at about 725–650 kya, when the climate system finds a three-state solution and strong 100-kyr climate cycles begin (Mudelsee and Stattegger 1997). These three states have more recent analogs and correspond to (1) full interglacial conditions, (2) the mild glacial conditions characteristic of Marine (oxygen) Isotope Stage (MIS) 3, and (3) maximum glacial conditions characteristic of MIS 2 (i.e., the Last Glacial Maximum) (Paillard 1998).

The MPR represents a significant shift in periodicity although there is not yet consensus on what this means. There are two views on the 100-kyr glacial–interglacial cycles and the role of eccentricity. The first is an old suggestion that there is some sort of nonlinear amplification in the climate system, which enhances the eccentricity signal and drives climate variations. The second is the more recent view that the other orbital parameters drive global climate change and that eccentricity acts as a pacing mechanism rather than as a driving force. We believe that this debate has not received sufficient attention in the paleoclimate and anthropology communities, with the result that some climatic interpretations still reinforce ideas dating from the early years of paleoclimatology. In many cases the last eight glacial–interglacial cycles are considered to be synonymous with "eccentricity forcing." This view or "myth" is fundamentally flawed and limits the interpretations of some excellent paleoclimatic records. Rather than rely on correlation of records to the approximately 100-kyr eccentricity cycles, Maslin and Ridgwell (2005) suggest instead that the critical influence of precession should be recognized. Raymo (1997) used precession to explain the rapid deglaciations that characterize the late Pleistocene period. It seems that every fourth or fifth precessional cycle, there is a rather weak cycle with low maxima and minima. Low maxima in Northern Hemisphere summer insolation allow ice sheets to maintain their size through these periods. During the following insolation minima, the ice sheets grow beyond their usual extent. A good example of this is the LGM when ice sheets were much larger than during the rest of the glacial period. These much larger ice sheets are very unstable and therefore with a slight change in orbital forcing they collapse rapidly and the whole climate system rebounds back into an interglacial period. Hence it is the excessive ice-sheet growth that produces such an extreme and rapid deglaciation, resulting in the sawtooth pattern of global climate change over the last 800 kya. Although eccentricity (and obliquity) determines the envelope of precessional amplitude and thus ultimately whether the minimum occurs on the fourth or fifth precessional cycle, the MPR does not represent the onset of nonlinear amplification of eccentricity to any meaningful interpretation of *nonlinear*. Instead, Maslin and Ridgwell (2005) suggest that the MPR could be thought of—not as a transition to a new mode of glacial–interglacial cycles per se—but simply the point at which a more intense and prolonged glacial state and associated subsequent rapid deglaciation becomes possible.

The MPR had a significant effect on African climate. Ségalen et al. (2007) concluded that C_4 grasses remained a relatively minor component of African environments until the late Pliocene and early Pleistocene. Pedogenic carbonate $\delta^{13}C$ data from existing localities in East Africa suggest that open ecosystems dominated by C_4 grass components emerged only during the MPR (i.e., after ~1 mya). Schefuß et al. (2003) have reconstructed the relative abundance of C_3 and C_4 plants using biomarkers in marine sediments off the coast of West

Africa. They find that the proportion of C_4 plants follows the enhanced glacial–interglacial cycles after the MPR and are directly influenced by sea-surface temperatures offshore of West Africa. They find no evidence that the climate of the Congo region is driven by precessional forcing. There is also growing evidence for the formation of large lakes between 1.1 and 0.9 mya in East Africa, such as in the Olorgesailie Formation, the Naivasha and Elementeita–Nakuru basins, and the Afar Basin (Trauth et al. 2005, 2007).

What We Do Not Know. The major difficulty in understanding the effects of these global climate transitions on African climate is the lack of high-resolution continental records. This problem is particularly acute in southern Africa. The terrestrial realm is severely restricted in the types of proxies that can be used, as well as the ability to accurately constrain the ages of the sediments, since in many cases the original record is removed through processes associated with subaerial exposure (e.g., Lowe and Walker 1984). The continental records that have been published generally do not provide the same level of continuous, detailed climate information as ocean records. Most of these continental records have a resolution greater than 10 kyr, which may cause the problem of climate-averaging (Hopley and Maslin 2010), whereby sediment and fauna from two very different climate regimes (say, two precession-scale periods) are combined. This means that many of the hominin habitat reconstructions are inaccurate as they combine two very different climate scenarios and their corresponding information about vegetation covers into one signal (Hopley and Maslin 2010).

At present only lakes and caves provide relatively continuous sections. Lake sediments are present in East Africa and the long-core drilling program at Lake Malawi has recovered a continuous sediment record spanning the last 145 kyr (Scholz et al. 2011, and references therein). Caves are present in southern Africa, and while the cave deposits have yielded abundant specimens, the stratigraphy of the caves is often highly complex (e.g., Scott 2000). Most of these sites were quarried first and analyzed later, thus severely impacting stratigraphic control. Finally, where there is limited stratigraphic control, records are sometimes dated based on an assumption of causal association with global events rather than through the development of an independent chronology. For example, the evidence for increased aridity associated with the iNHG was derived, as outlined in Partridge (1993), from geomorphological and biostratigraphic data sets that are not independent from one another and cannot be dated with precision.

SYNTHESIS OF LATE CENOZOIC AFRICAN CLIMATE CHANGE

On time-scales of more than 100 kyr, rift-related volcano–tectonic processes shaped the landscape of East Africa and profoundly influenced local climate and surface hydrology through the development of relief. Through the uplift of the Kenyan and Ethiopian plateaus, changes in orography and the associated rain shadow are believed to have been the major driving force for increased variability in moisture throughout East Africa. This increased sensitivity has resulted in a modern Rift Valley that hydrological modeling suggests could support lakes as deep as 150 meters through an increase in annual precipitation of only 15–30 percent (Bergner et al. 2003). Trauth et al. (2005, 2007) have documented three major late Cenozoic lake periods in East Africa (2.7–2.5 mya, 1.9–1.7 mya, and 1.1–0.9 mya), suggesting consistency in the moisture history of the Kenyan and Ethiopian rifts (Figure 4.7). Though preservation of East African lake records prior to 2.7 mya is patchy, there is also limited evidence for lake phases at about 4.7–4.3 mya, about 4.0–3.9 mya, about 3.4–3.3 mya, and about 3.20–2.95 mya. (Trauth et al. 2007). These lake phases correspond to drops in the East Mediterranean marine-dust abundance (Larrasoaña et al. 2003), which are thought to reflect the aridity of the eastern Algerian, Libyan, and western Egyptian lowlands north of the central Saharan watershed (Figure 4.14). The lake phases also correspond to an increased occurrence of sapropels in the Mediterranean Sea, which are thought to be caused by increased Nile River discharge (Lourens et al. 2004). The correspondence of the Mediterranean marine records with lake records of East Africa suggests a consistent moisture record for a region encompassing much of central and northern Africa over the last 3–5 myr.

East African extreme wet–dry phases correlate with significant intermediate-term decreases in aeolian-dust content from ocean-sediment cores adjacent to West Africa and Arabia (deMenocal 1995, 2004). Examination of these data demonstrates that both the lake and dust records are responding to precessional forcing and that these records are in-phase (deMenocal 1995, 2004; Deino et al. 2006; Kingston et al. 2007; Lepre et al. 2007; see Figure 4.12). Prior to 2.7–2.5 mya the extreme wet–dry "lake" phases appear every 400 kyr (see Figure 4.7); afterward they appear every 800 kyr, at 1.9–1.7 mya and 1.1–0.9 mya. Hence after 2.5 mya global climate changes seem to be required to cause an increased regional climate sensitivity to precessionally forced insolation and increased seasonality, which allows the climate to swing from developing large deep lakes to extreme aridity, resulting in the delivery of large dust loads to the adjacent oceans.

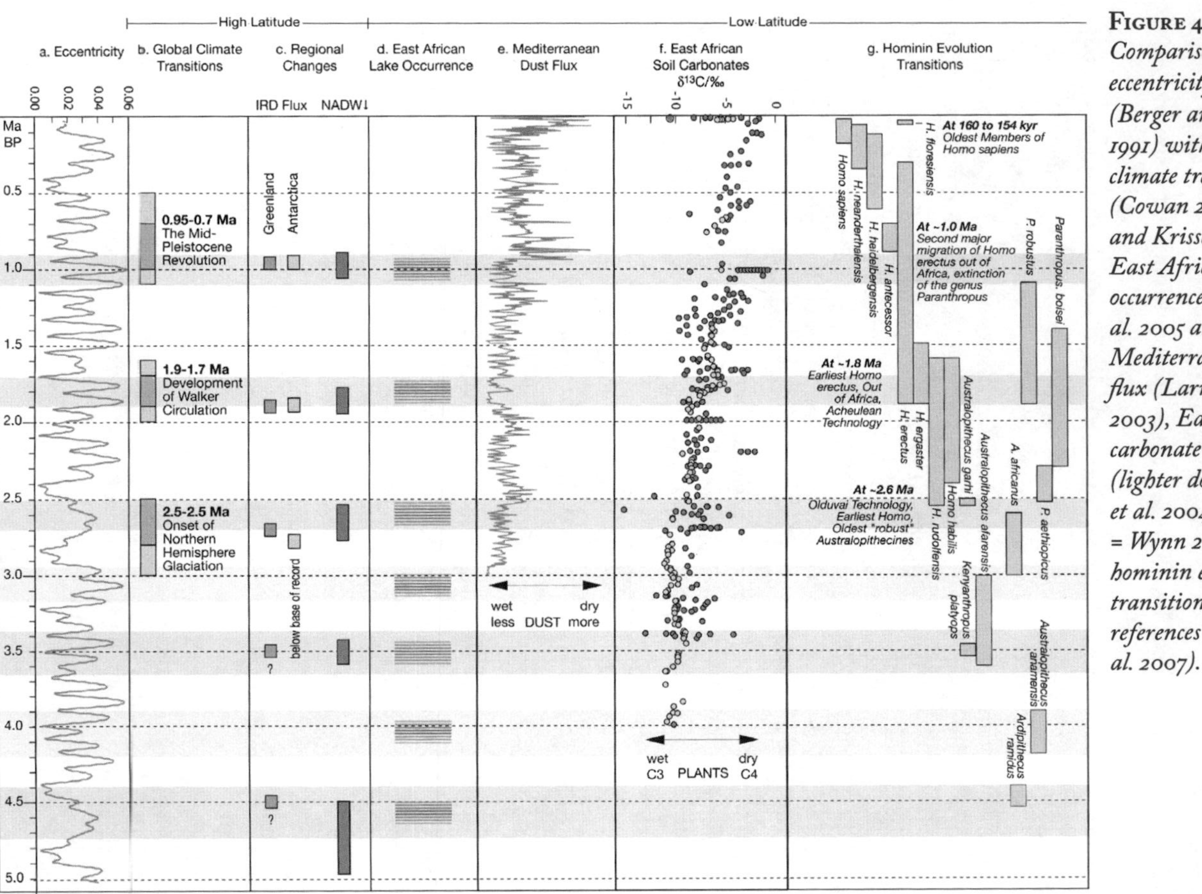

Figure 4.14. *Comparison of eccentricity variations (Berger and Loutre 1991) with high-latitude climate transitions (Cowan 2001; St. John and Krissek 2002), East African lake occurrence (Trauth et al. 2005 and this study), Mediterranean dust flux (Larrasoaña et al. 2003), East African soil carbonate carbon isotopes (lighter dots = Levin et al. 2004; darker dots = Wynn 2004), and hominin evolution transitions (see full references in Trauth et al. 2007).*

In contrast, prior to 2.5 mya eccentricity maxima alone were sufficient to produce regional sensitivity. The last three of the major Plio-Pleistocene extreme wet–dry phases all correspond to global climate transitions: iNHG, DWC, and MPR. Each of these transitions was accompanied by reduced NADW formation (Haug and Tiedemann 1998) and increased ice rafting from both Greenland and Antarctica (Cowan 2001; St. John and Krissek 2002; see Figure 4.14). Ice expansion in both hemispheres would have significantly increased the pole–equator thermal gradient, leading to north–south compression of the Intertropical Convergence Zone (ITCZ). A similar effect occurred during the LGM, where a strong compression of the ITCZ is observed both in paleoreconstructions of tropical hydrology (e.g., Peterson et al. 2000; Chiang et al. 2003) and via climate modeling (Lautenschlager and Herterich 1990; Bush and Philander 1999; Bush 2001). Trauth et al. (2007) have suggested that the compression of the ITCZ is essential for increasing the sensitivity of East Africa to precessional forcing of moisture availability; otherwise moisture is transported north and south away from the Rift Valley. Along the whole length of the rift, without this high-latitude climate control, East Africa cannot receive enough rainfall to fill large, deep, freshwater lakes during positive precessional periods. After 3 mya, both global climate forcing and eccentricity maxima are required to generate episodes of extreme precessionally forced climate change. Hence tectonics and high- and low-latitude forcing are required to explain the highly variable climate of Africa throughout the Plio-Pleistocene.

LINKING AFRICAN PALEOCLIMATE WITH EARLY HUMAN EVOLUTION

The relationship between climate and human evolution seems intuitive and indeed environmental factors have been suggested as a driving force in hominid evolution by many authors (see Kingston 2007 for detailed history). As discussed above, Vrba (1985) first identified global climate change as a cause of African mammalian evolution by documenting radiations in bovid species at about 2.5 mya coincident with the iNHG, which led to the development of the turnover-pulse hypothesis. However, with greater knowledge of African paleoclimates and mammalian fossil records, significant issues have developed regarding the TPH. When the TPH was first developed, the paleoclimate field was in its infancy, but through subsequent scientific ocean drilling it has become clear that (1) the iNHG was a long-term intensification beginning much earlier than 2.5 mya (Tiedemann et al. 1989), and (2) connections between high latitudes and low-to-middle latitudes are not as straightforward as originally thought.

The concept that significant global climate change forced major evolutionary changes in hominins remains pertinent, but both the idea of a turnover and the timing of such events have been challenged. It seems that the iNHG had less of an impact on the region (e.g., Behrensmeyer et al. 1997) than the DWC at about 2 mya (Ravelo et al. 2004). With greater understanding of African paleoclimates has also come new thinking about human evolution. The Variability Selection Hypothesis (Potts 1998; Potts et al. 1999) suggests that the complex intersection of orbitally forced changes in insolation and earth-intrinsic feedback mechanisms result in extreme, inconsistent environmental variability, selecting for behavioral and morphological mechanisms that enhance adaptive variability. The latest paleoclimate records described in this chapter have provided the basis of a new hypothesis referred to as the "Pulsed Climate Variability Hypothesis" (Maslin and Trauth 2009), which is discussed below.

There is now evidence, presented above, of extreme alternating wet and dry periods during the Plio-Pleistocene. These periods of extreme climate variability would have had a profound affect on the climate and vegetation of East Africa. Both lake- and marine-dust records show that these alternating wet–dry periods are forced by precession. It has been argued that it may not be appropriate or possible to link macroevolutionary change with orbitally forced climatic oscillations, as they may be too long term to have significant effects on biota. However, this is based on a misunderstanding of orbital forcing. Orbital forcing is not by definition always a long-term process, as all the orbital parameters are sinusoidal, which means that there are periods of little or no change followed by periods of large change. For example, imagine a pendulum swinging. At the end of its swing, when it is farthest from the middle, it is very slow and may even stop moving briefly when it reaches its maximum height. The pendulum then accelerates through the middle of its swing and moves at its fastest through the middle point of its swing, then it slows down on the other side. Precessional forcing is exactly the same. For example, the sinusoidal precessional forcing at the equator consists of periods of less than 2 kyr years, during which 60 percent of the total variation in daily insolation and seasonality occurs. These are followed by approximately 8 kyr when there is relatively little change in daily insolation (see Figure 4.15). Hence, precession does not result in smooth forcing, but rather produces rapid, strong forcing episodes that are combined with longer periods of relatively weak forcing. Rapid stratigraphic transitions from deep lacustrine to fluvial deposition associated with the diatomite deposits from Pliocene-aged lakes in the Baringo–Bogoria Basin suggest that this sinusoidal precessional forcing caused lakes to appear rapidly, remain part of the landscape for thousands of years, then disappear in a highly variable and erratic

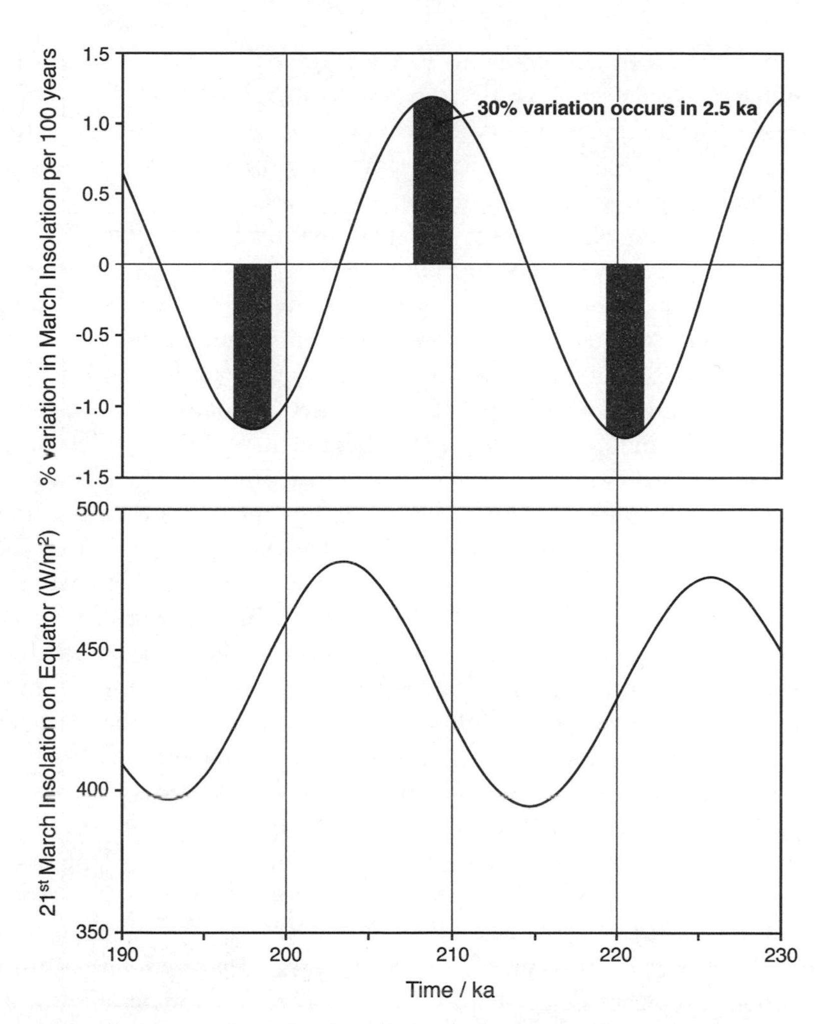

Figure 4.15. *Bottom panel: calculated insolation for March 21 at the equator between 230and 190 kya (Laskar 1990), which is controlled primarily by precessional variations. Top panel: calculated percentage variation in the March 21 insolation per hundred years. Note that 30% of the total variation of precession occurs in less than 2.5 kyr. So that within a single precessional cycle, 60% of the total variation occurs in two short periods of less than 2.5 kyr approximately 11 kyr apart.*

manner (Deino et al. 2006; Wilson 2011). In fact, the absence of shallow-water (littoral) diatom species at these key Plio-Pleistocene lake deposits (Kingston et al. 2007; Wilson 2011) suggests that lakes dried out completely in less than a few hundred years. This has important implications for the speciation and dispersal of mammals and hominins in East Africa. Trauth et al. (2005) have shown that between 5.0 and 0.5 mya, the periods of highly variable East African climate oscillats from very dry to very wet conditions occupied less than a third of the total 4.5 myr (Figure 4.14). Significantly, they show that twelve out of the fifteen hominin species (80%) that first appeared during this interval originated during one of these extreme "wet–dry" periods (Reed 1997; Dunsworth and Walker 2002; McHenry 2002; White 2002, White et al. 2006). Even taking into account the great difficulty in dating the first appearance of African hominins, and the problem of pseudospeciation events, this is compelling evidence for the preferential evolution of hominins during periods of extreme climate variability.

The difficulty in invoking orbital forcing arises not out of the question of scale, but out of timing—which part of these climate variations caused speciation and extinction events? Figure 4.16 presents three different models of the lake response to local orbital forcing. The first model suggests that there is a relatively smooth and gradual transition between periods with deep lakes and periods without lakes. If this "smooth" model is correct, there may have been prolonged periods of wet or arid conditions, which may invoke the Red Queen Hypothesis or the TPH as possible causes of evolution. The Red Queen Hypothesis suggests that continued adaptation is needed in order for a species to maintain its relative fitness among coevolving systems (Pearson 2001), and that biotic interactions rather than climate are the driving evolutionary forces. It is based on the Red Queen's race in Lewis Carroll's *Through the Looking-Glass*, when the Queen says "It takes all the running you can do, to keep in the same place" (see Barnosky 2001). However for this to occur, a fairly high-energy environment has to exist so that competition—not resources—is the dominant control. The extreme dry periods would support the TPH (Vrba 1985, 1995, 2000), which suggests that during arid conditions, selective evolution toward larger, more juvenilized descendants in the bovid lineages and hominins would have occurred. Alternatively, there could have been nonlinear dynamic changes related to the complex interactions of precipitation, temperature, and seasonality patterns to produce threshold changes in the ocurrence of the lakes and thus local vegetation, which drove evolutionary change. The second model envisages a "threshold" scenario whereby ephemeral lakes expand and contract extremely rapidly, producing the very rapid onset of extremely dry conditions required by deMenocal's (1995, 2004) "aridity hypothesis." The third model is an elaboration of the threshold model in which there is "extreme climate variability" during the rapid transition between deep-lake and no-lake states. Such a model would invoke extreme short-term variability that could drive speciation and extinction events, especially if this climate change occurred over a large geographic region.

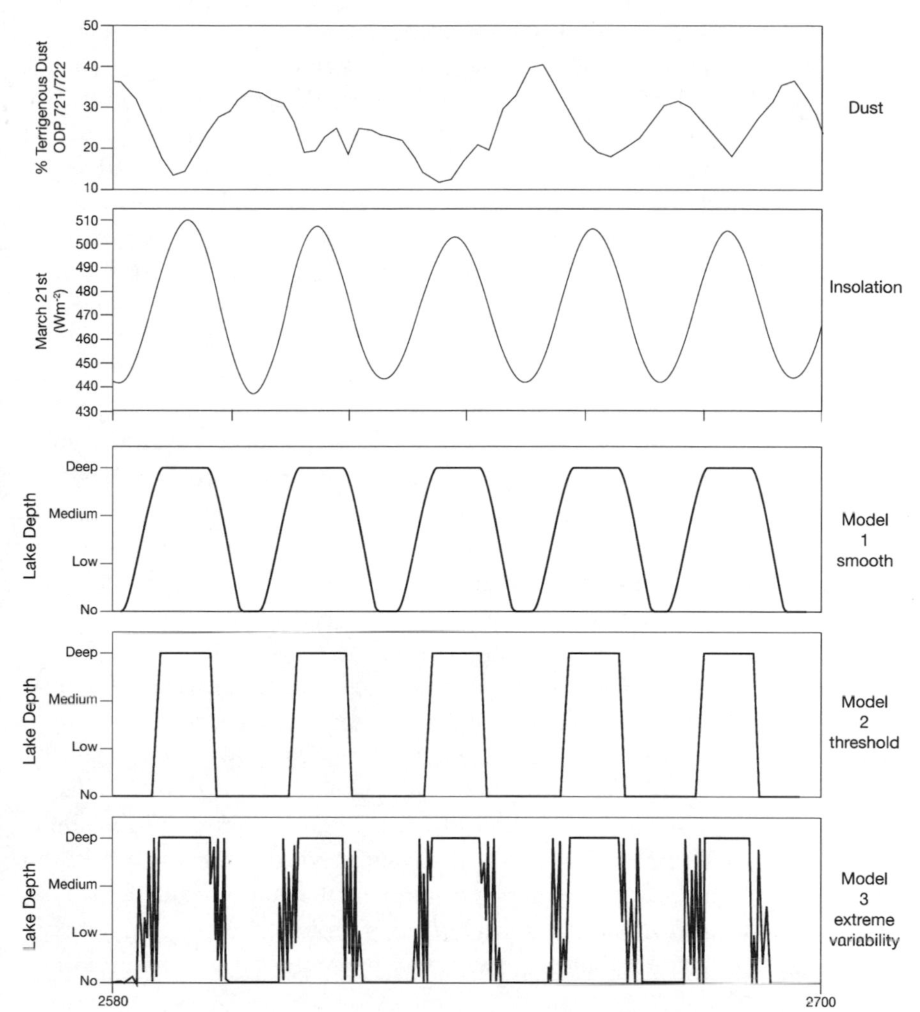

Figure 4.16. *Three theoretical models of possible lake changes in East Africa during the Plio-Pleistocene and their implications for the causes of human evolution (see main text). Model 1: "smooth" and relatively slow transitions from full-lake to no-lake conditions, which would imply that either high-energy wet conditions or prolonged aridity may have influenced human evolution. Model 2: "threshold" rapid transitions from full-lake to no-lake conditions may have influenced human evolution or it still may have been the high energy wet conditions or prolonged aridity (see Model 1) that influenced human evolution. Model 3: "extreme variability" with high variability during the transitions between full-lake and no-lake conditions, which may have influenced human evolution.*

This would produce the widespread environmental variability as required by the "Variability Selection Hypothesis" of human evolution (Potts 1998).

Data from Lake Baringo (Kingston et al. 2007; Wilson 2011) suggest that diatomites are typically bracketed by 20–30 centimeters of fine sand and silt horizons containing fish fossils that grade into high-energy terrestrial facies, indicating relatively rapid cycling between deep-lake and fully subaerial conditions. This suggests, for at least this region, that the third model for extreme climate variability is the most likely scenario, but we cannot rule out the Red Queen Hypothesis stimulated by prolonged periods of enhanced wet conditions.

CONCLUSIONS

Over the last two decades intense work on African paleoclimate and tectonic history has allowed us to begin to put together a coherent picture of how the environments of eastern and southern Africa have changed over the last 10 myr. The landscape of eastern Africa has been altered dramatically over this period of time. It changed from a relatively flat homogenous region covered with tropical mixed forest to a spectacularly heterogeneous region, with mountains over 4 kilometers high and vegetation ranging from desert to cloud forest. Added to this there were three major global climate transitions toward a cooler global climate and cyclic precessionally forced variations in rainfall. Critical to human evolution may have been the DWC at about 2 mya, as this may have produced the first El Niño/ ENSO events in Africa. The importance of ENSO is that it can produce extreme climates in Africa lasting a year or more, and thus influence an individual's survival, which is the scale at which evolution works. So, in many ways, with all this dramatic environmental change occurring in East Africa, it is not surprising that it was a hotbed of evolution, producing not only an ape that walks upright, but one that than can also ponder its own evolution.

REFERENCES

Ashley, G. M., and R. L. Hay. 2002. "Sedimentation Patterns in an Plio-Pleistocene Volcaniclastic Rift-Margin Basin, Olduvai Gorge, Tanzania." *Sedimentation in Continental Rifts, SEPM Special Publication* 73: 107–122.

Bagdasaryan, G. P., V. I. Gerasimovskiy, A. I. Polyakov, and R. K. Gukasyan. 1973. "Age of Volcanic Rocks in the Rift Zones of East Africa." *Geochemistry International* 10:66–71.

Baker, B. H., J. G. Mitchell, and L.A.J. Williams. 1988. "Stratigraphy, Geochronology and Volcano-Tectonic Evolution of the Kedong-Naivasha-Kinangop Region, Gregory Rift Valley, Kenya." *Journal of the Geological Society* 145 (1): 107–16. http://dx.doi.org/10.1144/gsjgs.145.1.0107.

Barnosky, A. D. 2001. "Distinguishing the Effects of the Red Queen and Court Jester on Miocene Mammal Evolution in the Northern Rocky Mountains." *Journal of Vertebrate Paleontology* 21 (1): 172–85. http://dx.doi.org/10.1671/0272-4634(2001)021[0172:DTEOTR]2.0.CO;2.

Bartoli, G., M. Sarnthein, M. Weinelt, H. Erlenkeuser, D. Garbe-Schonberg, and D. W. Lea. 2005. "Final Closure of Panama and the Onset of Northern Hemisphere Glaciation." *Earth and Planetary Science Letters* 237 (1-2): 33–44. http://dx.doi.org/10.1016/j.epsl.2005.06.020.

Behrensmeyer, A. K., R. Potts, A. Deino, and P. Ditchfield. 2002. "Sedimentation in Continental Rifts." In *SEPM Special Publication 73*, ed. R. W. Renaut and G. M. Ashley, 97–106.

Behrensmeyer, A. K., N. Todd, R. Potts, and G. McBrinn. 1997. "Late Pliocene Faunal Turnover in the Turkana Basin, Kenya and Ethiopia." *Science* 278: 1589–94.

Berger, A. 1988. "Milankovitch Theory and Climate." *Reviews of Geophysics* 26 (4): 624–57. http://dx.doi.org/10.1029/RG026i004p00624.

Berger, W. H., T. Bickert, H. Schmidt, and G. Wefer. 1993. "Quaternary Oxygen Isotope Record of Pelagic Foraminifera: Site 806, Ontong Java Plateau." In *Proceedings of the Ocean Drilling Program, Scientific Results, 130*, ed. W. H. Berger, L. W. Kroenke, and L. A. Mayer, 381–395. College Station: Ocean Drilling Program, Texas A&M University.

Berger, W. H., and E. Jansen. 1994. "Mid-Pleistocene Climate Chift: The Nansen Connection." In *The Polar Oceans and Their Role in Shaping the Global Environment*, ed. O. M. Johannessen, R. D. Muench, and J. E. Overland, 295–311. *AGU Geophysical Monographs* 85.

Berger, A., and M. F. Loutre. 1991. "Insolation Values for the Climate of the Last 10 Million Years." *Quaternary Science Reviews* 10 (4): 297–317. http://dx.doi.org/10.1016/0277-3791(91)90033-Q.

Berger, W. H., and G. Wefer. 1996. "Expeditions into the Past: Paleoceanographic Studies in the South Atlantic." In *The South Atlantic: Present and Past Circulation*, ed. G. Wefer, W. H. Berger, G. Siedler, and D. J. Webb, 363–410. Berlin: Spinger-Verlag. http://dx.doi.org/10.1007/978-3-642-80353-6_21

Berger, A., M. F. Loutre, and J. L. Mélice. 2006. "Equatorial Insolation: From Precession Harmonics to Eccentricity Frequencies." *Climate of the Past* 2 (2): 131–6. http://dx.doi.org/10.5194/cp-2-131-2006.

Bergner, A.G.N., M. H. Trauth, and B. Bookhagen. 3 2003. "Paleoprecipitation Estimates for the Lake Naivasha Basin (Kenya) During the Last 175 k.y. Using a Lake-Balance Model." *Global and Planetary Change* 36 (1-2):117–36. http://dx.doi.org/10.1016/S0921-8181(02)00178-9.

Bickert, T. A., G. H. Haug, and R. Tiedmann. 2004. "Late Neogene Benthic Stable Isotope Record of Ocean Drilling Program Site 999: Implications for Caribbean Paleoceanography, Organic Carbon Burial, and the Messinian Salinity Crisis." *Paleoceanography* 19 (1): PA1023. http://dx.doi.org/10.1029/2002PA000799.

Blisniuk, P., and M. R. Strecker. 1990. "Asymmetric Rift-Basin Development in the Central Kenya Rift." *TERRA Abstracts* 2: 51.

Bobe, R., and G. G. Eck. 2001. "Response of African Bovids to Pliocene Climatic Change." *Paleobiology* 27 (sp2): 1–48. http://dx.doi.org/10.1666/0094-8373(2001)027<0001:ROABTP>2.0.CO;2.

Bonnefille, R. 1983. "Evidence for a Cooler and Drier Climate in the Ethiopian Uplands Towards 2.5 Myr Ago." *Nature* 303 (5917): 487–91. http://dx.doi.org/10.1038/303487a0.

Bonnefille, R., and U. Mohammed. 1994. "Pollen-Inferred Climatic Fluctuations in Ethiopia during the Last 3000 Years." *Palaeogeography, Palaeoclimatology, Palaeoecology* 109 (2–4): 331–43. http://dx.doi.org/10.1016/0031-0182(94)90183-X.

Boos, W. R., and Z. Kuang. 2010. "Dominant Control of the South Asian Monsoon by Orographic Insulation versus Plateau Heating." *Nature* 463 (7278): 218–22. http://dx.doi .org/10.1038/nature08707. Medline:20075917

Boven, A. 1992. "The Applicability of the K-Ar and 40Ar/39Ar Techniques to Quaternary Volcanics: A Fundamental and Experimental Investigation." PhD thesis, Vrije Universiteit, Brussels.

Brown, F. H., and C .S. Feibel. 1991. "Stratigraphy, Depositional Environments and Palaeogeography of the Koobi Fora Formation." In *Koobi Fora Research Project*, vol. 3, ed. J.M. Harris, 1–30. Oxford: Clarendon.

Brown, N. J., C. A. Newell, S. Stanley, J. E. Chen, A. J. Perrin, K. Kajala, and J. M. Hibberd. 2011. "Independent and Parallel Recruitment of Preexisting Mechanisms underlying C4 Photosynthesis." *Science* 331 (6023): 1436–39. http://dx.doi.org/10.1126/science.1201248. Medline:21680841

Brown, R. W., K. Gallagher, and M. J. Duane. 1994. "A Quantitative Assessment of the Effects of Magmatism on the Thermal History of the Karoo Sedimentary Sequence." *Journal of African Earth Sciences* 18 (3): 227–43. http://dx.doi.org/10.1016/0899-5362 (94)90007-8.

Brown, R. W., K. Gallagher, A.J.W. Gleadow, and M. A. Summerfield. 2000. "Morphotectonic Evolution of the South Atlantic Margins of Africa and South America." In *Geomorphology and Global Tectonics*, ed. M. A. Summerfield, 257–283. Chichester: John Wiley and Sons Ltd.

Brown, R. W., M. A. Summerfield, and A.J.W. Gleadow. 2002. "Denudational History along a Transect across the Drakensberg Escarpment of Southern Africa Derived from Apatite Fission-Track Thermochronology." *Journal of Geophysical Research* 107 (B12): 2350. http://dx.doi.org/10.1029/2001JB000745.

Burton, K. W., H.-F. Ling, and K. R. O'Nions. 1997. "Closure of the Central American Isthmus and Its Effect on Deep-Water Formation in the North Atlantic." *Nature* 386 (6623): 382–5. http://dx.doi.org/10.1038/386382a0.

Bush, A.B.G. 2001. "Simulating Climates of the Last Glacial Maximum and of the Mid-Holocene: Wind Changes, Atmosphere-Ocean Interactions, and the Tropical Thermocline." In *The Oceans and Rapid Climate Change: Past, Present, and Future*, ed. D. Seidov, B. J. Haupt, and M. Maslin, 135–144. Washington, DC: AGU. http://dx.doi.org /10.1029/GM126p0135

Bush, A.B.G., and S.G.H. Philander. 1999. "The Climate of the Last Glacial Maximum: Results from a Coupled Atmosphere-Ocean General Circulation Model." *Journal of Geophysical Research* 104 (D20): 24509–25. http://dx.doi.org/10.1029/1999JD900447.

Bush, M. B., M. C. Miller, P. E. de Oliveira, and P. A. Colinvaux. 2002. "Orbital-Forcing Signal in Sediments of Two Amazonian Lakes." *Journal of Paleolimnology* 27 (3): 341–52. http://dx.doi.org/10.1023/A:1016059415848.

Butzer, K. W., F. W. Brown, and D. L. Thruber. 1969. "Horizontal Sediments of the Lower Omo Valley; the Kibish Formation." *Quaternaria* 11:15–29.

Cane, M. A., and P. Molnar. 2001. "Closing of the Indonesian Seaway as a Precursor to East African Aridification around 3-4 Million Years Ago." *Nature* 411 (6834): 157–62. http://dx.doi.org/10.1038/35075500. Medline:11346785

Chiang, J.C.H., M. Biasutti, and D. S. Battisti. 2003. "Sensitivity of the Atlantic Intertropical Convergence Zone to Last Glacial Maximum Boundary Conditions." *Paleoceanography* 18 (4): 1094. http://dx.doi.org/10.1029/2003PA000916.

Christensen, B. A., J. L. Kalbas, M. A. Maslin, and R. W. Murray. 2002. "Paleoclimatic Changes in Southern Africa during the Intensification of Northern Hemisphere Glaciation: Evidence from ODP Leg 175 Site 1085." *Marine Geology* 180 (1-4): 117–31. http://dx.doi.org/10.1016/S0025-3227(01)00209-2.

Clemens, S. C., D. W. Murray, and W. L. Prell. 8 1996. "Nonstationary Phase of the Plio-Pleistocene Asian Monsoon." *Science* 274 (5289): 943–8. http://dx.doi.org/10.1126/science.274.5289.943. Medline:8875928

Clemens, S. C., and W. L. Prell. 2003. "A 350,000-Year Summer-Monsoon Multiproxy Stack from the Owen Ridge, Northern Arabian Sea." *Marine Geology* 201 (1-3): 35–51. http://dx.doi.org/10.1016/S0025-3227(03)00207-X.

Clemens, S. C., and W. L. Prell. 2007. "The Timing of Orbital-Scale Indian Monsoon Changes." *Quaternary Science Reviews* 26 (3-4): 275–8. http://dx.doi.org/10.1016/j.quascirev.2006.11.010.

Clement, A. C., A. Hall, and A. J. Broccoli. 2004. "The Importance of Precessional Signals in the Tropical Climate." *Climate Dynamics* 22 (4): 327–41. http://dx.doi.org/10.1007/s00382-003-0375-8.

Cockburn, H.A.P., M. A. Seidl, and M. A. Summerfield. 1999. "Quantifying Denudation Rates on Inselbergs in the Central Namib Desert Using in Situ-Produced Cosmogenic ^{10}Be and ^{26}Al." *Geology* 27 (5): 399–402. http://dx.doi.org/10.1130/0091-7613(1999)027<0399:QDROII>2.3.CO;2.

Cockburn, H.A.P., R. W. Brown, M. A. Summerfield, and M. A. Seidl. 2000. "Quantifying Passive Margin Denudation and Landscape Development Using a Combined Fission-Track Thermochronology and Cosmogenic Isotope Analysis Approach." *Earth and Planetary Science Letters* 179 (3-4): 429–35. http://dx.doi.org/10.1016/S0012-821X(00)00144-8.

Copeland, P., T. M. Harrison, W.S.F. Kidd, X. Ronghua, and Z. Yuquan. 1987. "Rapid Early Miocene Acceleration of Uplift of the Gagdese Belt, Xizang (Southern Tibet), and

Its Bearing on Accomodation Mechanisms of the India-Asia Collision." *Earth and Planetary Science Letters* 86 (2-4): 240–52. http://dx.doi.org/10.1016/0012-821X(87)90224-X.

Cowan, E. A. 2001. "Identification of the Glacial Signal from the Antarctic Peninsula since 3.0 Ma at Site 1011 in a Continental Rise Sediment Drift." In *Proceedings of the Ocean Drilling Program, Scientific Results, 178,* ed. P. F. Barker, A. Camerlenghi, G. D. Acton, and A.T.S. Ramsay, 1–22. College Station: Ocean Drilling Program, Texas A&M University.

Crossley, R., and R. M. Knight. 1981. "Volcanism in the Western Part of the Rift Valley in Southern Kenya." *Bulletin of Volcanology* 44 (2): 117–28. http://dx.doi.org/10.1007/BF02597699.

Cruz, F. W., Jr., S. J. Burns, I. Karmann, W. D. Sharp, M. Vuille, A. O. Cardoso, J. A. Ferrari, P. L. Dias, and O. Viana. 2005. "Insolation-Driven Changes in Atmospheric Circulation over the Past 116,000 Years in Subtropical Brazil." *Nature* 434 (7029): 63–6. http://dx.doi.org/10.1038/nature03365. Medline:15744298

DeConto, R. M., and D. Pollard. 2003. "Rapid Cenozoic Glaciation of Antarctica Induced by Declining Atmospheric CO2." *Nature* 421 (6920): 245–9. http://dx.doi.org/10.1038/nature01290. Medline:12529638

Deino, A. L., J. D. Kingston, J. M. Glen, R. K. Edgar, and A. Hill. 2006. "Precessional Forcing of Lacustrine Sedimentation in the Late Cenozoic Chemeron Basin, Central Kenya Rift, and Calibration of the Gauss/Matuyama Boundary." *Earth and Planetary Science Letters* 247 (1-2): 41–60. http://dx.doi.org/10.1016/j.epsl.2006.04.009.

deMenocal, P. B. 1995. "Plio-Pleistocene African Climate." *Science* 270 (5233): 53–9. http://dx.doi.org/10.1126/science.270.5233.53. Medline:7569951

deMenocal, P. B. 2004. "African Climate Change and Faunal Evolution during the Pliocene-Pleistocene." *Earth and Planetary Science Letters* 220 (1-2): 3–24. http://dx.doi.org/10.1016/S0012-821X(04)00003-2.

deMenocal, P. B., and J. Bloemendal. 1995. "Plio-Pleistocene Climatic Variability in Subtropical Africa and the Palaeoenvironment of Hominin Evolution: A Combined Data-Model Approach." In *Paleoclimate and Evolution with Emphasis on Human Origins*, ed. E. S. Vrba, G. H. Denton, T. C. Partridge, and L. H. Burckle, 262–288. New Haven, CT: Yale University Press.

Denison, S. M., M. A. Maslin, C. Boot, R. D. Pancost, and V. J. Ettwein. 2005. "Precession-Forced Changes in South West African Vegetation during Marine Isotope Stages 101–100 (2.56–2.51 Ma)." *Palaeogeography, Palaeoclimatology, Palaeoecology* 220 (3-4): 375–86. http://dx.doi.org/10.1016/j.palaeo.2005.02.001.

de Wit, M.C.J., T. R. Marshal, and T. C. Partridge. 2000. "Fluvial Deposits and Drainage Evolution." In *The Cenozoic of Southern Africa*, ed. T. C. Partridge and R. R. Maud, 55–72. New York: Oxford University Press.

Diester-Haass, L., P. A. Meyers, and L. Vidal. 2002. "The Late Miocene Onset of High Productivity in the Benguela Current Upwelling System as Part of a Global Pattern." *Marine Geology* 180 (1-4): 87–103. http://dx.doi.org/10.1016/S0025-3227(01)00207-9.

Dunkley, P. M., M. Smith, D. J. Allen, and W. G. Darling. 1993. *The Geothermal Activity and Geology of the Northern Sector of the Kenya Rift Valley*. NERC Research Report SC/93/1. British Geological Survey, Keyworth.

Dunsworth, H., and A. Walker. 2002. "Early Genus Homo." In *The Primate Fossil Record*, ed. W. C. Hartwig, 401–406. New York: Cambridge University Press.

Dupont, L. M., B. Donner, L. Vidal, E. M. Pérez, and G. Wefer. 2005. "Linking Desert Evolution and Coastal Upwelling: Pliocene Climate Change in Namibia." *Geology* 33 (6): 461–4. http://dx.doi.org/10.1130/G21401.1.

Ebinger, C. J., T. Yemane, D. J. Harding, D. Tesfaye, S. Kelley, and D. C. Rex. 2000. "Rift Deflection, Migration, and Propagation: Linkage of the Ethiopian and Eastern Rifts, Africa." *Geological Society of America Bulletin* 112 (2): 163–76. http://dx.doi.org/10.1130/0016-7606(2000)112<163:RDMAPL>2.0.CO;2.

Edwards, E. J., C. P. Osborne, C.A.E. Strömberg, S. A. Smith, W J. Bond, P. A. Christin, A. B. Cousins, M. R. Duvall, D. L. Fox, R. P. Freckleton, et al., and the C4 Grasses Consortium. 2010. "The Origins of C4 Grasslands: Integrating Evolutionary and Ecosystem Science." *Science* 328 (5978): 587–91. http://dx.doi.org/10.1126/science.1177216. Medline:20431008

Einarsson, T., D. M. Hopkins, and R. R. Doell. 1967. "The Stratigraphy of Tjornes, Northern Iceland, and the History of the Bering Land Bridge." In *The Bering Land Bridge*, ed. D. M. Hopkins, 312–325. Stanford, CA: Stanford University Press.

Elenga, H., D. Schwartz, and A. Vincens. 1994. "Pollen Evidence of Late Quaternary Vegetation and Inferred Climate Changes in Congo." *Palaeogeography, Palaeoclimatology, Palaeoecology* 109 (2-4): 345–56. http://dx.doi.org/10.1016/0031-0182(94)90184-8.

Evernden, J. G., and G. H. Curtis. 1965. "The Potassium-Argon Dating of Late Cenozoic Rocks in East Africa and Italy." *Current Anthropology* 6 (4): 343–63. http://dx.doi.org/10.1086/200619.

Feakins, S. J., P. B. deMenocal, and T. I. Eglinton. 2005. "Biomarker Records of Late Neogene Changes in Northeast African Vegetation." *Geology* 33 (12): 977–80. http://dx.doi.org/10.1130/G21814.1.

Feakins, S. J., T. I. Eglinton, and P. B. deMenocal. 2007. "A Comparison of Biomarker Records of Northeast African Vegetation from Lacustrine and Marine Sediments ca. 3.40 Ma." *Organic Geochemistry* 38 (10): 1607–24. http://dx.doi.org/10.1016/j.orggeochem.2007.06.008.

Ficken, K. J., M. J. Wooller, D. L. Swain, F. A. Street-Perrott, and G. Eglinton. 2002. "Reconstruction of a Subalpine Grass-Dominated Ecosystem, Lake Rutundu, Mount Kenya: A Novel Multi-Proxy Approach." *Palaeogeography, Palaeoclimatology, Palaeoecology* 177 (1-2): 137–49. http://dx.doi.org/10.1016/S0031-0182(01)00356-X.

Fleming, A., M. A. Summerfield, J. O. Stone, L. K. Fifield, and R. G. Cresswell. 1999. "Denudation Rates for the Southern Drakensberg Escarpment, SE Africa, Derived from In-Situ-Produced Cosmogenic Cl–36: Initial Results." *Journal of the Geological Society* 156 (2): 209–12. http://dx.doi.org/10.1144/gsjgs.156.2.0209.

Foster, A., C. Ebinger, E. Mbede, and D. Rex. 1997. "Tectonic Development of the Northern Tanzanian Sector of the East African Rift System." *Journal of the Geological Society* 154 (4): 689–700. http://dx.doi.org/10.1144/gsjgs.154.4.0689.

Frank, M., B. C. Reynolds, and R. K. O'Nions. 1999. "Nd and Pb Isotopes in Atlantic and Pacific Water Masses before and after Closure of the Panama Gateway." *Geology* 27 (12): 1147–50. http://dx.doi.org/10.1130/0091-7613(1999)027<1147:NAPIIA>2.3.CO;2.

Gallagher, K., and R. W. Brown. 1999. "Denudation and Uplift at Passive Margins: The Record on the Atlantic Margin of Southern Africa." *Philosophical Transactions of the Royal Society of London. Series B, Biological Sciences* 357:835–59.

Gasse, F. 1990. "Lacustrine Basin Exploration, Case Studies, and Modern Analogs." In *AAPG Memoir* 50, ed. B.J. Katz, 1–340.

George, R.M.M., N. W. Rogers, and S. Kelley. 1998. "Earliest Magmatism in Ethiopia: Evidence for Two Mantle Plumes in One Flood Basalt Province." *Geology* 26 (10): 923–6. http://dx.doi.org/10.1130/0091-7613(1998)026<0923:EMIEEF>2.3.CO;2.

Giraudeau, J., P. A. Meyers, and B. A. Christensen. 2002. "Accumulation of Organic and Inorganic Carbon in Pliocene-Pleistocene Sediments along the Southwestern African Margin." *Marine Geology* 180 (1-4): 49–69. http://dx.doi.org/10.1016/S0025-3227(01)00205-5.

Harrison, T. M., P. Copeland, W.S.F. Kidd, and A. Yin. 1992. "Raising Tibet." *Science* 255 (5052): 1663–70. http://dx.doi.org/10.1126/science.255.5052.1663. Medline:17749419

Haug, G. H., and R. Tiedemann. 1998. "Effect of the Formation of the Isthmus of Panama on Atlantic Ocean Thermohaline Circulation." *Nature* 393 (6686): 673–76. http://dx.doi.org/10.1038/31447.

Hay, W. 1992. "The Cause of the Late Cenozoic Northern Hemisphere Glaciations: A Climate Change Enigma." *Terra Nova* 4 (3): 305–11. http://dx.doi.org/10.1111/j.1365-3121.1992.tb00819.x.

Hays, J. D., J. Imbrie, and N. J. Shackleton. 1976. "Variations in the Earth's Orbit: Pacemaker of the Ice Ages." *Science* 194 (4270): 1121–32. http://dx.doi.org/10.1126/science.194.4270.1121. Medline:17790893

Hillaire-Marcel, C., O. Carro, and J. Casanova. 1986. "^{14}C and Th/U Dating of Pleistocene and Holocene Stromatolites from East African Paleolakes." *Quaternary Research* 25 (3): 312–29. http://dx.doi.org/10.1016/0033-5894(86)90004-9.

Hopley, P. J., J. D. Marshall, G. P. Weedon, A. G. Latham, A. I. Herries, and K. L. Kuykendall. 2007. "Orbital Forcing and the Spread of C4 Grasses in the Late Neogene: Stable Isotope Evidence from South African Speleothems." *Journal of Human Evolution* 53 (5): 620–34. http://dx.doi.org/10.1016/j.jhevol.2007.03.007. Medline:17942141

Hopley, P. J., and M. A. Maslin. 2010. "Climate-Averaging of Terrestrial Faunas: An Example from the Plio-Pleistocene of South Africa." *Palaeobiology* 36 (1): 32–50. http://dx.doi.org/10.1666/0094-8373-36.1.32.

Huang, Y., K. H. Freeman, T. I. Eglinton, and F. A. Street-Perrott. 1999. "δ^{13}C Analyses of Individual Lignin Phenols in Quaternary Lake Sediments: A Novel Proxy for

Deciphering Past Terrestrial Vegetation Changes." *Geology* 27 (5): 471–4. http://dx.doi.org/10.1130/0091-7613(1999)027<0471:CAOILP>2.3.CO;2.

Huber, M., H. Brinkhuis, C. E. Stickley, K. Doos, A. Sluijs, J. Warnaar, S. A. Schellenberg, and G. L. Williams. 2004. "Eocene Circulation of the Southern Ocean: Was Antarctica Kept Warm by Subtropical Waters?" *Paleoceanography* 19 (4): PA4026. http://dx.doi.org/10.1029/2004PA001014.

Imbrie, J., A. Berger, E. A. Boyle, S. C. Clemens, A. Duffy, W. A. Howard, G. Kukla, J. Kutzbach, D. G. Martinson, A. McIntyre, et al. 1993. "On the Structure and Origin of Major Glaciation Cycles 2: The 100,000 Year Cycle." *Paleoceanography* 8 (6): 699–735. http://dx.doi.org/10.1029/93PA02751.

Imbrie, J., E. Boyle, S. C. Clemens, A. Duffy, W. A. Howard, G. Kukla, J. Kutzbach, D. G. Martinson, A. McIntyre, A. C. Mix, et al. 1992. "On the Structure and Origin of Major Glaciation Cycles 1: Linear Responses to Milankovitch Forcing." *Paleoceanography* 7 (6): 701–38. http://dx.doi.org/10.1029/92PA02253.

Keigwin, L. D. 1978. "Pliocene Closing of the Isthmus of Panama, Based on Biostratigraphic Evidence from nearby Pacific Ocean and Caribbean Cores." *Geology* 6 (10): 630–4. http://dx.doi.org/10.1130/0091-7613(1978)6<630:PCOTIO>2.0.CO;2.

Keigwin, L. D. 1982. "Isotopic Paleoceanography of the Caribbean and East Pacific: Role of Panama Uplift in Late Neogene Time." *Science* 217 (4557): 350–3. http://dx.doi.org/10.1126/science.217.4557.350. Medline:17791515.

Keller, G., C. E. Zenker, and S. M. Stone. 1989. "Late Neogene History of the Pacific-Caribbean Gateway." *Journal of South American Earth Sciences* 2 (1): 73–108. http://dx.doi.org/10.1016/0895-9811(89)90028-X.

Kennett, J. P. 1996. "A Review of Polar Climatic Evolution during the Neogene, Based on the Marine Sediment Record." In *Palaeoclimate and Evolution, with Emphasis on Human Origins*, ed. E. Vrba, G. H. Denton, T. C. Partridge, and L. H. Burckle, 49–64. New Haven, CT: Yale University Press.

Kingston, J. D. 2007. "Shifting Adaptive Landscapes: Progress and Challenges in Reconstructing Early Hominid Environments." *American Journal of Physical Anthropology: Yearbook of Physical Anthropology* 134 (Suppl 45): 20–58. http://dx.doi.org/10.1002/ajpa.20733. Medline:18046753

Kingston, J. D., A. L. Deino, R. K. Edgar, and A. Hill. 2007. "Astronomically Forced Climate Change in the Kenyan Rift Valley 2.7-2.55 Ma: Implications for the Evolution of Early Hominin Ecosystems." *Journal of Human Evolution* 53 (5): 487–503. http://dx.doi.org/10.1016/j.jhevol.2006.12.007. Medline:17935755

Kingston, J. D., A. Hill, and B. D. Marino. 1994. "Isotopic Evidence for Neogene Hominid Paleoenvironments in the Kenya Rift Valley." *Science* 264 (5161): 955–9. http://dx.doi.org/10.1126/science.264.5161.955. Medline:17830084

Krijgsman, W., C. G. Langereis, W. J. Zachariasse, M. Boccaletti, G. Moratti, R. Gelati, S. Iaccarino, G. Papani, and G. Villa. 1999. "Late Neogene Evolution of the Taza-Guercif Basin (Rifian Corridor, Morocco) and Implications for the Messinian Salinity Crisis." *Marine Geology* 153 (1-4): 147–60. http://dx.doi.org/10.1016/S0025-3227(98)00084-X.

Larrasoaña, J. C., A. P. Rohling, M. Winklhofer, and R. Wehausen. 2003. "Three Million Years of Monsoon Variability over the Northern Sahara." *Climate Dynamics* 21:689–98. http://dx.doi.org/10.1007/s00382-003-0355-z.

Laskar, J. 1990. "The Chaotic Motion of the Solar System: A Numerical Estimate of the Chaotic Zones." *Icarus* 88 (2): 266–91. http://dx.doi.org/10.1016/0019-1035(90)90084-M.

Laskar, J., P. Robutel, F. Joutel, M. Gastineau, A.C.M. Correia, and B. Levrard. 2004. "A Long-Term Numerical Solution for the Insolation Quantities of the Earth." *Astronomy & Astrophysics* 428 (1): 261–85. http://dx.doi.org/10.1051/0004-6361:20041335.

Lautenschlager, M., and K. Herterich. 1990. "Atmospheric Response to Ice Age Conditions: Climatology near the Earth's Surface." *Journal of Geophysical Research* 95 (D13): 22547–57. http://dx.doi.org/10.1029/JD095iD13p22547.

Lea, D. W., D. K. Pak, and H. J. Spero. 2000. "Climate Impact of Late Quaternary Equatorial Pacific Sea Surface Temperature Variations." *Science* 289 (5485): 1719–24. http://dx.doi.org/10.1126/science.289.5485.1719. Medline:10976060

Leakey, M. G., C. S. Feibel, I. McDougall, and A. Walker. 1995. "New Four-Million-Year-Old Hominid Species from Kanapoi and Allia Bay, Kenya." *Nature* 376 (6541): 565–71. http://dx.doi.org/10.1038/376565a0. Medline:7637803

Leakey, M. G., C. S. Feibel, I. McDougall, C. Ward, and A. Walker. 1998. "New Specimens and Confirmation of an Early Age for *Australopithecus anamensis*." *Nature* 393 (6680): 62–6. http://dx.doi.org/10.1038/29972. Medline:9590689

Lee-Thorp, J. A., M. Sponheimer, and J. M. Luyt. 2007. "Tracking Changing Environments Using Stable Carbon Isotopes in Fossil Tooth Enamel: An Example from the South African Hominin Sites." *Journal of Human Evolution* 53 (5): 595–601. http://dx.doi.org/10.1016/j.jhevol.2006.11.020. Medline:17920103

Lepre, C. J., R. L. Quinn, J.C.A. Joordens, C. C. Swisher, 3rd, and C. S. Feibel. 2007. "Plio-Pleistocene Facies Environments from the KBS Member, Koobi Fora Formation: Implications for Climate Controls on the Development of Lake-Margin Hominin Habitats in the Northeast Turkana Basin (Northwest Kenya)." *Journal of Human Evolution* 53 (5): 504–14. http://dx.doi.org/10.1016/j.jhevol.2007.01.015. Medline:17919684

Levin, N. E., J. Quade, S. W. Simpson, S. Semaw, and M. Rogers. 2004. "Isotopic Evidence for Plio-Pleistocene Environmental Change at Gona, Ethiopia." *Earth and Planetary Science Letters* 219 (1-2): 93–110. http://dx.doi.org/10.1016/S0012-821X(03)00707-6.

Lezine, A. 1991. "West African Paleoclimates during the Last Climatic Cycle Inferred from an Atlantic Deep-Sea Pollen Record." *Quaternary Research* 35 (3): 456–63. http://dx.doi.org/10.1016/0033-5894(91)90058-D.

Li, X. S., A. Berger, M. F. Loutre, M. A. Maslin, G. H. Haug, and R. Tiedemann. 1998. "Simulating Late Pliocene Northern Hemisphere Climate with the LLN 2-D Model." *Geophysical Research Letters* 25 (6): 915–8. http://dx.doi.org/10.1029/98GL00443.

Lourens, L. J. 1994. "Chapter 9: Long-Period Orbital Variations and Their Relation to Third-Order Eustatic Cycles and the Onset of Major Glaciations, 3.0 Million Years Ago." In *Astronomical Forcing of Mediterranean Climate during the Last 5.3 Million Years*, 199–206. PhD thesis, University of Utrecht, Holland.

Lourens, L. J., F. Hilgen, N. J. Shackleton, J. Laskar, and D. Wilson. 2004. "The Neogene Period." In *A Geologic Time Scale*, ed. F. Gradstein, J.G. Ogg, and G. Smith, 409–440. Cambridge: Cambridge University Press.

Lowe, J. J., and M.J.C. Walker. 1984. *Reconstructing Quaternary Environments*. NY: Prentice-Hall, Longman.

Mann, P., and J. Corrigan. 1990. "Model for Late Neogene Deformation in Panama." *Geology* 18 (6): 558–62. http://dx.doi.org/10.1130/0091-7613(1990)018<0558:MFLNDI>2.3.CO;2.

Maslin, M. A. 2004. "Atmosphere. Ecological versus Climatic Thresholds." *Science* 306 (5705): 2197–8. http://dx.doi.org/10.1126/science.1107481. Medline:15622563

Maslin, M. A., and B. Christensen. 2007. "Tectonics, Orbital Forcing, Global Climate Change, and Human Evolution in Africa: Introduction to the African Paleoclimate Special Volume." *Journal of Human Evolution* 53 (5): 443–64. http://dx.doi.org/10.1016 /j.jhevol.2007.06.005. Medline:17915289

Maslin, M. A., G. Haug, M. Sarnthein, R. Tiedemann, H. Erlenkeuser, and R. Stax. 1995. "Northwest Pacific Site 882: The Initiation of Major Northern Hemisphere Glaciation." *Proceedings of the Ocean Drilling Program, Scientific Results* 145: 315–329.

Maslin, M. A., X. S. Li, M. F. Loutre, and A. Berger. 1998. "The Contribution of Orbital Forcing to the Progressive Intensification of Northern Hemisphere Glaciation." *Quaternary Science Reviews* 17 (4-5): 411–26. http://dx.doi.org/10.1016/S0277-3791 (97)00047-4.

Maslin, M. A., R. D. Pancost, K. E. Wilson, J. Lewis, and M. H. Trauth. 2012. "Three and Half Million Year History of Moisture Availability of South West Africa: Evidence from ODP Site 1085 Biomarker Records." *Palaeogeography, Palaeoclimatology, Palaeoecology* 317–318:41–7. http://dx.doi.org/10.1016/j.palaeo.2011.12.009.

Maslin, M. A., and A. Ridgwell. 2005. "Mid-Pleistocene Revolution and the Eccentricity Myth." *Special Publication - Geological Society of London* 247 (1): 19–34. http://dx.doi.org /10.1144/GSL.SP.2005.247.01.02.

Maslin, M.A., D. Seidov, and J. J. Lowe. 2001. "Synthesis of the Nature and Causes of Sudden Climate Transitions during the Quaternary." In *The Oceans and Rapid Climate Change: Past, Present, and Future*, ed. D. Seidov, B. J. Haupt, and M. A. Maslin, 9–52. Am. Geophys. Union Geophys. Monogr. Series 126, AGU, Washington, DC.

Maslin, M. A., and M. H. Trauth. 2009. "Plio-Pleistocene East African Pulsed Climate Variability and Its Influence on Early Human Evolution." In *The First Humans: Origin and Early Evolution of the Genus Homo*, ed. F. E. Grine, R. E. Leakey, and J. G. Fleagle, 151–158. New York: Springer Verlag. http://dx.doi.org/10.1007/978-1-4020-9980-9_13

McClymont, E. L., and A. Rosell-Melé. 2005. "Links between the Onset of Modern Walker Circulation and the Mid-Pleistocene Climate Transition." *Geology* 33 (5): 389–92. http://dx.doi.org/10.1130/G21292.1.

McDougall, I., F. H. Brown, P. M. Vasoncelos, B. E. Cohen, D. S. Thiede, and M. J. Buchanan. 2012. "New Single Crystal ^{40}Ar/^{39}Ar Ages Improve Time Scale for Deposition of the Omo Group, Omo–Turkana Basin, East Africa." *Journal of the Geological Society* 169 (2): 213–26. http://dx.doi.org/10.1144/0016-76492010-188.

McDougall, I., and R. T. Watkins. 1988. "Potassium-Argon Ages of Volcanic Rocks from Northeast of Lake Turkana, Northern Kenya." *Geological Magazine* 125 (01): 15–23. http://dx.doi.org/10.1017/S001675680000933X.

McHenry, H. M. 2002. "Introduction to the Fossil Record of Human Ancestry." In *The Primate Fossil Record*, ed. W. C. Hartwig, 401–406. New York: Cambridge University Press.

Milankovitch, M. M. 1949. *Kanon der Erdbestrahlung und seine Anwendung auf das Eiszeitenproblem*. Royal Serbian Sciences, Spec. pub. 132, Section of Mathematical and Natural Sciences, 33, Belgrade, 633 pp. (*Canon of Insolation and the Ice Age Problem*, English translation by Israel Program for Scientific Translation and published for the US Department of Commerce and the National Science Foundation, Washington DC, 1969.)

Molnar, P., and P. England. 1990. "Late Cenozoic Uplift of Mountain Ranges and Global Climate Change: Chicken or Egg?" *Nature* 346 (6279): 29–34. http://dx.doi.org/10.1038/346029a0.

Morgan, M. E., J. D. Kingston, and B. D. Marino. 1994. "Carbon Isotopic Evidence for the Emergence of C4 Plants in the Neogene from Pakistan and Kenya." *Nature* 367 (6459): 162–5. http://dx.doi.org/10.1038/367162a0.

Mudelsee, M., and K. Stattegger. 1997. "Exploring the Structure of the Mid-Pleistocene Revolution with Advance Methods of Time-Series Analysis." *Geologische Rundschau* 86 (2): 499–511. http://dx.doi.org/10.1007/s005310050157.

Owen, R. B. 2002. "Sedimentological Characteristics and Origins of Diatomaceous Deposits in the East African Rift System." *Sedimentation in Continental Rifts: SEPM Special Publication*, ed. R. W. Renaut and G. M. Ashley, 73: 233–246.

Pagani, M., K. H. Freeman, and M. A. Arthur. 1999. "Late Miocene Atmospheric CO(2) Concentrations and the Expansion of C(4) Grasses." *Science* 285 (5429): 876–9. http://dx.doi.org/10.1126/science.285.5429.876. Medline:10436153

Paillard, D. 1998. "The Timing of Pleistocene Glaciations from a Simple Multiple-State Climate Model." *Nature* 391 (6665): 378–81. http://dx.doi.org/10.1038/34891.

Paillard, D., L. Labeyrie, and P. Yiou. 1996. "Macintosh Program Performs TimeSeries Analyses (AnalySeries 1.2 updated 2000)." *Eos, Transactions, American Geophysical Union* 77:379. http://dx.doi.org/10.1029/96EO00259.

Partridge, T. C. 1993. "Warming Phases in Southern Africa during the Last 150,000 Years: An Overview." *Palaeogeography, Palaeoclimatology, Palaeoecology* 101 (3-4): 237–44. http://dx.doi.org/10.1016/0031-0182(93)90016-C.

Partridge, T. C., P. B. deMenocal, S. A. Lorentz, M. J. Paiker, and J. C. Vogel. 1997. "Orbital Forcing of Climate over South Africa: A 200,000-Year Rainfall Record from the Pretoria Saltpan." *Quaternary Science Reviews* 16 (10): 1125–33. http://dx.doi.org/10.1016/S0277-3791(97)00005-X.

Pearson, P. N. 2001. Red Queen Hypothesis. *Encyclopedia of Life Sciences*. http://www.els.net.

Pearson, P. N., and M. R. Palmer. 2000. "Atmospheric Carbon Dioxide Concentrations over the Past 60 Million Years." *Nature* 406 (6797): 695–9. http://dx.doi.org/10.1038/35021000. Medline:10963587

Peterson, L. C., G. H. Haug, K. A. Hughen, and U. Röhl. 2000. "Rapid Changes in the

Hydrologic Cycle of the Tropical Atlantic during the Last Glacial." *Science* 290 (5498): 1947–51. http://dx.doi.org/10.1126/science.290.5498.1947. Medline:11110658
Petit, J. R., I. Basile, A. Leruyuet, D. Raynaud, C. Lorius, J. Jouzel, M. Stievenard, V. Y. Lipenkov, N. I. Barkov, B. B. Kudryashov, et al. 1997. "Four Climate Cycles in Vostok Ice Core." *Nature* 387 (6631): 359–60. http://dx.doi.org/10.1038/387359a0.
Pisias, N. G., and T. C. Moore, Jr. 1981. "The Evolution of Pleistocene Climate: A Time Series Approach." *Earth and Planetary Science Letters* 52 (2): 450–8. http://dx.doi.org/10.1016/0012-821X(81)90197-7.
Potts, R. 1998. "Environmental Hypothesis of Hominin Evolution." *Yearbook of Physical Anthropology* 41 (Suppl. 27): 93–136. http://dx.doi.org/10.1002/(SICI)1096-8644(1998)107:27+<93::AID-AJPA5>3.0.CO;2-X.
Potts, R., A. K. Behrensmeyer, and P. Ditchfield. 1999. "Paleolandscape Variation and Early Pleistocene Hominid Activities: Members 1 and 7, Olorgesailie Formation, Kenya." *Journal of Human Evolution* 37 (5): 747–88. http://dx.doi.org/10.1006/jhev.1999.0344. Medline:10536090.
Prell, W. L. 1984. "Covariance Patterns of Foraminiferal δ18O: An Evaluation of Pliocene Ice Volume Changes Near 3.2 Million Years Ago." *Science* 226 (4675): 692–4. http://dx.doi.org/10.1126/science.226.4675.692. Medline:17774947.
Qiang, X. K., Z. X. Li, C. McA. Powell, and H. B. Zheng. 2001. "Magnetostratigraphic Record of the Late Miocene Onset of East Asian Monsoon, and Pliocene Uplift of Northern Tibet." *Earth and Planetary Science Letters* 187 (1-2): 83–93. http://dx.doi.org/10.1016/S0012-821X(01)00281-3.
Quade, J., T. E. Cerling, and J. R. Bowman. 1989. "Development of Asian Monsoon Revealed by Marked Ecological Shift during the Latest Miocene in Northern Pakistan." *Nature* 342 (6246): 163–66. http://dx.doi.org/10.1038/342163a0.
Raab, M. J., R. W. Brown, K. Gallagher, A. Carter, and K. Weber. 2002. "Late Cretaceous Reactivation of Major Crustal Shear Zones in Northern Namibia: Constraints from Apatite Fission Track Analysis." *Tectonophysics* 349 (1-4): 75–92. http://dx.doi.org/10.1016/S0040-1951(02)00047-1.
Raab, M. J., R. W. Brown, K. Gallagher, K. Weber, and A.J.W. Gleadow. 2005. "Denudational and Thermal History of the Early Cretaceous Brandberg and Okenyenya Igneous Complexes on Namibia's Atlantic Passive Margin." *Tectonics* 24 (3): TC3006. http://dx.doi.org/10.1029/2004TC001688.
Rahmstorf, S., J. Marotzke, and J. Willebrand. 1996. "Stability of the Thermohaline Circulation." In *The Warmwatersphere of the North Atlantic Ocean*, ed. W. Kraus, 129–157. Berlin: Gebrüder Bornträger.
Ravelo, A. C., D. H. Andreasen, M. Lyle, A. Olivarez Lyle, and M. W. Wara. 2004. "Regional Climate Shifts Caused by Gradual Global Cooling in the Pliocene Epoch." *Nature* 429 (6989): 263–7. http://dx.doi.org/10.1038/nature02567. Medline:15152244
Raymo, M. E. 1991. "Geochemical Evidence Supporting T. C. Chamberlin's Theory of Glaciation." *Geology* 19 (4): 344–7. http://dx.doi.org/10.1130/0091-7613(1991)019<0344:GESTCC>2.3.CO;2.

Raymo, M. E. 1994a. "The Himalayas, Organic Carbon Burial and Climate in the Miocene." *Paleoceanography* 9 (3): 399–404. http://dx.doi.org/10.1029/94PA00289.

Raymo, M. E. 1994b. "The Initiation of Northern Hemisphere Glaciation." *Annual Review of Earth and Planetary Sciences* 22 (1): 353–83. http://dx.doi.org/10.1146/annurev.ea.22.050194.002033.

Raymo, M. E. 1997. "The Timing of Major Climate Terminations." *Paleoceanography* 12 (4): 577–85. http://dx.doi.org/10.1029/97PA01169.

Raymo, M. E., and K. Nisancioglu. 2003. "The 41 Kyr World: Milankovitch's Other Unsolved Mystery." *Paleoceanography* 18 (1): 1011. http://dx.doi.org/10.1029/2002PA000791.

Raymo, M. E., and W. F. Ruddiman. 1992. "Tectonic Forcing of Late Cenozoic Climate." *Nature* 359 (6391): 117–22. http://dx.doi.org/10.1038/359117a0.

Raymo, M. E., W. F. Ruddiman, and P .N. Froelich. 1988. "Influence of Late Cenozoic Mountain Building on Ocean Geochemical Cycles." *Geology* 16 (7): 649–53. http://dx.doi.org/10.1130/0091-7613(1988)016<0649:IOLCMB>2.3.CO;2.

Reed, K. E. Feb-Mar 1997. "Early Hominid Evolution and Ecological Change through the African Plio-Pleistocene." *Journal of Human Evolution* 32 (2-3): 289–322. http://dx.doi.org/10.1006/jhev.1996.0106. Medline:9061560

Reed, K. E., and J. L. Fish. 2005. "Tropical and Temperate Seasonal Influences on Human Evolution." In *Seasonality in Primates*, ed. D. Brockman and C. van Schaik, 491–520. Cambridge: Cambridge University Press. http://dx.doi.org/10.1017/CBO9780511542343.018

Richardson, J. L., and R. A. Dussinger. 1986. "Paleolimnology of Mid-Elevation Lakes in the Kenya Rift Valley." *Hydrobiologia* 143 (1): 167–74. http://dx.doi.org/10.1007/BF00026659.

Richardson, J. L., and A. E. Richardson. 1972. "History of an African Rift Lake and Its Climatic Implications." *Ecological Monographs* 42 (4): 499–534. http://dx.doi.org/10.2307/1942169.

Ridgwell, A. J., A. J. Watson, M. A. Maslin, and J. O. Kaplan. 2003. "Implications of Coral Reef Build-Up for the Controls on Atmospheric CO_2 since the Last Glacial Maximum." *Palaeoceanography* 18 (4): 1083. http://dx.doi.org/10.1029/2003PA000893.

Roveri, M., S. Lugli, V. Manzi, and B. C. Schreiber. 2008. "The Messinian Sicilian Stratigraphy Revisited: New Insights for the Messinian Salinity Crisis." *Terra Nova* 20 (6): 483–8. http://dx.doi.org/10.1111/j.1365-3121.2008.00842.x.

Ruddiman, W. F., and J. E. Kutzbach. 1990. "Late Cenozoic Plateau Uplift and Climate Change." *Transactions of the Royal Society of Edinburgh. Earth Sciences* 81 (04): 301–14. http://dx.doi.org/10.1017/S0263593300020812.

Ruddiman, W. F., and J. E. Kutzbach. 1991. "Plateau Uplift and Climatic Change." *Scientific American* 264 (3): 66–75. http://dx.doi.org/10.1038/scientificamerican0391-66. Medline:1711238

Ruddiman, W. F., and M. E. Raymo. 2003. "A Methane-Based Time Scale for Vostok Ice: Climatic Implications." *Quaternary Science Reviews* 22 (2-4): 141–55. http://dx.doi.org/10.1016/S0277-3791(02)00082-3.

Ruddiman, W. F., M. E. Raymo, H. H. Lamb, and J. T. Andrews. 1988. "Northern Hemisphere Climate Regimes during the Past 3 Ma: Possible Tectonic Connections." *Philosophical Transactions of the Royal Society of London. Series B, Biological Sciences* 318 (1191): 411–30. http://dx.doi.org/10.1098/rstb.1988.0017.

Ruddiman, W. F., M. Sarnthein, J. Backman, J. G. Baldauf, W. Curry, L. M. Dupont, T. Janecek, E. M. Pokras, M. E. Raymo, B. Stabell, et al. 1989. "Late Miocene to Pleistocene Evolution of Climate in Africa and the Low-Latitude Atlantic—Overview of Leg 108 results." In *Proceedings of the Ocean Drilling Program, Scientific Results, 108*, ed. W. F. Ruddiman, M. Sarnthein, and J. Baldauf, shipboard scientific party, 463–487. Ocean Drilling Program, College Station, TX.

Saltzman, B., K. A. Maasch, and M. Y. Verbitsky. 1993. "Possible Effects of Anthropogenically-Increased CO_2 on the Dynamics of Climate: Implications for Ice Age Cycles." *Geophysical Research Letters* 20 (11): 1051–4. http://dx.doi.org/10.1029/93GL01015.

Schefuß, E., S. Schouten, J.H.F. Jansen, and J. S. Sinninghe Damsté. 2003. "African Vegetation Controlled by Tropical Sea Surface Temperatures in the Mid-Pleistocene Period." *Nature* 422 (6930): 418–21. http://dx.doi.org/10.1038/nature01500. Medline:12660780

Schefuß, E., S. Schouten, and R. R. Schneider. 2005. "Climatic Controls on Central African Hydrology during the Past 20,000 Years." *Nature* 437 (7061):1003–6. http://dx.doi.org /10.1038/nature03945. Medline:16222296

Scholz, C. A., A. S. Cohen, and T. C. Johnson. 2011. "Southern Hemisphere Tropical Climate over the Past 145 Ka: Results of the Lake Malawi Scientific Drilling Project, East Africa." *Palaeogeography, Palaeoclimatology, Palaeoecology* 303 (1-4): 1–2. http://dx.doi .org/10.1 016/j.palaeo.2011.01.001.

Scott, L. 2000. "Vegetation History and Climate in the Savanna Biome South Africa since 190,000 Ka: A Comparison of Pollen Data from the Tswaing Crater (the Pretoria Saltpan) and Wonderkrater." *Quaternary International* 57–58:215–23.

Ségalen, L., J. A. Lee-Thorp, and T. Cerling. 2007. "Timing of C4 Grass Expansion across Sub-Saharan Africa." *Journal of Human Evolution* 53 (5): 549–59. http://dx.doi.org /10.1016/j.jhevol.2006.12.010. Medline:17905413

Sepulchre, P., G. Ramstein, F. Fluteau, M. Schuster, J. J. Tiercelin, and M. Brunet. 2006. "Tectonic Uplift and Eastern Africa Aridification." *Science* 313 (5792): 1419–23. http:// dx.doi.org/10.1126/science.1129158. Medline:16960002

Shackleton, N. J. 2000. "The 100,000-Year Ice-Age Cycle Identified and Found to Lag Temperature, Carbon Dioxide, and Orbital Eccentricity." *Science* 289 (5486): 1897–1902. http://dx.doi.org/10.1126/science.289.5486.1897. Medline:10988063

Shackleton, N. J., J. Imbrie, and N. G. Pisias. 1988. "The Evolution of Oceanic Oxygen-Isotope Variability in the North Atlantic over the Past Three Million Years [and Discussion]." *Philosophical Transactions of the Royal Society of London. Series B, Biological Sciences* 318:679–86. http://dx.doi.org/10.1098/rstb.1988.0030.

Stickley, C. E., H. Brinkhuis, S. A. Schellenberg, A. Sluijs, U. Röhl, M. Fuller, M. Grauert, M. Huber, J. Warnaar, and G L. Williams. 2004. "Timing and Nature of the Deepening of the Tasmanian Gateway." *Paleoceanography* 19 (4): PA4027. http://dx.doi.org/10.1029/2004PA001022.

St. John, K., and L. Krissek. 2002. "The Late Miocene to Pleistocene Ice Rafting History of Southeast Greenland." *Boreas* 31 (1): 28–35. http://dx.doi.org/10.1080/030094802100651.

Strecker, M. R. 1991. Das zentrale und südliche Kenia-Rift unter besonderer Berücksichtigung der neotektonischen Entwicklung. Habilitation thesis, University of Karlsruhe, Germany.

Strecker, M. R., P. M. Blisniuk, and G. H. Eisbacher. 1990. "Rotation of Extension Direction in the Central Kenya Rift." *Geology* 18 (4): 299–302. http://dx.doi.org/10.1130/0091-7613(1990)018<0299:ROEDIT>2.3.CO;2.

Sturchio, N. C., P.N. Dunkley, and M. Smith. 1993. "Climate-Driven Variations in Geothermal Activity in the Northern Kenya Rift Valley." *Nature* 362 (6417): 233–4. http://dx.doi.org/10.1038/362233a0.

Sundquist, E. T., and K. Visser. 2004. "The Geologic History of the Carbon Cycle." In *Treatise on Geochemistry*, Volume 8: *Biogeochemistry*, ed. W. H. Schlesinger, 425–72. Amsterdam: Elsevier.

Teaford, M. F., and P. S. Ungar. 2000. "Diet and the Evolution of the Earliest Human Ancestors." *Proceedings of the National Academy of Sciences of the United States of America* 97 (25): 13506–11. http://dx.doi.org/10.1073/pnas.260368897. Medline:11095758.

Tiedemann, R., M. Sarnthein, and N. J. Shackleton. 1994. "Astronomic Timescale for the Pliocene Atlantic $\delta^{18}O$ and Dust Flux Records of Ocean Drilling Program Site 659." *Paleoceanography* 9:619–38. http://dx.doi.org/10.1029/94PA00208.

Tiedemann, R., M. Sarnthein, and R. Stein. 1989. "Climatic Changes in the Western Sahara: Aeolo-Marine Sediment Record of the Last 8 Million Years." *Proceedings of the Ocean Drilling Program, Scientific Results*, 108: 241–278. Ocean Drilling Program, College Station, TX.

Tiercelin, J. J., and K. E. Lezzar. 2002. "A 300 Million Year History of Rift Lakes in Central and East Africa: An Updated Broad Review." In *The East African Great Lakes: Limnology, Paleolimnology and Biodiversity*, ed. E. O. Odada and D. O. Olago, 3–60. Dordrecht, Netherlands: Kluwer Academic Publishers. http://dx.doi.org/10.1007/0-306-48201-0_1

Tierney, J. E., J. M. Russell, Y. Huang, J. S. Damsté, E. C. Hopmans, and A. S. Cohen. 2008. "Northern Hemisphere Controls on Tropical Southeast African Climate during the Past 60,000 Years." *Science* 322 (5899): 252–5. http://dx.doi.org/10.1126/science.1160485. Medline:18787132

Trauth, M. H., A. L. Deino, A.G.N. Bergner, and M .R. Strecker. 2003. "East African Climate Change and Orbital Forcing during the Last 175 kyr BP." *Earth and Planetary Science Letters* 206 (3-4): 297–313. http://dx.doi.org/10.1016/S0012-821X(02)01105-6.

Trauth, M. H., M. A. Maslin, A. L. Deino, A. G. Bergner, M. Dühnforth, and M. R. Strecker. 2007. "High- and Low-Latitude Forcing of Plio-Pleistocene East African Climate and Human Evolution." *Journal of Human Evolution* 53:475–86. http://dx.doi.org/10.1016/j.jhevol.2006.12.009. Medline:17959230

Trauth, M. H., M. A. Maslin, A. L. Deino, A. Junginger, M. Lesoloyia, E. O. Odada, D. O. Olago, L. A. Olaka, M. R. Strecker, and R. Tiedemann. 2010. "Human Evolution in a Variable Environment: The Amplifier Lakes of Eastern Africa." *Quaternary Science Reviews* 29 (23-24): 2981–8. http://dx.doi.org/10.1016/j.quascirev.2010.07.007.

Trauth, M. H., M. A. Maslin, A. L. Deino, and M. R. Strecker. 2005. "Late Cenozoic Moisture History of East Africa." *Science* 309 (5743): 2051–3. http://dx.doi.org/10.1126/science.1112964. Medline:16109847

Tyson, P. D., and T. C. Partridge. 2000. "Evolution of Cenozoic Climates." In *The Cenozoic of Southern Africa*, ed. T. C. Partridge and R.R. Maud, 371–387. New York: Oxford University Press.

Tyson, P. D., and R. A. Preston-Whyte. 2000. *The Weather and Climate of Southern Africa, Cape Town, South Africa*. Oxford: Oxford University Press.

van der Beek, P., M. A. Summerfield, J. Braun, R. W. Brown, and A. Flemming. 2002. "Modelling Postbreakup Landscape Development and Denudational History across the Southeast African (Drakensberg Escarpment) Margin." *Journal of Geophysical Research* 107. http://dx.doi.org/10.1029/2001JB000744.

Verardo, D. J., and W. F. Ruddiman. 1996. "Late Pleistocene Charcoal in Tropical Atlantic Deep-Sea Sediments: Climatic and Geochemical Significance." *Geology* 24 (9): 855–7. http://dx.doi.org/10.1130/0091-7613(1996)024<0855:LPCITA>2.3.CO;2.

Vermeij, G. J. 1997. "Strait Answers from a Twisted Isthmus." *Paleobiology* 23:263–9.

Verschuren, D., J. S. Sinninghe Damsté, J. Moernaut, I. Kristen, M. Blaauw, M. Fagot, G. H. Haug, B. van Geel, M. De Batist, P. Barker, et al., and the CHALLACEA project members. 2009. "Half-Precessional Dynamics of Monsoon Rainfall near the East African Equator." *Nature* 462 (7273): 637–41. http://dx.doi.org/10.1038/nature08520. Medline:19956257

Vrba, E. S. 1985. "Environment and Evolution: Alternative Causes of the Temporal Distrubution of Evolutionary Events." *South African Journal of Science* 81:229–36.

Vrba, E. S. 1995. "The Fossil Record of African Antelopes (Mammalia, Bovidae) in Relation to Human Evolution and Paleoclimate." In *Paleoclimate and Evolution with Emphasis on Human Origins*, ed. E. S. Vrba, G. Denton, L. Burckle, and T. Partridge, 385–424. New Haven, CT: Yale University Press.

Vrba, E. S. 2000. "Major Features of Neogene Mammalian Evolution in Africa." In *The Cenozoic of Southern Africa*, ed. T. C. Partridge and R. R. Maud, 277–304. New York: Oxford University Press.

Walter, R. C., P. C. Manega, R. L. Hay, R. E. Drake, and G. H. Curtis. 1991. "Laser-Fusion ^{40}Ar/^{39}Ar Dating of Bed I, Olduvai Gorge, Tanzania." *Nature* 354 (6349): 145–49. http://dx.doi.org/10.1038/354145a0.

Wang, X., A. S. Auler, R. L. Edwards, H. Cheng, P. S. Cristalli, P. L. Smart, D. A. Richards, and C. C. Shen. 2004. "Wet Periods in Northeastern Brazil over the Past 210 kyr Linked to Distant Climate Anomalies." *Nature* 432 (7018): 740–3. http://dx.doi.org/10.1038/nature03067. Medline:15592409

Washbourn-Kamau, C. K. 1970. "Late Quaternary Chronology of the Nakuru-Elmenteita Basin, Kenya." *Nature* 226 (5242): 253–4. http://dx.doi.org/10.1038/226253c0. Medline:16057194

Washbourn-Kamau, C. K. 1975. "Late Quaternary Shorelines of Lake Naivasha, Kenya." *Azania* 10 (1): 77–92. http://dx.doi.org/10.1080/00672707509511614.

Washbourn-Kamau, C. K. 1977. "The Ol Njorowa Gorge, Lake Naivasha Basin, Kenya." In *Desertic Terminal Lakes*, ed. D. C. Greer, 297–307. Utah Water Research Laboratory.

Wesselman, H. B. 1985. "Fossil Micromammals as Indicators of Climatic Change about 2.4 Myr Ago in the Omo Valley, Ethiopia." *South African Journal of Science* 81:260–1.

White, T. 2002. "Earliest Hominids." In *The Primate Fossil Record*, ed. W. C. Hartwig, 406–417. New York: Cambridge University Press.

White, T. D., G. WoldeGabriel, B. Asfaw, S. Ambrose, Y. Beyene, R. L. Bernor, J. R. Boisserie, B. Currie, H. Gilbert, Y. Haile-Selassie, et al. 2006. "Asa Issie, Aramis and the Origin of *Australopithecus*." *Nature* 440 (7086): 883–9. http://dx.doi.org/10.1038/nature04629. Medline:16612373

Williams, M.A.J., R. Macdonald, and P. T. Leat. 1983. "Proceedings of Regional Seminar on Geothermal Energy in Eastern and Southern Africa." *UNESCO/USAID, Nairobi* 61–67.

Williams, M.A.J., F. M. Williams, F. Gasse, G. H. Curtis, and D. A. Adamson. 1979. "Pliocene-Pleistocene Environments at Gadeb Prehistoric Site, Ethiopia." *Nature* 282 (5734): 29–33. http://dx.doi.org/10.1038/282029a0.

Wilson, K. E. 2011. "Plio-Pleistocene Reconstruction of East African and Arabian Sea Palaeoclimate." PhD thesis, University College London.

Wilson, R.C.L. 2000. *The Great Ice Age*. London: Routledge.

WoldeGabriel, G., G. Heiken, T. D. White, B. Asfaw, W. K. Hart, and P. R. Renee. 2000. "Volcanism, Tectonism, Sedimentation, and the Paleoanthropological Record in the Ethiopian Rift System." *GSA Special Paper* 345:83–99.

Wolf-Welling, T.C.W., J. Thiede, A. M. Myhre, and Leg 151 Shipboard scientific party. 1995. "Bulk Sediment Parameter and Coarse Fraction Analysis: Paleoceanographic Implications of Fram Strait Sites 908 and 909, ODP Leg 151 (NAAG)." *Eos Transactions, Supplement* 76 (17): 166.

Wright, J. D., and K. G. Miller. 1996. "Control of North Atlantic Deep Water Circulation by the Greenland-Scotland Ridge." *Paleoceanography* 11 (2): 157–70. http://dx.doi.org/10.1029/95PA03696.

Wynn, J. G. 2004. "Influence of Plio-Pleistocene Aridification on Human Evolution: Evidence from Paleosols of the Turkana Basin, Kenya." *American Journal of Physical Anthropology* 123 (2): 106–18. http://dx.doi.org/10.1002/ajpa.10317. Medline:14730645

Zachos, J. C., M. Pagani, L. Sloan, E. Thomas, and K. Billups. 2001. "Trends, rhythms, and aberrations in global climate 65 Ma to present." *Science* 292 (5517): 686–93. http://dx.doi.org/10.1126/science.1059412. Medline:11326091

Ziegler, M., E. Tuenter, and L. J. Lourens. 2010. "The Precession Phase of the Boreal Summer Monsoon as Viewed from the Eastern Mediterranean (ODP Site 968)." *Quaternary Science Reviews* 29 (11-12): 1481–90. http://dx.doi.org/10.1016/j.quascirev.2010.03.011.

PART 2

Hominin Adaptations and Behavior

5

Early Hominin Posture and Locomotion

CAROL V. WARD

SINCE THE DISCOVERY OF THE TAUNG CHILD (Dart 1925) paleoanthropologists have understood that the key behavioral and anatomical shift that characterized early human evolution was a change in locomotor adaptation in which a group of hominins was selected to be habitually bipedal. Terrestrial bipedality has served as a significant preadaptation to the acquisition of other key human characteristics, such as toolmaking, life-history changes, and even intelligence (see Flinn et al. 2005). Understanding the nature and timing of the transition to terrestrial bipedality is key to an accurate interpretation of how and why humans evolved.

This chapter reviews the issues and evidence for *Australopithecus* locomotor behavior and evolution because it is *Australopithecus*, and in particular *A. afarensis*, that forms the basis for our understanding of the origins and early evolution of hominin locomotion. Recently, four new hominoid fossil species have been announced, all of which have been claimed to be hominins: *Sahelanthropus tchadensis* (Brunet et al. 2002) known from Chad and dated to 6–7 mya (million years ago), *Orrorin tugenensis* from about 6 mya in Kenya (Senut et al. 2001), *Ardipithecus kadabba* known from deposits dated to 5.8 mya in Ethiopia (Haile-Selassie 2001), and *Ardipithecus ramidus* from 4.4 mya, also in Ethiopia (White et al. 1994, 1995, 2009). However, the data published so for them are limited, so the nature of these hominins' locomotor adaptations cannot be fully ascertained, nor are their relationships with later

DOI: 10.5876/9781607322252:c05

hominins clear. There is evidence that *O. tugenensis* was bipedal, based on its femur (Richmond and Jungers 2008), suggesting that bipedality was established long before *Australopithecus*. *Ardipithecus ramidus* is suggested to have been bipedal when on the ground, although it still retained significant climbing abilities and it has been argued to have been a pronograde quadruped when in the trees (Lovejoy et al. 2009a, b). Still, *Australopithecus* remains the best-known early australopith and the model to which new fossils will be compared in order to assess the trajectory of early hominin locomotion.

In this chapter, I review the evidence for the posture and locomotion of *Australopithecus* and these earlier taxa. In doing so, I consider the issues of whether australopiths retained adaptations to arboreality, whether they still climbed trees, whether their bipedality itself differed from that of later hominins, and what sort of locomotor diversity is apparent among *Australopithecus* species.

WAS *AUSTRALOPITHECUS* FULLY BIPEDAL AND/OR STILL PARTLY ARBOREAL?

There is extensive and unequivocal evidence that australopiths were terrestrial bipeds (Lovejoy et al. 1973; Lovejoy 1975, 1978, 1988; Day and Wickens 1980; White 1980; Latimer 1983, 1991; Ohman 1986; Latimer et al. 1987; Latimer and Lovejoy 1989, 1990a, 1990b; Crompton et al. 1998; Kramer 1999; Haile-Selassie et al. 2010a; Ward et al. 2012; reviews in Stern 2000; Ward 2002) (Figures 5.1–5.3). Derived trait lists and discussions of their significance can also be found in McHenry (1994), Stern (2000), and Ward (2002) and also are reviewed in Aiello and Dean (1990).

Both species for which vertebral remains are known, *A. afarensis* and *A. africanus,* had the sinusoidal vertebral curvatures that allow the torso to balance over the hindlimbs efficiently and effectively in bipedal posture, including the posterior curvature, or lordosis, in the lumbar region (Robinson 1972; Ward and Latimer 1991, 2005; but see Sarmiento 1998) (Figure 5.1A). Apes and monkeys, on the other hand, have a fairly uniform anterior curvature to their spines. This demonstrates that not only were australopiths upright, their trunks were not inclined anteriorly with some partially upright posture on flexed lower limbs.

Australopiths had a short pelvis with laterally rotated iliac blades, allowing an effective hip abduction to balance the torso in a coronal plane during single-limb support phase of gait, and reduced torque about the hip joints for anterior–posterior and mediolateral balance (Johanson et al. 1982; Tague and Lovejoy 1986; Lovejoy et al. 1999) (Figure 5.1A and B). This pelvic restructuring also involves a broad sacrum (Robinson 1972; Leutenegger and Kelly 1977), a feature that also accommodates the widening intervertebral facet spacing found at the inferior end of the australopith vertebral column that allows hominins to obtain a lordosis (Ward and Latimer 2005).

The femoral neck of australopiths was ringed by thin cortical bone and an expanded trabecular region, reflecting increased shock-absorption capacity, and allowed by the hip abductors, which neutralize bending of the neck (Figure 5.1C) (Lovejoy et al. 1973; Johanson et al. 1982).

One key feature is the pronounced femoral bicondylar angle (Tardieu and Trinkaus 1994; Duren and Ward 1995; Duren 1999), which positions the body's center of gravity over the knee and ankle during single-limb support (Figure 5.1C). This angle is present when the knee is in extended posture and only develops with the onset of bipedal locomotion, demonstrating that australopiths indeed walked bipedally with a nearly extended lower limb at heel strike during gait, as do humans. The femoral condyles are flattened distally (Figure 5.1C), increasing chondral contact area during knee extension, and also reflecting a humanlike limb posture (review in Lovejoy 1988).

The proximal and distal tibia exhibit marked metaphyseal flaring adjacent to the joint surfaces, providing expanded cancellous bone volume for shock absorption during gait (see Lovejoy 1988, also Latimer and Lovejoy 1989, 1990a, 1990b) (Figure 5.2A and B). The tibia is also vertically oriented, lacking the valgus angle typical of apes that habitually invert their feet during arboreal climbing (Latimer et al. 1987; DeSilva 2009) (Figure 5.2B).

The expanded calcaneal tuberosity (Latimer and Lovejoy 1989) provides added cancellous bone area as well. The australopith hallux was adducted, as evidenced by the footprints at Laetoli, Tanzania (Leakey and Hay 1979), but also by the distal position of the first tarsometatarsal joint, the large insertion site for the peroneus longus muscle on the calcaneus that would prohibit this hallucal adductor in apes from functioning the same as it does in australopiths, and the lack of a smooth surface on the proximal first metatarsal that would permit rotation at this joint (Figure 5.2D, E and F) (Latimer and Lovejoy 1990b).

The foot of *Australopithecus* was also stiff, lacking the midfoot flexibility characteristic of great apes that facilitates pedal grasping, and providing effective propulsion during terrestrial bipedal gait (DeSilva 2010; Ward et al. 2011). This is apparent in the flat, dorsoplantarly expanded fourth metatarsal joints on all hominins, and in the articulation between lateral cuneiform and fourth metatarsal, something that apes lack (Figure 5.3A) (Ward et al. 2011). There also were permanent humanlike longitudinal and transverse arches of the foot, also

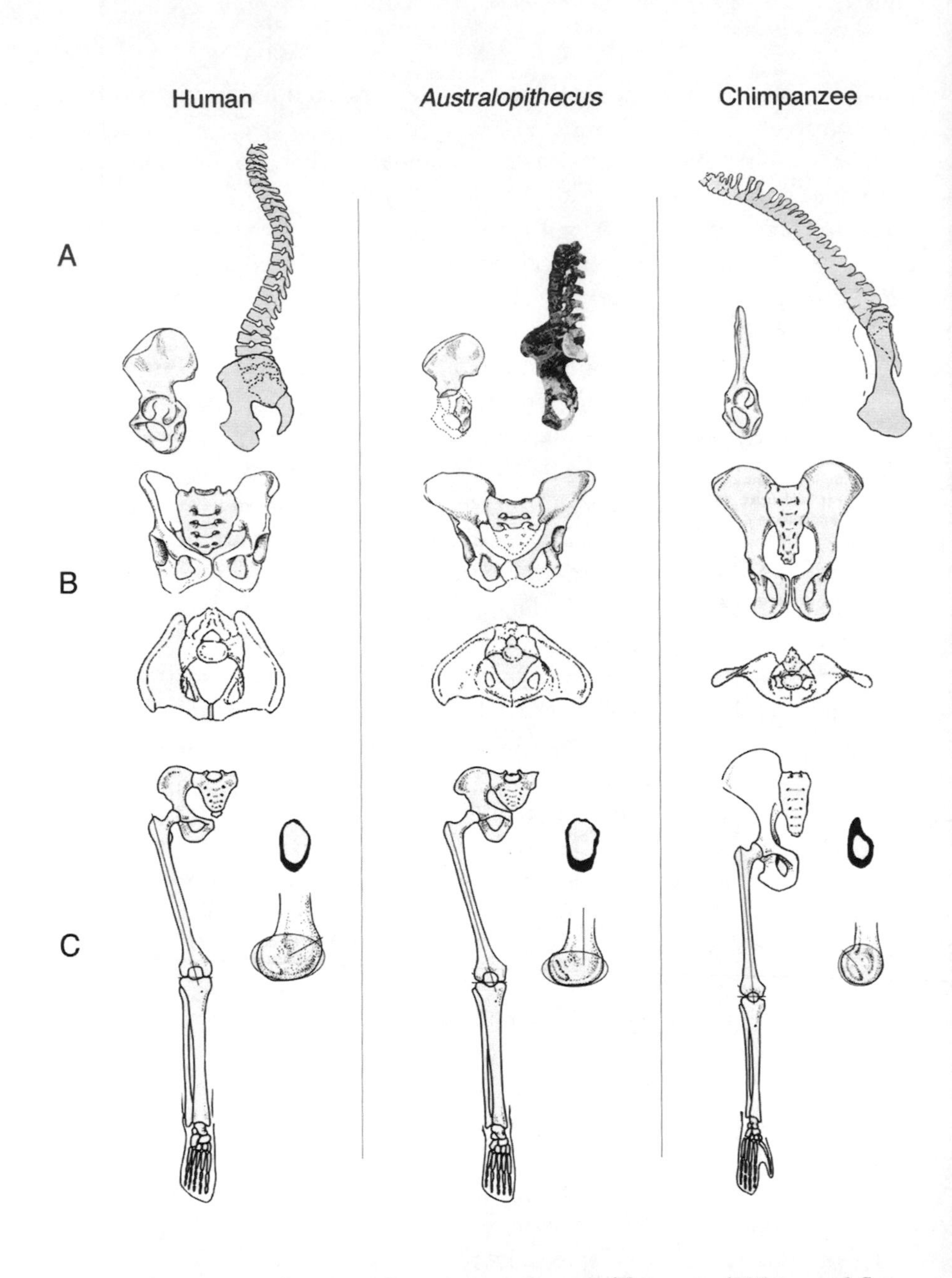

seen in the 3.6-mya hominin footprints at Laetoli, Tanzania (White and Suwa 1987). The fourth metatarsal is twisted about its long axis, so that when the metatarsal heads contact the ground the base is angled medially, reflecting a transverse arch (Ward et al. 2011). The distal ends of the metatarsals are distally

Figure 5.1 (facing page). *Morphological features of the australopith skeleton reflecting adaptation to habitually upright bipedal posture and locomotion, compared with comparable elements of humans and chimpanzees. This is not intended to assume that the ancestral condition was exactly like that of a chimpanzee; the chimpanzee was chosen merely as a representative, closely related nonhominin for purposes of the comparisons discussed in the text. There are almost certainly aspects of morphology—for example, the craniocaudally elongate pelvis—that differ from the morphology of the chimp–human ancestor. (A) Left sides: left hipbones in lateral view. Right sides: left hipbones of extant species in lateral view (redrawn from Klein 2001), and right side of* Australopithecus *(STS 14,* A. africanus *photo from Robinson 1972) in medial view with vertebral columns (illustration by Luba Gudz). Note the laterally facing iliac blades and spinal curvatures of the hominins in contrast with the chimpanzee. (B) Pelvis in anterior (above, redrawn from Klein 2001) and cranial (below, redrawn from Stern 2000) views. Again, note iliac orientation and sacral width. (C) Lower-limb skeleton in anterior view (redrawn from Klein 2001), with femoral neck cross-sectional outline upper right (redrawn from Stern 2000) and lateral view of lateral femoral condyle lower right (redrawn from Klein 2001). Note the distinct femoral bicondylar angle of the hominins, the thin femoral neck cortex, and the elliptical femoral condyle of the hominins in contrast to the ape.*

and not plantarly inclined (Lovejoy et al. 2009a; Ward et al. 2011), reflecting a longitudinal arch. In addition, the metatarsal heads are "domed" dorsally reflecting habitual dorsiflexion of metatarsophalangeal joints during bipedal striding (Figure 5.3B, C, and D) (Latimer and Lovejoy 1990a).

It is clear that not only had australopiths evolved adaptations for bipedality, they reduced their arboreal efficacy by sacrificing traits such as a flexible foot with opposable hallux, high intermembral index (Figure 5.4), relatively long fingers and toes, and grasping feet (Latimer and Lovejoy 1989; Latimer 1991; Ward 2002) (Figure 5.5). Thus, the vector of morphological change (defined as the magnitude and direction of morphological character transformation throughout a lineage, Simpson 1953) leading to australopiths was toward anatomy that enhanced terrestrial bipedality and diminished arboreal competence (Latimer 1991). This suggests that morphologies were indeed adaptations shaped by natural selection (Weishampel 1995; Lauder 1996) for bipedal posture and locomotion.

While this is indeed clear, it is also true that the australopith locomotor skeleton is not identical to that of modern humans, nor even to earlier species of *Homo*. It is in these differences that the challenges to our interpretations lie. While nearly all researchers agree that bipedality was the adaptively most significant mode of locomotion in early hominins, some contend that *A. afarensis* retained significant adaptations to arboreality and thus was partly arboreal (e.g.,

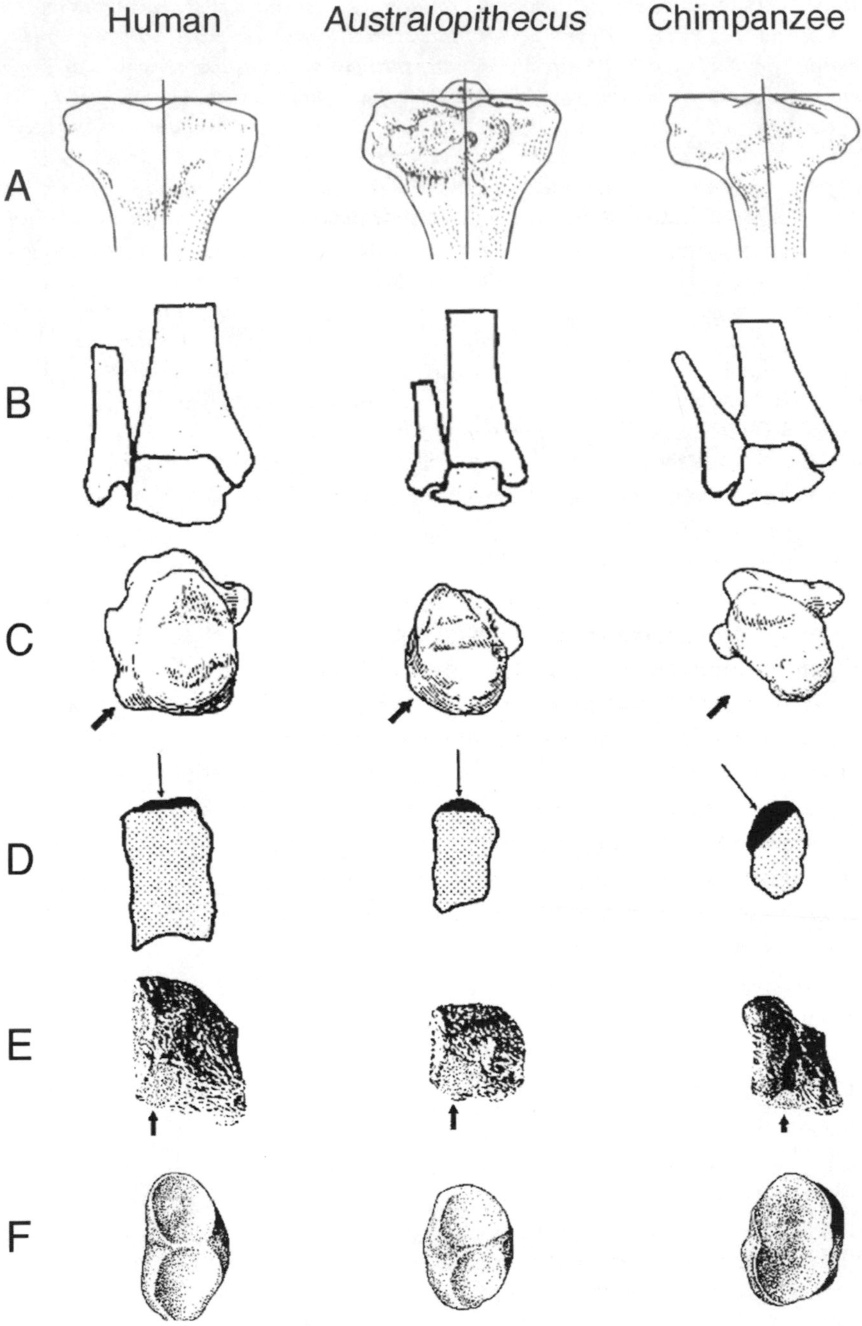

	Human	*Australopithecus*	Chimpanzee
A			
B			
C			
D			
E			
F			

Figure 5.2 (facing page). *More morphological features of the australopith skeleton reflecting adaptation to habitually upright bipedality, with same qualifier as in Figure 5.1. (A) Proximal tibias in anterior view. Note the strong metaphyseal flare in the hominins, but the narrow, hollowed out morphology of the chimpanzee. (B) Distal tibia and fibula articulated with talus in posterior view. Note the valgus angle of the chimpanzee leg, but the vertical orientation in the hominins (redrawn from Latimer et al. 1987). (C) Left calcaneal tuberosity in posterior view. Note the expanded lateral plantar process of the hominins in contrast with the ape (redrawn from Latimer et al. 1987). (D) Horizontal section through left medial cuneiforms. Note the distal orientation and flatter contours of the hallucal tarsometatarsal joint in the hominins, reflecting an adducted hallux, but the medial orientation and curved shape of the chimpanzee's joint (redrawn from Latimer and Lovejoy 1989). (E) Medial cuneiform, medial view. Note the large insertion area for the peroneus longus muscle (redrawn from Latimer and Lovejoy 1990b). (F) Proximal end of hallucal metatarsal, proximal view. Note the two distinct concavities of the hominins, at least partly separated by a transverse ridge that would limit rotation and translation at this joint, compared with the smoothly continuous morphology of the chimpanzee's joint surface that permits opposition and rotation (redrawn from Latimer and Lovejoy 1990b).*

Senut 1980; Stern and Susman 1981, 1983, 1991; Feldesman 1982; Jungers 1982, 1991; Jungers and Stern 1983; Schmid 1983; Rose 1984, 1991; Susman et al. 1984; Deloison 1985, 1991, 1992; Tardieu 1986a, 1986b; Susman and Stern 1991; Duncan et al. 1994; Stern 2000; Harcourt-Smith and Aiello 2004), perhaps with a compromised form of bipedal progression stemming from these retained arboreal characters (Susman et al. 1984; Preuschoft and Witte 1991; Rak 1991; Susman and Demes 1994; Cartmill and Schmitt 1996; MacLatchy 1996; Schmitt et al. 1996; Ruff 1998; Schmitt et al. 1999; Stern 1999).

Testing this hypotheses and determining the extent of arboreal adaptations in *Australopithecus* relies on our ability to interpret the functional and adaptive significance of its primitive characters (review in Ward 2002). Derived traits are relatively easy to interpret. Since natural selection is the only force of evolution capable of producing long-term directional morphological change, derived morphologies are almost always the result of selection on behaviors enhanced by those morphologies. Thus, derived characters that facilitate upright posture and bipedal progression are easy to understand as the result of selection for bipedality. Primitive traits, however, are more complicated, as they could be retained by stabilizing selection for a particular behavior, or simply because they were selectively neutral.

Therefore, understanding the significance of those features in which *Australopithecus* differs from humans relies on determining whether they were, in fact, derived relative to the ancestral condition. If so, we can construct hypotheses to explain their origin.

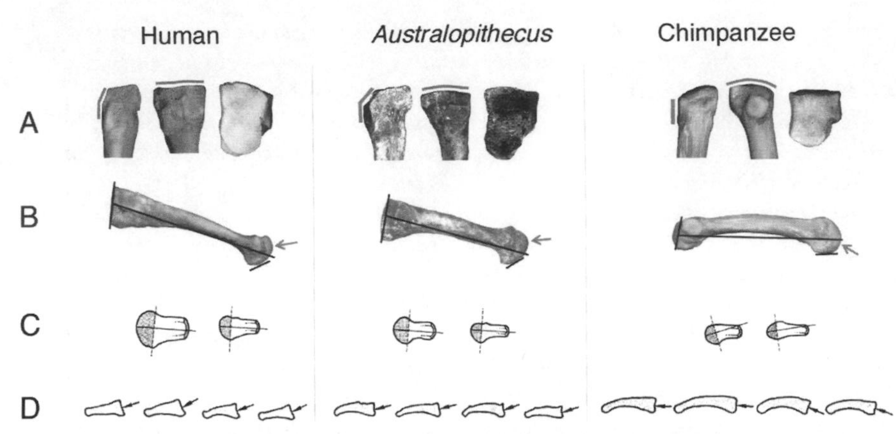

FIGURE 5.3. *More morphological features of the australopith skeleton reflecting adaptation to habitually upright bipedality, with same qualifier as in Figure 5.1 (redrawn from Stern 2000 and Ward et al. 2011). (A) Proximal end of left fourth metatarsals in (left to right) dorsal, medial, and proximal views, all showing morphologies associated with a stiff midfoot and lack of midfoot flexibility in the hominins as compared with apes (Ward et al. 2011). Dorsal view shows the obliquely oriented facet for the lateral cuneiform in humans and* Australopithecus, *not seen in apes. Medial view illustrates the flat contour of the base in hominins, as opposed to the convex surface of apes that facilitates midfoot flexion and extension (see also DeSilva 2010). Proximal view illustrates the dorsoplantarly elongate fourth tarsometatarsal joint characteristic of hominins (see also Lovejoy et al. 2009a), but not seen in apes. (B) Medial view of fourth metatarsals. Plantar inclination of the base and distal inclination of the plantar portion of the head articular surface, along with dorsal "doming" of the head, all reflect the presence of longitudinal arches of the foot. Combined with the clear torsion of the bone reflecting a transverse arch (Ward et al. 2011), which arcs dorsally from the fourth metatarsal medially, this also indicates a strong medial arch. (C) Side view of distal end of first and third metatarsals. Note the distal orientation of this surface perpendicular to the long axis of the diaphyses in hominins in contrast with the chimpanzee, in which the surface is plantarly oriented (Latimer and Lovejoy 1990a). (D) Pedal phalanges, side view. Note the dorsal orientation of the proximal articular surface (arrows reflect normals to the subchondral surface) in hominins in contrast to the plantar orientation seen in the chimpanzee (Latimer and Lovejoy 1990a).*

If not, and the traits represent primitive retentions, we need to creatively consider the reasons why. If a trait compromises the derived function, we can infer that stabilizing selection retained it for an alternate function. If not, we are left with the inability to discriminate between the hypotheses that the trait was retained for a reason, or just because there was no reason for selection to eliminate it.

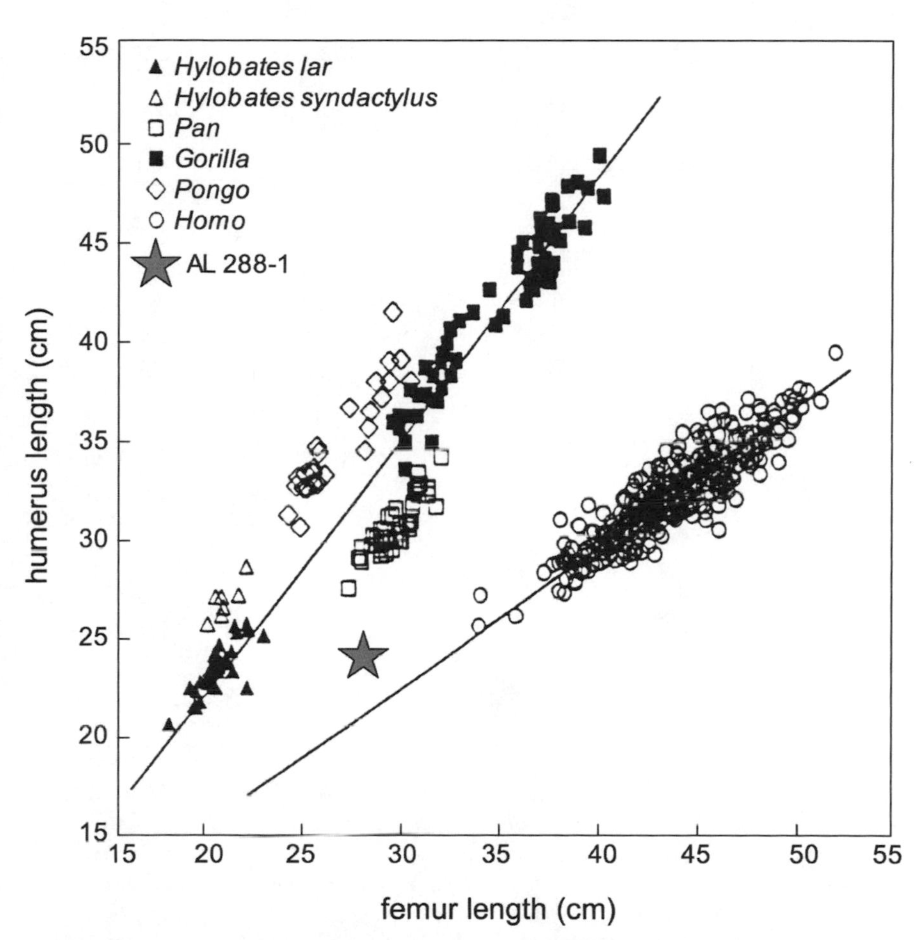

Figure 5.4. *Scatterplot of humeral length to femur length in extant hominoids and* Australopithecus afarensis *(AL 288–1), redrawn from Latimer (1991). Apes all have humeri that are longer relative to femur length than do humans. AL 288–1 does have a longer humerus for its femur length than do modern humans, but much shorter than that of an ape, suggesting that her humerofemoral index was reduced from the primitive condition, suggesting selection for more nearly even proximal-limb lengths.*

Another complication is that sometimes form precedes function, and anatomy continues to improve animals' abilities to engage in behaviors that enhance their reproductive success. The best illustration of this in the paleoanthropological literature is the work of Kay and Ungar (1997), in which they demonstrated that although the molar-cusp morphology of Miocene apes was more like extant frugivores with fairly low crowns and short shearing crests, dental

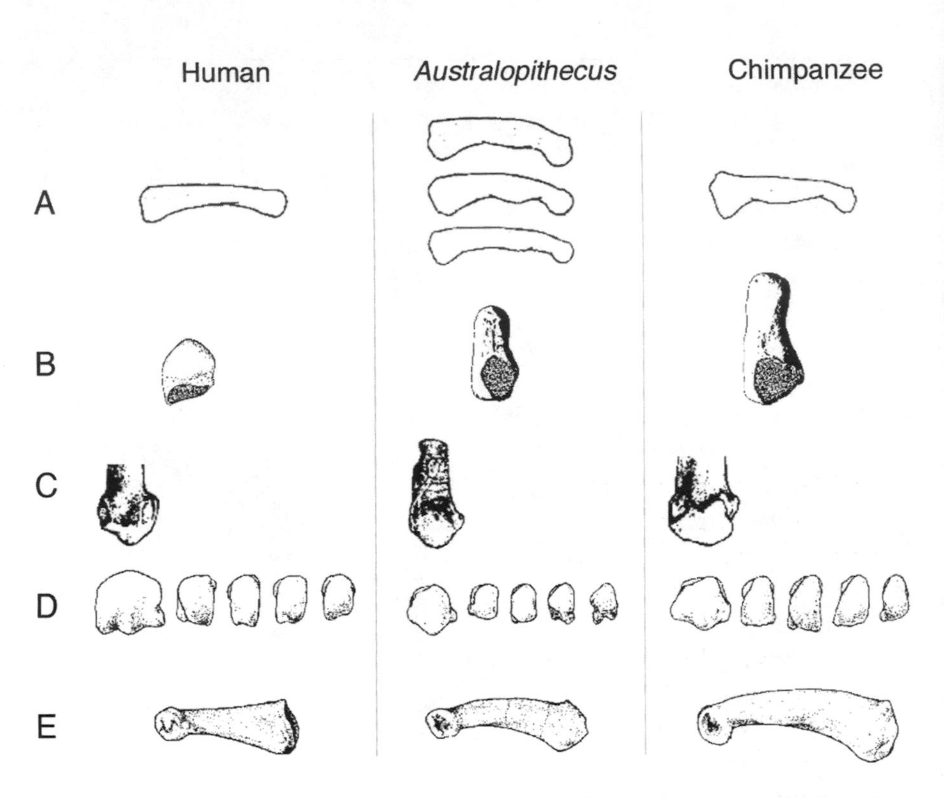

Figure 5.5. *Features of australopiths dissimilar to those of humans and more like those of apes (redrawn from Stern 2000; originally from Stern and Susman 1983). (A) Manual proximal phalanges, lateral view. Australopiths have relatively longer and more curved phalanges with stronger flexor ridges than do humans, although not as long as those of extant apes. (B) Right pisiforms, dorsolateral view. Australopiths have a longer pisiform than do humans, but not as long as that of chimpanzees. (C) Distal end of hallucal metatarsal, plantar view. The distal articular surface is not as flattened distally in australopiths as in humans, although more so than in chimpanzees. (D) Distal view of metatarsal heads 1–5 (left to right), showing how the distal portions of the articluar surfaces of australopiths are not expanded as in humans. (E) Proximal pedal phalanges, side view. As with the hand, pedal phalanges of australopiths are longer and more curved than those of humans, but not as long and curved as those of chimpanzees.*

microwear analysis indicates a similar range of diets to those of extant apes, including high degrees of folivory. Thus, we must seriously consider the possibility that even though australopiths were not anatomically identical to modern humans, their locomotor behavior was equivalent. Precise empirical methods to identify such effects are not readily available for postcranial anatomy, although phenotypically labile traits have the best hope of reflecting behavior for comparison with less labile traits that more closely track genetic change, or history of selection. We must also consider other reasons for potential changes from *Australopithecus* to *Homo* that may involve selective factors other than locomotion, such as throwing or manipulating objects, and so on.

The rest of this chapter considers which traits in *Australopithecus* are indeed primitive retentions, as far as we can tell at the present time, and what these retentions may mean for interpreting *Australopithecus* locomotion.

FROM WHAT KIND OF ANCESTOR DID *AUSTRALOPITHECUS* EVOLVE?

All four great-ape species share a suite of morphologies that enhance their abilities to engage in vertical climbing and below-branch arboreal activities given their relatively large body sizes. With no further information, it would seem clear that these shared features were indisputably homologous, and that hominins evolved from an ancestor that was postcranially much like extant great apes (Figure 5.6). However, by incorporating fossil-ape data into our phylogenetic schemes, the picture is not so straightforward (e.g., Begun, 1994; Begun et al. 1997a; Rose 1997; C. Ward 1997).

Fossil data suggest that extant great apes' postcranial similarities may not all be homologous, however, and so the last common ancestor of chimpanzees and humans may not have closely resembled extant apes in postcranial form. The oft-cited example is of *Sivapithecus*, probably a member of the *Pongo* clade based on striking craniofacial similarities (Figure 5.6) (review in S. Ward 1997). However, its humerus has less torsion than any extant hominoid (Larson 1996, 1998; Pilbeam et al. 1990; Madar 1994; S. Ward 1997; but see Moyà-Solà et al. 1999). Low humeral torsion is presumed to reflect a ventral shoulder joint orientation placed on a narrow thorax (Benton 1965, 1976; Ward 1993, Ward et al. 1993; Churchill 1996; Judd 2005; but see Chan 1997), unlike the broad thorax and laterally oriented shoulders typical of extant hominoids (Evans and Krahl 1945; Le Gros Clark and Thomas 1951; Martin and Saller 1959; Napier and Davis 1959; Larson 1996, 1998; Pilbeam et al. 1990; Churchill 1996). Furthermore, the *Sivapithecus* postcrania also reveal that members of this genus had less well-developed halluces than do extant orangutans or African apes and shorter, less-curved phalanges (Rose 1986, 1997; Spoor et al. 1991; Begun 1993). These *Sivapithecus* postcranial fossils imply that *Pongo* is convergent upon African apes in many of its more specialized climbing features (Pilbeam et al. 1990; S. Ward 1997).

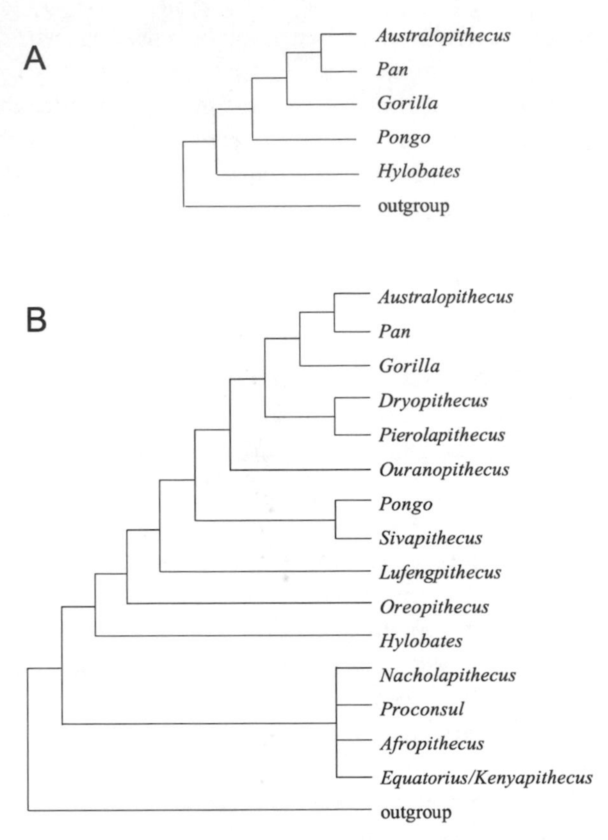

FIGURE 5.6. *(A) Commonly accepted cladogram of extant hominoids and* Australopithecus. *(B) Cladogram of well-known living and fossil hominoids, modified from Begun et al. (1997a). When only extant taxa are considered, parsimony would suggest that the common ancestor of* Australopithecus *and apes was postcranially much like that of an extant great ape. However, many Miocene apes differ significantly from their extant relatives, and including these data alters the balance of parsimony to suggest that the common* Australopithecus–*chimpanzee ancestor may have differed from the morphology seen in modern chimps.*

Early hominins also share features with fossil "basal" hominoids (Begun and Kordos 1997; Begun et al. 1997a, 1997b) such as *Sivapithecus*, *Proconsul*, and/or *Equatorius*, and with *Hylobates*, that are not found among extant great apes, such as shorter pelves with longer lumbar vertebral columns, less-curved forelimb bones, longer thumbs, shorter medial rays of the hand and foot, and less humeral torsion (e.g. Larson 1996, 1998; Rose 1997; C. Ward 1997; MacLatchy et al. 2000). Foot length is not known from the fossil record after *Proconsul*, whose feet were not as long as those of extant apes (Walker and Teaford 1988). Either hominins have reverted to the primitive condition in these features, or extant apes are homoplastic in these ways. Either scenario involves substantial homoplasy, making polarity determinations among hominoids difficult. Even the suspensory later Miocene apes *Pierolapithecus, Hispanopithecus,* and *Oreopithecus* have vertebrae more closely resembling those of hylobatids

The likely Asian–African ape homoplasies open the possibility that chimps and gorillas may also be homoplastic in some features. This hypothesis is supported by some researchers studying kinematic and anatomical correlates of knuckle-walking in chimps and gorillas (Inouye 1994; Dainton and Macho 1999; Kivell and Schmitt 2009). If so, some of these hominin characters may be primitive, which would indicate less change from the last common ancestor than we now suppose. This hypothesis is supported by the apparent low position of the upper thorax and horizontal orientation of the shoulder in *Australopithecus* (Ohman 1986), along with the presence of five lumbar vertebrae as in humans and hylobatids, rather than extant great apes.

Australopithecus metacarpals are the same length relative to the forearms as in all extant hominoids except *Pan* (Drapeau 2001). This discovery highlights the potential error in assuming that *Pan* reflects the morphology of the last common ancestor. Apparent differences in hand proportions between the late Miocene apes *Oreopithecus* and *Hispanopithecus* also point to variation within hominoids in hand structure and proportions (Moyà-Solà et al. 1999).

Recently, however, arguments have been made in favor of the hypothesis that the last common ancestor was a knuckle-walker, and African ape postcranial similarities are indeed homologous (Richmond et al. 2001). This perspective is the most widely accepted at this time. The fact that *Ardipithecus* has been argued to lack features associated with knuckle-walking, and that it lacks some key apomorphies of great apes, such as a narrow sacrum, flexible midfoot, elongate pelvis, and expanded midcarpal joint (Lovejoy et al. 2009a,b,c,d), could also indicate that these are hominin synapomorphies. Thus, the debate continues.

Many *A. afarensis* traits appear to be primitive even when the fossil evidence is considered. A long pisiform (Figure 5.5B), dorsoplantarly narrow navicular, and metatarsals lacking expanded dorsal margins of their head (Figure 5.5C and D) are known for *Oreopithecus* (Sarmiento 1987), *Ardipithecus* (Lovejoy et al. 2009a), and extant apes. *Australopithecus* lacks the expanded vertebral bodies of humans, resembling all taxa for which there are data (e.g., McHenry 1992): *Proconsul* (Napier and Davis 1959; Walker and Pickford 1983; Walker and Teaford 1988; Ward et al. 1993), *Nacholapithecus* (Rose et al. 1996; Nakatsukasa et al. 1998; Nakatsukasa et al. 2000), *Equatorius* (Le Gros Clark and Leakey 1951), *Morotopithecus* (Walker and Rose 1968; Gebo et al. 1997; MacLatchy et al.

2000), *Oreopithecus* (Harrison 1986; Sarmiento 1987), *Hispanopithecus* (Moyà-Solà and Köhler 1996), and extant primates. Within the hand, a pronounced third metacarpal styloid process, found only in *Homo* and some gorillas, is absent in *Proconsul* (Napier and Davis 1959) and *Oreopithecus* (Moyà-Solà et al. 1999). Also, the first metacarpal base is highly concavo-convex in *Proconsul* (personal observation), *Sivapithecus* (Spoor et al. 1991), *Oreopithecus* (Moyà-Solà et al. 1999), and extant apes. Small apical tufts are found on the distal phalanges of *Afropithecus* (Leakey et al. 1988), *Proconsul* (Napier and Davis 1959; Begun 1994), *Nacholapithecus* (Nakatsukasa et al. 1998), *Oreopithecus* (Moyà-Solà et al. 1999), *Hispanopithecus* (Moyà-Solà and Köhler 1996), and extant primates. The mediolaterally broad knee joint typical of extant apes and seen to a lesser extent in *A. afarensis* (Tardieu 1979) is also developed to some degree in *Proconsul* (Walker and Pickford 1983), and more so in *Morotopithecus*, *Sivapithecus*, and *Oreopithecus* (Harrison 1986; Sarmiento 1987; S. Ward 1997).

Within the forelimb, *A. afarensis* also appears to be primitive in arm, forearm, and metacarpal segment lengths (Drapeau 2001; Drapeau et al. 2005). *Australopithecus* has a slightly higher brachial index compared with extant apes (see Kimbel et al. 1994) and appears to represent the primitive large hominoid condition. *A. afarensis* does not appear to have shortened its upper limbs from the primitive condition, nor lengthened them, showing less directional change than often assumed (Jungers 1982, 1984; Wolpoff 1983; White et al. 1993; White 1994; Haile-Selassie et al. 2010a). There is thus no evidence that *A. afarensis* is derived in forelimb proportions, supporting the hypothesis that their lower limbs are elongated rather than their forelimbs reduced (Wolpoff 1983). Instead, *Pan* and *Pongo*, and to a lesser extent *Gorilla*, seem to have independently elongated their forearms from the likely ancestral condition.

Many primitive traits, however, are derived toward a humanlike condition from the apelike one no matter which phylogeny one considers. *Australopithecus* appears to have had reduced manual and pedal phalanges that were relatively shorter and straighter than in any primate except *Homo* (Bush et al. 1982; White 1994) (Figure 5.5A and E). The hamate hamulus is proximodistally shorter than that of extant apes and humans (Sarmiento 1994, 1998; Ward et al. 1999b), perhaps suggesting a shorter carpal tunnel, related to a shorter hand. The high intermembral index of African apes appears to have been present in at least *Oreopithecus* (Hürzeler 1960; Moyà-Solà and Köhler 1996) and *Hispanopithecus* (Moyà-Solà and Köhler 1996), and probably in *Nacholapithecus* (Nakatsukasa et al. 2000), although not in *Proconsul* (Walker and Pickford 1983; Walker and Teaford 1988), but was reduced in *Australopithecus* (Figure 5.4). *Australopithecus* had a shorter ulna relative to its femur length than any known nonhominin member of the extant ape clade, suggesting a reduction in forelimb to hindlimb proportions, a conclusion also supported by its humeral to femoral proportions (Jungers 1982, 1984; Wolpoff 1983; White et al. 1993; White 1994) (Figure 5.4).

A number of primitive characters of *A. afarensis* appear to reflect soft-tissue differences from humans, and suggest a somewhat more extant great ape–like muscular configuration in these early hominins. Some of the *Australopithecus* hip and thigh muscle-attachment sites appear to have differed from those of *Homo* (Ward 2002; Harmon 2008). The anterior fibers of the gluteus minimus, or scansorius, muscle has a more extensive attachment in australopiths than in modern humans (Ward 2002; Harmon 2008; see also Häusler 2001), perhaps related to their slightly less sagittally oriented iliac blades (Robinson 1972). The conformation of the *A. afarensis* ischial tuberosity also differs from those of humans, being longer with the adductor magnus origin site set at an angle to the rest of the tuberosity. This different conformation may reflect other primitive musculature. Similarly, the sartorius, gracilis, and semitendinosus insert directly into the bone along the tibial tuberosity in *Australopithecus* (Ward 2002).

Latimer and Lovejoy (1990b) suggest that the convex first metatarsal head of *A. afarensis*, unlike the flatter configuration in humans, may reflect more idiosyncratic loading in *A. afarensis* than in *Homo*. They argue this may be due to the fact that the triceps surae may not yet have undergone the changes in muscle belly length and pennation seen in humans that likely accompanied limb elongation (see also Bramble and Lieberman 2004), and so relatively more plantarflexion was accomplished with the peroneal muscles. The large peroneal muscles are suggested by the large peroneal groove on the fibula and peroneal trochlea on the calcaneus (Latimer and Lovejoy 1989, 1990b). This hypothesis is difficult to test, as muscles do not often fossilize. Given the lack of capacity for hallucal abduction in any *Australopithecus*, it is highly unlikely that the rounded head of the first metatarsal reflects opposability, but whether it somehow reflects arboreality is unspecified and unknown.

Taken together, though, these indicators of a somewhat primitive muscle configuration in *Australopithecus* suggest that these early hominins were not fully modern in their muscular anatomy. This is an important point, because when muscles contract to respond to ground reaction forces or to produce movement, they generate strain within bones that can trigger modeling responses during growth and remodeling after skeletal maturity (review and discussion in Martin et al. 1998; Carter and Beaupré 2001). This means that some bony differences between *Australopithecus* and humans could be related to one of two factors, or both. Either these hominins were behaving differently, or that they were behaving in the same way but with slightly different muscle vectors due to the primitive muscle arrangement. The latter situation could produce a slightly different pattern of bone strain, influencing traits such as the geometry of the acetabulum, proximal femoral shaft, patellar groove, or hallucal tarsometatarsal joint shape. Indeed, Berge (1991) hypothesized that *A. afarensis* only could have moved bipedally efficiently if it had a more apelike arrangement of the hip musculature due to its slightly different bone structure. Although these ideas have not been tested, they represent a plausible alternate hypothesis that should be considered when evaluating skeletal differences among hominins.

Another feature in which *A. afarensis* has been considered significantly different from humans is in the anteroposteriorly compressed sacral ala (Figure 5.1). However, given the mediolaterally extensive and anteroposteriorly compressed nature of the *A. afarensis* pelvis, and the broadly flaring blades, this feature may simply reflect iliac form and not be a relevant variable itself (see Lovejoy et al. 2000; Lovejoy 2005).

A final line of evidence we can use to assess character polarities comes from data for the earliest hominins *Sahelanthropus*, *Orrorin,* and *Ardipithecus*. All three genera have been described as probable bipeds, but the nature of this bipedality and their morphological similarity to later hominins is still unknown or undescribed. *Sahelanthropus* is known only from craniodental remains. Because it has a short basicranium with basion intersected by the bicarotid cord, like other hominins, its discoverers tentatively suggest it was upright, barring further evidence (Brunet et al. 2002). A nearly complete femur is known for *Orrorin,* which its discoverers say displays bipedal features (Senut et al. 2001). None of the characters they mention in their article, however, are found exclusively in bipeds. It could be that when the internal contours of the femoral neck are revealed, and when proximal- to middle-shaft dimensions are compared, those data will be sufficient to make a judgment. *Ardipithecus* has also been described as a biped, based on a short basicranium in *Ar. ramidus* (White et al. 1994; Suwa et al. 2009) as in *Sahelanthropus* (Brunet et al. 2002), a broad low pelvis with wide sacrum, a wide subpubic angle, and a large anterior inferior iliac spine for attachment of the iliofemoral ligament (which limits extension and is tightened in extended limb postures; Lovejoy et al. 2009d), a stiff foot (Lovejoy et al. 2009a), a presumed flexible spine (Lovejoy and McCollum 2010), and a proximal phalanx of *Ar. kadabba* that resembles those at Hadar (Haile-Selassie 2001). *Ar. kadabba* also appears to have had a curved ulna, but this is found in hominin ulnas from the Omo river valley (Omo L40–19) and Olduvai Gorge (OH 36). The forearm remains of *Ar. ramidus* appear more like those of *A. afarensis* than they do like African apes with a relatively high olecranon and anteriorly facing trochlea (White et al. 1994). However, *Ar. ramidus* retained a large grasping hallux and other features suggesting that it also retained significant climbing abilities as well, a unique pattern among hominoids.

So, in summary, most features of *A. afarensis* that are generally interpreted to be primitive probably are, although the prevalence of homoplasy among hominoids should suggest caution in making this assumption. The likely homoplasies between Asian and African apes raise the possibility that African apes may be convergent upon one another in ways other than forelimb length, making some apparently derived hominin features possible homoplasies, like relatively short forelimbs, short pelves, 5–6 lumbar vertebrae, and thumb–finger proportions. This could be true even if the last common ancestor of African apes and humans was a knuckle-walker. Some features that appeared to be derived in *A. afarensis*, like the short metacarpals or forelimb proportions, appear to be primitive, with extant apes often displaying a derived condition.

It is clear that australopiths had undergone selection for terrestrial bipedality to the extent that they sacrificed features that would have facilitated climbing trees, so there is no question as to which locomotor behavior most strongly affected the survival and reproductive success of their ancestors. Even in the case of most of the features that remain incompletely humanlike, and more apelike (Figure 5.5), australopiths still may be derived in the direction of humans and away from an apelike condition. Yet, australopiths retained primitive traits from their arboreal forbears. At present, it is not possible to determine the reason for these retentions. If these traits compromised their ability to be effective bipeds, we could infer selection for a certain level of arboreal competence, evidence for or against which I discuss below.

WAS *AUSTRALOPITHECUS* BIPEDALITY DIFFERENT FROM HUMAN BIPEDALITY?

Some researchers have implied that the gait of *A. afarensis* differed from that of humans because they had slightly different limb proportions, muscle attachment sites, and a pelvis with less coronally oriented iliac blades and a smaller anterior horn of the acetabulum than do humans (Figure 5.1). Some have suggested that this indicates the bipedal gait of *A. afarensis* was more energetically expensive than that of humans, or at least different (Berge 1991, 1994; Cartmill and Schmitt 1996; Preuschoft and Witte 1991; Rak 1991; Rose 1984, 1991; Ruff 1998; Schmitt et al. 1996,1999; Stern 1999; Susman et al. 1984; Tardieu 1991). Even if it is true that *Australopithecus* might have looked different when it walked than do extant humans (Rak 1991; Susman and Demes 1994; Schmitt et al. 1996, 1999; Ruff 1998), it remains to be demonstrated that they were less capable bipeds, or, if they differed from us, exactly how. If *Australopithecus* could not walk as fast, as far, or as well as do humans, and this was because of the retention of traits that enhanced arboreality, then this would support the hypothesis that stabilizing selection had acted to maintain some level of arboreal competence.

Some researchers have proposed that *A. afarensis* gait was kinematically distinct from that of humans, and that they walked with their hips and knees in a more flexed posture (Susman et al. 1984; Preuschoft and Witte 1991; Stern 1999; but see Lovejoy and McCollum 2010). This is not well supported by anatomical evidence, however; *A. afarensis* certainly had a significant lumbar lordosis (Ward et al., 2012; Ward and Latimer 1991, 2005) (Figure 5.1) reflecting habitually upright posture, which is in fact reduced or eliminated when the torso is anteriorly inclined for balance over flexed lower limbs. The pronounced femoral bicondylar angle indicates that *A. afarensis* actually engaged in bipedal walking or running with the knee in extended posture (Tardieu and Trinkaus 1994; Duren 1999) (Figure 5.1).

A bent-knee posture is also inconsistent the distal femoral morphology of *A. afarensis*. The expanded calcaneal tuberosity of *A. afarensis* indicates that initial contact during gait was a heel strike, and that weight was borne on a single support limb (Latimer and Lovejoy 1989) (Figure 5.2). Heel strike would have occurred with the knee at or near full extension, as it is in humans, and the knee would have been straightened further through midstance and late stance. The lateral lip of the patellar groove is higher than the medial one in both *A. afarensis* specimens AL 129-1 and AL 288-1, and differs from the situation seen in apes, despite the mediolaterally broad distal femur retained in *A. afarensis*. *Australopithecus afarensis* also exhibits elliptical lateral femoral condyles with the flattest region on the distal surface, as do modern humans and unlike the rounded shape seen in apes, although this flattening is not as pronounced as humans perhaps because of its small size (Tardieu 1998) (Figure 5.1). This morphology serves to increase tibiofemoral articular contact at or near full extension of the joint, indicating that peak transarticular loads occurred at or near full knee extension, rather than in a bent-knee posture, during bipedal progression, as it does in humans, as joints appear to be shaped during growth by response to hydrostatic pressure in articular cartilage (Frost 1990a; Hamrick 1999). A bent-knee gait also is unlikely from a kinematic perspective (Crompton et al. 1998; Kramer and Eck 2000).

Others have argued for a different gait pattern in australopiths as well. The small femoral head (McHenry 1991; Ruff 1998) and strong diaphysis of AL 288-1, coupled with a longer femoral neck, has been suggested by Ruff (1998) to mean that *A. afarensis* pelvic movements during gait would have had to be greater due to higher predicted hip and femoral strains. However, the fact that *A. afarensis* seems to have had proportionally stronger bones throughout its skeleton (Ruff 1998; Coffing 1998) could confound Ruff's comparison. It is also unclear why *A. afarensis* would have a very large bicondylar angle in this situation. Rak (1991) also posited greater pelvic movements, although in pelvic rotation rather than lateral elevation.

The relatively longer toes of *A. afarensis* (Stern and Susman 1983; Susman et al. 1984; White and Suwa 1987) have been suggested to imply a different gait pattern. Whether or not the gait of *A. afarensis* would have looked unusual due to having larger feet is uncertain. There are no associated pedal and lower-limb skeletons to assess relative foot length empirically. Further, variation among extant humans with different relative foot and toe lengths is considered normal, and few researchers would argue that any humans are more or less effective bipeds as a consequence.

In summary, while several researchers have posited a less-efficient form of bipedality for *A. afarensis*, none yet has falsified the hypothesis that lower-limb posture was not within the range of modern humans. Furthermore, the various analyses fail to support a single mechanism by which they differ, instead positing either different hip postures or movement during locomotion. Data are unclear as to whether the apelike aspects of the *A. afarensis* skeleton diminished their efficiency as terrestrial bipeds. While the detailed kinematic profile of *A. afarensis* individuals might differ in some ways from that of humans, there is no certainty that they would have been less capable bipeds as a consequence of their retained primitive traits. Data suggesting different distributions of subchondral surface at the *A. afarensis* hip joint (MacLatchy 1996) may reflect the different patterns of musculoskeletal anatomy surrounding the hip joint rather than different limb posture (Ward 2002). This then raises the question of why *Homo* altered the primitive hominin pattern, with its apparent mechanical improvements for bipedality. Increasing body size may have played a part, along with other selective pressures to modify pelvic form to accommodate childbirth for larger-brained infants (Tague and Lovejoy 1986; Ruff 1998). At present, there is no clear refutation of the hypothesis that the primitive traits in the *A. afarensis* skeleton compromised their bipedal ability.

WHETHER IT WAS OR WAS NOT ADAPTIVE TO DO SO, DID *AUSTRALOPITHECUS* CLIMB TREES?

Knowing what australopiths actually did, regardless of what they were adapted to do, would be relevant for reconstructing the behavior of these creatures.

Relatively few studies so far have investigated this possibility. Data about actual behaviors will come from traits that are phenotypically plastic and that reflect an individual's behavior pattern (Churchill 1996; Lieberman 1997).

One type of data that has been linked to climbing is muscularity, or interpretation of muscularity based on bone robusticity and muscle insertion-site morphology. It appears that skeletal indicators of overall muscularity were greater in *A. afarensis* than modern humans. Indicators of muscle size and strength in *A. afarensis* tend to be large relative to modern humans, such as the deltoid tuberosity (Aiello and Dean 1990), flexor-muscle insertion sites manually and pedally (Marzke 1983; Stern and Susman 1983; Susman et al. 1984), the peroneal groove on the fibula, and the peroneal trochlea on the calcaneus (Latimer and Lovejoy 1989, 1990b). However, there is no evidence that this was to improve climbing, nor that it did improve climbing, nor that it reflects adaptations to other behaviors such as foraging, defense, and so on, nor that it was part of a more global biologic or hormonal situation. Indicators of relative muscularity in all regions of the *A. afarensis* skeleton are generally greater than in modern humans, as with all premodern hominins, most of which were clearly not arboreally adapted (Ruff et al. 1993, 1999; Coffing 1998).

Richmond (1998, 2003) and Paciulli (1995) have suggested that phalangeal curvature reflects the ontogeny of locomotor behavior (see also Richmond 2007). They argue that phalangeal curvature is greatest during the juvenile period when juveniles spend the greatest proportional amount of time engaging in arboreal behaviors, and decreases later in life. If their results are confirmed by other analysis, particularly longitudinal ones, this would be a character indicating that at least juvenile *A. afarensis* spent time in the trees.

However, Tardieu and Preuschoft (1996) inferred that the morphology of developing epiphyseal plates was affected by behavior, and the immature Hadar femora resemble those of humans and not apes. This suggested to them that *A. afarensis* individuals were not climbing, as they lacked the apparent adaptations for stabilizing the epiphysis during arboreality. Duren (2001) has also found significant links between epiphyseal orientation and form and locomotion among catarrhines.

Finally, when assessing the likely behaviors of *Australopithecus*, we can consider the capabilities evident in their morphology. Even if *A. afarensis* spent time in the trees, without a flexible foot and grasping hallux (White and Suwa 1987; Latimer and Lovejoy 1990b; but see Clarke and Tobias 1995; Ward et al. 2012) they would have been significantly less agile than are apes and monkeys, especially in the case of females holding infants. Despite the apparently slightly longer and more curved toes than is typical for modern humans, their toes were still shorter and straighter than those of apes (Bush et al. 1982; Latimer and Lovejoy 1990b; Stern and Susman 1983; White and Suwa 1987) and the midfoot stiff and inflexible (Ward et al. 2011), and so the size of supports they could have grasped would be less. Even if their manual grasp was greater than that of humans, and their pedal grasp like that of the manual grasp of modern 2-year-old humans (Susman et al. 1984; Stern 2000), we still cannot be sure that this was retained for adaptively significant arboreal behaviors. They would have been more competent in the trees than are typical humans, as they were much stronger and had very slightly better mechanics with longer arms, fingers, and toes. Any arboreal locomotion would still necessarily have been fairly slow and deliberate, however—more for something like sleeping in trees than for hunting or fleeing from predators. Certainly females with infants would have difficulty climbing trees, because without grasping halluces they would have to hold onto their infant with at least one arm. Still, sleeping in trees for safety is a potentially significant behavior that, at least in theory, could have selected for some relatively weak climbing capabilities.

Thus, although *Australopithecus* may have climbed trees, there is no evidence that this behavior had any adaptive advantages for them to the extent that it influenced their morphology. The strongest evidence for retained selection for arboreal competence is probably the retention of some primitive traits for almost 3 myr (million years) in australopith evolution. However, the fact that so many features improving arboreal competence were reduced or eliminated indicates that climbing was of little or no significance for shaping the australopith skeleton.

DID ALL *AUSTRALOPITHECUS* SPECIES SHARE THE SAME LOCOMOTOR ADAPTATION?

It appears that the general pattern of postcranial morphology exhibited within *Australopithecus* was essentially constant, and taxa vary little from one another, although there may have been minor modifications of the basic pattern in different species. Broadly speaking, there is an *Australopithecus* pattern of morphology that changes appreciably only with the advent of *Homo erectus* around 1.8 mya.

The earliest *Australopithecus* species identified so far is *A. anamensis* (Leakey et al. 1995, 1998; Ward et al. 1999a, 2001). Postcranially, *A. anamensis* is known from only the proximal and distal thirds of a tibia, a distal humerus, a nearly complete radius, a capitate, and a partial proximal manual phalanx, as well as a partial femur and some other undescribed fragments (White et al. 2006). In most ways, these bones closely resemble those of *A. afarensis*. From the tibia, we can infer that *A. anamensis* was a habitual biped, because like all other known hominins it displays a diaphysis that is oriented normal to the talocrural joint surface, rather than at the varus angle found in apes (Latimer et al. 1987; Ward et al. 1999a). The humerus, discovered long before it was attributed to *A. anamensis* (Patterson and Howells 1967), and despite suggestions that it is more like that of *Homo* (Senut and Tardieu 1985; Baker et al. 1998), is indistinguishable from that of *A. afarensis* (Feldesman 1982, Hill and S. Ward 1988, Lague and Jungers 1996; Ward et al. 2001), as is the phalanx. The radius is similar as well, and belonged to a forearm that was longer than the longest one preserved for *A. afarensis* (Heinrich et al. 1993). This suggests that the forearms of *A. anamensis* were at least as long as those of *A. afarensis*, and almost assuredly longer for their body size than those of *Homo*. The capitate is poorly preserved, but suggests a more laterally facing second metacarpal facet than seen in any other hominins, and instead resembles the condition found in great apes. The only place in which these taxa differ is the capitate facet reflecting rotational ability of the second carpometacarpal joint (Leakey et al. 1998). Because it is likely that *Australopithecus anamensis* and *A. afarensis* represent a single lineage (Kimbel et al. 2006; Haile-Selassie 2010; Haile-Selassie et al. 2010b), there is evidence of postcranial change only in the hand, but that bipedality had been established by 4.2 mya. However, too little fossil evidence is available to test these ideas with any certainty.

Australopithecus afarensis and *A. africanus* are similar postcranially in overall morphological pattern, although there are some comparatively minor differences (Häusler 2001 and references therein). When they differ, *A. africanus* generally has slightly more humanlike morphology than does *A. afarensis*, but not always. Although their hands are strikingly similar morphologically, *A. africanus* appears to have more gracile metacarpal shafts and slightly straighter phalanges with less-well-developed flexor ridges than does *A. afarensis* (Ricklan 1987). There also seems to be a large apical tuft on the pollical distal phalanx. Although this bone is not known for *A. afarensis*, the other *A. afarensis* distal phalanges have narrow apical tufts (Bush et al. 1982). The *A. africanus* pelvis appears to have had a slightly more anteroposteriorly expanded inlet with more sagittally oriented iliac blades with better-developed cranial angles of the sacrum. This suggests a different overall shape of the lower torso in each species. Häusler (2001) reports an indistinct iliofemoral ligament attachment in *A. afarensis* compared with a distinct one in *A. africanus*, but given the prevalence of an intertrochanteric line on the femur in *A. afarensis*, it seems that this ligament was strong in both species. Häusler (2001) also shows that the latissimus dorsi attachment site on the iliac crest is more medially restricted in *A. africanus*, like modern humans, than in *A. afarensis*, suggesting less powerful upper limbs in the Sterkfontein hominins.

There are several other ways in which *A. africanus* has been interpreted to be more primitive and arboreal, however. Clarke and Tobias (1995) argue that STW 573 had an abductable hallux, but this was not confirmed by a geometric morphometric analysis of these bones (Harcourt-Smith and Aiello 2004). Observation of the original specimen reveals that this specimen does not differ from *A. afarensis* or from the joint morphology and orientation of OH8 (Leakey 1960, 1961), which is attributed to *Homo/Australopithecus habilis* (Leakey et al. 1964). *A. africanus* also has been suggested to have more apelike limb proportions than does *A. afarensis* (McHenry and Berger 1998;). However, Häusler (2001) notes that upper-limb length to acetabulum size does not differ between two partial skeletons of *A. afarensis* (Al 288–1) and *A. africanus* (STW 431). Differences noted by McHenry and Berger (1998) in associated skeletons reflect the relatively large radial head of *A. africanus* (Häusler 2001).

In summary, *A. africanus* may have been slightly more derived than was *A. afarensis* in most features where the species differ. Because *A. africanus* is later in time (McKee 1993; Walter 1994; McKee et al. 1995), this could indicate a directional trend. If these taxa represent a single lineage, this might signal disaptation in the traits that change and their related soft tissues.

Australopithecus boisei is too poorly known postcranially to compare (see Grausz et al. 1988). *A. robustus*, on the other hand, is known from a number of postcranial bones, although many cannot be attributed to *Australopithecus* with certainty, as *Homo* is also known from the same sites. The putative *A. robustus* hand bones are described as more humanlike, with broader apical tufts on the terminal phalanges and flat first metacarpal bases (Susman 1988, 1989). Its ilium (SK 3155) has been described as having a relatively wider and more cranially extensive iliac blade with a larger auricular surface than do those of *A. africanus* and *A. afarensis* (Brain et al. 1974; McHenry 1975), but variation among species may reflect body size and/or individual variation (McHenry 1975). Other *A. robustus* postcranial bones are not notably different from those of *A. afarensis* or other *Australopithecus* species (Broom and Robinson 1949; Napier 1959, 1964; Robinson 1970; Day and Scheuer 1973; Brain et al. 1974; Susman 1989; McHenry 1994; Susman et al. 2001). Suggested variation certainly does not approach the condition seen in *Homo erectus/ergaster*.

No other *Australopithecus* species is known postcranially. Postcrania from the Hata Member of the Bouri Formation in Ethiopia cannot be definitively attributed to *A. garhi*, the only hominin identified at the site (Asfaw et al. 1999).

However, these fossils display a primitive, *A. afarensis*–like brachial index but with proportionately longer lower limbs, suggesting selection for increased lower-limb length if this hominin was a descendent of *A. afarensis*.

Thus, the basic pattern of postcranial anatomy exhibited by *A. afarensis* appears to persist for over 3 myr, suggesting that its locomotor adaptation was stable, and not undergoing ongoing selection for improved terrestrial competence. The slight differences between *A. africanus* and *A. afarensis* might signal slight directional selection toward a *Homo*-like condition if *A. africanus* descended from *A. afarensis*, which is uncertain. Similarly, if the postcrania from the *A. garhi* site of Bouri Hata (Asfaw et al. 1999) represent a descendent of *A. afarensis*, lower-limb elongation would have occurred within the genus. These observations are speculative and are meant only to point out possibilities. If they were eventually supported by further evidence, this might document selection against primitive traits within the genus *Australopithecus*.

The length of time over which the basic *Australopithecus* pattern appears to have been retained seems to be the strongest evidence that primitive morphology retained an adaptive advantage for *Australopithecus* individuals, although it does not falsify the null hypothesis that these primitive traits were simply selectively neutral. If bipedality began with late Miocene hominins like *Sahelanthropus* and *Orrorin*, both about 6 mya, it would have been around for 3 myr before the appearance of *A. afarensis*. Even so, it had been around for almost 1 myr since *A. anamensis*. The general australopith body plan remained for perhaps up to 1 mya until the disappearance of this genus. If the primitive *Australopithecus* traits were disadvantageous, selection would have had ample time to eliminate or alter them. If they were selectively neutral, we need to explain why they never changed until the appearance of *Homo*.

SUMMARY AND CONCLUSIONS

It is clear from available fossil evidence that the immediate ancestors of *Australopithecus* had undergone selection to be terrestrial bipeds. This hypothesis is supported by the numerous derived skeletal modifications from the likely ancestral condition visible in the preserved parts of their skeletons. Australopiths were fully upright and would have walked with a similar extended limb posture to modern humans, given their fully modern lumbar lordosis and femoral bicondylar angle.

The hypothesis that arboreal competence continued to confer selective advantages for australopiths beyond the relatively awkward levels of which humans are capable remains difficult to test. It is almost certain that australopiths would have been more capable climbers than are most modern humans. They were smaller and stronger, with slightly longer forearms, hands, and feet, but none of these features is nearly as well developed as in any ape. Particularly in light of the strong directional signal away from arboreality toward bipedality, it is difficult to disprove the null hypothesis that these traits were adaptively valuable. These traits do not appear to have compromised effective bipedality. They were, apparently, retained throughout the history of *Australopithecus* evolution, although evidence is slim, given the spotty fossil record. Although there are the occasional features that may differ among species, at present there is little fossil evidence for significant locomotor diversity within *Australopithecus*, unless the Bouri Hata fossils do indeed turn out to belong to *A. garhi*.

Moreover, the massive rearrangement of bone and joint morphology and orientation evident in the *Australopithecus* skeleton reveals that substantial loads were incurred while traveling bipedally. Extant apes are capable facultative bipeds, being able to travel short distances during food-gathering episodes on two feet (Hunt 1994). Their ability to climb trees is more important to them in terms of reproductive success, however, so they have not been selected to modify their skeletons to become better bipeds. The adaptive consequences of problems that might stem from a musculoskeletal system poorly designed for bipedality, such as back problems or joint pain, are minimal in very short distance locomotor bouts. On the other hand, the fitness of an animal traveling longer distances more regularly, at high speed, or for particularly reproductively valuable reasons (inter- or intraspecific competition, carrying, etc.), would be in jeopardy with any such maladies. For this reason, the only scenario that would select for the major alterations hominins underwent is if bipedality was employed for more than simply standing and feeding, or walking very short distances. The adaptive importance of bipedal travel is highlighted by the loss of the many advantages of arboreal traits, in particular a grasping hallux, useful for climbing trees with speed and agility for feeding, hunting, or predator escape, or for enabling infants to cling to their mothers' fur.

In conclusion, if australopiths and other early hominins were partly arboreal, postcranial changes seen in *Homo erectus/ergaster* might be due to a shift toward full bipedality. On the other hand, if australopiths were indeed fully bipedal, the changes in the *Homo* skeleton must have been due to other factors, such as an increase in efficiency due to walking longer distances or running, and/or selection for body size, throwing, tool use, and transport, or any number of other possibilities. Continued efforts toward better understanding early hominin evolution are critical to understanding the origin and early evolution of our clade.

ACKNOWLEDGMENTS

I thank Matt Sponheimer, Julia Lee-Thorp, Kaye Reed, and Peter Ungar for inviting me to contribute to this volume, and for their patience. I thank Lee Berger, Ron Clarke, Don Johanson, Bill Kimbel, Bruce Latimer, Meave Leakey, Steve Leigh, Brian Richmond, Philip Tobias, and the staffs of the Cleveland Museum of Natural History, National Museums of Kenya, Transvaal Museum, National Museums of Ethiopia, University of the Witswatersrand, and Institute of Human Origins for access to fossil, comparative skeletal, and cadaver specimens. My research was supported by NSF SBR 9601025 and the University of Missouri Research Council and Research Board.

REFERENCES

Aiello, L. C., and C. Dean. 1990. *An Introduction to Human Evolutionary Anatomy*. London: Academic Press.

Asfaw, B., T. D. White, O. Lovejoy, B. Latimer, S. Simpson, and G. Suwa. 1999. "*Australopithecus garhi*: A New Species of Early Hominid from Ethiopia." *Science* 284 (5414): 629–35. http://dx.doi.org/10.1126/science.284.5414.629. Medline:10213683

Baker, E. W., A. A. Malyango, and T. Harrison. 1998. "Phylogenetic Relationships and Functional Morphology of the Distal Humerus from Kanapoi, Kenya." *American Journal of Physical Anthropology* 26 (Suppl. 66)

Begun, D. R. 1993. "New Catarrhine Phalanges from Rudabanya (Northeastern Hungary) and the Problem of Parallelism and Convergence in Hominoid Postcranial Morphology." *Journal of Human Evolution* 24 (5): 373–402. http://dx.doi.org/10.1006/jhev.1993.1028.

Begun, D. R. 1994. "Relations among the Great Apes and Humans: New Interpretations Based on the Fossil Great Ape *Dryopithecus*." *Yearbook of Physical Anthropology* 37 (S19): 11–63. http://dx.doi.org/10.1002/ajpa.1330370604.

Begun, D. R., and L. Kordos. 1997. "Phyletic Affinities and Functional Convergence in *Dryopithecus* and Other Miocene and Living Hominids." In *Function, Phylogeny, and Fossils: Miocene Hominoid Evolution and Adaptations*, ed. D. R. Begun, C. V. Ward, and M. D. Rose, 291–316. New York: Plenum Press.

Begun, D. R., C. V. Ward, and M. D. Rose. 1997a. "Events in Hominoid Evolution." In *Function, Phylogeny, and Fossils: Miocene Hominoid Evolution and Adaptations*, ed. D. R. Begun, C. V. Ward, and M. D. Rose, 389–415. New York: Plenum Press.

Begun, D. R., C. V. Ward, and M. D. Rose. 1997b. *Function, Phylogeny and Fossils: Miocene Hominoid Evolution and Adaptations*. New York: Plenum Press.

Benton, R. 1965. "Morphological Evidence for Adaptations within the Epaxial Region of the Primates." In *The Baboon in Medical Research*, ed. H. Vagtborg, 10–20. Austin: University of Texas Press.

Benton, R. S. 1976. "Structural Patterns in the Pongidae and Cercopithecidae." *Yearbook of Physical Anthropology* 18:65–88.

Berge, C. 1991. "Quelle est la signification fonctionnelle du pelvis très large d'*Australopithecus afarensis* (AL 288–1)?" In *Origine(s) de la Bipédie chez les Hominidés*, ed. B. Senut and Y. Coppens, 113–120. Paris: CNRS.

Berge, C. 1994. "How Did the Australopithecines Walk? A Biomechanical Study of the Hip and Thigh of *Australopithecus afarensis*." *Journal of Human Evolution* 26 (4): 259–73. http://dx.doi.org/10.1006/jhev.1994.1016.

Brain, C. K., E. S. Vrba, and J. T. Robinson. 1974. "A New Hominid Innominate Bone from Swartkrans." *Annals of the Transvaal Museum* 29:55–66.

Bramble, D. M., and D. E. Lieberman. 2004. "Endurance Running and the Evolution of Homo." *Nature* 432 (7015): 345–52. http://dx.doi.org/10.1038/nature03052. Medline:15549097

Broom, R., and J. T. Robinson. 1949. "A New Type of Fossil Man." *Nature* 164 (4164): 322–3. http://dx.doi.org/10.1038/164322a0. Medline:18137042

Brunet, M., F. Guy, D. Pilbeam, H. T. Mackaye, A. Likius, D. Ahounta, A. Beauvilain, C. Blondel, H. Bocherens, J.-R. Boisserie, et al. 2002. "A New Hominid from the Upper Miocene of Chad, Central Africa." *Nature* 418: 145–51. Medline:12097880.

Bush, M. E., C. O. Lovejoy, D. C. Johanson, and Y. Coppens. 1982. "Hominid Carpal, Metacarpal and Phalangeal Bones Recovered from the Hadar Formation: 1974–1977 Collections." *American Journal of Physical Anthropology* 57 (4): 651–78. http://dx.doi.org/10.1002/ajpa.1330570410.

Carter, D. R., and G. S. Beaupré. 2001. *Skeletal Function and Form: Mechanobiology of Skeletal Development, Aging and Regeneration*. Cambridge: Cambridge University Press.

Cartmill, M., and D. Schmitt. 1996. "Pelvic Rotation in Human Walking and Running: Implications for Early Hominid Bipedalism." *American Journal of Physical Anthropology* Suppl. 22: 81.

Chan, L.-K. 1997. "Thoracic Shape and Shoulder Biomechanics in Primates." PhD dissertation, Duke University, Durham, NC.

Churchill, S. E. 1996. "Particulate versus Integrated Evolution of the Upper Body in Late Pleistocene Humans: A Test of Two Models." *American Journal of Physical Anthropology* 100 (4): 559–83. http://dx.doi.org/10.1002/(SICI)1096-8644(199608)100:4<559::AID-AJPA9>3.0.CO;2-L. Medline:8842328

Clarke, R. J., and P. V. Tobias. 1995. "Sterkfontein Member 2 Foot Bones of the Oldest South African Hominid." *Science* 269 (5223): 521–4. http://dx.doi.org/10.1126/science.7624772. Medline:7624772

Coffing, K. E. 1998. "The Metacarpals of *Australopithecus afarensis*: Locomotor and Behavioral Implications of Cross-Sectional Geometry." PhD dissertation, Johns Hopkins University, Baltimore.

Crompton, R. H., L. Yu, W. Weijie, M. Günther, and R. Savage. 1998. "The Mechanical Effectiveness of Erect and "Bent-Hip, Bent-Knee" Bipedal Walking in *Australopithecus afarensis*." *Journal of Human Evolution* 35 (1): 55–74. http://dx.doi.org/10.1006/jhev.1998.0222. Medline:9680467

Dainton, M., and G. A. Macho. 1999. "Did Knuckle Walking Evolve Twice?" *Journal of Human Evolution* 36 (2): 171–94. http://dx.doi.org/10.1006/jhev.1998.0265. Medline:10068065

Dart, R. A. 1925. "*Australopithecus africanus:* The Man-Ape of South Africa." *Nature* 115 (2884): 195–9. http://dx.doi.org/10.1038/115195a0.
Day, M. H., and J. L. Scheuer. 1973. "SKW 14147: A New Hominid Metacarpal from Swartkrans." *Journal of Human Evolution* 2 (6): 429–38. http://dx.doi.org/10.1016/0047-2484(73)90121-8.
Day, M. H., and E. H. Wickens. 1980. "Laetoli Pliocene Hominid Footprints and Bipedalism." *Nature* 286 (5771): 385–7. http://dx.doi.org/10.1038/286385a0.
Deloison, Y. 1985. "Comparative Study of Calcanei of Primates and *Pan-Australopithecus-Homo* Relationships." In *Hominid Evolution: Past, Present and Future*, ed. P. V. Tobias, 143–147. New York: Alan R. Liss.
Deloison, Y. 1991. "Les Australopithèques Marchaient-ils Comme Nous?" In *Origine(s) de la Bipédie chez les Hominidés*, ed. Y. Coppens and B. Senut, 177–186. Paris: CNRS.
Deloison, Y. 1992. "Articulation cunéométatarsienne de l'hallux considérée comme un des éléments déterminants de la forme de locomotion à partir de son anatomie osseuse: Comparison entre l'australopithèque, l'homme et le chimpanzé." *Comptes rendus de l'Académie des Sciences Paris*, Sèr. II, 314: 1379–85.
DeSilva, J. M. 2009. "Functional Morphology of the Ankle and the Likelihood of Climbing in Early Hominins." *Proceedings of the National Academy of Sciences of the United States of America* 106 (16): 6567–72. http://dx.doi.org/10.1073/pnas.0900270106. Medline:19365068
DeSilva, J. M. 2010. "Revisiting the 'Midtarsal Break.'" *American Journal of Physical Anthropology* 141 (2): 245–58. Medline:19672845
Drapeau, M. S. 2001. "Functional Analysis of the Associated Partial Forelimb Skeleton from Hadar, Ethiopia (AL 438–1)." PhD Dissertation, University of Missouri, Columbia.
Drapeau, M. S., C. V. Ward, W. H. Kimbel, D. C. Johanson, and Y. Rak. 2005. "Associated Cranial and Forelimb Remains Attributed to *Australopithecus afarensis* from Hadar, Ethiopia." *Journal of Human Evolution* 48 (6): 593–642. http://dx.doi.org/10.1016/j.jhevol.2005.02.005. Medline:15927662
Duncan, A. S., J. Kappelman, and L. J. Shapiro. 1994. "Metatarsophalangeal Joint Function and Positional Behavior in *Australopithecus afarensis*." *American Journal of Physical Anthropology* 93 (1): 67–81. http://dx.doi.org/10.1002/ajpa.1330930105. Medline:8141243
Duren, D. L. 1999. "Developmental Determinants of Femoral Morphology." MA thesis, Kent State University, Kent, OH.
Duren, D. L. 2001. "Physeal Orientation, Form and Function: Relationships with Primate Locomotor Behavior." PhD Dissertation, Kent State University, Kent, OH.
Duren, D .L., and C. V. Ward. 1995. "Femoral Physeal Plate Angles and Their Relationship to Bipedality." *American Journal of Physical Anthropology* Suppl. 20:86.
Evans, F. G., and V. E. Krahl. 1945. "The Torsion of the Humerus: A Phylogenetic Survey from Fish to Man." *American Journal of Anatomy* 76 (3): 303–37. http://dx.doi.org/10.1002/aja.1000760303.
Feldesman, M. R. 1982. "Morphometric Analysis of the Distal Humerus of Some Cenozoic Catarrhines: the Late Divergence Hypothesis Revisited." *American Journal of Physical Anthropology* 59 (1): 73–95. http://dx.doi.org/10.1002/ajpa.1330590108. Medline:6814259

Flinn, M. V., D. C. Geary, and C. V. Ward. 2005. "Ecological Dominance, Social Competition, and Coalitionary Arms Races: Why Humans Evolved Extraordinary Intelligence." *Evolution and Human Behavior* 26 (1): 10–46. http://dx.doi.org/10.1016/j.evolhumbehav.2004.08.005.

Frost, H. M. 1990a. "Skeletal Structural Adaptations to Mechanical Usage (SATMU): 1. Redefining Wolff's Law: The Bone Modeling Problem." *Anatomical Record* 226 (4): 403–13. http://dx.doi.org/10.1002/ar.1092260402. Medline:2184695

Gebo, D. L., L. MacLatchy, R. Kityo, A. Deino, J. Kingston, and D. Pilbeam. 1997. "A Hominoid Genus from the Early Miocene of Uganda." *Science* 276 (5311): 401–4. http://dx.doi.org/10.1126/science.276.5311.401. Medline:9103195

Grausz, H. M., R. E. Leakey, A. Walker, and C. V. Ward. 1998. "Associated Cranial and Post-Cranial Bones of *Australopithecus boisei*." In *Evolutionary History of the Robust Australopithecines*, ed. F. E. Grine, 127–32. Chicago: Aldine.

Grausz, H. M., R. E. Leakey, A. Walker, and C. V. Ward. 2008. "Associated cranial and post-cranial bones of *Australopithecus boisei*." In *Evolutionary History of the Robust Australopithecines*. F. E. Grine (ed). pp. 127-132. Aldine, Chicago.

Haile-Selassie, Y. 2001. "Late Miocene Hominids from the Middle Awash, Ethiopia." *Nature* 412 (6843): 178–81. http://dx.doi.org/10.1038/35084063. Medline:11449272

Haile-Selassie, Y. 2010. "Phylogeny of Early *Australopithecus*: New Fossil Evidence from the Woranso-Mille (Central Afar, Ethiopia)." *Philosophical Transactions of the Royal Society of London* 365 (1556): 3323–31. http://dx.doi.org/10.1098/rstb.2010.0064. Medline:20855306

Haile-Selassie, Y., B. M. Latimer, M. Alene, A. L. Deino, L. Gibert, S. M. Melillo, B. Z. Saylor, G. R. Scott, and C. O. Lovejoy. 2010a. "An Early *Australopithecus Afarensis* Postcranium from Woranso-Mille, Ethiopia." *Proceedings of the National Academy of Sciences of the United States of America* 107 (27): 12121–6. http://dx.doi.org/10.1073/pnas.1004527107. Medline:20566837

Haile-Selassie, Y., B. Z. Saylor, A. Deino, M. Alene, and B. M. Latimer. 2010b. "New Hominid Fossils from Woranso-Mille (Central Afar, Ethiopia) and Taxonomy of Early *Australopithecus*." *American Journal of Physical Anthropology* 141 (3): 406–17. Medline:19918995

Hamrick, M. W. 1999. "A Chondral Modeling Theory Revisited." *Journal of Theoretical Biology* 201 (3): 201–8. http://dx.doi.org/10.1006/jtbi.1999.1025. Medline:10600363

Harcourt-Smith, W.E.H., and L. C. Aiello. 2004. "Fossils, Feet and the Evolution of Human Bipedal Locomotion." *Journal of Anatomy* 204 (5): 403–16. http://dx.doi.org/10.1111/j.0021-8782.2004.00296.x. Medline:15198703

Harmon, E. H. 2008. "Early Hominin Greater Trochanter Shape." *American Journal of Physical Anthropology* 135 (Suppl.): 112–3.

Harrison, T. 1986. "A Reassessment of the Phylogenetic Relationships of *Oreopithecus Bambolii* Gervais." *Journal of Human Evolution* 15:541–83. http://dx.doi.org/10.1016/S0047-2484(86)80073-2.

Häusler, M. F. 2001. "New Insights into the Locomotion of *Australopithecus africanus:* Implications of the Partial Skeleton STW 431 (Sterkfontein, South Africa)." PhD dissertation, Universität Zürich, Zurich.

Heinrich, R. E., M. D. Rose, R. E. Leakey, and A. C. Walker. 1993. "Hominid Radius from the Middle Pliocene of Lake Turkana, Kenya." *American Journal of Physical Anthropology* 92 (2): 139–48. http://dx.doi.org/10.1002/ajpa.1330920203. Medline:8273826

Hill, A., and S. Ward. 1988. "Origin of the Hominidae: The Record of African Large Hominoid Evolution between 14 My and 4 My." *Journal of Human Evolution* 31:49–83.

Hunt, K. D. 1994. "The Evolution of Hominid Bipedality: Ecology and Functional Morphology." *Journal of Human Evolution* 26 (3): 183–202. http://dx.doi.org/10.1006/jhev.1994.1011.

Hürzeler, J. 1960. "The Significance of *Oreopithecus* in the Genealogy of Man." *Triangle* 4:164–74. Medline:14405578

Inouye, S. E. 1994. "Ontogeny of Knuckle-Walking Hand Postures in African Apes." *Journal of Human Evolution* 26 (5-6): 459–85. http://dx.doi.org/10.1006/jhev.1994.1028.

Johanson, D. C., C. O. Lovejoy, W. H. Kimbel, T. D. White, S. C. Ward, M. E. Bush, B. M. Latimer, and Y. Coppens. 1982. "Morphology of the Pliocene Partial Hominid Skeleton (A. L. 288–1) from the Hadar Formation, Ethiopia." *American Journal of Physical Anthropology* 57 (4): 403–51. http://dx.doi.org/10.1002/ajpa.1330570403.

Judd, A. 2005. "Humeral Torsion and Thoracic Shape in Anthropoid Primates." MA thesis, University of Missouri, Columbia.

Jungers, W. 1984. "Aspects of Size and Scaling in Primate Biology with Special Reference to the Locomotor Skeleton." *Yearbook of Physical Anthropology* 27 (S5): 73–97. http://dx.doi.org/10.1002/ajpa.1330270505.

Jungers, W. L. 1982. "Lucy's Limbs: Skeletal Allometry and Locomotion in *Australopithecus afarensis*." *Nature* 297 (5868): 676–8. http://dx.doi.org/10.1038/297676a0.

Jungers, W. L. 1991. "A Pygmy Perspective on Body Size and Shape in *Australopithecus afarensis* (A. L. 288–1, "Lucy")." In *Origine(s) de la Bipédie Chez les Hominidés*, ed. Y. Coppens and B. Senut, 215–224. Paris: CNRS.

Jungers, W. L., and J. T. Stern, Jr. 1983. "Body Proportions, Skeletal Allometry and Locomotion in the Hadar Hominids: A Reply to Wolpoff." *Journal of Human Evolution* 12 (7): 673–84. http://dx.doi.org/10.1016/S0047-2484(83)80007-4.

Kay, R. F., and P. S. Ungar. 1997. "Dental Evidence for Diet in Some Miocene Catarrhines with Comments on the Effects of Phylogeny on the Interpretation of Adaptation." In *Function, Phylogeny, and Fossils: Miocene Hominoid Evolution and Adaptations*, ed. D. R. Begun, C. V. Ward, and M. D. Rose, 131–151. New York: Plenum Press.

Kimbel, W. H., D. C. Johanson, and Y. Rak. 1994. "The First Skull and Other New Discoveries of *Australopithecus afarensis* at Hadar, Ethiopia." *Nature* 368 (6470): 449–51. http://dx.doi.org/10.1038/368449a0. Medline:8133889

Kimbel, W. H., C. A. Lockwood, C. V. Ward, M. G. Leakey, Y. Rak, and D. C. Johanson. 2006. "Was *Australopithecus anamensis* Ancestral to *A. afarensis*? A Case of Anagenesis in the Hominin Fossil Record." *Journal of Human Evolution* 51 (2): 134–52. http://dx.doi.org/10.1016/j.jhevol.2006.02.003. Medline:16630646

Kivell, T. L., and D. Schmitt. 2009. "Independent Evolution of Knuckle-Walking in African Apes Shows that Humans Did Not Evolve from a Knuckle-Walking Ancestor."

Proceedings of the National Academy of Sciences of the United States of America 106 (34): 14241–6. http://dx.doi.org/10.1073/pnas.0901280106. Medline:19667206

Klein, R. G. 2001. *The Human Career*. 2nd ed. Chicago: University of Chicago Press.

Kramer, P. A. 1999. "Modelling the Locomotor Energetics of Extinct Hominids." *Journal of Experimental Biology* 202 (Pt 20): 2807–18. Medline:10504316

Kramer, P. A., and G. G. Eck. 2000. "Locomotor Energetics and Leg Length in Hominid Bipedality." *Journal of Human Evolution* 38 (5): 651–66. http://dx.doi.org/10.1006/jhev.1999.0375. Medline:10799258

Lague, M. R., and W. L. Jungers. 1996. "Morphometric Variation in Plio-Pleistocene Hominid Distal Humeri." *American Journal of Physical Anthropology* 101 (3): 401–27. http://dx.doi.org/10.1002/(SICI)1096-8644(199611)101:3<401::AID-AJPA8>3.0.CO;2-0. Medline:8922185

Larson, S. G. 1996. "Estimating Humeral Torsion on Incomplete Fossil Anthropoid Humeri." *Journal of Human Evolution* 31 (3): 239–57. http://dx.doi.org/10.1006/jhev.1996.0059.

Larson, S. G. 1998. "Parallel Evolution in the Hominoid Trunk and Forelimb." *Evolutionary Anthropology* 6 (3): 87–99. http://dx.doi.org/10.1002/(SICI)1520-6505(1998)6:3<87::AID-EVAN3>3.0.CO;2-T.

Latimer, B. 1983. "The Anterior Foot Skeleton of *Australopithecus afarensis*." *American Journal of Physical Anthropology* 60:217.

Latimer, B. 1991. "Locomotor Adaptations in *Australopithecus afarensis:* The Issue of Arboreality." In *Origine(s) de la Bipédie Chez les Hominidés*, ed. Y. Coppens and B. Senut, 169–176. Paris: CNRS.

Latimer, B., and C. O. Lovejoy. 1989. "The Calcaneus of *Australopithecus afarensis* and Its Implications for the Evolution of Bipedality." *American Journal of Physical Anthropology* 78 (3): 369–86. http://dx.doi.org/10.1002/ajpa.1330780306. Medline:2929741

Latimer, B., and C. O. Lovejoy. 1990a. "Metatarsophalangeal Joints of *Australopithecus afarensis*." *American Journal of Physical Anthropology* 83 (1): 13–23. http://dx.doi.org/10.1002/ajpa.1330830103. Medline:2221027

Latimer, B., and C. O. Lovejoy. 1990b. "Hallucal Tarsometatarsal Joint in *Australopithecus afarensis*." *American Journal of Physical Anthropology* 82 (2): 125–33. http://dx.doi.org/10.1002/ajpa.1330820202. Medline:2360609

Latimer, B., J. C. Ohman, and C. O. Lovejoy. 1987. "Talocrural Joint in African Hominoids: Implications for *Australopithecus afarensis*." *American Journal of Physical Anthropology* 74 (2): 155–75. http://dx.doi.org/10.1002/ajpa.1330740204. Medline:3122581

Lauder, G. V. 1996. "The Argument from Design." In *Adaptation*, ed. G. V. Lauder and M. R. Rose, 55–92. San Diego: Academic Press.

Le Gros Clark, W. E., and L.S.B. Leakey. 1951. *The Miocene Hominoidea of East Africa*. Fossil Mammals of Africa, No. 1. London: British Museum of Natural History.

Le Gros Clark, W. E., and D. P. Thomas. 1951. "Associated Jaws and Limb Bones of *Limnopitheucs macinnesi*." *Fossil Mammals of Africa* 3:1–27.

Leakey, L.S.B. 1960. "Recent Discoveries at Olduvai Gorge." *Nature* 188 (4755): 1050–2. http://dx.doi.org/10.1038/1881050a0.

Leakey, L.S.B. 1961. "New Finds at Olduvai Gorge." *Nature* 189 (4765): 649–50. http://dx.doi.org/10.1038/189649a0. Medline:13759921

Leakey, L.S.B., P. V. Tobias, and J. R. Napier. 1964. "A New Species of the Genus Homo from Olduvai Gorge." *Nature* 202 (4927): 7–9. http://dx.doi.org/10.1038/202007a0. Medline:14166722

Leakey, M. D., and R. L. Hay. 1979. "Pliocene Footprints in the Laetolil Beds at Laetoli, Northern Tanzania." *Nature* 278 (5702): 317–23. http://dx.doi.org/10.1038/278317a0.

Leakey, M. G., C. S. Feibel, I. McDougall, and A. Walker. 1995. "New Four-Million-Year-Old Hominid Species from Kanapoi and Allia Bay, Kenya." *Nature* 376 (6541): 565–71. http://dx.doi.org/10.1038/376565a0. Medline:7637803

Leakey, M. G., C. S. Feibel, I. McDougall, C. V. Ward, and A. Walker. 1998. "New Specimens and Confirmation of an Early Age for *Australopithecus anamensis*." *Nature* 393 (6680): 62–6. http://dx.doi.org/10.1038/29972. Medline:9590689

Leakey, R. E., M. G. Leakey, and A. C. Walker. 1988. "Morphology of *Afropithecus turkanensis* from Kenya." *American Journal of Physical Anthropology* 76 (3): 289–307. http://dx.doi.org/10.1002/ajpa.1330760303. Medline:3137824

Leutenegger, W., and J. T. Kelly. 1977. "Relationship of Sexual Dimorphism in Canine Size and Body Size to Social, Behavioral, and Ecological Correlates in Anthropoid Primates." *Primates* 18 (1): 117–36. http://dx.doi.org/10.1007/BF02382954.

Lieberman, D .E. 1997. "Making Behavioral and Phylogenetic Inferences from Hominid Fossils: Considering the Developmental Influence of Mechanical Forces." *Annual Review of Anthropology* 26 (1): 185–210. http://dx.doi.org/10.1146/annurev.anthro.26.1.185.

Lovejoy, C. O. 1975. "Biomechanical Perspectives on the Lower Limb of Early Hominids." In *Primate Functional Morphology and Evolution*, ed. R.H. Tuttle, 291–306. Paris: Mouton. http://dx.doi.org/10.1515/9783110803808.291

Lovejoy, C. O. 1978. "A Biomechanical Review of the Locomotor Diversity of Early Hominids." In *Early Hominids of Africa*, ed. C. J. Jolly, 403–39. New York: St. Martin's Press.

Lovejoy, C. O. 1988. "Evolution of Human Walking." *Scientific American* 259 (5): 118–25. http://dx.doi.org/10.1038/scientificamerican1188-118. Medline:3212438

Lovejoy, C. O. 2005. "The Natural History of Human Gait and Posture. Part 1. Spine and Pelvis." *Gait & Posture* 21 (1): 95–112. http://dx.doi.org/10.1016/S0966-6362(04)00014-1. Medline:15536039

Lovejoy, C. O., M. J. Cohn, and T. D. White. 1999. "Morphological Analysis of the Mammalian Postcranium: A Developmental Perspective." *Proceedings of the National Academy of Sciences of the United States of America* 96 (23): 13247–52. http://dx.doi.org/10.1073/pnas.96.23.13247. Medline:10557306

Lovejoy, C. O., M J. Cohn, and T. D. White. 2000. "The Evolution of Mammalian Morphology: A Developmental Perspective." In *Development, Growth and Evolution: Implications for the Study of the Hominid Skeleton*, ed. P. O'Higgens and M. J. Cohn, 41–56. San Diego: Academic Press.

Lovejoy, C. O., K. G. Heiple, and A. H. Burstein. 1973. "The Gait of *Australopithecus*." *American Journal of Physical Anthropology* 38 (3): 757–79. http://dx.doi.org/10.1002/ajpa.1330380315. Medline:4735528

Lovejoy, C. O., B. Latimer, G. Suwa, B. Asfaw, and T. D. White. 2009a. "Combining Prehension and Propulsion: The Foot of *Ardipithecus ramidus*." *Science* 326 (5949): 72e1–8. http://dx.doi.org/10.1126/science.1175832. Medline:19810198

Lovejoy, C. O., and M. A. McCollum. 2010. "Spinopelvic Pathways to Bipedality: Why No Hominids Ever Relied on a Bent-Hip, Bent-Knee Gait." *Philosophical Transactions of the Royal Society of London* 365 (1556): 3289–99. http://dx.doi.org/10.1098/rstb.2010.0112. Medline:20855303

Lovejoy, C. O., S. W. Simpson, T. D. White, B. Asfaw, and G. Suwa. 2009b. "Careful Climbing in the Miocene: The Forelimbs of *Ardipithecus ramidus* and Humans Are Primitive." *Science* 326 (5949): 70e1–8, e78. http://dx.doi.org/10.1126/science.1175827. Medline:19810196

Lovejoy, C. O., G. Suwa, S. W. Simpson, J. H. Matternes, and T. D. White. 2009c. "The Great Divides: *Ardipithecus ramidus* Reveals the Postcrania of Our Last Common Ancestors with African Apes." *Science* 326 (5949): 100–6. http://dx.doi.org/10.1126/science.1175833. Medline:19810199

Lovejoy, C. O., G. Suwa, L. Spurlock, B. Asfaw, and T. D. White. 2009d. "The Pelvis and Femur of *Ardipithecus ramidus*: The Emergence of Upright Walking." *Science* 326 (5949): 72e1–6. http://dx.doi.org/10.1126/science.1175831. Medline:19810197

MacLatchy, L. M. 1996. "Another Look at the Australopithecine Hip." *Journal of Human Evolution* 31 (5): 455–76. http://dx.doi.org/10.1006/jhev.1996.0071.

MacLatchy, L. M., D. Gebo, R. Kityo, and D. Pilbeam. 2000. "Postcranial Functional Morphology of *Morotopithecus bishopi*, with Implications for the Evolution of Modern Ape Locomotion." *Journal of Human Evolution* 39 (2): 159–83. http://dx.doi.org/10.1006/jhev.2000.0407. Medline:10968927

Madar, S. 1994. "Humeral Shaft Morphology of *Sivapithecus*." *American Journal of Physical Anthropology* Suppl. 20:140.

Martin, R., and K. Saller. 1959. *Lehrbuch der Anthropologie*, vol. 2. Stuttgart: Gustav Fischer.

Martin, R. B., D. B. Burr, and N. A. Sharkey. 1998. *Skeletal Tissue Mechanics*. New York: Springer-Verlag.

Marzke, M. W. 1983. "Joint Functions and Grips of the *Australopithecus afarensis* Hand, with Special Reference to the Region of the Capitate." *Journal of Human Evolution* 12 (2): 197–211. http://dx.doi.org/10.1016/S0047-2484(83)80025-6.

McHenry, H. M. 1975. "A New Pelvic Fragment from Swartkrans and the Relationship between the Robust and Gracile Australopithecines." *American Journal of Physical Anthropology* 43 (2): 245–61. http://dx.doi.org/10.1002/ajpa.1330430211. Medline:810037

McHenry, H. M. 1991. "Sexual Dimorphism in *Australopithecus afarensis*." *Journal of Human Evolution* 20 (1): 21–32. http://dx.doi.org/10.1016/0047-2484(91)90043-U.

McHenry, H. M. 1992. "Body Size and Proportions in Early Hominids." *American Journal of Physical Anthropology* 87 (4): 407–31. http://dx.doi.org/10.1002/ajpa.1330870404. Medline:1580350

McHenry, H. M. 1994. "Early Hominid Postcrania: Phylogeny and Function." In *Integrative Paths to the Past: Paleoanthropological Advances in Honor of F. Clark Howell*, ed. R. S. Corriccini and R. L. Ciochon, 251–268. Englewood Cliffs, NJ: Prentice Hall.

McHenry, H. M., and L. R. Berger. 1998. "Body Proportions in *Australopithecus afarensis* and *A. africanus* and the Origin of the Genus Homo." *Journal of Human Evolution* 35 (1): 1–22. http://dx.doi.org/10.1006/jhev.1997.0197. Medline:9680464

McKee, J. K. 1993. "Faunal Dating of the Taung Hominid Fossil Deposit." *Journal of Human Evolution* 25 (5): 363–76. http://dx.doi.org/10.1006/jhev.1993.1055.

McKee, J. K., J. F. Thackeray, and L. R. Berger. 1995. "Faunal Assemblage Seriation of Southern African Pliocene and Pleistocene Fossil Deposits." *American Journal of Physical Anthropology* 96 (3): 235–50. http://dx.doi.org/10.1002/ajpa.1330960303. Medline:7785723

Moyà-Solà, S., and M. Köhler. 1996. "A *Dryopithecus* Skeleton and the Origins of Great-Ape Locomotion." *Nature* 379 (6561): 156–9. http://dx.doi.org/10.1038/379156a0. Medline:8538764

Moyà-Solà, S., M. Köhler, and L. Rook. 1999. "Evidence of Hominid-Like Precision Grip Capability in the Hand of the Miocene Ape Oreopithecus." *Proceedings of the National Academy of Sciences of the United States of America* 96 (1): 313–7. http://dx.doi.org/10.1073/pnas.96.1.313. Medline:9874815

Nakatsukasa, M., Y. Kunimatsu, D. Shimizu, and H. Ishida. 2000. "A New Skeleton of the Large Hominoid from Nachola, Northern Kenya." *American Journal of Physical Anthropology* Suppl. 30:235.

Nakatsukasa, M., A. Yamanaka, Y. Kunimatsu, D. Shimizu, and H. Ishida. 1998. "A Newly Discovered *Kenyapithecus* Skeleton and Its Implications for the Evolution of Positional Behavior in Miocene East African Hominoids." *Journal of Human Evolution* 34 (6): 657–64. http://dx.doi.org/10.1006/jhev.1998.0228. Medline:9650105

Napier, J. R. 1959. *Fossil Metacarpals from Swartkrans*. Fossil Mammals of Africa, No. 17. London: British Museum of Natural History.

Napier, J. R. 1964. "The Evolution of Bipedal Walking in the Hominids." *Archives de Biologie* 77 (Suppl.): 673–708.

Napier, J. R., and P. Davis. 1959. "The Forelimb Skeleton and Associated Remains of *Proconsul africanus*." *Fossil Mammals of Africa* 16:1–70.

Ohman, J. C. 1986. "The First Rib of Hominoids." *American Journal of Physical Anthropology* 70 (2): 209–29. http://dx.doi.org/10.1002/ajpa.1330700208. Medline:3090892

Paciulli, L. M. 1995. "Ontogeny of Phalangeal Curvature and Positional Behavior in Chimpanzees." *American Journal of Physical Anthropology* Suppl. 20:165.

Patterson, B., and W. W. Howells. 1967. "Hominid Humeral Fragment from Early Pleistocene of Northwestern Kenya." *Science* 156 (3771): 64–6. http://dx.doi.org/10.1126/science.156.3771.64. Medline:6020039

Pilbeam, D., M. D. Rose, J. C. Barry, and S.M.I. Shah. 1990. "New *Sivapithecus* Humeri from Pakistan and the Relationship of *Sivapithecus* and *Pongo*." *Nature* 348 (6298): 237–9. http://dx.doi.org/10.1038/348237a0. Medline:2234091

Preuschoft, H., and H. Witte. 1991. "Biomechanical Reasons for the Evolution of Hominid Body Shape." In *Origine(s) de la Bipédie Chez les Hominidés*, ed. Y. Coppens and B. Senut, 59–77. Paris: CNRS.

Rak, Y. 1991. "Lucy's Pelvic Anatomy: Its Role in Bipedal Gait." *Journal of Human Evolution* 20 (4): 283–90. http://dx.doi.org/10.1016/0047-2484(91)90011-J.

Richmond, B. G. 1998. "Ontogeny and Biomechanics of Phalangeal Form in Primates." PhD dissertation, State University of New York, Stony Brook.

Richmond, B. G. 2003. "Early Hominin Locomotion and the Ontogeny of Phalangeal Curvature in Primates." *American Journal of Physical Anthropology* 36 (Suppl.): 178–179.

Richmond, B. G. 2007. "Biomechanics of Phalangeal Curvature." *Journal of Human Evolution* 53 (6): 678–90. http://dx.doi.org/10.1016/j.jhevol.2007.05.011. Medline:17761213

Richmond, B .G., D. R. Begun, and D. S. Strait. 2001. "Origin of Human Bipedalism: The Knuckle-Walking Hypothesis Revisited." *Yearbook of Physical Anthropology* 44 (Suppl 33): 70–105. http://dx.doi.org/10.1002/ajpa.10019. Medline:11786992

Richmond, B. G., and W .L. Jungers. 2008. "*Orrorin tugenensis* Femoral Morphology and the Evolution of Hominin Bipedalism." *Science* 319 (5870): 1662–5. http://dx.doi.org /10.1126/science.1154197. Medline:18356526

Ricklan, D. E. 1987. "Functional Anatomy of the Hand of *Australopithecus africanus*." *Journal of Human Evolution* 16 (7-8): 643–64. http://dx.doi.org/10.1016/0047-2484(87)90018-2.

Robinson, J. T. 1970. "Two New Early Hominid Vertebrae from Swartkrans." *Nature* 225 (5239): 1217–9. http://dx.doi.org/10.1038/2251217a0. Medline:5435351

Robinson, J. T. 1972. *Early Hominid Posture and Locomotion*. Chicago: University of Chicago Press.

Rose, M. D. 1984. "A Hominine Hip Bone, KNM-ER 3228, from East Lake Turkana, Kenya." *American Journal of Physical Anthropology* 63 (4): 371–8. http://dx.doi.org/10.1002 /ajpa.1330630404. Medline:6428239

Rose, M. D. 1986. "Further Hominoid Postcranial Specimens from the Late Miocene Nagri Formation of Pakistan." *Journal of Human Evolution* 15 (5): 333–67. http://dx.doi.org /10.1016/S0047-2484(86)80016-1.

Rose, M. D. 1991. "The Process of Bipedalization in Hominids." In *Origine(s) de la Bipédie Chez les Hominidés*, ed. Y. Coppens and B. Senut, 37–48. Paris: CNRS.

Rose, M. D. 1997. "Functional and Phylogenetic Features of the Forelimb in Miocene Hominoids." In *Function, Phylogeny, and Fossils: Miocene Hominoid Evolution and Adaptations*, ed. D .R. Begun, C. V. Ward, and M. D. Rose, 79–100. New York: Plenum Press.

Rose, M. D., Y. Nakano, and H. Ishida. 1996. "*Kenyapithecus* Postcranial Specimens from Nachola, Kenya." *African Study Monographs* Suppl. 24: 3–56.

Ruff, C. B. 1998. "Evolution of the Hominid Hip." In *Primate Locomotion: Recent Advances*, ed. E. Strasser, J. Fleagle, A. L. Rosenberger, and H. M. McHenry, 449–69. New York: Plenum Press.

Ruff, C. B., H. M. McHenry, and J. F. Thackeray. 1999. "Cross-Sectional Morphology of the SK 82 and 97 Proximal Femora." *American Journal of Physical Anthropology* 109 (4): 509–21.

http://dx.doi.org/10.1002/(SICI)1096-8644(199908)109:4<509::AID-AJPA7>3.0.CO;2-X. Medline:10423266

Ruff, C. B., E. Trinkaus, A. Walker, and C. S. Larsen. 1993. "Postcranial Robusticity in Homo. I: Temporal Trends and Mechanical Interpretation." *American Journal of Physical Anthropology* 91 (1): 21–53. http://dx.doi.org/10.1002/ajpa.1330910103. Medline:8512053

Sarmiento, E. E. 1987. "The Phylogenetic Position of *Oreopithecus* and Its Significance in the Origin of the Hominoidea." *American Museum Novitates* 2881:1–44.

Sarmiento, E. E. 1994. "Terrestrial Traits in the Hands and Feet of Gorillas." *American Museum Novitates* 3091:1–56.

Sarmiento, E. E. 1998. "Generalized Quadrupeds, Committed Bipeds and the Shift to Open Habitats: An Evolutionary Model of Hominid Divergence." *American Museum Novitates* 3250:1–78.

Schmid, P. 1983. "Eine Rekonstruktion des Skelettes von A.L. 288–1 (Hadar) und deren Konsequenzen." *Folia Primatologica* 40 (4): 283–306. http://dx.doi.org/10.1159/000156111.

Schmitt, D., P. Lemelin, and A. C. Trueblood. 1999. "Shock Wave Transmission through the Human Body during Normal and Compliant Walking." *American Journal of Physical Anthropology* Suppl. 28:243–4.

Schmitt, D., J. T. Stern, and S. G. Larson. 1996. "Compliant Gait in Humans: Implications for Substrate Reaction Forces during Australopith Bipedalism." *American Journal of Physical Anthropology* Suppl. 22:209.

Senut, B. 1980. "New Data on the Humerus and Its Joints in Plio-Pleistocene Hominids." *Collegium Anthropologicum* 1:87–93.

Senut, B., M. Pickford, D. Gommery, P. Mein, K. Cheboi, and Y. Coppens. 2001. "First Hominid from the Miocene (Lukeino Formation, Kenya)." *Comptes rendus de l'Académie des Sciences Paris* 332:137–44.

Senut, B., and C. Tardieu. 1985. "Functional Aspects of Plio-Pleistocene Hominid Limb Bones: Implications for Taxonomy and Phylogeny." In *Ancestors: The Hard Evidence*, ed. E. Delson, 193–201. New York: Alan R. Liss.

Simpson, G. G. 1953. *The Major Features of Evolution*. New York: Columbia University Press.

Spoor, C. F., P. Y. Sondaar, and S. T. Hussain. 1991. "A New Hominoid Hamate and First Metacarpal from the Late Miocene Nagri Formation of Pakistan." *Journal of Human Evolution* 21 (6): 413–24. http://dx.doi.org/10.1016/0047-2484(91)90092-A.

Stern, J. T. 2000. "Climbing to the Top: A Personal Memoir of *Australopithecus afarensis*." *Evolutionary Anthropology* 9 (3): 113–33. http://dx.doi.org/10.1002/1520-6505(2000)9:3<113::AID-EVAN2>3.0.CO;2-W.

Stern, J. T., Jr. 1999. "The Cost of Bent-Knee, Bent-Hip Bipedal Gait. A Reply to Crompton et al." *Journal of Human Evolution* 36 (5): 567–70. http://dx.doi.org/10.1006/jhev.1999.0290. Medline:10222170

Stern, J. T., and R. L. Susman. 1981. "Electromyography of the Gluteal Muscles in *Hylobates, Pongo* and *Pan*: Implications for the Evolution of Hominid Bipedality." *American Journal of Physical Anthropology* 55 (2): 153–66. http://dx.doi.org/10.1002/ajpa.1330550203.

Stern, J. T., Jr., and R. L. Susman. 1983. "The Locomotor Anatomy of *Australopithecus afarensis*." *American Journal of Physical Anthropology* 60 (3): 279–317. http://dx.doi.org/10.1002/ajpa.1330600302. Medline:6405621

Stern, J. T., and R. L. Susman. 1991. "'Total Morphological Pattern' versus the 'Magic Trait': Conflicting Approaches to the Study of Early Hominid Bipedalism." In *Origine(s) de la Bipédie Chez les Hominidés*, ed. Y. Coppens and B. Senut, 99–112. Paris: CNRS.

Susman, R. L. 1988. "Hand of *Paranthropus robustus* from Member 1, Swartkrans: Fossil Evidence for Tool Behavior." *Science* 240 (4853): 781–4. http://dx.doi.org/10.1126/science.3129783. Medline:3129783

Susman, R. L. 1989. "New Hominid Fossils from the Swartkrans Formation (1979-1986 Excavations): Postcranial Specimens." *American Journal of Physical Anthropology* 79 (4): 451–74. http://dx.doi.org/10.1002/ajpa.1330790403. Medline:2672829

Susman, R. L., and A. B. Demes. 1994. "Relative Foot Length in *Australopithecus afarensis* and Its Implications for Bipedality." *American Journal of Physical Anthropology* 62:187–98.

Susman, R L., D. de Ruiter, and C. K. Brain. 2001. "Recently Identified Postcranial Specimens of *Paranthropus* and Early *Homo* from Swartkrans Cave, South Africa." *American Journal of Physical Anthropology* 41:607–30.

Susman, R. L., and J. T. Stern. 1991. "Locomotor Behavior of Early Hominids: Epistemology and Fossil Evidence." In *Origine(s) de la Bipédie chez les Hominidés*, ed. Y. Coppens and B. Senut, 121–132. Paris: CNRS.

Susman, R. L., J. T. Stern, Jr., and W. L. Jungers. 1984. "Arboreality and Bipedality in the Hadar Hominids." *Folia Primatologica* 43 (2-3): 113–56. http://dx.doi.org/10.1159/000156176. Medline:6440837

Suwa, G., B. Asfaw, R. T. Kono, D. Kubo, C. O. Lovejoy, and T. D. White. 2009. "The *Ardipithecus ramidus* Skull and Its Implications for Hominid Origins." *Science* 326 (5949): 68e1–7, e7. http://dx.doi.org/10.1126/science.1175825. Medline:19810194

Tague, R. G., and C. O. Lovejoy. 1986. "The Obstetric Pelvis of A. L. 288–1 (Lucy)." *Journal of Human Evolution* 15 (4): 237–55. http://dx.doi.org/10.1016/S0047-2484(86)80052-5.

Tardieu, C. 1979. "Aspects bioméchaniques de l'articulation du genou chez les primates." *Bulletins de la Société Anatomique de Paris* 4:66–86.

Tardieu, C. 1986a. "The Knee Joint in Three Hominoid Primates: Application to Plio-Pleistocene Hominids and Evolutionary Implications." In *Current Perspectives in Primate Biology*, ed. D. M. Taub and T. A. King, 182–192. New York: Van Norstrand Reinhold.

Tardieu, C. 1986b. "Evolution of the Knee Intra-Articular Menisci in Primates and Some Hominids." In *Primate Evolution*, ed. J. G. Else and P. C. Lee, 183–190. Cambridge: Cambridge University Press.

Tardieu, C. 1998. "Short Adolescence in Early Hominids: Infantile and Adolescent Growth of the Human Femur." *American Journal of Physical Anthropology* 107 (2): 163–78. http://dx.doi.org/10.1002/(SICI)1096-8644(199810)107:2<163::AID-AJPA3>3.0.CO;2-W. Medline:9786331

Tardieu, C. 1991. "Etude comparative des déplacements du centre de gravité du corps pendant la marche par une nouvelle méthode d'analyse tridimensionnelle: Mise à l'épreuve d'une hypothèse évolutive." In *Origine(s) de la Bipédie Chez les Hominidés*, ed. Y. Coppens and B. Senut, 49–58. Paris: Editions du CNRS.

Tardieu, C., and H. Preuschoft. 1996. "Ontogeny of the Knee Joint in Humans, Great Apes and Fossil Hominids: Pelvi-Femoral Relationships during Postnatal Growth in Humans." *Folia Primatologica* 66 (1-4): 68–81. http://dx.doi.org/10.1159/000157186. Medline:8953751

Tardieu, C., and E. Trinkaus. 1994. "Early Ontogeny of the Human Femoral Bicondylar Angle." *American Journal of Physical Anthropology* 95 (2): 183–95. http://dx.doi.org/10.1002/ajpa.1330950206. Medline:7802095

Walker, A., and M. Pickford. 1983. "New Postcranial Fossils of *Proconsul africanus* and *Proconsul nyanzae*." In *New Interpretations of Ape and Human Ancestry*, ed. R. L. Ciochon and R. S. Corruccini, 325–351. New York: Plenum Press.

Walker, A., and M. D. Rose. 1968. "Fossil Hominoid Vertebra from the Miocene of Uganda." *Nature* 217 (5132): 980–1. http://dx.doi.org/10.1038/217980a0. Medline:5642859

Walker, A., and M. F. Teaford. 1988. "The Kaswanga Primate Site: An Early Miocene Hominoid Site on Rusinga Island, Kenya." *Journal of Human Evolution* 17 (5): 539–44. http://dx.doi.org/10.1016/0047-2484(88)90041-3.

Walter, R. C. 1994. "Age of Lucy and the First Family: Single-Crystal ^{40}Ar/^{39}Ar Dating of the Denen Dora and Lower Kada Hadar Members of the Hadar Formation, Ethiopia." *Geology* 22 (1): 6–10. http://dx.doi.org/10.1130/0091-7613(1994)022<0006:AOLATF>2.3.CO;2.

Ward, C. V. 1993. "Torso Morphology and Locomotion in Proconsul nyanzae." *American Journal of Physical Anthropology* 92 (3): 291–328. http://dx.doi.org/10.1002/ajpa.1330920306. Medline:8291620

Ward, C. V. 1997. "Functional Anatomy and Phyletic Implications of the Hominoid Trunk and Hindlimb." In *Function, Phylogeny, and Fossils: Miocene Hominoid Evolution and Adaptations*, ed. D. R. Begun, C. V. Ward, and M. D. Rose, 101–130. New York: Plenum Press.

Ward, C. V. 2002. "Early Hominid Posture and Locomotion: Where Do We Stand?" *Yearbook of Physical Anthropology* 45:185–215.

Ward, C. V., W. H. Kimbel, and D. C. Johanson. 2011. "Complete Fourth Metatarsal and Arches in the Foot of *Australopithecus afarensis*." *Science* 331 (6018): 750–3. http://dx.doi.org/10.1126/science.1201463. Medline:21311018.

Ward, C. V., W. K. Kimbel, D. C. Johanson, and Y. Rak. 2012. "New Postcranial Fossils Attributed to *Australopithecus afarensis* from Hadar, Ethiopia." *Journal of Human Evolution* 63:1–57.

Ward, C. V., and B. Latimer. 1991. "The Vertebral Column of *Australopithecus*." *American Journal of Physical Anthropology* Suppl. 12:180.

Ward, C. V., and B. Latimer. 2005. "Upright Posture and Vertebral Anatomy in *Australopithecus*." *American Journal of Physical Anthropology* S40:218.

Ward, C. V., M. G. Leakey, B. Brown, F. Brown, J. Harris, and A. Walker. 1999b. "South Turkwel: A New Pliocene Hominid Site in Kenya." *Journal of Human Evolution* 36 (1): 69–95. http://dx.doi.org/10.1006/jhev.1998.0262. Medline:9924134

Ward, C. V., M. G. Leakey, and A. Walker. 2001. "Morphology of *Australopithecus anamensis* from Kanapoi and Allia Bay, Kenya." *Journal of Human Evolution* 41 (4): 255–368. http://dx.doi.org/10.1006/jhev.2001.0507. Medline:11599925

Ward, C. V., A. Walker, and M. G. Leakey. 1999a. "The New Hominid Species *Australopithecus anamensis*." *Evolutionary Anthropology* 7 (6): 197–205. http://dx.doi.org/10.1002/(SICI)1520-6505(1999)7:6<197::AID-EVAN4>3.0.CO;2-T.

Ward, C. V., A. Walker, M .F. Teaford, and I. Odhiambo. 1993. "Partial skeleton of *Proconsul nyanzae* from Mfangano Island, Kenya." *American Journal of Physical Anthropology* 90 (1): 77–111. http://dx.doi.org/10.1002/ajpa.1330900106. Medline:8470757

Ward, S. 1997. "The Taxonomy and Phylogenetic Relationships of *Sivapithecus* Revisited." In *Function, Phylogeny, and Fossils: Miocene Hominoid Evolution and Adaptations*, ed. D. R. Begun, C. V. Ward, and M. D. Rose, 269–290. New York: Plenum Press.

Weishampel, D. B. 1995. "Fossils, Function and Phylogeny." In *Functional Morphology in Vertebrate Paleontology*, ed. J. J. Thomason, 34–54. Cambridge: Cambridge University Press.

White, T. D. 1980. "Evolutionary Implications of Pliocene Hominid Footprints." *Science* 208 (4440): 175–6. http://dx.doi.org/10.1126/science.208.4440.175. Medline:17745537

White, T. D. 1994. "Ape and Hominid Limb Length." *Nature* 369 (6477): 194. http://dx.doi.org/10.1038/369194b0.

White, T. D., B. Asfaw, Y. Beyene, Y. Haile-Selassie, C. O. Lovejoy, G. Suwa, and G. WoldeGabriel. 2009. "*Ardipithecus ramidus* and the Paleobiology of Early Hominids." *Science* 326 (5949): 75–86. http://dx.doi.org/10.1126/science.1175802. Medline:19810190

White, T. D., and G. Suwa. 1987. "Hominid Footprints at Laetoli: Facts and Interpretations." *American Journal of Physical Anthropology* 72 (4): 485–514. http://dx.doi.org/10.1002/ajpa.1330720409. Medline:3111270

White, T. D., G. Suwa, and B. Asfaw. 1994. "*Australopithecus ramidus*, A New Species of Early Hominid from Aramis, Ethiopia." *Nature* 371 (6495): 306–12. http://dx.doi.org/10.1038/371306a0. Medline:8090200

White, T. D., G. Suwa, and B. Asfaw. 1995. "*Australopithecus ramidus*, A New Species of Early Hominid from Aramis, Ethiopia." *Nature* 375 (6526): 88. http://dx.doi.org/10.1038/375088a0. Medline:7677838

White, T. D., G. Suwa, W. K. Hart, R. C. Walter, G. WoldeGabriel, J. de Heinzelin, J. D. Clark, B. Asfaw, and E. Vrba. 1993. "New Discoveries of *Australopithecus* at Maka in Ethiopia." *Nature* 366 (6452): 261–5. http://dx.doi.org/10.1038/366261a0. Medline:8232584

White, T. D., G. WoldeGabriel, B. Asfaw, S. Ambrose, Y. Beyene, R. L. Bernor, J.-R. Boisserie, B. Currie, H. Gilbert, Y. Haile-Selassie, et al. 2006. "Asa Issie, Aramis and the Origin of *Australopithecus*." *Nature* 440 (7086): 883–9. http://dx.doi.org/10.1038/nature04629. Medline:16612373

Wolpoff, M. H. 1983. "Lucy's Little Legs." *Journal of Human Evolution* 12 (5): 443–53. http://dx.doi.org/10.1016/S0047-2484(83)80140-7.

6

The Functional Morphology of Jaws and Teeth: Implications for Understanding Early Hominin Dietary Adaptations

Peter S. Ungar
and David J. Daegling

The famed French politician and gourmet Jean-Anthelme Brillat-Savarin (1826) once wrote "tell me what you eat, I'll tell you who you are." Our diet choices are such an important part of defining us that an entire discipline, nutritional anthropology, has developed to study how and why we are what we eat. Paleoanthropologists take this one step further, suggesting that an understanding of the evolution of human diets can even inform us on how we came to be the way we are as a species.

Diet is an important key to defining an animal species, its ecological role, and its relationships within a biological community. Primatologists, for example, recognize that "diet is the single most important parameter underlying behavioral and ecological differences among living primates" (Fleagle 1999). The same must have been true for extinct primate species, including our own ancestors, the early hominins. The aim of this chapter is to consider one of the most common approaches that paleontologists take to reconstructing the diets of fossil mammals: the study of relationships between jaw and tooth form on the one hand, and diet on the other. How do we do this? The most common approach is the comparative method. We observe that a given structure in living animals is used for a specific function. This allows us to hypothesize a relationship between that structure and its function. If all living animals that have this structure use it in the same way, we can "predict" (or retrodict) that fossil animals with that

DOI: 10.5876/9781607322252:c06

structure would have used it the same way (Kay and Cartmill 1975; Anthony and Kay 1993). Studies of the mechanics of anatomical structures and theoretical models further help us understand what a trait is best suited for, which can be very useful for evaluating hypothesized form–function relationships and for trying to figure out the function of a structure that has no modern counterpart.

This chaper presents a comparative study of the shapes of jaws and teeth of living primates, particularly of those aspects that relate to food processing, with an eye toward reconstructing the diets of fossil species. Teeth and jaws are the most common elements found in most vertebrate fossil assemblages, including the early hominins. As durable parts of the digestive system, they seem ideally suited to being studied for the reconstruction of diet. Here we offer examples of how the process works for two of our hominin forebears, *Australopithecus africanus* and *Paranthropus robustus*. Craniodental adaptations of these two hominins are often compared and contrasted, and these suggest differences in the diets of these species.

AUSTRALOPITHECUS AFRICANUS AND *PARANTHROPUS ROBUSTUS*

Researchers have focused attention on the dietary adaptations of South African Plio-Pleistocene hominins for more than a half century, ever since Robinson's (1954) classic study comparing *Australopithecus africanus* and *Paranthropus robustus*. Robinson noted that *P. robustus* had larger cheek teeth and smaller incisors, larger chewing muscles, more enamel chipping on their cheek teeth, and other features that suggested to him a specialized herbivorous diet compared with a more generalized, omnivorous *A. africanus*.

Many subsequent workers have focused on differences in the sizes of the teeth of these two hominins (e.g., Tobias 1967; Groves and Napier 1968; Jolly 1970; Pilbeam and Gould 1974; Szalay 1975; Kay 1975a; Wood and Stack 1980; Walker 1981; Peters 1981; Kay 1985; Lucas et al. 1985; Wood and Ellis 1986; Demes and Creel 1988; McHenry 1988; Ungar and Grine 1991; Teaford and Ungar 2000). Many of these researchers suggested that the larger molars and smaller incisors of *Paranthropus*, compared with *Australopithecus*, reflect a diet that was dominated by large quantities of low-quality foods requiring little incisal preparation but more heavy chewing. While relationships between tooth size and diet are still not well understood (Teaford et al. 2002; Ungar et al. 2006), this interpretation does appear to be consistent with craniofacial functional morphology of these hominins (Du Brul 1977; Ward and Molnar 1980; Rak 1983; Sakka 1984; Daegling and Grine 1991). Most would agree that the facial architectures and jaw shapes of these hominin species differ from one another and from modern primates in a manner that relates to their dietary adaptations.

The contrasts between the "robust" and "gracile" australopiths also hold when examining dental microwear. Grine and colleagues (Grine 1981; Grine and Kay 1987; Ungar and Grine 1991) demonstrated significant differences in both molar and incisor microwear between these species. Grine (1986) observed, for example, that *P. robustus* had more microwear pits on their occlusal surfaces than did *A. africanus*. In contrast, *A. africanus* had relatively more microscopic striations on their facets, which were longer and thinner, and showed more of a preferred orientation than those of the "robust" australopiths. He interpreted these results to suggest that *A. africanus* consumed a greater variety of foods including larger, more abrasive items than eaten by *P. robustus*.

In this chapter we review current understandings of how jaw and tooth form relate to diet, and take another look at the functional morphology of these structures in *A. africanus* and *P. robustus*. Differences in dental morphology have been particularly useful to this end. Tooth-form evidence suggests that these hominins differed in the mechanical properties of the foods to which they were adapted, but that the story is rather complex. While researchers have built a scenario suggesting a fundamental dietary dichotomy between these two species over the past half century, there is, in fact, little evidence to suggest that these species differed fundamentally in the foods they preferred to eat. It is more likely that the differences between these australopiths reflect selection for different fallback strategies when preferred foods were less available.

BIOMECHANICS AND FOOD FRACTURE

Before we can discuss jaw and tooth functional morphology, we must know something about the biomechanics of food fracture. After all, most mammalian teeth and jaws function principally to break foods. Digestion begins in the mouth. Chewing fractures food particles, increasing exposed surface area for digestive enzymes in the gastrointestinal tract to free nutrients. This is important for most mammals because of the unusually high levels of energy we require to maintain our high metabolic rates. It stands to reason then, that nature should select for tooth and jaw shapes that are efficient for processing the foods that a species has evolved to eat. As Spears and Crompton (1996) have noted, dental morphology affects the nature, magnitude, and distribution of stresses on food particles. Likewise, the physical properties of food influence the nature, magnitude, and distribution of forces that act on the jaws during chewing.

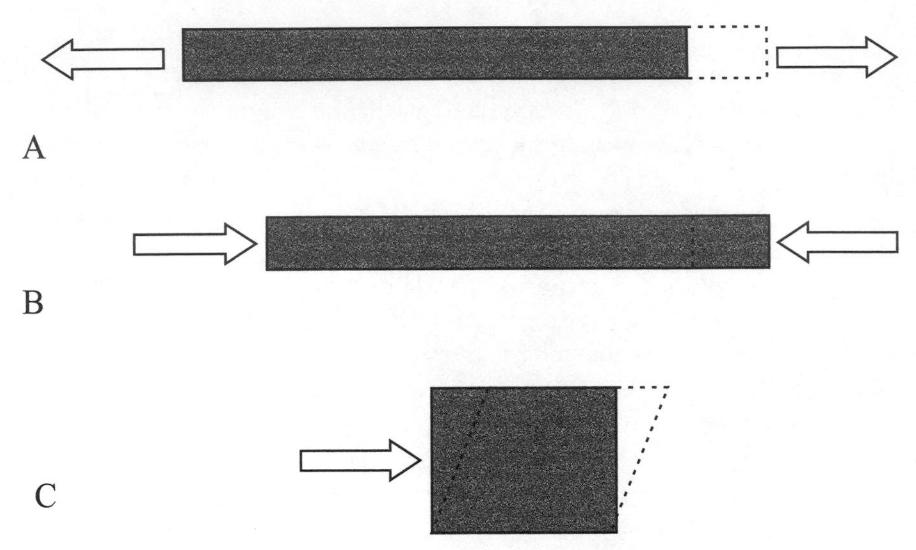

Figure 6.1. *Stress and strain. (A) Tensile stress results in tensile strain. Elongation is reckoned as positive strain. The stress is determined by the force applied and the cross-sectional area of the loaded structure. (B) Compressive stress results in compressive strain, which in this case is negative. (C) Shear strain is reckoned in angular terms, and can be perceived as the tendency of adjacent bits of material to slide past one another. In a sheared structure, tension and compression occur at 45° to the applied force.*

If we understand the mechanisms of food failure, we should be better able to understand craniodental functional morphology. A comprehensive review of all the relevant concepts and more is offered by Lucas (2004). We first need to consider how solids react when loaded. Strength is key to understanding food fracture, and the concepts of stress and strain are key to understanding strength. Stress (σ) can be thought of as analogous to "pressure." It is force normalized by the area over which it acts (F/A). Strain (ε), in contrast, refers to the deformation of a material under load. It can be simply calculated as a change in length as a fraction of original length ($\Delta L/L$). Normal strain along a single axis may be negative (as in compression), or positive (as in elongation or tensile strain). In real life, however, deformation is often more complicated than straight tension or compression along a single axis. Shear strain occurs as angular deformation, such that a rectangle is deformed into a parallelogram (Figure 6.1). Mastication usually brings a combination of tensile, compressive and shear strains.

While stress and strain are different entities, the effects of stress are manifested as strain, and the two variables are proportional to one another for given materials, at least over the interval of load known as the elastic range (Figure

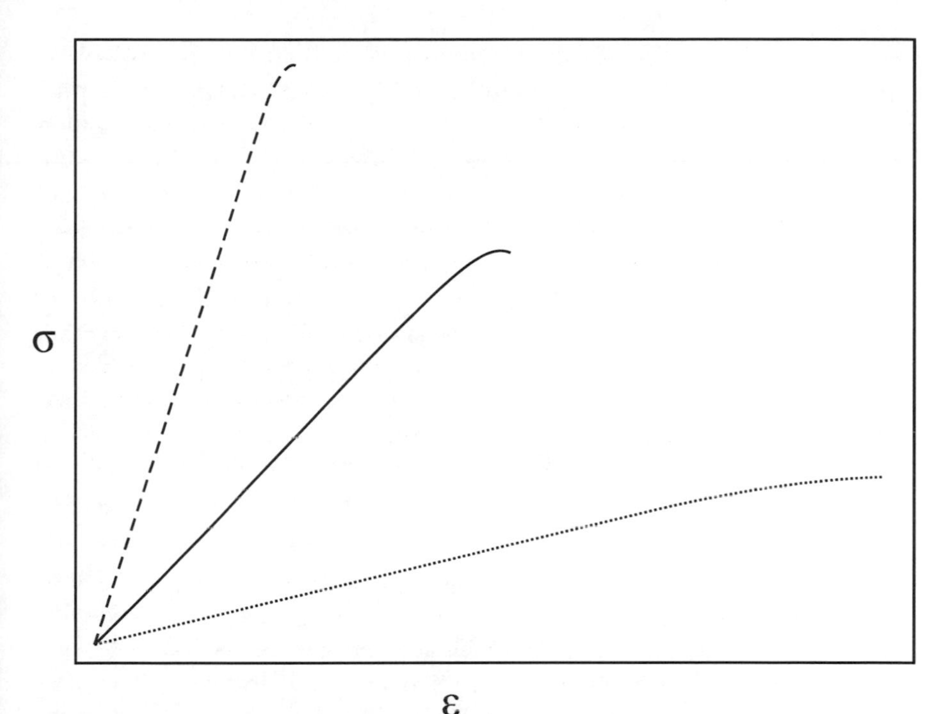

Figure 6.2. *Stress–strain curves for three hypothetical materials (σ, stress; ε, strain). The yield point is found where the slope changes; points beyond this are said to lie in the plastic range. Recovery of the original shape is expected when the load is removed somewhere along the linear portion of the curve (termed the elastic range). Beyond the yield point, deformation is permanent. A rubbery material is represented by the dotted line; lots of strain under little stress is typical of compliant materials. The dashed line represents a stiff material like steel which shows little strain despite high stress. Materials like bone are in between (solid line).*

6.2). This linear relationship is part of a stress–strain curve, and a given material has a unique slope describing its elastic range. Elastic deformation occurs when a strained solid returns to its original shape after a load is removed from it (think of a rubber band being stretched and released). Stiff foods have steep stress–strain slopes, whereas compliant items display a relatively flat stress–strain curve.

After a point, the linear relation between stress and strain breaks down and a material begins to yield. Failure can involve either fracture (think of that rubber band snapping if it is stretched too far) or plastic deformation (think of a wad of modeling clay being pounded into a table). The key properties of foods in this context are hardness (resistance to plastic deformation), and toughness (resistant to fracture). The shapes of mammalian teeth seem to be particularly influenced by the fracture toughness of those foods to which they are adapted. Tougher foods are more resistant to fracture because they resist propagation of cracks through their substance. Brittle solids, in contrast, require less work to propagate a crack through them, though initiating that crack may require considerable force depending on how hard that item is. Diets that include hard objects are associated with increased jaw dimensions in primates and marsupials (Hogue 2008), presumably because the additional effort needed to break these items also requires buttressing of the jaw to maintain bone stress within tolerable limits.

MANDIBULAR FUNCTIONAL MORPHOLOGY AND BIOMECHANICS

How natural selection has influenced jaw form

The sizes, shapes, and structural compositions of the upper and lower jawbones are governed by a number of distinct functional requirements. These include providing structural support for the teeth, space for the developing dentition, and bony processes for attachment of masticatory muscles. Food size and behavior displays of the canines can also place demands on the jaws, requiring a minimum amount of gape (Smith 1984; Hylander 2013). There are less obvious requirements that impact jaw shape as well; the respiratory airway needs to remain open, and the tongue (a structure that is itself essential for food processing) must have enough space in the mouth to function effectively. In the end, jaw shape is difficult to understand because it represents several functional compromises.

We can start with certain assumptions about how natural selection has influenced jaw form. The first is that selection favors jaws that are efficient at producing bite forces. The second is that the jaws are strong enough to withstand these occlusal forces, the muscle forces that produce them, and the resultant forces that act across the jaw joint (called the joint-reaction force). These assumptions lead to the unremarkable conclusion that natural selection has produced jaws that are capable of breaking down the foods a species eats while endowing them with the strength to stay intact after hundreds if not thousands of chewing cycles per day.

This conclusion may not be earth shattering, but it does lead to certain valuable predictions for understanding how diet influences mandibular morphology. The overall configuration of the vertebrate jaw is, from a purely mechanical perspective, surprisingly poorly suited to producing biting forces. The relative positions of the jaw joint (the temporomandibular joint, or TMJ, in mammals), the masticatory muscles, and the teeth mean that only a fraction of muscular effort is realized as bite force (see discussion on levers below). Still, all primates have to live with this configuration, and foods that require high bite forces should select for jaws with the most favorable placements possible for the muscles, TMJ, and teeth.

The second premise of jaw evolution is that jaws should not break, even after years of processing food. These bones must be strong enough and tough enough to perform their functions while maintaining their structural integrity. We know from experience that spontaneous fracture of the jaw is not something even an avid gum-chewer need worry about, and animals tend not to starve to death because they have chewed their mandibles into mush. The formula for a strong jaw is not complicated: make it bigger. Larger bones are harder to break. On the other hand, they are also more difficult to move. This leads to an engineering conundrum—you want a jaw that will not fracture, but you do not want one so heavy that it cannot be moved effectively by the muscles attached to it. Thus, mandible size is a compromise. A jaw must be heavy enough to withstand forces created during the chewing cycle but light enough for the muscles to accelerate it effectively to produce necessary bite forces ($F = ma$, after all).

Approaches to studying jaw form and function

The varied methods for linking diet to jaws can be collapsed into three categories: comparative, theoretical, and experimental. These are not mutually exclusive; in fact, most effective studies combine two or all three of these approaches to clarify the relationships between morphological variation and dietary adaptations. Even so, it is useful to consider these perspectives separately as each makes certain fundamental assumptions about how one relates form to function.

The comparative approach relies on the consistency of relationships between form and function across species. For example, large incisors in primate species are usually related to processing large, husked fruits (Hylander 1975a; Ungar 1996). The front teeth are used to reduce such foods to bits small enough to be chewed. The co-occurrence of large incisors and large, husked fruits in the diet are thus explained by feeding behavior. Once form–function relationships are established using extant species, these can be used to infer function from form in fossil species. Paleoanthropologists rely on the comparative method constantly.

Such studies contrast with classic biomechanical analyses, which link morphology to diet on theoretical grounds. This latter approach usually begins by trying to figure out how an anatomical complex works (e.g., how occluding teeth reduce a food bolus, or how the mandible moves the teeth through occlusion), and then looks at performance under different conditions (e.g., how different mandibular proportions influence reaction forces at the jaw joint). The biomechanical perspective attempts to reduce an anatomical system to a series of objective physical principles. The resulting numerical summaries facilitate comparisons and give a clear indication of efficiency, however that is defined. But the biomechanical perspective is only as accurate as our ability to describe the forces that act in an anatomical system, and correct identification of these forces can be frustratingly difficult.

A theoretical approach growing in popularity for describing craniofacial and mandibular mechanics is finite element analysis. Finite element analysis is a purely mathematical technique ideally suited for solving complex mechanical problems, such as determining the physical behavior of a materially and structurally irregular object, such as a tooth or bone. This mathematical approach offers a very powerful tool for describing the distribution of stress and strain throughout an anatomical structure. As biomechanical models go, finite element analysis is a computational bulldozer; it takes on the seemingly insoluble structural and material complexity of a whole bone, separating it into constituent parts and solving loading problems for each element.

Anthropological applications of this approach are still in their infancy, but show considerable promise (Marinescu et al. 2005; Richmond et al. 2005; Strait et al. 2005, 2009). While the technique represents jaw geometry reasonably well, the challenge of producing a realistic model that mimics all the forces and processes associated with chewing is likely to persist into the foreseeable future (Korioth and Versluis 1997; Daegling and Hylander 1998). The main problem is that boundary conditions of a finite element model, such as loading, location and orientation of restraints, and material property specifications are all defined by the user, so the model is only as good as the numbers used to generate it (see also Grine et al. 2010).

The experimental approach, in contrast, measures the variables of interest directly, as opposed to inferring them computationally. Some experimental designs do not rely on simplifying assumptions, such as the activity of a certain muscle, the sharpness of a tooth cusp, or the shape of a bone. Examples of this approach include electromyographic investigations (e.g., Weijs and Dantuma 1975; Hylander and Johnson 1994; Ross and Hylander 2000; Hylander et al. 2005) and strain gauge studies (Weijs and DeJongh 1977; Hylander 1979b, 1981, 1984). Other experimental approaches, such as photoelastic stress analysis (Ward and Molnar 1980) and considerations of masticatory efficiency (Lucas and Luke 1984) necessarily depend on certain assumptions about structure and function to permit interpretation. The principal limitation of experimental approaches is that most protocols are limited in their ability to analyze more than a small set of variables at a time. Thus, the data collected rarely provide a global accounting of either muscle activity or strain patterns. In vivo experimental approaches (i.e., those performed in living subjects), however, allow the most direct means available for inferring functional behavior, and in vitro studies (i.e., those performed on biological tissues but not in living subjects) can also be designed to evaluate assumptions made in comparative or theoretical investigations (cf. Hylander 1979a, 1979b, 1984; Demes et al. 1984; Wolff 1984; Daegling and Hylander 1998; Daegling and Hotzman 2003).

Reconstructing diet from mandibular functional morphology: a brief history

While researchers have begun to understand relationships between maxillary shape and the functional demands of chewing (Hylander 1977; Spencer and Demes 1993; Antón 1996; Rafferty et al. 2003; Vinyard et al. 2003; Lieberman et al. 2004), most studies relating jaw form to diet in mammals have focused on the mandible. Such studies extend back more than half a century (Arendsen de Wolf Exalto 1951; Becht 1953; Smith and Savage 1959; Alexander 1968; Turnbull 1970). Early comparative analyses established beyond doubt that jaw geometry relates to feeding behavior, and subsequent work has confirmed this for higher primates (Hylander 1979a, 1979b; Smith 1983; Bouvier 1986a,b; Daegling 1992; Cole 1992; Ravosa 1990, 1996; Spencer 1999; Taylor 2002).

Attempts to establish correlations between mandible form and aspects of diet (e.g., hard nuts and seeds, leaves) have met with mixed success. Some comparisons bear out predictions, others do not. For example, comparative study of capuchin species that eat different foods in the wild reveals that the species that processes hard nuts and fruits in its diet (*Cebus apella*) has predictably larger mandibles than its counterpart that is less specialized in its diet (*Cebus capucinus*) (Daegling 1992). However, a similar study of two African colobine monkeys (Daegling and McGraw 2001) failed to establish important functional differences in their mandibles despite the frequent processing of tough objects in one species but not the other.

Most comparative studies of extant species are undertaken with some thought of paleontological application. The logic in these cases is straightforward; if diet can be correlated with specific aspects of mandible size and shape, then the presence of these features in fossil specimens provides a reliable window into feeding behaviors in the past. Comparisons among modern taxa are thus used as analogs for fossils. A problem with the hominin fossil record is that we have few useful living analogs. Indeed, australopiths have mandibular proportions unlike those of any living mammal let alone any higher primate. The literature reveals three approaches to dealing with this challenge: (1) draw broad ecological analogies based on general aspects of mandible size and shape, (2) address the uniqueness of jaw form by trying to understand size and proportions in the context of tooth or body size variation, or (3) reduce the unique morphology to a finite set of biomechanical variables (measuring strength or leverage, for example) to allow direct comparison to other fossil and extant forms.

While researchers have tried to distinguish the diets of early hominins using the comparative approach (Jolly 1970; Szalay 1975; Cachel 1975; DuBrul 1977) there have been very few biomechanical data used in such analyses. Such studies are thought provoking and well argued, but their conclusions depend on the analog chosen for comparison. Thus, the baboon analogy (Jolly 1970) suggests that early hominins were seed eaters, the hyena analogy (Szalay 1975) suggested that more robust early hominins crushed bone, and the carnivore analogy (Cachel 1975) suggested that different-sized early hominins ate meat torn from their different-sized prey. These studies considered little more than the general robustness of the early hominin jaws as a basis for argument. The connection between diet and jaw form was drawn in only the most general terms. DuBrul's (1977) study, however, invoked specific mechanical principles to illustrate the functional significance of different skull morphologies. His bear analogy suggested that differences in skull form between *Australopithecus* and *Paranthropus* resembled that between grizzlies and pandas. He argued that the herbivorous panda (and by analogy, *Paranthropus*) has a retracted face and enlarged areas for the masseter and medial pterygoid muscles to attach in order to maximize biting forces at the expense of gape. This contrasts with the craniofacial morphology of more omnivorous grizzlies (and by analogy, *Australopithecus*).

Other researchers have looked to the size and scaling of mandibles to assess form and function. The null hypothesis is that differences in size and shape are explained by differences in body size, so that *Australopithecus* and *Paranthropus* jaws are essentially scaled versions of one another (Pilbeam and Gould 1974; Wolpoff 1977), despite the fact that mandibular proportions, such as the ratio of corpus breadth to corpus height, are apparently correlated with differences in size (Chamberlain and Wood 1985). Chamberlain and Wood (1985) suggested that corpus proportions for the australopiths were best explained as a more or less simple correlate of body size, meaning that dietary adaptation did not necessarily underlie differences in jaw morphology among species. Other investigations suggest that scaling of corpus proportions among fossil samples seems to depend, at least in part, on ancestry (cf. Wolpoff 1977; Chamberlain and Wood 1985; Wood and Aiello 1998), as Smith (1983) found for higher primates in general. Smith (1983) in fact warned that "neither size nor shape seem to be related to diet in a way that would be useful for paleontological inference." Hylander's (1979b, 1984) in vivo investigations of bone-strain patterns have, in contrast, offered useful functional interpretations of corpus size and shape variation. He examined loads inferred from the strain-gauge studies in efforts to determine those features of corpus geometry best suited for resisting the loads created by masticatory forces. This work inspired a generation of studies that used a "strength of materials" approach to mandibular function (reviewed in Taylor 2002). This approach allowed for meaningful biomechanical comparisons and for interpretation of the functional significance of unique morphologies in the fossil record.

Bone mechanics: material and structural considerations

We must know something about mandibular mechanics in order to understand the second premise of jaw evolution (that the mandible must be strong and rigid enough to accommodate the forces of mastication). We know, for example, that other things being equal, a larger mandible will experience less stress and strain than a smaller one. This conclusion follows from the fact that if we apply the same amount of muscular force to a jaw with more bone in it, the ratio of force to area is less (relative to the smaller jaw) and the stress is therefore lower. Since stress is proportional to strain for a given tissue over a range of loads, it also follows that the strain is less as well.

We also know that jaw shape (especially its cross-sectional geometry) influences the distribution and magnitude of masticatory stresses (Hylander 1979a; Smith 1983; Daegling 1993; Ravosa 1991, 1996, 2000). Chewing bends and twists the jaws in predictable ways, and theory suggests that certain cross-sectional shapes are more effective than others at resisting these types of forces. For example, bending stress (which you can produce in a wood pencil by pulling downward on each end) depends not only on cross-sectional size but also geometrical shape of the cross section. One reason primate mandibles are generally deeper than they are broad is that most bending stress is created by vertically directed forces, and the distribution of bone serves to limit the amount of stress that would otherwise arise in a shallower jaw. Morphologists have borrowed a concept from engineering, the *cross-sectional moment of inertia* (or the *second moment of area*), which considers both the amount of material in a cross section as well as its location relative to the axis about which bending occurs. These variables summarize both size and shape in a mechanically meaningful way, because larger values quantify better stress-bearing capabilities. For example, a plastic ruler is much easier to bend by holding it flat than it is holding it on edge. The reason is that when the ruler is held on edge, more of the plastic is in line with the bending force to resist the accompanying stress. Conversely, the ruler held flat deflects more, evincing greater strain owing to higher stress. Because the flat ruler has most of its material aligned with the axis about which it is being bent (rather than perpendicular to it), it has a very low cross-sectional moment of inertia.

Jaws are also twisted by the forces of chewing. Twisting (or *torsion*) can be visualized by holding a cardboard tube at both ends and twisting one hand clockwise and the other counterclockwise. Twisting results in shear strains that are—as with bending—minimized when there is more material present to resist it. The ideal shape for resisting a twisting load is a cylindrical one, and hollow tubes represent effective designs for resisting torsional stress. The fact that primate mandible cross-sections are almost always deeper than they are broad suggests that vertical bending stress is more important (i.e., potentially dangerous) than twisting stress during mastication. That said, the appropriate variable for measuring torsional stiffness in the primate jaw remains a matter of uncertainty (Daegling and Hylander 1998).

Comparative studies consider the mandible to be structurally variable but materially comparable among samples. In addition, approaches using linear measures or cross-sectional moments of inertia assume that the material being considered is homogeneous. The assumption, simply stated, is that bone is bone. This assumption permits us to do comparative analysis without worrying about whether the behavior of bone tissue is identical throughout a given jaw or in a sample of different ones. In truth though, bone stiffness is not uniform within and between jaws. The assumption that an orangutan jaw is identical in its elasticity to a human mandible has not been well tested (Schwartz-Dabney and Dechow 2003; Bhatavadekar et al. 2006), despite clear evidence that variations in bone density and mineralization affect material properties and ultimately stress–strain profiles. While we must keep this in mind, it need not necessarily derail work that is based on the assumption of material homogeneity. The real issue is whether variations between species are of a magnitude that they might change our interpretations.

Is the mandible a beam or a lever?

Mandibles can be modeled as a third-class lever (Hylander 1975b; Smith 1978; Spencer 1999) or as a loaded beam (Hylander 1979b; Smith 1983). The difference

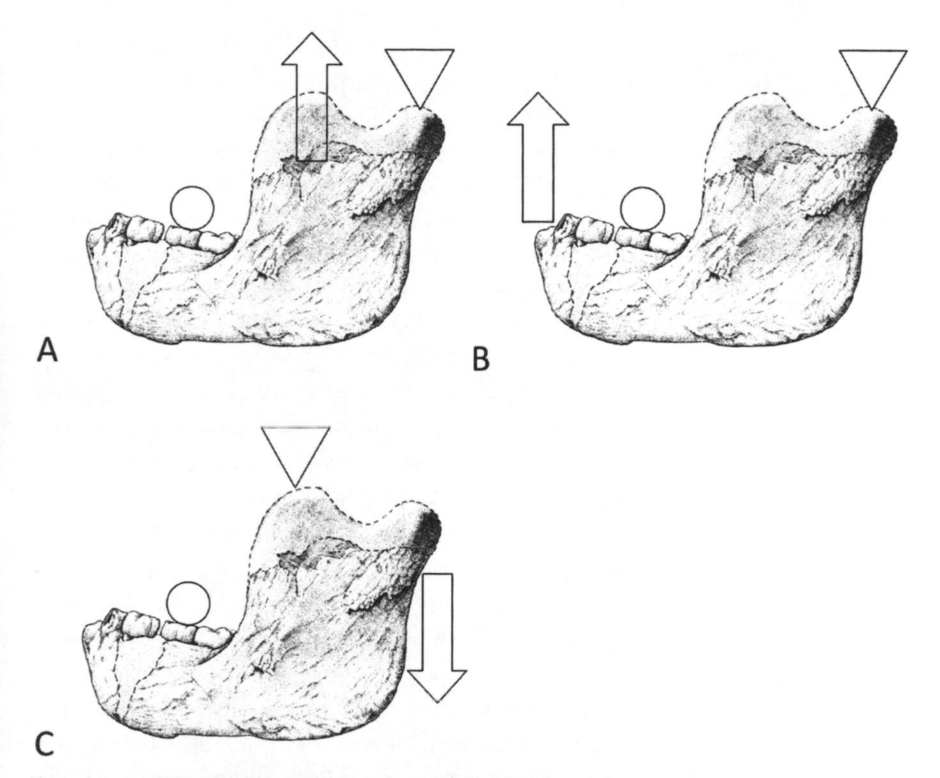

Figure 6.3. *The SKW 5 mandible of* Paranthropus robustus *illustrates the three classes of levers, only one of which (third class, A) corresponds to the primate configuration. The triangle is the fulcrum, the circle is the load (the thing we are trying to move), and the arrow is the force we apply to do the work of moving the load. If we could magically move the muscle force forward from the ramus to the front of the jaw (B), we would have a more efficient mechanical arrangement (a second-class lever), and if we could move the muscles back and the fulcrum (= jaw joint) forward (C), we would have the best arrangement of all (first-class lever). Obviously, there are very good functional and developmental reasons why B and C, despite their mechanical efficacy, have never evolved. (Adapted from Daegling and Grine 1991.)*

between the approaches is that a lever model assumes the mandible behaves as a rigid body, while the beam model assumes that the mandible is deformable and that the distribution of stresses can be predicted by considering basic dimensions. While neither model is correct in the strictest sense, both are reasonable approximations and greatly simplify both theoretical and comparative analyses. The choice of model to apply to a particular problem depends on the questions of interest. Lever models provide a readily interpretable accounting of the effects of variation in external forces on masticatory function. On the other hand, beam models are ideal for providing an approximation of the internal forces (i.e., stress) acting on the jaw during feeding activity.

One implication of describing the mandible as a third-class lever is that there are two classes of levers that might be preferred (Figure 6.3). The primate mandible is not well designed for chewing. The muscles that act on it have relatively little leverage with which to produce bite forces because the resistance arm is greater than the lever arm. In addition, because the fulcrum of the lever lies to the same side of both the bite points and the resultant muscle forces, much muscular effort is wasted as joint reaction force.

In the currency of natural selection, however, the relative inefficiency of mandibular leverage is irrelevant, since the third-class arrangement is common to all primates. There are probably very sound developmental and integrative reasons why first- and second-class jaw levers have not evolved among mammals, but the most convenient explanation is the historical one: primates inherited the basic mammalian jaw apparatus. Still, if selection for increased bite force has operated during primate evolution, then one outcome we might expect to see is modification of the leverage system to enhance the conversion of muscle force to bite force.

Comparative studies that focus on the strength of the mandible usually begin with beam theory. While lever models are concerned with how the overall configuration of the jaw produces bite force, beam models focus on how differences among species influence the magnitude and distribution of internal stresses in mandibular bone. The difference then is one of focus, how masticatory forces are generated and imparted to food items versus how the forces are resisted within the jaw itself.

Dietary inferences from theoretical and experimental studies

Experimental studies have demonstrated that foods that resist plastic deformation result in elevated strain levels in mandibular bone (Hylander 1979b; Weijs and de Jongh 1977) and that dietary consistency influences morphology in terms of geometry, mass, and remodeling during growth. For example, monkeys fed hard diets exhibit greater degrees of remodeling and greater bone mass in their mandibles than monkeys raised on soft diets (Bouvier and Hylander 1981). In addition, it appears that masticatory strain is a necessary component for normal craniofacial development; animals raised without the benefit of normal masticatory forces (i.e., lacking a diet of resistant foods) exhibit reduced muscle mass, altered mandibular geometry and facial dimensions, as well as serious malocclusions (Beecher and Corruccini 1981; Corruccini and Beecher 1982, 1984; Ciochon et al. 1997). These latter studies serve to emphasize that a certain amount of stress and strain is actually desirable for the maintenance of proper masticatory function.

These experimental results have been invoked to support the idea that over evolutionary time, dietary shifts to more mechanically challenging foodstuffs should result in predictable changes in morphology. If we recognize that the growth process has a heritable (i.e., genetic) basis, and that the normal reaction of bone cells varies within populations, then the linking of developmental and evolutionary processes is not conceptually difficult. That primates have species-specific jaw form is self-evident, but the important point is that this genetically determined baseline form is not infinitely malleable during growth, despite the plasticity of bone modeling and remodeling and the obvious possibility of dietary modification. A macaque monkey raised on the diet of a gorilla will still grow up to have a jaw that is unmistakably monkeylike in its proportions. At the same time, morphological features are clearly affected by behavior during life.

The types of loads identified in vivo (Hylander 1977, 1979b, 1981, 1984) can be analyzed independently to assess the stress-resisting capacity and ultimately the strength of different morphological configurations. These studies have served to focus comparative studies, in that differences in jaw form can be linked to diet by understanding how different foods or feeding behaviors change the relative importance of different loading regimes. Hylander's work established that leaf consumption, or folivory, which involves chewing fibrous items, is often associated with stronger jaws and a more efficient apparatus for producing bite forces (Hylander 1979a; Ravosa 1990; Spencer 1999; Taylor 2002). Hard-object feeding is also associated with stronger jaws (Hylander 1979a; Daegling 1992; Anapol and Lee 1994). In both cases, strength is not enhanced by altering bone distribution or changing cross-sectional shape; rather, relative size differences seem to account for variation in biomechanical competence (Figure 6.4).

It is important to recognize the underlying implication of what a "stronger" jaw means. Harder items require greater bite forces, which are in turn produced by recruiting more muscle force. These forces entail larger joint-reaction forces and require more stress-resisting material in the mandible. In addition, bending stresses tend to be higher in longer jaws because of the added leverage of muscle and reaction forces (this is not true in the case of torsion). In effect, the implicit argument is that stress levels are equivalent in different jaws. Folivores experience neither more nor less stress in their mandibles than do other primates as their jaws differ so as to maintain some optimal level of stress. It is

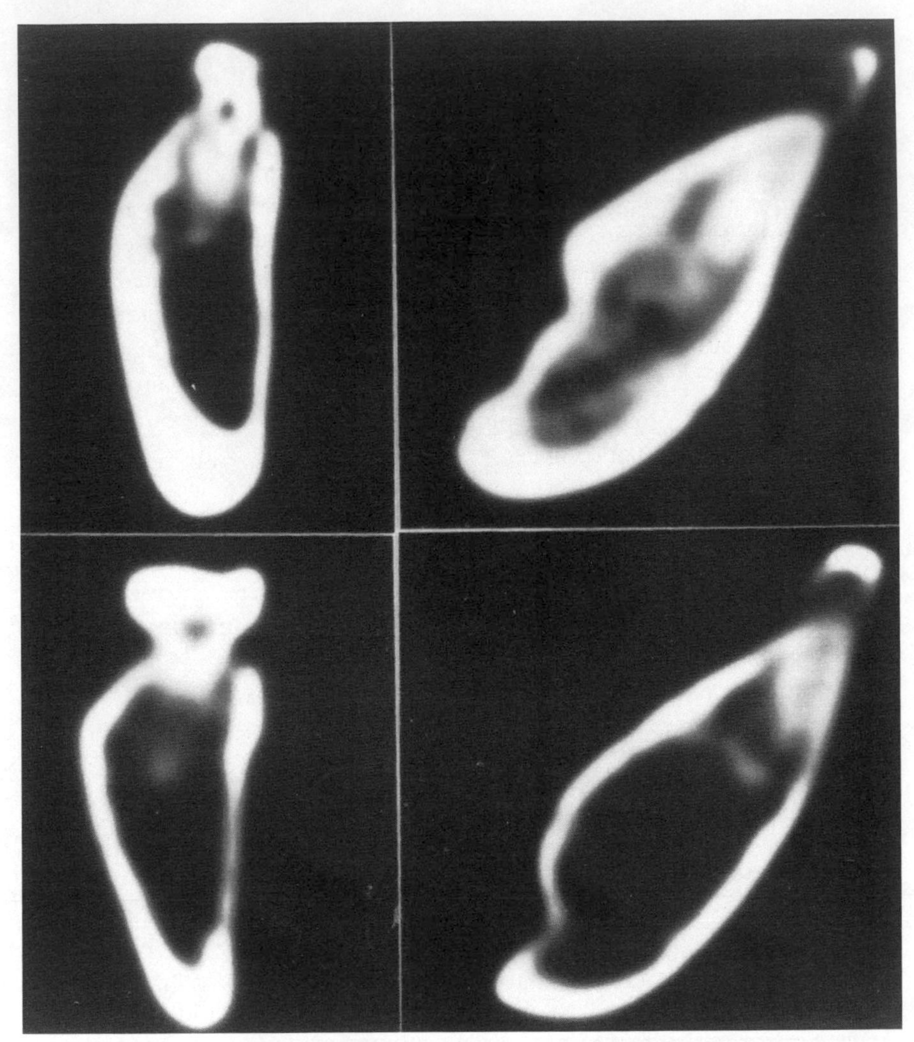

Figure 6.4. *Computed tomography (CT) scans through molar and symphyseal (midline) sections of two species of capuchin monkey.* Cebus apella *(top row) is famous for its consumption of palm nuts and other hard items;* Cebus capucinus *(bottom row) is not known to process such hard objects in the wild. Cross-sectional moments of inertia are larger in* C. apella *about any axis of bending and their jaws are unquestionably stronger. Note that in terms of external geometry the shapes of the sections are very similar between the two species; in addition, the amount of bone used (relative to the total subperisoteal area) is only marginally greater in* C. apella *(Daegling 1992). This means that mechanical strength is achieved by using about the same proportion of bone in a larger overall section.*

very difficult to specify what this optimal stress is, however, given that strains (and most likely stresses) vary considerably throughout the mandible (Daegling 1993, 2004).

Two other observations complicate attempts to link diet directly to jaw form. First, some studies fail to show predicted differences in biomechanical properties despite marked dietary differences between species (e.g., Bouvier 1986a; Anapol and Lee 1994; Daegling and McGraw 2001). Second, increases in biomechanical robusticity (i.e., relatively stronger mandibles) are correlated with larger body sizes, which in turn relate to decreased food quality (Ravosa 2000). Thus, caloric yield per unit of masticatory effort is often less in larger animals than in smaller ones. Moreover, the available evidence suggests that larger primates might have relatively less available muscle force (Cachel 1984), which may require, from the standpoint of both bite-force production and stress resistance, less favorable patterns of muscle recruitment. The connections between diet and mandibular form can be seen in this context to be direct effects of body-size variation.

Craniomandibular evidence for diet in *Australopithecus* and *Paranthropus*

Australopithecus africanus and *P. robustus* have very distinctive and unique jaws. From a biomechanical perspective, the two species are capable of withstanding higher masticatory loads than are extant large-bodied primates (Hylander 1988; Daegling and Grine 1991. Hylander (1979a, 1988) has noted, for example, that *Paranthropus* has large transverse dimensions of the corpus. He related these to the resistance of twisting forces resulting from large transverse components of biting and muscular forces. In fact, whatever type of masticatory load is considered, these early hominins were better suited to resist the resultant stresses than are living apes (Table 6.1; Daegling 1990). Others have suggested that mandibular dimensions in early hominins are instead related to the co-occurrence of reduced canines and enlarged cheek teeth (Wolpoff 1975; Chamberlain and Wood 1985). The idea here is that thick jaws are required to support large teeth, and smaller canines have eliminated the need for a deep jaw to support a long canine root. Comparative tooth and jaw data among living primates (Plavcan and Daegling 2006) only weakly support the idea that jaw proportions are determined by the size of the dentition.

The mandibles of *A. africanus* and *P. robustus* are actually quite similar in functional terms (Daegling and Grine 1991). Their cross–sectional shapes are more or less identical (Figure 6.5). Both utilize cortical bone in ways that are

Table 6.1. Stress resistance at M_1 sections in mandibles of great apes and early hominins

Load	*Taxon*	*Mean*	*S.d.*	*Range*
Vertical bending[1]				
	Pan	216.7	22.3	171–254
	Pongo	237.6	22.1	183–279
	Gorilla	240.9	25.4	204–286
	Australopithecus (N = 3)	337.7	24.9	309–353
	Paranthropus (N = 4)	332.5	79.4	253–442
Torsion[2]				
	Pan	549.2	46.2	434–607
	Pongo	511.0	66.3	398–626
	Gorilla	599.7	47.6	490–677
	Australopithecus (N = 1)	722.0	–	–
	Paranthropus (N = 4)	798.5	107.7	645–891
Lateral bending[3]				
	Pan	99.1	9.7	79–113
	Pongo	96.7	7.7	83–110
	Gorilla	93.1	8.1	76–109
	Australopithecus (N = 2)	142.5	40.3	114–171
	Paranthropus (N = 4)	128.3	18.8	102–145
Shear[4]				
	Pan	90.7	8.5	75–105
	Pongo	89.2	7.6	70–99
	Gorilla	94.0	6.6	83–105
	Australopithecus (N = 1)	108.0	–	–
	Paranthropus (N = 4)	120.0	14.6	102–137

1. Calculated as the ratio of the fourth root of the appropriate area moment of inertia to the moment arm in vertical bending $\times 10^4$ (Daegling 1992, 1993). The ratio is dimensionless, with larger values representing greater strength. Great ape samples N = 20 for each taxon with males and females equally represented.
2. Calculated as the ratio of the third root of Bredt's formula for torsional resistance to mandibular length $\times 10^4$ (Daegling 1992, 1993). The ratio is dimensionless, with larger values representing greater strength given mandibular size. The denominator is not an estimate of the twisting moment arm, but should be viewed as a size proxy. Great ape samples N = 20 for each taxon with males and females equally represented.
3. Calculated as the ratio of the fourth root of the appropriate area moment of inertia to the moment arm in lateral bending $\times 10^4$ (Daegling 1992, 1993). The ratio is dimensionless, with larger values representing greater strength. Great ape samples N = 20 for each taxon with males and females equally represented.
4. Calculated as the ratio of the square root of cortical bone area to mandibular length $\times 10^4$ (Daegling 1992, 1993). The ratio is dimensionless, with larger values representing greater shear resistance given mandibular size. The denominator is not an estimate of the applied force. Great ape samples N = 20 for each taxon with males and females equally represented.

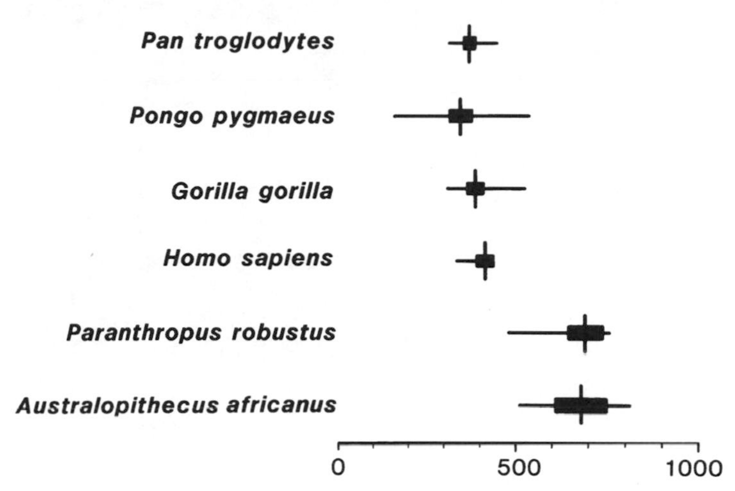

FIGURE 6.5. *Cross-sectional proportions of early hominin mandibles in comparison with modern great apes and humans. The ratio is of the largest and smallest area moments of inertia (by definition perpendicular to one another) determined through the centroid of the sections. This bending index provides a measure of shape based on the amount and distribution of bone within the section. The hominins are distinct from modern taxa and yet are indistinguishable from one another.*

reminiscent of modern apes (Figure 6.6), but because of their large mandibular corpus dimensions, they are much stronger (biomechanically speaking) than are modern analogs (Figure 6.7). Yet, despite the similarity in their mandibles, their craniofacial skeletons are very different. There has been some debate in the literature concerning the functional implications of these differences. Many have argued that hypertrophy of the facial skeleton in the genus *Paranthropus* can be related to masticatory function (e.g., Rak 1983). At least some features though, such as a thickened palate, might reflect developmental processes rather than resistance of masticatory stresses (McCollum 1997). Indeed, palatal thickening may have the effect of *increasing* strains elsewhere in the facial skeleton (Richmond et al. 2005). What seems clear from this finding is that the overall configuration of the facial skeleton—rather than specific osseous features—is the paramount consideration for assessing the mechanical effects of masticatory loads. Modeling stress profiles in facial bones is also considerably more problematic than in the mandible, owing primarily to the irregular geometry, presence of sutures, and the voids created by nasal and orbital cavities as well as paranasal sinuses. Beam models are probably inadequate; finite element analysis

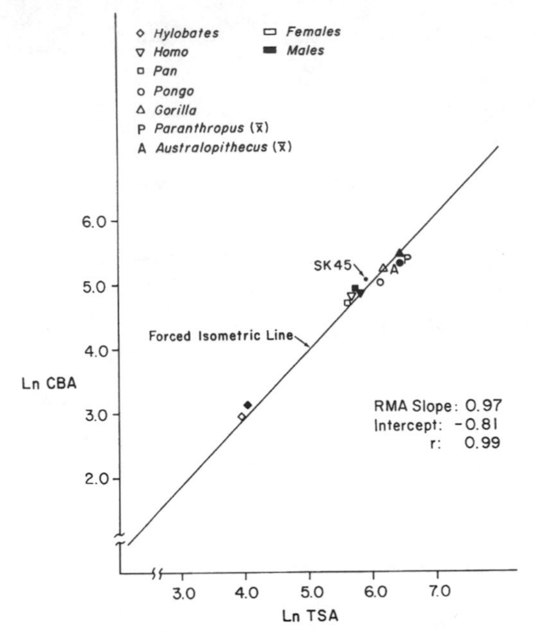

Figure 6.6. *Reduced major axis regression of cortical bone area (CBA) relative to the total subperiosteal area (TSA) from samples of modern hominoids and* Paranthropus robustus *and* Australopithecus africanus. *The isometric line represents the slope of proportionality. The proximity of the data points to this line suggests all hominoids utilize the same proportion of bone tissue relative to overall cross-sectional dimensions.*

represents a viable modeling option (Richmond et al. 2005), with the drawback that it is currently not suited to assess the effects of population variation. Finally, it is also prudent to recognize that developmental and adaptive biomechanical explanations are not necessarily mutually exclusive.

What jaws tell us about masticatory loads

If primates load their jaws differently to process different diets, differences in bending and twisting loads should result in differences in corpus geometry. Nevertheless, attempts to associate mandibular cross-sectional shape with diet have not been very successful, even with an improving understanding of masticatory mechanics enabled by in vivo studies. When these *in vivo* data are mapped onto morphological variation in the mandible, it becomes clear

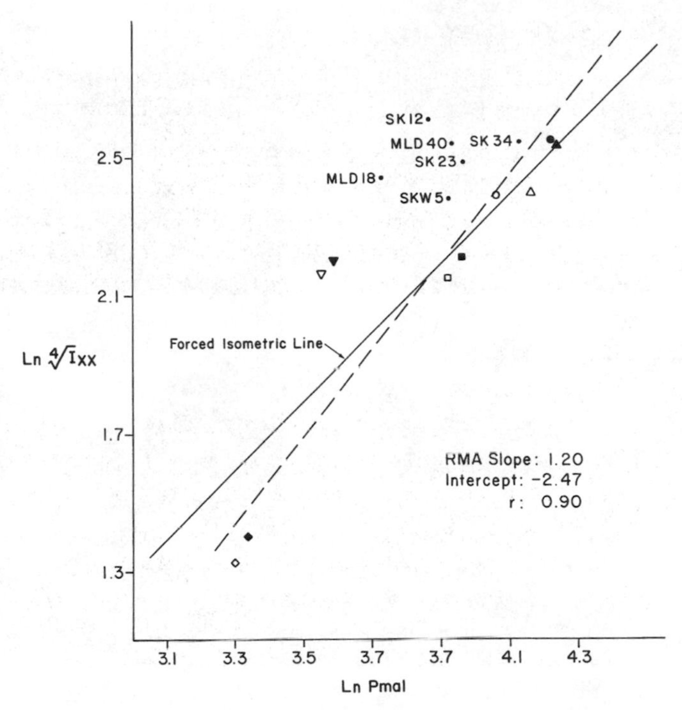

Figure 6.7. *Reduced major axis regression of the cross-sectional moment of inertia (I_{XX}) that resists vertical bending (among the more important masticatory loads that the jaws must resist) against an estimate of the parasagittal moment arm (Pmal) that reflects the leverage of the bending force. Symbols as in Figure 6.6. MLD specimens represent* Australopithecus *fossils, SK and SKW represent jaws of Paranthropus. The position of the early hominins above both isometric and best-fit lines is indicative of their enhanced strength under this load. Note that these forms achieve this strength in two ways: relative to orangutans and gorillas, they are strong by virtue of their relatively short jaws (reduced bending leverage), while relative to chimps and humans, they are strong by virtue of their absolutely large corpus dimensions.*

that stress histories cannot be inferred from cross-sectional geometry alone (Daegling 1993). Perhaps one problem is our difficulty distinguishing whether relative differences in jaw size reflect greater masticatory loads per occlusal event or a greater number of loading cycles per day. There is no obvious morphological feature of the mandible that we can use to address this, presumably because we do not yet understand the long-term effects of higher loads compared with those of more frequent loads (Figure 6.8). We cannot even assume

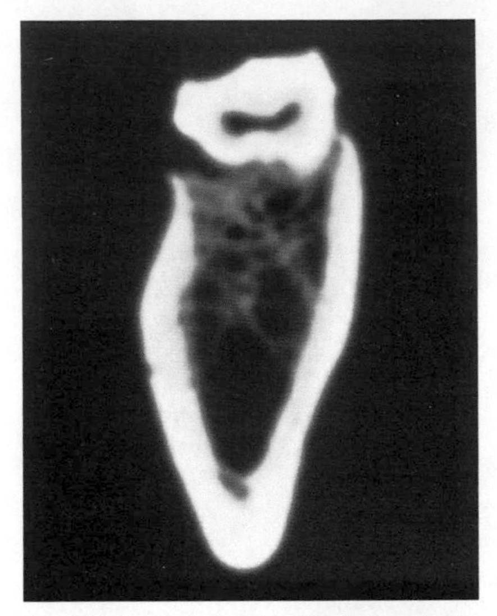
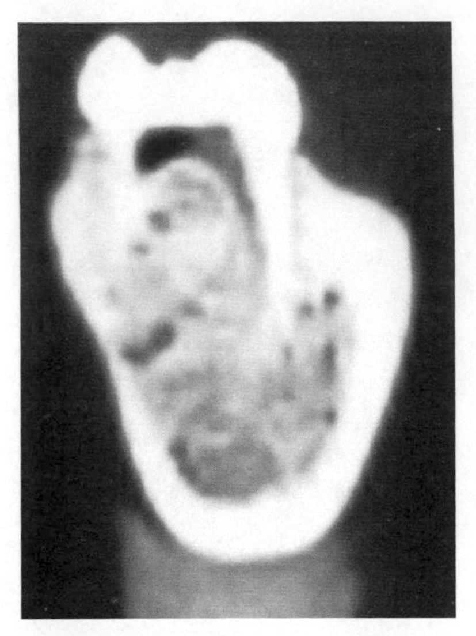

FIGURE 6.8. *CT scans of* Gorilla *and* Paranthropus *mandibles through the* M_2 *(not to scale). The clear differences in corpus shape and cortical bone distribution are functionally significant, but whether we can link these differences to the physical properties of the food items consumed remains a topic for further exploration.*

that different regions of mandibular bone will respond similarly to changes in functional loading (Rubin et al. 1990).

TEETH

RECONSTRUCTING DIET FROM DENTAL FUNCTIONAL MORPHOLOGY: A BRIEF HISTORY

Relationships between teeth and diet are fortunately more direct and therefore easier to demonstrate and understand, and such studies have a long and celebrated history (Hunter 1771; Owen 1840; Cuvier 1863; Cope 1883; Osborn 1907; Gregory 1922; Simpson 1933). While researchers have recognized for centuries that tooth form reflects function, it was still not really until the 1970s that this work began to come of age, as researchers developed mechanical models of how dental features related to masticatory movements (Crompton and Sita-Lumsden 1970; Kay and Hiiemae 1974). Primates that have sharp blades or long crests on their molar teeth typically have opposing teeth that come into occlusion at steep angles for shearing and slicing. Tough foods are sheared between the leading edges of crown crests. Primate shearing blades are typically recip-

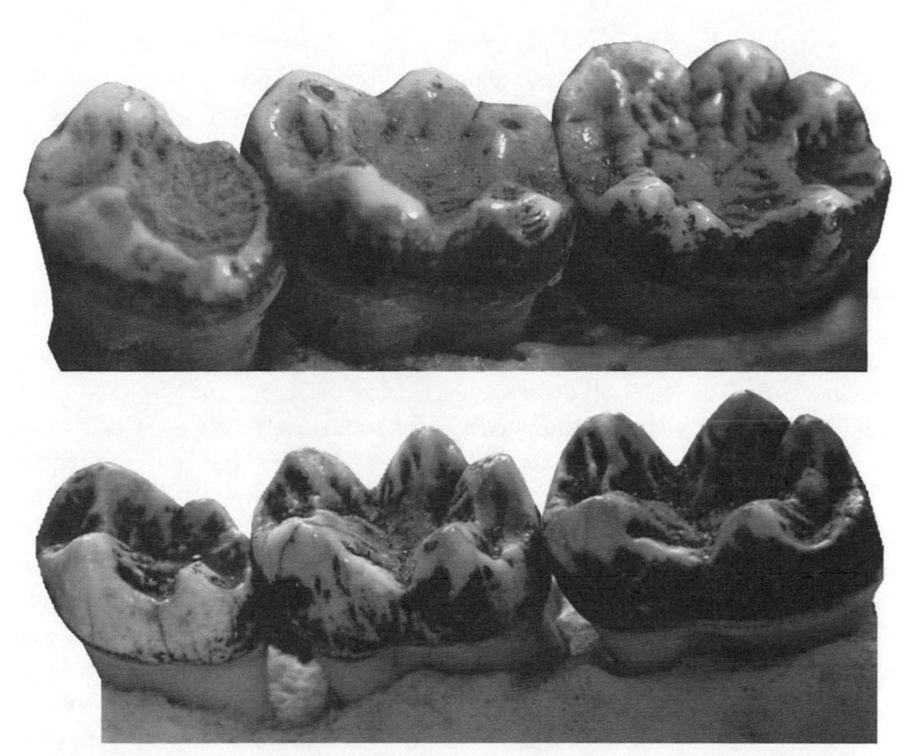

FIGURE 6.9. *Dentitions of a chimpanzee (above) and a gorilla (below). These teeth (P_4 through M_2) represent similarly worn specimens. Note the differences in the slopes and relief of the surfaces. These correspond to differences in the diets of these two species.*

rocally concave to minimize contact area (Figure 6.9). In contrast, mammals adapted to consume hard, brittle foods more often crush these foods between flatter "mortar and pestle" surfaces.

As some researchers worked on biomechanics of teeth and food fracture, others began to associate specific crown morphologies with specific diets. Rosenberger and Kinzey (1976), for example, contrasted the sharp, shearing teeth of New World monkeys that eat insects and leaves with the flat, crushing morphology of those that eat fruits. Seligsohn and Szalay (1978) likewise noted occlusal morphologies related to diet in lemurs, attributing sharp edges the molars of leaf-eating species, and puncturing cusp tips on the molars of those that habitually process bamboo.

It became clear by the late 1970s that relationships between tooth form and function in living primates could be used to infer the diets of fossil forms, but that a quantitative and comparative approach to dental functional morphology was needed. Richard Kay took on the challenge (Kay 1978; Kay and Hylander 1978; Kay and Covert 1984) and developed a shearing quotient (SQ) that could be used as a measure of the efficiency with which a tooth can be used to slice or shear tough foods. Shearing crests run from front to back, or mesiodistally, between the tips of cusps and the notches between them (Figure 6.10). The longer these crests are relative to the overall length of a tooth, the steeper the approach between opposing teeth and the more area is available to shear and slice tough foods.

Shearing-quotient calculation is straightforward. First, the lengths of all shearing crests on a series of unworn molar teeth are measured and summed. The mesiodistal lengths of these teeth are also recorded, and summed crest length is then plotted and regressed against tooth length for a set of closely related species with similar diets (e.g., fruit-eating ape). The resulting regression line allows us to determine the expected crest length for a closely related species (in this case, an ape) with a given diet (in this case, fruits) and a given tooth length (Figure 6.10). Shearing quotients are calculated as the difference between observed and expected crest length (computed from the regression-line formula) divided by expected crest length. In our example then, an ape with a high, positive SQ value will have longer crests than expected of a frugivore, and one with a negative SQ would have shorter crests. Numerous studies have shown that leaf- and insect-eating primates have relatively longer crests (higher SQs) than do fruit-eaters. Further, among frugivores, those that consume harder, more brittle fruits or nuts have the shortest crests (and lowest SQ values). Studies have shown that SQs accurately track diet within all major primate groups, from lemurs and lorises to New World monkeys, Old World monkeys, and apes (Kay and Covert 1984; Anthony and Kay 1993; Strait 1993; Meldrum and Kay 1997; Ungar 1998).

Given the relationships established for living primates then, it should be possible to plot extinct primate teeth on a graph of crest length over tooth length for living primates with known diets to infer the diets of the fossil forms. And indeed, researchers have used SQs to reconstruct the diets of fossil primates from every epoch in which they are well known (Kay 1977; Kay and Simons 1980; Anthony and Kay 1993; Strait 1993; Williams and Covert 1994; Ungar and Kay 1995; Fleagle et al. 1997; Teaford et al. 2002; Ungar et al. 2004).

Dental functional morphology of the early hominins

Despite the SQ studies of all of these fossil primates, there has been remarkably little work published on the dental functional morphology of fossil homi-

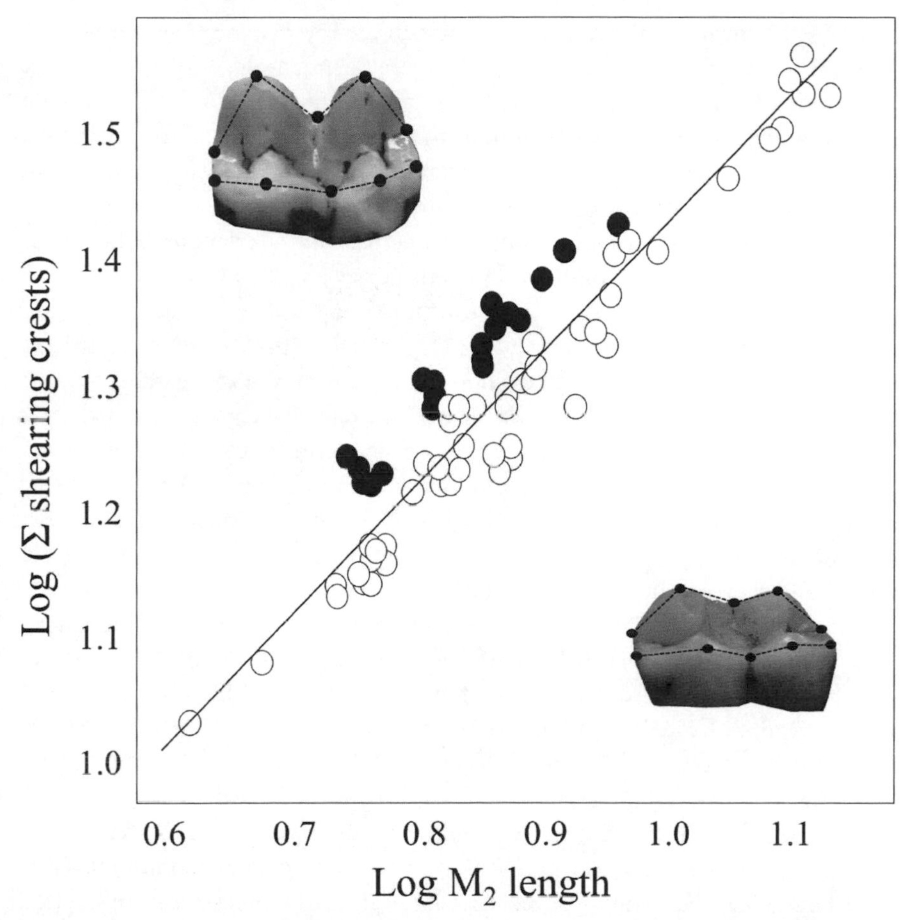

Figure 6.10. *Shearing-crest analysis of Old World monkeys. The black circles represent species of colobines, which tend to be more folivorous, whereas the white circles indicate more frugivorous cercopithecines. The line through the data is a least-squares regression line. Notice the clear separation of Old World monkeys by diet. The tooth images show measurements used to compute shearing crest lengths. The upper-left tooth surface shows relatively longer crests than does the lower-right tooth. (Data presented come from Kay 1978.)*

nins. Even Kay (1985), the architect of the shearing quotient, reported no SQ values in his comprehensive review of dental evidence for diet in the australopiths. One problem is that many of the early hominins have flat, bulbous cusps, with little in the way of discernable mesiodistal cresting on their molar teeth. Another limitation is the very small sample of unworn fossil hominin teeth for analysis. As molars begin to wear, cusp tips quickly become obliterated, making it impossible to measure crest lengths (which are the distances between those tips and the notches between them). There are, for example, only nine unworn lower second molars (M_2s, the tooth most often used in SQ studies) among the hundreds of published early hominin teeth known from South Africa. We have little hope of using SQ studies to understand dental functional morphology for most early hominin species in the near future.

Nevertheless, there have been some clues concerning the dental–dietary adaptations of the early hominins. More than three decades ago, Grine (1981) noted that *A. africanus* evinced more inclined facets on their cheek teeth than did *P. robustus* (especially from Swartkrans). He opined that *A. africanus* engaged in more shearing, wherein facet faces slid past one another nearly parallel to their planes of contact while *P. robustus* had less inclined facets, suggesting a shallower approach into and out of centric occlusion and a more catholic masticatory repertoire with both perpendicular and parallel components to occlusal events. He proposed that this, along with other lines of evidence including allometry, microwear fabrics, and craniofacial morphology, implied that *P. robustus* ate harder and more fibrous foods than did *A. africanus.*

Shearing quotient values have been computed for those few unworn *A. africanus* and *P. robustus* (see below) M_2s that are available (Ungar et al. 1999). Results of this limited study, based on four *A. africanus* and five *P. robustus* specimens, support Grine's (1981) interpretations, while reaffirming that the australopiths as a group had flat, blunt molar teeth. Although the *A. africanus* and *P. robustus* ranges overlap, the average SQ value for the "gracile" australopiths is higher than that for the "robust" hominins (Figure 6.11). Still, the averages for both species are lower than the means reported for all extant hominoids (Kay 1977; Ungar and Kay 1995). Given that blunter, flatter molars are well suited to crush hard foods (Lucas et al. 1994), and that hard-object-feeding primates show lower SQs than closely related soft-fruit specialists (Fleagle et al. 1997), it is reasonable to assume that these australopiths were admirably equipped to process hard, brittle objects. On the other hand, *A. africanus*, and especially *P. robustus*, would have had difficulty shearing and slicing tough, pliant foods (Teaford and Ungar 2000).

Dental topographic analysis of living primates

While shearing-quotient studies are effective for unworn teeth, they do not work well when landmarks used for crest-length measurement (i.e., cusp tips) become obliterated by wear. Restricting our studies of dental functional morphology to unworn teeth is not a great idea, however. First, as described above,

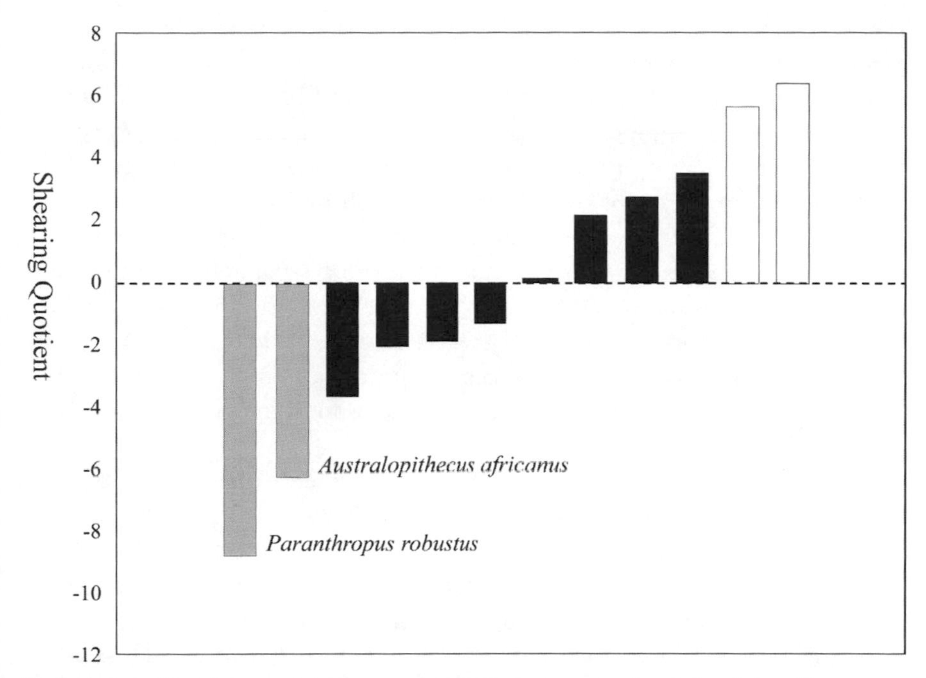

Figure 6.11. *Shearing quotient mean values for living apes and early hominins. The white bars represent the more folivorous apes (siamang and gorilla), whereas the black bars represent more frugivorous apes (gibbons and chimpanzees). The hatched lines represent* Australopithecus africanus *and* Paranthropus robustus. *The zero value reflects the regression line of shearing-crest lengths over tooth length (see Figure 6.10) computed for the frugivorous apes. More-positive values are residuals indicating longer crests, whereas more-negative values indicate less occlusal relief. Notice that the more folivorous species have higher SQs and are clearly separated from the frugivores. Also, the australopiths are distinct from the living apes with lower SQ values. (Data from Ungar et al. 1999.)*

this limits sample sizes because most primate molar teeth, whether in the mouth, museum collections, or the ground, are worn. Second, natural selection does not stop when teeth begin to wear—probably quite the opposite in fact (King et al. 2005). We have to consider that natural selection should favor shapes that wear in a manner that keep teeth mechanically efficient for fracturing the foods to which they are adapted (Kay 1981; Teaford 1983; Ungar and M'Kirera 2003). If we do not include studies of worn teeth, we are getting an incomplete picture of the dental form–function relationship.

Some researchers have tried to include both unworn and worn teeth in functional analyses. Smith (1999), for example, used a technique modified from

Wood, Abbott, and Graham (1983) to measure 2D planimetric areas of individual cusps as seen in occlusal view. These values should not change with wear so long as individual cusp boundaries can be identified and outlined. Smith's (1999) study of cusp proportions group chimpanzees and gibbons to the exclusion of gorillas, suggesting to her that there may be a functional signal. Still, such studies depend on being able to distinguish individual cusp boundaries, which can go quickly on thin-enameled molars. A more important limitation of this is that the fact that the two frugivores group separately from the more folivorous ape does not necessarily imply a causal relationship between relative planimetric cusp area and diet, and there are no clear biomechanical explanations for any functional signal in the cusp proportions identified. However, such explanations are a prerequisite to the inference of form–function relationships (Kay and Cartmill 1977).

The main problem with functional studies of planimetric cusp area is that chewing occurs in a 3D environment, and two teeth with similar projected 2D areas may differ greatly in cusp relief. Cusp relief is critical to the angle of approach between the lower and upper teeth as facets come into occlusion during chewing. This, in turn, determines whether foods are fractured by shearing or crushing, which reflects the fracture properties of those foods (see above). What is needed is a 3D approach to characterizing functional aspects of tooth form. This approach should allow us to include both unworn and worn teeth in analyses.

This is where dental topographic analysis comes in. A cloud of three-dimensional points representing the occlusal surface of a tooth is collected and analyzed using geographic information systems (GIS) software. Several types of instrument can be used to collect data points, depending on the resolution and work envelope required for the job. The electromagnetic digitizer (Zuccotti et al. 1998), laser scanner (Ungar and Williamson 2000), confocal microscope (Jernvall and Selänne 1999), and piezo touch-probe scanner (King et al. 2005) have all been used with success for dental topographic analysis.

GIS software packages collect, store, manipulate, analyze, and display geographically referenced information. The GIS approach was first developed to model and examine the surface of the Earth. Dental topographic analysis takes advantage of these tools, characterizing teeth as landscapes—cusps become mountains, fissures become valleys. The resulting virtual tooth surface can then be measured and analyzed using conventional GIS tools. Data on many different aspects of topography, from cusp slope and relief to surface area, angularity or jaggedness, aspect, basin volume "drainage patterns," and many other topographic attributes can be collected quickly and objectively (Zuccotti et al. 1998; Ungar and Williamson 2000).

One of the principal advantages of this approach is that it does not rely on a set of fixed landmarks for measurement, so that tooth wear is not a problem. Occlusal surface shape attributes can be compared independent of degree of tooth wear. Thus, we can compare and contrast variably worn teeth from different individuals, and even document functional changes in occlusal shape for individuals as their teeth wear (Dennis et al. 2004; King et al. 2005).

Here we illustrate the dental topographic analysis approach by summarizing a comparison of variably worn molars of chimpanzees, gorillas, and orangutans (M'Kirera and Ungar 2003; Ungar and M'Kirera 2003; Merceron et al. 2006; Ungar 2007), and a comparative analysis of occlusal shape in the early hominins *A. africanus* and *P. robustus* (Ungar 2007).

The first study involved an analysis of lower second-molar teeth (M_2s) of wild-shot museum specimens of the Central African chimpanzee *Pan troglodytes troglodytes* (n = 54), the western lowland gorilla *Gorilla gorilla gorilla* (n = 47), and the Bornean orangutan *Pongo pygmaeus pygmaeus* (n = 51). These species were chosen for two reasons. First, they make a good baseline series for comparisons with our Plio-Pleistocene ancestors. The great apes are our closest living relatives and their teeth all follow the same "bauplan" or general structure as ours and those of the early hominins. Second, these species vary subtly in their diets in ways expected to show slight to moderate differences in their occlusal morphology. *P. troglodytes* and *G. g. gorilla* both prefer soft, succulent fruits when they are available. Reports suggest that fruit flesh makes up some 70–80 percent and 45–55 percent of their diets respectively (Williamson et al. 1990; Kuroda 1992; Nishihara 1992; Tutin et al. 1997), with gorillas falling back on tough, fibrous foods like stems and leaves when fruit is scarce. Bornean orangutan food preferences depend greatly on seasonal availability, but their annual fruit consumption comes in at about 55–65 percent (Rodman 1977; MacKinnon 1977), intermediate between the two African apes—with due note of caveats concerning differences in data-collection protocols (Doran et al. 2002).

The occlusal surfaces of high-resolution replicas were scanned using a laser scanner to generate clouds of points representing those surfaces. These points were sampled at an interval of 25.4 μm (1/1,000th of an inch) along the x- and y-axes. GIS software was then used to interpolate the surface, or make a virtual model of the occlusal table of each tooth. Details can be found in M'Kirera and Ungar (2003) and Ungar and M'Kirera (2003).

Occlusal surfaces are complicated structures that can be described by several different variables. Here we review results for one of these: average occlusal slope. This attribute is easy to visualize, it is of obvious functional significance, and it separates species with differing diets. Occlusal slope is defined as the

Table 6.2. Dental topographic analysis slope-descriptive statistics (data from Ungar 2004; and Ungar 2006)1.

A. The early hominin data set (50 µm resolution point cloud from piezo scanner).

	Stage 1			*Stage 2*			*Stage3*		
	Mean	*S.d.*	*N*	*Mean*	*S.d.*	*N*	*Mean*	*S.d.*	*N*
Australopithecus	42.85	2.774	6	36.17	4.581	4	27.89	3.926	4
Paranthropus	35.79	6.03	7	35.61	4.511	2	25.99	2.834	2

B. The extant ape data set (25.4µm resolution point cloud from laser scanner).

	Stage 1			*Stage 2*			*Stage3*		
	Mean	*S.d.*	*N*	*Mean*	*S.d.*	*N*	*Mean*	*S.d.*	*N*
Gorilla	37.75	5.036	7	36.29	2.665	10	36.29	2.665	10
Pan	32.88	5.859	5	30.15	5.771	28	26.48	4.680	18
Pongo	36.29	2.665	10	32.86	2.274	16	30.32	2.170	6

1. The early hominin and living-ape data should not be compared directly because these data sets were collected from digital elevation models of different resolutions. Data for individual taxa should be compared only within the studies.

average change in elevation between adjacent points across a surface. Results for other attributes, such as surface relief and angularity, can be found in the literature (M'Kirera and Ungar 2003; Ungar and M'Kirera 2003) and provide complementary information on molar functional morphology.

Once the dental topography data were collected, each specimen was scored for gross wear according to Scott's (1979) protocol, and teeth were grouped by wear stage (see Ungar 2004 for details) and analyzed using a ranked data two-way ANOVA model, to see whether species differed in occlusal slope, whether differently worn teeth differed in occlusal slope, and whether any differences observed between the species remained the same at different stages of gross tooth wear. Pairwise comparisons tests were used to determine the sources of significant variation for taxon and wear stage differences.

The results of this study are summarized in Figure 6.12 (top) and Table 6.2. There are three noteworthy findings. First, as one might expect, slope values tend to be lower in more-worn specimens—teeth become flatter as they wear down. Second, the species differed significantly in mean occlusal slope such that gorillas had the steepest sloped surfaces, followed by orangutans. Chimpanzees had the least average change in elevation between adjacent points on the surface. This corresponds to expected differences given fracture properties, especially toughness, of the diets of these apes (see above). Third, the differences between

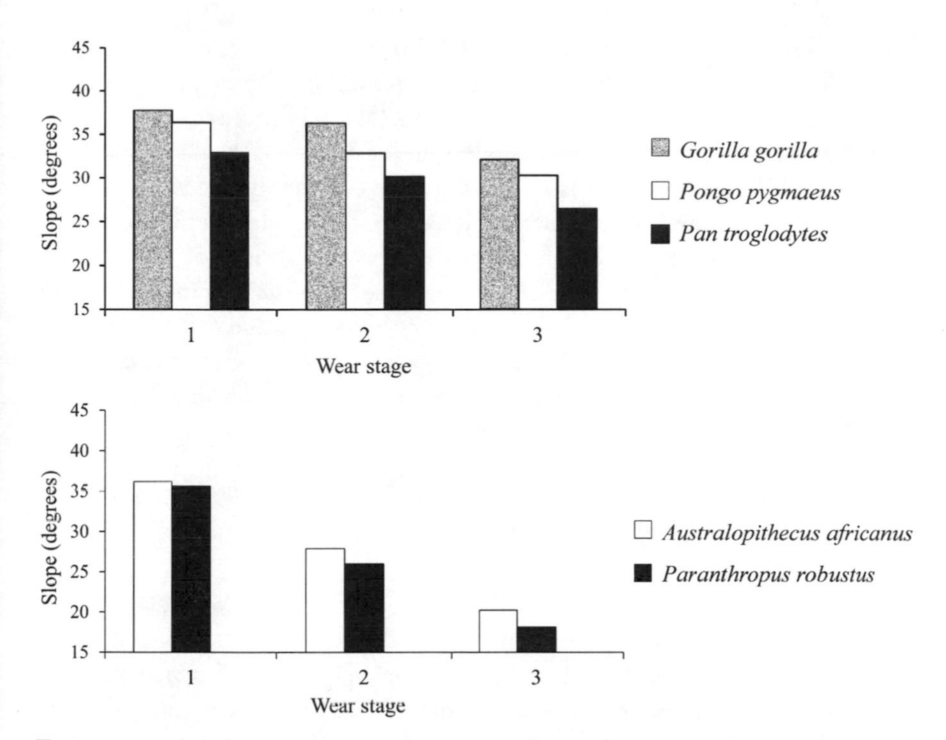

Figure 6.12. *Dental topographic analysis of average occlusal-slope data for living apes (above) and fossil hominins (below). Wear stages are as defined in Ungar (2004). Note that all species show lower average slope values with greater wear. Note also that differences between the species remain consistent at each stage of wear. (Data from Ungar 2006 and Ungar and M'Kirera 2003.) The living apes and fossil hominins are considered separately because the data for each were collected at different resolutions using different types of instruments. The extant ape data were collected at 25.4µm x,y,z resolution using the Surveyor 500 laser scanner (see text) and the fossil-hominin data were collected at 50 µm x,y,z resolution using the Roland DGA PICZA piezo scanner.*

the species were similar at each of the three stages of wear, as suggested by the lack of interaction in the model between species and wear stage.

Why is this important? It suggests that variably worn teeth *can* be used in dental topography analyses so long as we control for degree of wear. Therefore, we can make dental functional morphology studies of fossil species represented only by worn teeth, so long as there is a baseline of comparative data for specimens with similar degrees of wear. This dramatically increases the number and variety of fossil species that can be studied and the number of specimens that can be included in analyses.

Food Preferences or Fallback Resource Exploitation? These data bring up an important sidenote. While it is evident that dental differences between these apes reflect differences in their diets, it not so clear whether these differences relate more to food preferences or to selection for occasional but important fallback foods. While many have assumed that more commonly eaten foods have a greater selective influence on molar "design" (e.g., Kay 1975b), primate teeth can also reflect adaptations for less frequently eaten but still important resources (Kinzey 1978; Lambert et al. 2004). Apes are a case in point. Hominoids seem to prefer succulent, sugar-rich foods that are high in energy and easy to consume and digest—this is a legacy of the ancestral catarrhine dietary adaptation (Ross 2000; Ungar 2005). These resources tend to be easy to digest and offer a low cost–benefit ratio, and they may not require a specialized dental morphology for efficient digestion.

On the other hand, morphological specializations may actually reflect selection for foods that are less desirable and more difficult to process, especially those seasonally critical resources that get an animal through lean times. This phenomenon leads to Liem's Paradox, which was first noted for fish (Robinson and Wilson 1998). The paradox reflects an adaptation to less-favored foods. In other words, some species actually avoid the very foods to which they are adapted unless more preferred resources are unavailable. This seems to be the case with gorillas, for example, which choose foods high in nonstarch sugars and sugar-to-fiber ratios when these are available (Remis 2002). Still, more than a dozen field studies have demonstrated that gorillas can tolerate more fibrous diets than can sympatric chimpanzees (Wrangham 2005). The differences between chimpanzee and gorilla diets become clear at "crunch times." Western lowland gorillas and central African chimpanzees at Lopé, for example, have 60–80 percent overlap in plant species they eat (Williamson et al. 1990; Tutin and Fernandez 1993), with both species eating soft fruits much of the year. When preferred fruits become scarce, however, the gorillas fall back more on leaves and other fibrous plant parts than do chimpanzees. The same even holds true for the mountain gorillas at Bwindi, where they are found alongside chimpanzees (Stanford and Nkurunungi 2003).

Dental Topography of South African Early Hominins

Dental topographic analysis has been used to study early hominin species (e.g., Ungar 2004). Data for *A. africanus* and *P. robustus* (from Ungar 2007) are summarized here as an example. A total of thirty-three undamaged M_2s, including *A. africanus* specimens from Makapansgat and Sterkfontein (n =

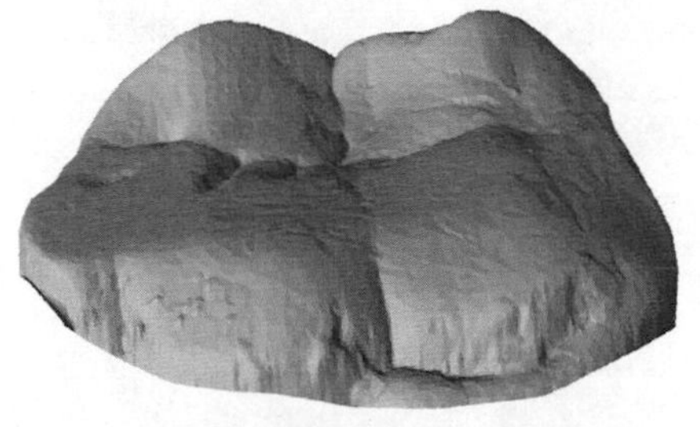

Australopithecus africanus (Stw 404)

Paranthropus robustus (SK 23)

Figure 6.13. *Triangulated irregular network models for early hominins. These models were constructed from digital elevation data for representative M_2s of* Australopithecus africanus *(Stw 404) and* Paranthropus robustus *(SK 23). Specimens used to generate these models are both in wear stage 2.*

18) and *P. robustus* from Drimolen, Swartkrans and Kromdraai (n = 15) were included in this analysis. This compares with a total of nine specimens suitable for SQ study (see above).

Point-cloud data for this study were collected using a piezo touch-probe scanner. The resolution of this instrument is fairly low (50 μm), but still sufficient for comparing occlusal slope in hominins. This scanner is becoming increasingly popular for modeling teeth because of its low price and ease of use (e.g., Archer and Sanson 2002; Martin-de las Heras et al. 2005; King et al. 2005). These data were again analyzed using GIS software as described above for the living great apes. The most-worn hominin specimens, those that preserved little occlusal morphology, were not included in the statistical analyses because recent work has suggested that at this point functional efficiency drops markedly and "dental senescence" begins to set in (see, for example, King et al. 2005).

Results for these hominins are presented in Figures 6.12 (bottom) and 6.13, and Table 6.2. As with the living-ape study, there is significant variation between species and wear stages but no interaction between the two factors. First, less-worn specimens have significantly higher mean occlusal slope values than do more-worn specimens. Second, *A. africanus* has higher average occlusal slope than does *P. robustus*. And third, the difference between the hominins remains consistent at each wear stage, as suggested by the lack of interaction between the factors.

These results suggest that *A. africanus* molars could more efficiently shear and slice tough foods than could comparably worn *P. robustus* cheek teeth. In contrast, *P. robustus* could more efficiently crush hard, brittle foods than could *A. africanus*. This result is consistent with both the SQ study (see above) and the observations by Grine (1981) that *A. africanus* had more-inclined occlusal facets than did most *P. robustus* specimens. These results also make sense given most studies of dental allometry, microwear, and craniofacial morphology for these hominins (see Ungar 2007 for review).

LIEM'S PARADOX AND EARLY HOMININ DIETARY ADAPTATIONS

So, did *Paranthropus* prefer to eat harder, brittler foods than did *A. africanus*, or do the differences in occlusal morphology between these species reflect selection for different fallback foods? Liem's Paradox offers an interesting and provocative model. As we noted above, gorillas have digestive anatomy that allows them to subsist entirely on tough, fibrous foods when necessary, whereas chimpanzees cannot. Did the craniodental toolkit of *Paranthropus* likewise afford it a more flexible diet (in this case, the ability to efficiently fracture hard, brittle foods rather than tough, elastic ones) compared with *Australopithecus*?

If so, perhaps Liem's Paradox can be applied to early hominins too. Researchers have tended to think of *Paranthropus* as a *specialized* hominin because of its highly derived adaptive morphology (Teaford and Ungar 2000). It may be, however, that the derived "robust" australopith morphology was actually capable of efficiently processing a *broader range* of foods, including both extremely hard, brittle items and the less mechanically challenging ones eaten by the less-specialized "gracile" forms. Perhaps then, *Paranthropus* was more of a generalist than many have assumed (see Wood and Strait 2000). This has some very fundamental implications for natural selection in early hominins and human evolution.

This idea was recently proposed by Scott et al. (2005) on the basis of microwear evidence. They noted that while *A. africanus* and *P. robustus* differed in their central tendencies for microwear surface complexity and anisotropy, the two species overlapped considerably in both attributes. This suggests that while the two australopiths often ate foods with similar fracture properties, they ate foods with different fracture properties at least on occasion. *A. africanus* and *P. robustus* showed more heterogeneity in anisotropy and surface complexity respectively. This hints that the "gracile" australopiths ate foods with a greater range of toughness, whereas the "robust" forms ate foods with greater ranges of hardness and brittleness. The authors concluded that *A. africanus* and *P. robustus* probably often ate similar foods, and differed mostly in fallback resources exploited.

This is also consistent with results for jaw functional morphology described above and for isotope analyses for the early hominins. The fact that jaw-shape data do not clearly separate *A. africanus* and *P. robustus* might then not reflect limitations in our data sets, but rather might actually suggest that loading regimes for these hominins were similar most of the time. Likewise, a lack of difference between carbon isotope ratios for *A. africanus* and *P. robustus* and their intermediate position between C_3 and C_4 resource specialists (Lee-Thorp 2002; van der Merwe et al. 2003) is certainly consistent with two hominins with generalized diets and similar food preferences. Occasional differences in fallback-food exploitation would likely become swamped given time-averaging and specimen sampling that spans the months during which the enamel crown developed. Perhaps then, we might think of jaw shape and tooth chemistry results not as limited in their abilities to distinguish hominins with very different diets, but rather, as showing similarities in food preferences. This interpretation, however, must be reconciled with the observation that the upper facial skeletons of *Australopithecus* and *Paranthropus* are morphologically distinct.

Can the dental topography results for the South African australopiths offer any further clues to the degree of difference between the diets of these hominins? While different instruments were used to collect the data for living apes and the australopiths, it is clear that the degree of difference in mean occlusal slope between the early hominins is no greater than, and probably somewhat less than, that between western lowland gorillas and central African chimpanzees. Because the difference between the African apes can be explained by differences in fallback-food exploitation, the same may well be true for *A. africanus* and *P. robustus*. All of the lines of evidence (tooth chemistry, jaw shape, dental microwear, and occlusal morphology) taken together are consistent with the hypothesis that *A. africanus* and *P. robustus* had similar, broad-based diets that differed mostly in fallback foods taken during times of resource stress. A fallback explanation is also parsimonious with what is known about primate feeding ecology today; however, it is at present not possible to directly test a fallback hypothesis in the fossil record. For example, a competing hypothesis would be that the South African australopiths had subtle differences in dietary profiles that were not necessarily seasonally driven or determined by periodic resource scarcity. This implicates *P. robustus* as having a broader dietary repertoire than *A. africanus*, consistent with the principle articulated by Liem's Paradox. While both the morphological and wear data do not readily permit discriminating between these alternatives in the paleontological context, there is no reason to suspect that seasonal differences in resource availability did not characterize the Pliocene landscape. Whether *Paranthropus* was periodically compelled to rely on poor-quality foods and/or opportunistically seek out mechanically challenging but nutritionally rich items is probably beyond our resolution for the time being. New approaches and new technologies are allowing us to get more-and-more precise pictures of the dietary adaptations of past species (e.g., see Sponheimer et al. 2006), and we should not conclude that what we do not know today will remain unknown tomorrow. It is an exciting time for studies of the evolution of human diet, and all lines of evidence, including mandibular and dental functional morphology have an important role to play in our understanding of the adaptations of our hominin predecessors.

ACKNOWLEDGMENTS

We thank Matt Sponheimer for organizing the symposium that led to this volume and Luci Betti for drawing Figure 6.3. The work described here was funded by the US National Science Foundation and the LSB Leakey Foundation. This chapter is a publication of the Stony Brook Alumni Craniodental Research Team.

REFERENCES

Alexander, R. M. 1968. *Animal Mechanics*. London: Sidgwick & Jackson.

Anapol, F., and S. Lee. 1994. "Morphological Adaptation to Diet in Platyrrhine Primates." *American Journal of Physical Anthropology* 94 (2): 239–61. http://dx.doi.org/10.1002/ajpa.1330940208. Medline:8085615

Anthony, M.R.L., and R. F. Kay. 1993. "Tooth Form and Diet in Ateline and Alouattine Primates: Reflections on the Comparative Method." *American Journal of Science* 293 (A): 356–82. http://dx.doi.org/10.2475/ajs.293.A.356.

Antón, S. C. 1996. "Cranial Adaptation to a High Attrition Diet in Japanese Macaques." *International Journal of Primatology* 17 (3): 401–27. http://dx.doi.org/10.1007/BF02736629.

Archer, D., and G. Sanson. 2002. "Form and Function of the Selenodont Molar in Southern African Ruminants in Relation to Their Feeding Habits." *Journal of Zoology (London, England)* 257 (1): 13–26. http://dx.doi.org/10.1017/S0952836902000614.

Arendsen de Wolf Exalto, E. 1951. "On Differences in the Lower Jaw of Animalivorous and Herbivorous Mandibles." *Proceedings of the Koninklijke Nederlandse Academie van Wetenschappen. Series C: Biological and Medical Sciences*. 54: 237–246: 405–410.

Becht, G. 1953. "Comparative Biologic-Anatomical Researches on Mastication in Some Mammals." *Proceedings of the Koninklijke Nederlandse Academie van Wetenschappen. Series C: Biological and Medical Sciences* 56:508–27.

Beecher, R. M., and R. S. Corruccini. 1981. "Effects of Dietary Consistency on Craniofacial and Occlusal Development in the Rat." *Angle Orthodontist* 51 (1): 61–9. Medline:6939354

Bhatavadekar, N. B., D. J. Daegling, and A. J. Rapoff. 2006. "Application of an Image-Based Weighted Measure of Skeletal Bending Stiffness to Great Ape Mandibles." *American*

Journal of Physical Anthropology 131 (2): 243–51. http://dx.doi.org/10.1002/ajpa.20397. Medline:16596598

Bouvier, M. 1986a. "A Biomechanical Analysis of Mandibular Scaling in Old World Monkeys." *American Journal of Physical Anthropology* 69 (4): 473–82. http://dx.doi.org/10.1002/ajpa.1330690406.

Bouvier, M. 1986b. "Biomechanical Scaling of Mandibular Dimensions in New World Monkeys." *International Journal of Primatology* 7 (6): 551–67. http://dx.doi.org/10.1007/BF02736661.

Bouvier, M., and W. L. Hylander. 1981. "Effect of Bone Strain on Cortical Bone Structure in Macaques (*Macaca mulatta*)." *Journal of Morphology* 167 (1): 1–12. http://dx.doi.org/10.1002/jmor.1051670102. Medline:7241595

Brillat-Savarin, Jean-Anthelme. 1826. *Physiologie du Goût*. Paris: Feydeau.

Cachel, S. 1975. "A New View of Speciation in *Australopithecus*." In *Paleoanthropology, Morphology and Paleoecology*, ed. R. H. Tuttle, 183–201. The Hague: Mouton. http://dx.doi.org/10.1515/9783110810691.183

Cachel, S. 1984. "Growth and Allometry in Primate Masticatory Muscles." *Archives of Oral Biology* 29 (4): 287–93. http://dx.doi.org/10.1016/0003-9969(84)90102-X. Medline:6586125

Chamberlain, A. T., and B. A. Wood. 1985. "A Reappraisal of the Variation in Hominid Mandibular Corpus Dimensions." *American Journal of Physical Anthropology* 66 (4): 399–405. http://dx.doi.org/10.1002/ajpa.1330660408.

Ciochon, R. L., R. A. Nisbett, and R. S. Corruccini. Apr-Jun 1997. "Dietary Consistency and Craniofacial Development Related to Masticatory Function in Minipigs." *Journal of Craniofacial Genetics and Developmental Biology* 17 (2): 96–102. Medline:9224944

Cole, T. M., III. 1992. "Postnatal Heterochrony of the Masticatory Apparatus in *Cebus apella* and *Cebus albifrons*." *Journal of Human Evolution* 23 (3): 253–82. http://dx.doi.org/10.1016/S0047-2484(05)80003-X.

Cope, E. D. 1883. "On the Trituberculate Type of Molar Tooth in the Mammalia. Paleontological Bulletin Number 37." *Proceedings of the American Philosophical Society* 21:324–6.

Corruccini, R. S., and R. M. Beecher. 1982. "Occlusal Variation Related to Soft Diet in a Nonhuman Primate." *Science* 218 (4567): 74–6. http://dx.doi.org/10.1126/science.7123221. Medline:7123221.

Corruccini, R. S., and R. M. Beecher. 1984. "Occlusofacial Morphological Integration Llowered in Baboons Raised on Soft Diet." *Journal of Craniofacial Genetics and Developmental Biology* 4 (2): 135–42. Medline:6746878.

Crompton, A. W., and A. G. Sita-Lumsden. 1970. "Functional Significance of the Therian Molar Pattern." *Nature* 227 (5254): 197–9. http://dx.doi.org/10.1038/227197a0. Medline:5428418

Cuvier, G. 1863. *The Animal Kingdom*. London: Henry Bohm.

Daegling, D. J. 1990. "Geometry and Biomechanics of Hominoid Mandibles." PhD dissertation, State University of New York, Stony Brook.

Daegling, D. J. 1992. "Mandibular Morphology and Diet in the Genus *Cebus*." *International Journal of Primatology* 13 (5): 545–70. http://dx.doi.org/10.1007/BF02547832.

Daegling, D. J. 1993. "The Relationship of in Vivo Bone Strain to Mandibular Corpus Morphology in *Macaca fascicularis*." *Journal of Human Evolution* 25 (4): 247–69. http://dx.doi.org/10.1006/jhev.1993.1048.

Daegling, D. J. 2004. "Relationship of Strain Magnitude to Morphological Variation in the Primate Skull." *American Journal of Physical Anthropology* 124 (4): 346–52. http://dx.doi.org/10.1002/ajpa.10356. Medline:15252862

Daegling, D .J., and F. E. Grine. 1991. "Compact Bone Distribution and Biomechanics of Early Hominid Mandibles." *American Journal of Physical Anthropology* 86 (3): 321–39. http://dx.doi.org/10.1002/ajpa.1330860302. Medline:1746641

Daegling, D. J., and J. L. Hotzman. 2003. "Functional Significance of Cortical Bone Distribution in Anthropoid Mandibles: An *in vitro* Assessment of Bone Strain under Combined Loads." *American Journal of Physical Anthropology* 122 (1): 38–50. http://dx.doi.org/10.1002/ajpa.10225. Medline:12923903

Daegling, D. J., and W. L. Hylander. 1998. "Biomechanics of Torsion in the Human Mandible." *American Journal of Physical Anthropology* 105 (1): 73–88. http://dx.doi.org/10.1002/(SICI)1096-8644(199801)105:1<73::AID-AJPA7>3.0.CO;2-E. Medline:9537929

Daegling, D. J., and W. S. McGraw. 2001. "Feeding, Diet and Jaw Form in West African *Colobus* and *Procolobus*." *International Journal of Primatology* 22 (6): 1033–55. http://dx.doi.org/10.1023/A:1012021823076.

Demes, B., and N. Creel. 1988. "Bite Force, Diet, and Cranial Morphology of Fossil Hominids." *Journal of Human Evolution* 17 (7): 657–70. http://dx.doi.org/10.1016/0047-2484(88)90023-1.

Demes, B., H. Preuschoft, and J.E.A. Wolff. 1984. "Stress-Strength Relationships in the Mandibles of Hominoids." In *Food Acquisition and Processing in Primates*, ed. D. J. Chivers, B. A. Wood, and A. Bilsborough, 369–390. New York: Plenum.

Dennis, J. C., P. S. Ungar, M. F. Teaford, and K. E. Glander. 2004. "Dental Topography and Molar Wear in *Alouatta palliata* from Costa Rica." *American Journal of Physical Anthropology* 125 (2): 152–61. http://dx.doi.org/10.1002/ajpa.10379. Medline:15365981

Doran, D. M., A. McNeilage, D. Greer, C. Bocian, P. Mehlman, and N. Shah. 2002. "Western Lowland Gorilla Diet and Resource Availability: New Evidence, Cross-Site Comparisons, and Reflections on Indirect Sampling Methods." *American Journal of Primatology* 58 (3): 91–116. http://dx.doi.org/10.1002/ajp.10053. Medline:12454955

Du Brul, E. L. 1977. "Early Hominid Feeding Mechanisms." *American Journal of Physical Anthropology* 47 (2): 305–20. http://dx.doi.org/10.1002/ajpa.1330470211. Medline:410308.

Fleagle, J. G. 1999. *Primate Adaptations and Evolution*. 2nd ed. New York: Academic Press.

Fleagle, J. G., R. F. Kay, and M.R.L. Anthony. 1997. "Fossil New World Monkeys." In *Vertebrate Paleontology in the Neotropics*, ed. R. F. Kay, R. H. Madden, R. L. Cifelli, and J. J. Flynn, 473–495. Washington, DC: Smithsonian Institution Press.

Gregory, W. K. 1922. *The Origin and Evolution of Human Dentition*. Baltimore: Williams and Wilkins.

Grine, F. E. 1981. "Trophic Differences between "Gracile" and "Robust" Australopithecines: A Scanning Electron Microcope Analysis of Occlusal Events." *South African Journal of Science* 77:203–30.
Grine, F. E. 1986. "Dental Evidence for Dietary Differences in *Australopithecus* and *Paranthropus*: A Quantitative-Analysis of Permanent Molar Microwear." *Journal of Human Evolution* 15 (8): 783–822. http://dx.doi.org/10.1016/S0047-2484(86)80010-0.
Grine, F. E., S. Judex, D. J. Daegling, E. Ozcivici, P. S. Ungar, M. F. Teaford, M. Sponheimer, J. Scott, R. S. Scott, and A. Walker. 2010. "Craniofacial Biomechanics and Functional and Dietary Inferences in Hominin Paleontology." *Journal of Human Evolution* 58 (4): 293–308. http://dx.doi.org/10.1016/j.jhevol.2009.12.001. Medline:20227747.
Grine, F. E., and R. F. Kay. 1987. "Quantitative Analysis of Occlusal Microwear in *Australopithecus* and *Paranthropus*." *Scanning Microscopy* 1 (2): 647–56. Medline:3112937.
Groves, C. P., and J. R. Napier. 1968. "Dental Dimensions and Diet in Australopithecines." *Proceedings of the VIII International Congress of Anthropological and Ethnological Sciences* 3: 273–276.
Hogue, A. S. 2008. "Mandibular Corpus Form and Its Functional Significance: Evidence from Marsupials." In *Primate Craniofacial Function and Biology*, ed. C. J. Vinyard, M. J. Ravosa, and C. E. Wall, 329–356. New York: Springer. http://dx.doi.org/10.1007/978-0-387-76585-3_15
Hunter, J. 1771. *The Natural History of Human Teeth*. London: J. Johnson.
Hylander, W. L. 1975a. "Incisor Size and Diet in Anthropoids with Special Reference to Cercopithecidae." *Science* 189 (4208): 1095–8. http://dx.doi.org/10.1126/science.808855. Medline:808855
Hylander, W. L. 1975b. "The Human Mandible: Lever or Link?" *American Journal of Physical Anthropology* 43 (2): 227–42. http://dx.doi.org/10.1002/ajpa.1330430209. Medline:1101706
Hylander, W. L. 1977. "The Adaptive Significance of Eskimo Craniofacial Morphology." In *Orofacial Growth and Development*, ed. A. A. Dahlberg and T. M. Graber, 129–170. The Hague: Mouton. http://dx.doi.org/10.1515/9783110807554.129
Hylander, W. L. 1979a. "The Functional Significance of Primate Mandibular Form." *Journal of Morphology* 160 (2): 223–39. http://dx.doi.org/10.1002/jmor.1051600208. Medline:458862.
Hylander, W. L. 1979b. "Mandibular Function in *Galago crassicaudatus* and *Macaca fascicularis*: An *in vivo* Approach to Stress Analysis of the Mandible." *Journal of Morphology* 159 (2): 253–96. http://dx.doi.org/10.1002/jmor.1051590208. Medline:105147.
Hylander, W. L. 1981. "Patterns of Stress and Strain in the Macaque Mandible." In *Craniofacial Biology*, ed. D. S. Carlson, 1–37. Monograph 10, Craniofacial Growth Series. Ann Arbor: Center for Human Growth and Development, University of Michigan.
Hylander, W. L. 1984. "Stress and Strain in the Mandibular Symphysis of Primates: A Test of Competing Hypotheses." *American Journal of Physical Anthropology* 64 (1): 1–46. http://dx.doi.org/10.1002/ajpa.1330640102. Medline:6731608.
Hylander, W. L. 1988. "Implications of *In Vivo* Experiments for Interpreting the Functional Significance of "Robust" Australopithecine Jaws." In *Evolutionary History of the "Robust" Australopithecines*, ed. F. E. Grine, 55–83. New York: Aldine de Gruyter.

Hylander, W. L. 2013. "Functional Links Between Canine Height and Jaw Gape in Catarrhineswith Special Reference to Early Hominins." *American Journal of Physical Anthropology.* 150(2): 247–249. http://dx.doi.org/10.1002/ajpa.22195.

Hylander, W. L., and K. R. Johnson. 1994. "Jaw Muscle Function and Wishboning of the Mandible during Mastication in Macaques and Baboons." *American Journal of Physical Anthropology* 94 (4): 523–47. http://dx.doi.org/10.1002/ajpa.1330940407. Medline:7977678

Hylander, W. L., C. E. Wall, C. J. Vinyard, C. F. Ross, M. R. Ravosa, S. H. Williams, and K. R. Johnson. 2005. "Temporalis Function in Anthropoids and Strepsirrhines: An EMG Study." *American Journal of Physical Anthropology* 128 (1): 35–56. http://dx.doi.org/10.1002/ajpa.20058. Medline:15714512

Jernvall, J., and L. Selänne. 1999. "Laser Confocal Microscopy and Geographic Information Systems in the Study of Dental Morphology." *Paleontologica Electronica.* 2 (18): 1–17.

Jolly, C. J. 1970. "The Seed-Eaters: A New Model of Hominid Differentiation Based on a Baboon Analogy." *Man* 5 (1): 5–26. http://dx.doi.org/10.2307/2798801.

Kay, R. F. 1975a. "Allometry and Early Hominids (Comment)." *Science* 189: 63.

Kay, R. F. 1975b. "The Functional Adaptations of Primate Molar Teeth." *American Journal of Physical Anthropology* 43 (2): 195–215. http://dx.doi.org/10.1002/ajpa.1330430207. Medline:810034.

Kay, R. F. 1977. "Diets of Early Miocene African Hominoids." *Nature* 268 (5621): 628–30. http://dx.doi.org/10.1038/268628a0.

Kay, R. F. 1978. "Molar Structure and Diet in Extant Cercopithecidae." In *Development, Function, and Evolution of Teeth*, ed. P. M. Butler and K. A. Joysey, 309–339. New York: Academic Press.

Kay, R. F. 1981. "The Nut-Crackers: A New Theory of the Adaptations of the Ramapithecinae." *American Journal of Physical Anthropology* 55 (2): 141–51. http://dx.doi.org/10.1002/ajpa.1330550202.

Kay, R. F. 1985. "Dental Evidence for the Diet of *Australopithecus.*" *Annual Review of Anthropology* 14 (1): 315–41. http://dx.doi.org/10.1146/annurev.an.14.100185.001531.

Kay, R. F., and M. Cartmill. 1975. "Skull of *Palaechthon* and Comments on Ecological Adaptations of Plesiadapoidea." *American Journal of Physical Anthropology* 42: 311–311.

Kay, R. F., and M. Cartmill. 1977. "Cranial Morphology and Adaptations of *Palaechthon nacimienti* and Other Paromomyidae (Plesiadapoidea, Primates), with a Description of a New Genus and Species." *Journal of Human Evolution* 6 (1): 19–53. http://dx.doi.org/10.1016/S0047-2484(77)80040-7.

Kay, R. F., and H. H. Covert. 1984. "Anatomy and Behavior of Extinct Primates." In *Food Acquisition and Processing in Primates*, ed. D. J. Chivers, B. A. Wood, and A. Bilsborough, 467–508. New York: Plenum Press.

Kay, R. F., and K. M. Hiiemae. 1974. "Jaw Movement and Tooth Use in Recent and Fossil Primates." *American Journal of Physical Anthropology* 40 (2): 227–56. http://dx.doi.org/10.1002/ajpa.1330400210. Medline:4815136

Kay, R. F., and W. L. Hylander. 1978. "The Dental Structure of Mammalian Folivores with Special Reference to Primates and Phalangeroidea (Marsupialia)." In *The Ecology of Arboreal Folivores*, ed. G. G. Montgomery, 173–191. Washington, DC: Smithsonian Institution.

Kay, R. F., and E. L. Simons. 1980. "The Ecology of Oligocene African Anthropoidea." *International Journal of Primatology* 1 (1): 21–37. http://dx.doi.org/10.1007/BF02692256.

King, S. J., S. J. Arrigo-Nelson, S. T. Pochron, G. M. Semprebon, L. R. Godfrey, P. C. Wright, and J. Jernvall. 2005. "Dental Senescence in a Long-Lived Primate Links Infant Survival to Rainfall." *Proceedings of the National Academy of Sciences of the United States of America* 102 (46): 16579–83. http://dx.doi.org/10.1073/pnas.0508377102. Medline:16260727.

Kinzey, W. G. 1978. "Feeding Behavior and Molar Features in Two Species of Titi Monkey." In *Recent Advances in Primatology*, Volume 1: *Behavior*, ed. D. J. Chivers and J. Herbert, 373–85. New York: Academic Press.

Korioth, T.W.P., and A. Versluis. 1997. "Modeling the Mechanical Behavior of the Jaws and Their Related Structures by Finite Element (FE) analysis." *Critical Reviews in Oral Biology and Medicine* 8 (1): 90–104. http://dx.doi.org/10.1177/10454411970080010501. Medline:9063627.

Kuroda, S. 1992. "Ecological Interspecies Relationships between Gorillas and Chimpanzees in the Ndoki-Nouabale Reserve, Northern Congo." In *Topics in Primatology*, Volume 2: *Behavior, Ecology and Conservation*, ed. N. Itoigawa, Y. Sugiyama, G. P. Sackett, and R.K.R. Thompson, 385–94. Tokyo: University of Tokyo Press.

Lambert, J. E., C. A. Chapman, R. W. Wrangham, and N. L. Conklin-Brittain. 2004. "Hardness of Cercopithecine Foods: Implications for the Critical Function of Enamel Thickness in Exploiting Fallback Foods." *American Journal of Physical Anthropology* 125 (4): 363–8. http://dx.doi.org/10.1002/ajpa.10403. Medline:15386250.

Lee-Thorp, J. 2002. "Hominid Dietary Niches from Proxy Chemical Indicators in Fossils: The Swartkrans Evidence." In *Human Diet: Its Origin and Evolution*, ed. P. S. Ungar and M. F. Teaford, 123–141. Westport, CT: Bergin and Garvey.

Lieberman, D. E., G. E. Krovitz, F. W. Yates, M. Devlin, and M. St Claire. 2004. "Effects of Food Processing on Masticatory Strain and Craniofacial Growth in a Retrognathic Face." *Journal of Human Evolution* 46 (6): 655–77. http://dx.doi.org/10.1016/j.jhevol.2004.03.005. Medline:15183669.

Lucas, P. W. 2004. *Dental Functional Morphology: How Teeth Work*. New York: Cambridge University Press. http://dx.doi.org/10.1017/CBO9780511735011.

Lucas, P. W., R. T. Corlett, and D. A. Luke. 1985. "Plio-Pleistocene Hominid Diets: An Approach Combining Masticatory and Ecological Analysis." *Journal of Human Evolution* 14 (2): 187–202. http://dx.doi.org/10.1016/S0047-2484(85)80006-3.

Lucas, P. W., and D. A. Luke. 1984. "Optimum Mouthful for Food Comminution in Human Mastication." *Archives of Oral Biology* 29 (3): 205–10. http://dx.doi.org/10.1016/0003-9969(84)90056-6. Medline:6587841.

Lucas, P. W., C. R. Peters, and S. R. Arrandale. 1994. "Seed-Breaking Forces Exerted by Orang-Utans with Their Teeth in Captivity and a New Technique for Estimating Forces Produced in the Wild." *American Journal of Physical Anthropology* 94 (3): 365–78. http://dx.doi.org/10.1002/ajpa.1330940306. Medline:7943191.

MacKinnon, J. 1977. “A Comparative Ecology of Asian Apes.” *Primates* 18 (4): 747–72. http://dx.doi.org/10.1007/BF02382929.

Marinescu, R., D. J. Daegling, and A. J. Rapoff. 2005. “Finite-Element Modeling of the Anthropoid Mandible: The Effects of Altered Boundary Conditions.” *Anatomical Record, Part A: Discoveries in Molecular, Cellular, and Evolutionary Biology* 283 (2): 300–9. http://dx.doi.org/10.1002/ar.a.20166. Medline:15747352.

Martin-de las Heras, S., A. Valenzuela, C. Ogayar, A. J. Valverde, and J. C. Torres. 2005. “Computer-Based Production of Comparison Overlays from 3D-Scanned Dental Casts for Bite Mark Analysis.” *Journal of Forensic Sciences* 50 (1): 127–33. http://dx.doi.org/10.1520/JFS2004226. Medline:15831006.

McCollum, M. A. 1997. “Palatal Thickening and Facial Form in *Paranthropus*: Examination of Alternative Developmental Models.” *American Journal of Physical Anthropology* 103 (3): 375–92. http://dx.doi.org/10.1002/(SICI)1096-8644(199707)103:3<375::AID-AJPA7>3.0.CO;2-P. Medline:9261500

McHenry, H. J. 1988. “New Estimates of Body Weight in Early Hominids and Their Significance to Encephalization and Megadontia in “Robust” Australopithecines.” In *Evolutionary History of the Robust Australopithecines*, ed. F. E. Grine, 133–148. New York: Aldine.

Meldrum, D. J., and R. F. Kay. 1997. “*Nuciruptor rubricae*, a New Pitheciin Seed Predator from the Miocene of Colombia.” *American Journal of Physical Anthropology* 102 (3): 407–27. http://dx.doi.org/10.1002/(SICI)1096-8644(199703)102:3<407::AID-AJPA8>3.0.CO;2-R. Medline:9098507

Merceron, G., S. Taylor, R. Scott, Y. Chaimanee, and J. J. Jaeger. 2006. “Dietary Characterization of the Hominoid *Khoratpithecus* (Miocene of Thailand): Evidence from Dental Topographic and Microwear Texture Analyses.” *Naturwissen* 93 (7): 329–33. http://dx.doi.org/10.1007/s00114-006-0107-0.

M’kirera, F., and P. S. Ungar. 2003. “Occlusal Relief Changes with Molar Wear in *Pan troglodytes troglodytes* and *Gorilla gorilla gorilla*.” *American Journal of Primatology* 60 (2): 31–41. http://dx.doi.org/10.1002/ajp.10077. Medline:12784284.

Nishihara, T. 1992. “A Preliminary Report on the Feeding Habits of Western Lowland Gorillas (*Gorilla gorilla gorilla*) in the Ndoki Forest, Northern Congo.” In *Topics in Primatology*, Volume 2: *Behavior, Ecology and Conservation*, ed. N. Itoigawa, Y. Sugiyama, G. P. Sackett, and R.K.R. Thompson, 225–40. Tokyo: University of Tokyo Press.

Osborn, H. F. 1907. *Evolution of Mammalian Molar Teeth to and from the Triangular Type*. New York: MacMillan Company. http://dx.doi.org/10.5962/bhl.title.1571

Owen, R. 1840. *Odontography*. London: Hippolyte Bailliere.

Peters, C. R. 1981. “Robust vs. Gracile Early Hominid Masticatory Capabilities: The Advantages of the Megadonts.” In *The Perception of Human Evolution*, ed. L. L. Mai, E. Shanklin, and R. W. Sussman, 161–181. Los Angeles: University of California Press.

Pilbeam, D., and S. J. Gould. 1974. “Size and Scaling in Human Evolution.” *Science* 186 (4167): 892–901. http://dx.doi.org/10.1126/science.186.4167.892. Medline:4219964

Plavcan, J. M., and D. J. Daegling. 2006. "Interspecific and Intraspecific Relationships between Tooth Size and Jaw Size in Primates." *Journal of Human Evolution* 51 (2): 171–84. http://dx.doi.org/10.1016/j.jhevol.2006.02.005. Medline:16620914

Rafferty, K. L., S. W. Herring, and C. D. Marshall. 2003. "Biomechanics of the Rostrum and the Role of Facial Sutures." *Journal of Morphology* 257 (1): 33–44. http://dx.doi.org/10.1002/jmor.10104. Medline:12740894

Rak, Y. 1983. *The Australopithecine Face*. New York: Academic Press.

Ravosa, M. J. 1990. "Functional Assessment of Subfamily Variation in Maxillomandibular Morphology among Old World Monkeys." *American Journal of Physical Anthropology* 82 (2): 199–212. http://dx.doi.org/10.1002/ajpa.1330820209. Medline:2360614

Ravosa, M. J. 1991. "Structural Allometry of the Prosimian Mandibular Corpus and Symphysis." *Journal of Human Evolution* 20 (1): 3–20. http://dx.doi.org/10.1016/0047-2484(91)90042-T.

Ravosa, M. J. 1996. "Jaw Morphology and Function in Living and Fossil Old World Monkeys." *International Journal of Primatology* 17 (6): 909–32. http://dx.doi.org/10.1007/BF02735294.

Ravosa, M. J. Sep-Oct 2000. "Size and Scaling in the Mandible of Living and Extinct Apes." *Folia Primatologica* 71 (5): 305–22. http://dx.doi.org/10.1159/000021754. Medline:11093035

Remis, M. J. 2002. "Food Preferences among Captive Western Gorillas (*Gorilla gorilla gorilla*) and Chimpanzees (*Pan troglodytes*)." *International Journal of Primatology* 23 (2): 231–49. http://dx.doi.org/10.1023/A:1013837426426.

Richmond, B. G., B. W. Wright, I. R. Grosse, P. C. Dechow, C. F. Ross, M. A. Spencer, and D. S. Strait. 2005. "Finite Element Analysis in Functional Morphology." *Anatomical Record, Part A: Discoveries in Molecular, Cellular, and Evolutionary Biology* 283 (2): 259–74. http://dx.doi.org/10.1002/ar.a.20169. Medline:15747355.

Robinson, B. W., and D. S. Wilson. 1998. "Optimal Foraging, Specialization, and a Solution to Liem's Paradox." *American Naturalist* 151 (3): 223–35. http://dx.doi.org/10.1086/286113. Medline:18811353

Robinson, J. T. 1954. "Prehominid Dentition and Hominid Evolution." *Evolution* 8 (4): 324–34. http://dx.doi.org/10.2307/2405779.

Rodman, P. S. 1977. "Feeding Behaviour of Orangutans of the Kutai Nature Reserve, East Kalimantan." In *Primate Ecology: Studies of Feeding and Ranging Behaviour in Lemurs, Monkeys and Apes*, ed. T. H. Clutton-Brock, 383–413. London: Academic Press.

Rosenberger, A. L., and W. G. Kinzey. 1976. "Functional Patterns of Molar Occlusion in Platyrrhine Primates." *American Journal of Physical Anthropology* 45 (2): 281–97. http://dx.doi.org/10.1002/ajpa.1330450214. Medline:822731.

Ross, C. F. 2000. "Into the Light: The Origin of Anthropoidea." *Annual Review of Anthropology* 29 (1): 147–94. http://dx.doi.org/10.1146/annurev.anthro.29.1.147.

Ross, C. F., and W. L. Hylander. 2000. "Electromyography of the Anterior Temporalis and Masseter Muscles of Owl Monkeys (*Aotus trivirgatus*) and the Function of the Postorbital Septum." *American Journal of Physical Anthropology* 112 (4): 455–68.

http://dx.doi.org/10.1002/1096-8644(200008)112:4<455::AID-AJPA4>3.0.CO;2-4. Medline:10918124

Rubin, C. T., K. J. McLeod, and S. D. Bain. 1990. "Functional Strains and Cortical Bone Adaptation: Epigenetic Assurance of Skeletal Integrity." *Journal of Biomechanics* 23 (Suppl. 1): 43–54. http://dx.doi.org/10.1016/0021-9290(90)90040-A. Medline:2081744

Sakka, M. 1984. "Cranial Morphology and Masticatory Adaptations." In *Food Acquisition and Processing in Primates*, ed. D. J. Chivers, B. A. Wood, and A. Bilsborough, 415–427. New York: Plenum.

Schwartz-Dabney, C. L., and P. C. Dechow. 2003. "Variations in Cortical Material Properties throughout the Human Dentate Mandible." *American Journal of Physical Anthropology* 120 (3): 252–77. http://dx.doi.org/10.1002/ajpa.10121. Medline:12567378

Scott, E. C. 1979. "Dental Wear Scoring Technique." *American Journal of Physical Anthropology* 51 (2): 213–7. http://dx.doi.org/10.1002/ajpa.1330510208.

Scott, R. S., P. S. Ungar, T. S. Bergstrom, C. A. Brown, F. E. Grine, M. F. Teaford, and A. Walker. 2005. "Dental Microwear Texture Analysis Shows Within-Species Diet Variability in Fossil Hominins." *Nature* 436:693–5. http://dx.doi.org/10.1038/nature03822. Medline:16079844.

Seligsohn, D., and F. S. Szalay. 1978. "Relationship between Natural Selection and Dental Morphology: Tooth Function and Diet in *Lepilemur* and *Hapalemur*." In *Development, Function and Evolution of Teeth*, ed. P. M. Butler and K. A. Joysey, 289–307. New York: Academic Press.

Simpson, G. G. 1933. "Paleobiology of Jurassic Mammals." *Paleobiologica* 5:127–58.

Smith, E. 1999. "A Functional Analysis of Molar Morphometrics in Living and Fossil Hominoids Using 2-D Digitized Images." PhD dissertation, University of Toronto, Toronto.

Smith, J. M., and R.J.G. Savage. 1959. "The Mechanics of Mammalian Jaws." *School Science Review* 40:289–301.

Smith, R. J. 1978. "Mandibular Biomechanics and Temporomandibular Joint Function in Primates." *American Journal of Physical Anthropology* 49 (3): 341–9. http://dx.doi.org/10.1002/ajpa.1330490307. Medline:103437

Smith, R. J. 1983. "The Mandibular Corpus of Female Primates: Taxonomic, Dietary, and Allometric Correlates of Interspecific Variations in Size and Shape." *American Journal of Physical Anthropology* 61 (3): 315–30. http://dx.doi.org/10.1002/ajpa.1330610306. Medline:6614146

Smith, R. J. 1984. "Comparative Functional Morphology of Maximum Mandibular Opening (Gape) in Primates." In *Food Acquisition and Processing in Primates*, ed. D. J. Chivers, B. A. Wood, and A. Bilsborough, 231–55. New York: Plenum Press.

Spears, I. R., and R. H. Crompton. 1996. "The Mechanical Significance of the Occlusal Geometry of Great Ape Molars in Food Breakdown." *Journal of Human Evolution* 31 (6): 517–35. http://dx.doi.org/10.1006/jhev.1996.0077.

Spencer, M.A. 1999. "Constraints on Masticatory System Evolution in Anthropoid Primates." *American Journal of Physical Anthropology* 108 (4): 483–506. http://dx.doi

.org/10.1002/(SICI)1096-8644(199904)108:4<483::AID-AJPA7>3.0.CO;2-L. Medline:10229390

Spencer, M. A., and B. Demes. 1993. "Biomechanical Analysis of Masticatory System Configuration in Neandertals and Inuits." *American Journal of Physical Anthropology* 91 (1): 1–20. http://dx.doi.org/10.1002/ajpa.1330910102. Medline:8512051

Sponheimer, M., B. H. Passey, D. J. de Ruiter, D. Guatelli-Steinberg, T. E. Cerling, and J. A. Lee-Thorp. 2006. "Isotopic Evidence for Dietary Variability in the Early Hominin *Paranthropus robustus*." *Science* 314 (5801): 980–2. http://dx.doi.org/10.1126/science.1133827. Medline:17095699.

Stanford, C. B., and J. B. Nkurunungi. 2003. "Do Wild Chimpanzees and Mountain Gorillas Compete for Food?" *American Journal of Physical Anthropology* Suppl. 36: 198–99.

Strait, S. G. 1993. "Molar Morphology and Food Texture among Small Bodied Insectivorous Mammals." *Journal of Mammalogy* 74 (2): 391–402. http://dx.doi.org/10.2307/1382395.

Strait, D. S., Q. Wang, P. C. Dechow, C. F. Ross, B. G. Richmond, M. A. Spencer, and B. A. Patel. 2005. "Modeling Elastic Properties in Finite-Element Analysis: How Much Precision Is Needed to Produce an Accurate Model?" *Anatomical Record, Part A: Discoveries in Molecular, Cellular, and Evolutionary Biology* 283 (2): 275–87. http://dx.doi.org/10.1002/ar.a.20172. Medline:15747346.

Strait, D. S., G. W. Weber, S. Neubauer, J. Chalk, B. G. Richmond, P. W. Lucas, M. A. Spencer, C. Schrein, P. C. Dechow, C. F. Ross, et al. 2009. "The Feeding Biomechanics and Dietary Ecology of *Australopithecus africanus*." *Proceedings of the National Academy of Sciences of the United States of America* 106 (7): 2124–9. http://dx.doi.org/10.1073/pnas.0808730106. Medline:19188607

Szalay, F. S. 1975. "Hunting-Scavenging Protohominids: A Model for Hominid Origins." *Man* 10 (3): 420–9. http://dx.doi.org/10.2307/2799811.

Taylor, A. B. 2002. "Masticatory form and function in the African apes." *American Journal of Physical Anthropology* 117 (2): 133–56. http://dx.doi.org/10.1002/ajpa.10013. Medline:11815948

Teaford, M. F. 1983. "The Morphology and Wear of the Lingual Notch in Macaques and Langurs." *American Journal of Physical Anthropology* 60 (1): 7–14. http://dx.doi.org/10.1002/ajpa.1330600103. Medline:6869505

Teaford, M. F., and P. S. Ungar. 2000. "Diet and the Evolution of the Earliest Human Ancestors." *Proceedings of the National Academy of Sciences of the United States of America* 97 (25): 13506–11. http://dx.doi.org/10.1073/pnas.260368897. Medline:11095758

Teaford, M. F., P. S. Ungar, and F. E. Grine. 2002. "Paleontological Evidence for the Diets of African Plio-Pleistocene Hominins with Special Reference to early *Homo*." In *Human Diet: Its Origin and Evolution*, ed. P. S. Ungar and M. F. Teaford, 143–166. Westport, CT: Bergin and Garvey.

Tobias, P. V. 1967. *The Cranium and Maxillary Dentition of Australopithecus (Zinjanthropus) boisei*. Cambridge: Cambridge University Press. http://dx.doi.org/10.1017/CBO9780511897795

Wolpoff, M. H. 1977. "Systematic Variation in Early Hominid Corpus Dimensions." *Anthropologischer Anzeiger* 36 (1): 3–6. Medline:411415.

Wood, B. A., S. A. Abbott, and S. H. Graham. 1983. "Analysis of the Dental Morphology of Plio-Pleistocene Hominids, II: Mandibular Molars—Study of Cusp Areas, Fissure Pattern and Cross Sectional Shape of the Crown." *Journal of Anatomy* 137 (Pt 2): 287–314. Medline:6415025.

Wood, B. A., and L. C. Aiello. 1998. "Taxonomic and Functional Implications of Mandibular Scaling in Early Hominins." *American Journal of Physical Anthropology* 105 (4): 523–38. http://dx.doi.org/10.1002/(SICI)1096-8644(199804)105:4<523::AID-AJPA9>3.0.CO;2-O. Medline:9584893

Wood, B. A., and M. Ellis. 1986. "Evidence for Dietary Specialization in the 'Robust' Australopithecines." *Anthropos (Brno)* 23:101–24.

Wood, B. A., and C. G. Stack. 1980. "Does Allometry Explain the Differences between Gracile and Robust Australopithecines?" *American Journal of Physical Anthropology* 52 (1): 55–62. http://dx.doi.org/10.1002/ajpa.1330520108.

Wood, B. A., and D. S. Strait. 2000. "*Paranthropus boisei*: A Derived Eurytope?" *American Journal of Physical Anthropology* Suppl. 30: 326.

Wrangham, R. W. 2005. "The Delta Hypothesis: Hominoid Ecology and Hominin Origins." In *Interpreting the Past: Essays on Human, Primate and Mammal Evolution in Honor of David Pilbeam*, ed. D. E. Lieberman, R. J. Smith, and J. Kelley, 231–242. Boston: Brill Academic Publishers.

Zuccotti, L. F., M. D. Williamson, W. F. Limp, and P. S. Ungar. 1998. "Technical Note: Modeling Primate Occlusal Topography Using Geographic Information Systems Technology." *American Journal of Physical Anthropology* 107 (1): 137–42. http://dx.doi.org/10.1002/(SICI)1096-8644(199809)107:1<137::AID-AJPA11>3.0.CO;2-1. Medline:9740307

7

Dental Microwear and Paleoecology

Mark F. Teaford,
Peter S. Ungar, and
Frederick E. Grine

Over the past fifty years, analyses of microscopic wear patterns on teeth, or dental microwear analyses, have shed light on diet and tooth use in living and fossil animals (e.g., Walker 1976; Walker et al. 1978; Rensberger 1978; Puech and Prone 1979; Grine 1981; Rose et al. 1981; Ryan 1981; Walker 1981; Puech 1984a; Teaford and Walker 1984; Rensberger 1986; Teaford 1988; Ryan and Johanson 1989; Ungar 1990; Teaford and Glander 1991; Solounias and Moelleken 1992a; Strait 1993; Daegling and Grine 1994; Ungar 1994; Danielson and Reinhard 1998; King et al. 1999a; Silcox and Teaford 2002; Solounias and Semprebon 2002; Leakey et al. 2003; Nystrom et al. 2004; Semprebon et al. 2004; Merceron et al. 2005b; Scott et al. 2005; Grine et al. 2006a, 2006b; Teaford et al. 2008; Ungar et al. 2008, 2010; Rodrigues et al. 2009; Williams et al. 2009; Solounias et al. 2010; Firmat et al. 2011). However, studies of dental microwear are limited by the fact that most foods are not hard enough to scratch enamel (Peters 1982; Lucas 1991). Thus many microwear patterns are actually caused by abrasives on foods rather than by the foods themselves. Still, since any such evidence represents the direct effect of an item in an animal's environment on that animal's teeth, this raises interesting possibilities for insights into paleoecology. In essence, what can analyses of diet, and abrasives in or on foods, tell us about the environments in which animals once lived?

Surprisingly, relatively few studies of dental microwear have addressed this question. A major reason for this

DOI: 10.5876/9781607322252:c07

shortcoming is that most workers have focused on primates, and these taxa have very eclectic diets as compared with many other animals. Thus, it is hard to extract paleoecological insights from microwear evidence when the animal in question feeds on a variety of foods in a variety of substrates. Moreover, despite the fact that primates range from snowy regions of Japan to the deserts of the Ethiopian highlands, modern taxa are, for the most part, found in tropical forests and woodlands, with few in African savannas (Richard 1985; Campbell et al. 2006). This further limits the usefulness of modern primate microwear evidence for paleoecological interpretations. Fortunately, nonprimate mammals inhabit other habitats and have more restricted diets and/or narrower ranges of feeding and foraging behaviors, so that they may yield additional ecological insights. The purpose of this chapter is to examine what has been done in the way of dental microwear analysis, and to see what it might tell us about paleoecology.

HISTORY OF DENTAL MICROWEAR ANALYSES

Initial analyses of dental microwear were qualitative in nature and based on light-microscope assessments of tooth surfaces, a state to which some workers are only now returning after more than half a century of methodological advancements. In the first studies, Butler (1952) and Mills (1955, 1963) noticed characteristic orientations of scratches on the teeth of different mammals, providing the initial evidence for the existence of two "phases" of masticatory jaw movement in primates. Similarly, Dahlberg and Kinzey (1962) observed the possibility of documenting differences in diet based on (among other things) differences in the amount of microscopic scratching on teeth. Subsequent work by a number of researchers rekindled interest in the topic (Walker 1976; Puech 1977; Rensberger 1978; Walker et al. 1978; Puech and Prone 1979; Ryan 1979; Puech et al. 1980; Walker 1980, 1981; Grine 1981; Puech et al. 1981; Ryan 1981; Rensberger 1982), with a methodological shift to use of the scanning electron microscope (SEM) because of its superior depth of focus and resolution of detail.

Of course, finer microscopic resolution raised the possibility of finer dietary/ecological distinctions—as long as that information could be put to efficient use. This led to the initial quantification of dental microwear, using the incidence of scratches and pits, measurements of the lengths and widths of these features, and assessment of feature orientation, employing various forms of computer-controlled digitizers or calipers in conjunction with SEM micrographs (e.g., Fine and Craig 1981; Gordon 1982, 1984b, c; Teaford and Walker 1984; Teaford 1985; Grine 1986; Kelley 1986; Solounias et al. 1988; Young and Robson 1987).

However, as the number of studies began to grow, it quickly became apparent that some workers were using different and sometimes incomparable methods to quantify dental microwear. This presented researchers with an array of methodological difficulties (Covert and Kay 1981; Gordon 1982, Gordon and Walker 1983; Kay and Covert 1983; Gordon 1984b, 1988; Teaford 1988). One attempt to circumvent the subjectivity associated with the standard techniques of microwear quantification was the use of Fourier transforms and image processing introduced by Kay (1987). This method was applied in the analysis of the same micrographs of *Australopithecus africanus* and *Paranthropus robustus* occlusal facets that had been quantified earlier by more "traditional" methods (Grine and Kay 1988; Kay and Grine 1988). Their conclusions corresponded to the differences observed earlier (Grine 1986).

In order to standardize microwear quantification, Ungar (Ungar et al. 1991; Ungar 1995) developed a "semi-automated" procedure that is still used by many researchers today. To use it, images must be stored in digital format so they can be opened by the freeware package "Microware" (http://comp.uark.edu/~pungar/). Each microwear feature in the micrograph has to be identified and measured using a mouse and cursor on a computer screen. While the technique provides a standardized series of measurements for analysis, and stores them in a format readily accessible to most statistical packages, the work is still time-consuming and somewhat subjective. Nevertheless, the availability of this standardized technique prompted work on a wide range of taxa. Most studies, however, focused on what animals were eating, or how their teeth were being used, rather than on other factors with ecological implications, such as the environment in which the foods were found, the aridity of these environments, feeding heights in the canopy, and so on.

REVIEW OF STUDIES OF MODERN MATERIAL

Interpretations of fossils, of course, should ultimately be based on an appreciation of living animals. Thus, we here review the different approaches to dental microwear analysis of extant taxa before returning to the topic of paleoecology.

Analyses of museum material

Museum collections of modern mammalian teeth have proven to be an invaluable source of data for rare animals, and for sample sizes that cannot be obtained through work on living animals in the wild. Analyses of museum specimens with good background data (i.e., exact locality and time of year of collection) can permit the establishment of correlations between certain diets or abrasives (or patterns of tooth use), and certain microwear patterns. These correlations depend on which teeth are analyzed, because anterior teeth are used differently than posterior teeth, with the incisors and canines being used to *ingest* food, and the premolars and molars being used to *chew* food once it has been ingested.

Analyses of incisor microwear have yielded two basic conclusions. Both are intuitively concordant. First, animals that use their incisors very heavily in the ingestion of food show higher densities of incisal microwear features (Ryan 1981; Kelley 1986, 1990; Ungar 1990, 1994; Krueger and Ungar 2010). Second, the orientation of striations on the incisors reflects the direction of preferred movement of food (or other items) across these teeth (Walker 1976; Rose et al. 1981; Ryan 1981; Ungar 1994; Bax and Ungar 1999; Rivals and Semprebon 2010; Krueger and Ungar 2010). Thus, for example, the orangutan, which generally uses its incisors a great deal in preparing food, shows more scratches on its incisors than does the gibbon, and the scratches often run in a mesiodistal direction, reflecting the direction that branches are pulled between the front teeth (Ungar 1994). More interestingly for the purposes of this chapter, analyses of incisor microwear have also yielded insights that may be more generally applicable to paleoecological reconstructions, because there is evidence that the particle sizes of abrasives are reflected in the sizes of microscopic scratches on the teeth. This, in turn, may be indicative of feeding height in the canopy, as phytoliths in plants are generally larger than the abrasive particles that dominate clay-based tropical soils (Ungar 1990, 1994).

Analyses of molar microwear have demonstrated additional points of interest. As noted above, correlations between the orientation of jaw movement and scratches on wear facets have yielded insights into the mechanics of mastication in a variety of mammalian species (Gordon 1984c; Rensberger 1986; Young and Robson 1987; Teaford and Byrd 1989; Young et al. 1990; Hojo 1996). In addition, grazers tend to show more microscopic scratches on their molars compared to browsers (Solounias and Moelleken 1992a, b; Solounias and Hayek 1993; Solounias and Moelleken 1994; MacFadden et al. 1999; Solounias and Semprebon 2002; Merceron et al. 2004a, 2004b, 2005a, 2005b; Schubert et al. 2006; Ungar et al. 2007). Animals that eat hard objects usually show large pits on their molars, while leaf-eaters tend to exhibit relatively more scratches than pits (Teaford and Walker 1984; Teaford 1988). Those "hard objects" can evidently include nuts, but also smaller items like insect exoskeletons (Strait 1993; Silcox and Teaford 2002). Microwear may also be encountered on nonocclusal (buccal and lingual) molar surfaces, and this may give additional indications of the abrasiveness of the diet, the size of food items, or even the degree of terrestriality (Puech 1977; Lalueza Fox 1992; Lalueza Fox and Pérez-Pérez 1993; Pérez-Pérez et al. 2003; Ungar and Teaford 1996). However, studies of nonocclusal microwear published to date have yet to demonstrate how microwear might be formed on the sides of teeth, where applied forces are minimal to nonexistent. Thus, the precise causes of potential correlations between diet and nonocclusal microwear remain to be explained. Hopefully, ongoing work (Galbany et al. 2009a, b; Romero and DeJuan 2007; Romero et al. 2009) may provide some answers.

Studies of molar microwear have also yielded glimpses of even more subtle differences associated with variation in diet. These include differences between closely related genera (Solounias and Hayek 1993; Teaford 1993; Daegling and Grine 1999; Oliveira 2001; Merceron et al. 2004b, 2005a, 2005b; Calandra et al. 2008, 2010; DeMiguel et al. 2010; Merceron et al. 2010; Ramdarshan et al. 2010; Semprebon et al. 2011), differences between species (e.g., *Gorilla beringei* vs. *G. gorilla*) (King et al. 1999a), and differences between populations within species (Teaford and Robinson 1989; Mainland 1998, 2000, 2003, 2006; Merceron et al. 2004a).

Obviously, such analyses are only as good as the data that are recorded for museum samples, and the observed or published dietary information for those species or populations. Unfortunately, there are very few primate collections that provide the exact date and precise location of collection for each specimen. As a result, finer-resolution studies of diet and dental microwear based on museum samples of primates are relatively rare. Possibilities are notably better for other taxa, as some have been collected systematically in large numbers to help control populations (e.g., the roe deer analyzed by Merceron et al. 2004a), or have been slaughtered at specific times of the year (e.g., the sheep and goats analyzed by Mainland 1998, 2000, 2003).

Analyses of live animals

Unfortunately, obtaining fine-resolution correlations between dental microwear and diet in living animals, whether in the lab or the wild, can be a rather difficult task when culling animals at specific times of the year is not an option. Because living animals must be anesthetized to make molds of their teeth, the timing and type of anesthesia to be administered may lead to problems. For example, the common use of ketamine administered after 8–12 hours of fasting will leave the animal rigidly hard to work with, salivating excessively, and with a thick organic film built up on its teeth. As a result, it is perhaps not surprising that, to date, there has been only one successful study to have employed laboratory animals (Teaford and Oyen 1989a, 1989b). In that study, groups of vervet monkeys were raised on hard and soft diets to assess the effects of food properties on craniofacial growth. The hard diet consisted of monkey chow and apples, and the soft diet of water-softened monkey chow and pureed applesauce. Although both diets had the same basic ingredients, including a very abrasive monkey chow component, two rather surprising differences were observed. First, the *incisors* of the soft-food animals were more heavily worn than those of the hard-food animals, because the former routinely rubbed handfuls of soft, abrasive, food across their incisors, whereas the latter hardly used their incisors at all (Teaford 1988). Second, the *molars* of the monkeys on the soft diet showed smaller pits on their occlusal surfaces (Teaford and Oyen 1989b), perhaps due to adhesive wear caused by repeated tooth–tooth contacts in chewing (Teaford and Runestad 1992).

The in vivo study by Teaford and Oyen (1989b) also demonstrated that the turnover in dental microwear can be quite rapid in animals with an abrasive diet. Indeed, all of the microwear features in an area sampled by an SEM micrograph were observed to change in 1–2 weeks, depending on whether the animal was raised on the hard or soft diet (Teaford and Oyen 1989a). This observation is the basis for the idea of the "Last Supper" effect (Grine 1986), where occlusal dental microwear may only record the effects of the most-recently eaten foods on the teeth. Of course, microwear on the buccal side of the molar teeth will show far slower turnover (Pérez-Pérez et al. 1994), but then animals do not chew on the buccal surfaces so it is still not clear what buccal microwear actually reflects.

A pilot study by Noble and Teaford (1995) using human volunteers and foods normally thought to be hard or abrasive reaffirmed that at least some foods in the North American diet (e.g., popcorn kernels) can scratch enamel. Other work has focused on rates of microscopic wear (Teaford and Tylenda 1991), which can be used to gain insights into clinical problems (Raphael et al. 2003; Janal et al. 2007), or the applicability of restorative dental materials (Turssi et al. 2005; Wu et al. 2005).

Laboratory studies of living animals are necessarily limited in that laboratory diets are not nearly as diverse as those in the wild, where seasonal and geographic differences may have a huge impact on dental microwear patterns. Similarly, the range of abrasives in different environments, or microhabitats, raises the possibility of detecting surprisingly subtle ecological differences through dental microwear analyses. Thus, work with animals in the wild has the potential to provide a wealth of information that is relevant to diet and can be used in paleoecological reconstructions. This type of work was pioneered by

Walker et al. (1978) on hyraxes, where skulls were collected directly from the same area in which behavioral observations were recorded. Unfortunately, while studies of living primates in the wild have been attempted a number of times, they have usually met with limited success. Primates often live in forested habitats where they are hard to see, and even harder to catch. Even in open habitats (e.g., baboons in the East African savanna), the work is difficult. Thus, to date, only two studies have consistently yielded high-quality copies of the teeth of living mammals in the wild. The first is the ongoing study at La Pacifica in the Guanacaste region of Costa Rica (Teaford and Glander 1991; Ungar et al. 1995; Teaford and Glander 1996; Dennis et al. 2004). There, howler monkeys (*Alouatta palliata*) are regularly observed, captured, and released in a dry tropical-forest setting. That work has certainly verified some of the observations from museum specimens (e.g., the relationship between leaf-eating and scratches on teeth). It has also given us glimpses of other complicating factors, such as variation in the amount of molar microwear from season to season, between riverine and nonriverine microhabitats (Teaford and Glander 1996), and possible effects of other forms of abrasives in natural plant foods—for example, silica trichomes on leaves (Teaford et al. 2006), and dust in the canopy (Ungar et al. 1995). The studies at La Pacifica have also shown that tooth wear generally proceeds at a rapid pace in the wild—in fact, at about 8–10 times the pace of that in US dental patients (Teaford and Glander 1991). Thus the "Last Supper" effect has the potential to differ markedly between species.

More recently, a second long-term study has begun to yield high-resolution casts of primate teeth in the wild (Nystrom et al. 2004). The study populations, from the Anubis–hamadryas hybrid zone of Awash National Park, Ethiopia, have been the focus of multidisciplinary work for over thirty years (Nagel 1973; Phillips-Conroy 1978; Sugawara 1979; Phillips-Conroy and Jolly 1986; Phillips-Conroy et al. 1991, 2000; Szmulewicz et al. 1999; Dirks et al. 2002). Taking dental impressions before the heavy onset of new leaves and grasses in this seasonal environment allowed Nystrom et al. (2004) to implicate "small-caliber environmental grit" as the main cause of the observed microwear patterns. Moreover, these patterns showed no significant difference between sexes, age groups, or different troops.

Studies such as these provide some insights into the specific causes of dental microwear, such as the abrasives on leaves that can cause striations on teeth (Lucas and Teaford 1995; Danielson and Reinhard 1998; Reinhard et al. 1999; Gügel et al. 2001; Teaford et al. 2006; Reinhard and Danielson 2005), and acids that can etch enamel (Mannerberg 1960; Boyde 1964; Puech 1984b; Puech et al. 1986; Teaford 1988, 1994; Ungar 1994; King et al. 1999b; Ranjitkar et al. 2008).

Microwear patterns also can be caused by what might be termed the *indirect* effects of food (Puech 1984a; Teaford 1988; Pastor 1992, 1993; Teaford 1994; Ungar 1994; Ungar et al. 1995; Daegling and Grine 1999; Nystrom et al. 2004). For instance, certain food-preparation procedures (grinding, cooking, etc.) may introduce abrasives into foods, causing a high incidence of microscopic scratches on teeth—scratches not caused by the foods themselves, but by the methods with which they were prepared (Pastor 1992, 1993; Teaford and Lytle 1996). Similarly, animals may also eat soft foods, and still show many scratches on their teeth if the foods are coated with abrasives (e.g., earthworms coated with dirt) (Silcox and Teaford 2002). Finally, especially if an animal has a soft but tough diet, tooth-on-tooth wear can yield characteristic microwear patterns as enamel edges penetrate the food and grind past each other, yielding a high incidence of small pits that are probably caused by the adhesive wear of enamel on enamel (Puech et al. 1981; Walker 1984; Puech 1984a; Puech et al. 1986; Radlanski and Jäger 1989; Teaford and Runestad 1992; Rafferty et al. 2002).

In vitro laboratory studies

If studies of living animals have yielded a complex array of possibilities, why not conduct experimental studies of enamel abraded by different substances? Some studies have demonstrated that the orientation of scratches on a tooth's surface can indeed reflect the orientation of tooth–food–tooth movements (e.g., Ryan 1979; Teaford and Walker 1983; Gordon 1984c; Walker 1984; Teaford and Byrd 1989; Morel et al. 1991). Other studies have shown that certain agents, such as windblown sand, or various acids, can leave characteristic microwear patterns on teeth (Puech and Prone 1979; Puech, Prone, Kraatz 1980; Puech, Prone, Albertini 1981; Gordon 1984a; Puech et al. 1985; Rensberger and Krentz 1988; King et al. 1999b; Ranjitkar et al. 2008). However, there have been surprisingly few controlled studies of the wear patterns caused by different types of foods.

Peters (1982) used standard physical-property-testing equipment while examining the effects of a range of African foods on dental microwear, ultimately showing that few foods could actually scratch enamel, with extraneous abrasives being one of the prime culprits instead (see also Puech et al. 1986). Only with more detailed analyses did subsequent work (e.g., Gügel et al. 2001) begin to demonstrate the effects of specific foods on microwear patterns (e.g., "cereal-specific" microwear related to phytolith content in certain grains). At the present time, the closest thing we have to a natural experiment is work on herbivores raised on known diets (e.g., Mainland 1998, 2000, 2003; Merceron et al. 2004a; Vanpoucke et al. 2009). Granted, the analyses were done on dental material from slaughtered animals, thus the diets have not been monitored or changed during the course of study. However, the restricted diets of the animals in question, and the combination of dental microwear analyses and fecal analyses have allowed researchers to show that traditional distinctions between "browsers" and "grazers" are not as simple as we might imagine, with microwear patterns being influenced by a range of factors, including phytolith content, extraneous abrasives, food toughness, and so on (Mainland 1998, 2000, 2003).

What don't we know about dental microwear and diet in modern animals?

At first glance, with all the work that has been done, it might seem that we know a great deal about dental microwear and diet. In reality, however, all we have are tiny windows into a complex world. Studies of living animals have been conducted on only two primate species (*Alouatta palliata* and *Papio hamadryas*) in two habitats, the dry tropical forest of Costa Rica, and the thornbush and savanna grassland of Ethiopia. Would dental microwear patterns differ for howlers or baboons in other habitats? Undoubtedly. It is not only likely that they would eat somewhat different types of foods, but that extraneous factors such as the degree of arboreality and terrestriality and aridity would influence exogenous factors that impact microwear. Other questions that bear consideration include how might other species share a habitat with *Alouatta* or *Papio* (or other taxa), and how would that be reflected in differences in dental microwear? What is the magnitude of seasonal, annual, geographic, and interspecific differences in dental microwear for other types of animals elsewhere in the world? How does the incidence of dental microwear relate to specific abrasives and foods in the wild? Clearly, a massive amount of work has yet to be done on live animals in the wild if we are to push analyses of fossils to their current limits of resolution.

Similarly, laboratory studies have barely begun to sort through the intricacies of dental microwear formation. As noted earlier, the effects of specific food items have yet to be documented in any systematic fashion, and the effects of foods naturally consumed in the wild have yet to be examined in any detail. As diets in the wild can be quite variable, and as dental microwear features can change quite quickly (Teaford and Oyen 1989a; Teaford and Glander 1991), what are the effects of different food items on overall microwear patterns within a specific diet? Will items that are abrasive, hard, or acidic effectively swamp other microwear patterns? Will the "Last Supper" effect vary between species, populations, and/or individual behaviors?

In essence, we must not lose sight of the fact that the microwear pattern on a given tooth is essentially a complex summary of past events, often with multiple signals superimposed on each other. Some wear processes, like acid etching, leave fairly distinctive patterns, but can have varying effects on teeth depending on the length of exposure of the tooth to the acid (Teaford 1994) and the abrasiveness of the rest of the diet (Noble and Teaford 1995). Other wear processes, like abrasion, may yield evidence of the effects of foods themselves, or the effects of extraneous abrasives adhering to foods (Teaford 1988; Pastor 1993). Still other processes, like adhesion, may yield wear patterns similar to those caused by small hard objects (Radlanski and Jäger 1989; Teaford and Runestad 1992; Rafferty et al. 2002). When animals have varied diets that incorporate foods of different properties, the effects of each food may be hard to decipher.

All of the analyses discussed above have been conducted on enamel, but what about wear on dentin? Until now, investigators have eschewed dentin, mainly because it is so soft as to be almost impossible to clean without introducing artificial microwear patterns. However, because it is so soft, it might be an indicator of even subtler diet distinctions than enamel, especially among those foods that are too soft to cause enamel scratches or pits. Clearly, more laboratory work needs to be done to explore the possibility of employing dentin microwear.

With regard to the analysis of museum specimens, there are very few collections of material for which detailed dietary information has been documented before the animals were killed (e.g., portions of the Sumatran collections at The Academy of Natural Sciences in Philadelphia, and the Dian Fossey gorilla collection at the Smithsonian Institution). Thus, virtually all studies of museum samples are limited in their resolution by the resolution of published dietary information for the animals in question.

ANALYSES OF PALEONTOLOGICAL AND ARCHAEOLOGICAL SAMPLES

Given the limitations of dental microwear analyses of modern animals, what can analyses of extinct populations tell us about paleoecology? Paleontological analyses of dental microwear have included a wide variety of animals, but as with studies of living mammals, more work has been done on primates than on anything else. Analyses of Miocene hominoids have helped document an impressive array of dietary adaptations in these early apes (Teaford and Walker 1984; Daegling and Grine 1994; Ungar 1996; King et al. 1999a). By contrast, analyses of Plio-Pleistocene cercopithecoid specimens have suggested a surprisingly limited array of dietary adaptations in both East and South Africa

(El Zaatari et al. 2005; Lucas and Teaford 1994; Leakey et al. 2003; Teaford et al. 2008), with variation in the incidence of pitting on the molars being, at most, 30 percent of that seen in extant colobines and cercopithecines. Still, the degree of nonocclusal microwear on the molars has been used as putative support for the idea that certain fossil colobines (e.g., *Cercopithecoides*) led a more terrestrial lifestyle than do modern colobines (Ungar and Teaford 1996). Molar microwear analyses have also helped to document the effects of phylogenetic constraints in fossil apes by documenting similar functions in taxa that have undergone shifts in molar morphology through time (Kay and Ungar 1997; Ungar et al. 2004).

Analyses of early hominins have documented differences in the possible dietary habits of *A. africanus* and *P. robustus* (Grine 1981, 1986, 1987; Grine and Kay 1988; Kay and Grine 1988). Recent work has taken analyses a step further by making the distinction between dental *capabilities* and dental *use*, as the capability to process certain foods may well have been of critical importance in certain situations (Teaford and Ungar 2000; Teaford et al. 2002; Ungar et al. 2006, 2008). For instance, analyses of early *Homo* molar microwear have shown that specimens attributed to *Homo erectus/ergaster* show a higher incidence of pitting than do those assigned to *Homo habilis* (Ungar et al. 2006), suggesting the consumption of tougher or harder items by *H. erectus/ergaster* (possibly as a critical fallback food). Studies of *Paranthropus boisei* have shown that, despite its massive, "hyperrobust" masticatory apparatus, this early hominin probably avoided hard foods, or only rarely used them as a critical fallback food (Ungar et al. 2008). Similar dietary inferences were also drawn from analyses of the geochronologically earlier East African taxa, *Australopithecus anamensis* and *Australopithecus afarensis* (Grine et al. 2006a, 2006b; Ungar et al. 2010).

Analyses of Holocene human populations have also yielded results of interest to paleoecology, due to the fact that finer temporal and geographic resolution is possible in such analyses. Thus, the transition from hunting and gathering to agriculture has left a complex signal in the microwear record, because it seems to depend on which populations are examined, and in which habitats. (Bullington 1991, Pastor 1992; Pastor and Johnston 1992; Schmidt 2001; Teaford 1991, 2002; Teaford et al. 2001; Organ et al. 2005). For example, in central North America, there seems to be a transition in preagricultural populations to a slightly harder diet before the actual switch to agriculture (Schmidt 2001), with subsequent agricultural populations having a softer, less variable diet (Bullington 1991). Along the southeastern coast of the United States, however, microwear evidence for that dietary transition is complicated by significant local differences in microhabitat, most notably between coastal and inland sites, with the former evidently including significant amounts of large-grained abrasives in foods (Teaford 1991, 2002; Teaford et al. 2001). In other regions of the world (e.g., the Indian subcontinent), interpretations are further complicated by changes in food-processing techniques. Stone grinding tools introduced significant abrasives into what would otherwise have been a fairly nonabrasive diet (Molleson and Jones 1991; Pastor 1992; Pastor and Johnston 1992), while the boiling of foods, in other populations, led to a marked *decrease* in the amount of molar microwear (Molleson et al. 1993; Ma and Teaford 2010).

Dental microwear work on nonprimate fossils has ranged from rodents (Rensberger 1978, 1982; Lewis et al. 2000; Rodrigues et al. 2009) to ruminants (Solounias and Dawson-Saunders 1988; Solounias and Hayek 1993; Solounias and Moelleken 1992a, b, 1994; Mainland 1998, 2000; MacFadden et al. 1999; Solounias and Semprebon 2002; Rivals and Deniaux 2003, 2005; Merceron et al. 2004a, 2004b, 2005a, b; Schubert et al. 2006; Ungar et al. 2007; DeMiguel et al. 2010), and from carnivores (Van Valkenburgh et al. 1990; Ungar et al. 2010) to dinosaurs (Fiorillo 1998; Varriale 2004; Schubert and Ungar 2005; Williams et al. 2009). What insights have they given us into paleoecology?

To date, workers have generally taken one of three approaches. Most have tried to document the diets of species that might serve as indicators of specific habitats or ecological zones (e.g., browsers vs. grazers). Others have tried to gain insights into the season of death of animals in some collections. A few others have tried to shed light on the size of abrasives in paleoenvironments.

The most commonly cited dietary distinction used in such paleoecological interpretations is that between browsers and grazers. Interestingly, the most commonly cited microwear evidence in support of such a distinction is one that was noted over thirty years ago in the classic paper by Walker et al. (1978): grazers tend to have more microwear and especially more scratches on their teeth than do browsers. In essence, methods of analysis may have changed, but some basic discoveries have stood the test of time!

The presence of browsers within a paleontological sample has usually been used as an indicator of more closed or forested environments, while the presence of grazers has been used as an indicator of more open grassland environments (e.g., Rensberger 1978; Solounias and Semprebon 2002; Rivals and Deniaux 2003; Merceron et al. 2004b, 2005a, 2005b; Rodrigues et al. 2009; Solounias et al. 2010). Such distinctions will undoubtedly prove to be oversimplifications as we come to know more about ecology and dental microwear. However, they have already forced researchers to consider multiple lines of evidence to further refine their interpretations (e.g., isotope analyses used by Merceron et al. 2004b, Merceron and Ungar 2005; Schubert et al. 2006; Grine et al. 2012).

Documentation of the season of death has also proved useful in interpretations of past populations. Obviously, such distinctions would be difficult to make for most fossil samples, but with large enough samples from narrow enough time ranges, some might be possible (e.g., analyses of prey species have yielded evidence of hunting seasons in *Homo heidelbergensis*) (Rivals and Deniaux 2003, 2005). With more recent human populations, such insights might also prove valuable (e.g., seasonal overgrazing by sheep domesticated by historic Norse populations) (Mainland 2006).

Finally, if investigators are to push dental microwear analyses further, the relationship between the sizes and shapes of microwear features and the sizes and shapes of abrasives needs to be better understood. Correlations between the size of abrasives and dental microwear patterns were initially suggested by Ungar (1992, 1994). He noted that phytoliths in upper-canopy leaves were larger than clay-based soil particles on the ground, perhaps helping to explain patterns of microwear differences between species of primates. Walker and Hagen (1994) subsequently suggested that the angularity of abrasive particles might complicate such interpretations, as large jagged particles might cause smaller-than-expected microwear features. However, little else has been done. Danielson and Reinhard (1998; Reinhard and Danielson 2005) suggested that calcium oxalate phytoliths were a "ubiquitous" presence in the diets of human populations in the American Southwest—populations with heavy gross tooth wear and microwear. Merceron et al. (2004b) suggested that there may be phytoliths of different sizes in C_3 vis-à-vis C_4 plants. Still, no modern comparative analyses have followed up on these suggestions.

Methodological cautions in analysis of fossil material

In light of the array of possibilities that revolve around dental microwear research, it must be stressed that the biggest challenge facing such analyses is a methodological one. Because not all studies have followed the same protocol, not all results are strictly comparable. Even with the use of semiautomated, computerized, digitizing routines (Ungar et al. 1991), interobserver error rates of SEM-based microwear quantification may be quite high (Grine et al. 2002). When this is coupled with the fact that most analyses have used relatively small samples (even for species with variable diets), the net effect is that dental microwear analyses have barely begun to live up to their potential. This is especially true in applying dental microwear results to paleoecological interpretations.

Recently, workers have begun to address this issue through the use of two new approaches: lower magnification work by light microscopy (where features are categorized into a limited number of classes) (Solounias and Semprebon 2002; Semprebon et al. 2004), and a higher-magnification combination of confocal microscopy and scale-sensitive fractal analysis (Scott et al. 2005). Although the former holds the promise of quick, cheap analyses of larger samples, and although published tests of interobserver error rates hint at better replicability than standard SEM-based analyses (Semprebon et al. 2004), the authors failed to present measures of error rates in a form that could be compared with those documented by Grine et al. (2002).

The low-magnification technique also requires significant training to master, and interobserver error rates have yet to be checked for categorizations of features into different sizes—the very distinctions that hold the most promise of documenting subtle dietary differences (Godfrey et al. 2004). Moreover, because it entails the use of low magnifications, it certainly will miss information provided by finer microwear features (Teaford 2007), and it may even leave experienced workers unable to distinguish larger microwear features from dentin exposures (Walker 2007). Finally, the low-magnification technique is apparently incapable of differentiating antemortem wear from taphonomic artifacts in many cases. This is manifest from a comparison of the sample sizes of South African fossil monkeys employed in the SEM study by El Zaatari et al. (2005) and the "low-mag" study by Williams et al. (2006). While El Zaatari et al. (2005) found that only six molars attributed to *Papio robinsoni* from Swartkrans Member 1 preserved reliable microwear not subjected to postmortem taphonomic damage, Williams et al. (2006) included twenty-two specimens (virtually the entire available sample) from the same deposit. Similarly, El Zaatari et al. (2005) concluded that no specimen of *Papio angusticeps* from the sites of Cooper's A and B and Kromdraai A preserved microwear that had not been taphonomically altered, whereas Williams et al. (2006) encountered no such problem with eighteen specimens (again, virtually the entire collection available from these sites)! Thus, this technique may ultimately be restricted to making only the grossest of dietary distinctions, as the inclusion of specimens with postmortem wear will often make the documentation of significant antemortem differences difficult, if not impossible, leaving some interpretations effectively meaningless.

On the other hand, the confocal approach holds the promise of providing truly objective characterizations of entire wear surfaces—and in three dimensions—with no need to identify or categorize specific wear features. Thus, it would seem to be the best technique available for obtaining useful numbers. To emphasize this dramatic shift from standard landmark-based measurements, the developers have referred to it as "dental microwear texture analysis" (Ungar et al. 2003; Scott et al. 2005). This technique is still in the process of refinement, with new analytical routines continuing to be developed. A database for future interpretations is still being gathered, and comparisons with data generated by previous techniques are still being completed. Initial results, though, are very promising, yielding insights into diet variability in early hominins. Microwear patterns overlap markedly between *A. africanus* and *P. robustus*, with most showing simple, isotropic wear surfaces, though some "gracile" and "robust" specimens have extremely anisotropic and complex ones respectively. This pattern is comparable to that seen in extant primates that "fall back" on mechanically challenging foods at times of resource stress (Scott et al. 2005). Microwear texture analysis holds the potential for analyses of much larger samples in a truly objective fashion.

CONCLUSIONS

While dental microwear analyses have been conducted for over fifty years, paleoecological applications of this approach are still in their infancy. Clearly the potential for insight is there, as certain microwear patterns may ultimately be tied to the presence of certain abrasives in the environment and to differences in phytogeographic cover. However, broader differences in dental microwear, between habitats, or microhabitats, have yet to be examined in any detail. Only with recent methodological improvements are investigators finally beginning to analyze samples of sufficient size and diversity to adequately document the ecological causes of differences in dental microwear. This work will require a great deal of effort, on a wide range of fronts, including analyses of living animals in the wild and in the lab, and more analyses of fossil material. The net effect, however, may produce finer resolution of ecological differences between past populations.

REFERENCES CITED

Bax, J. S., and P. S. Ungar. 1999. "Incisor Labial Surface Wear Striations in Modern Humans and Their Implications for Handedness in Middle and Late Pleistocene Hominids." *International Journal of Osteoarchaeology* 9 (3): 189–98. http://dx.doi.org/10.1002/(SICI)1099-1212(199905/06)9:3<189::AID-OA474>3.0.CO;2-N.

Boyde, A. 1964. "The Structure and Development of Mammalian Enamel." PhD dissertation, University of London.

Bullington, J. 1991. "Deciduous Dental Microwear of Prehistoric Juveniles from the Lower Illinois River Valley." *American Journal of Physical Anthropology* 84 (1): 59–73. http://dx.doi.org/10.1002/ajpa.1330840106. Medline:2018101

Butler, P. M. 1952. "The Milk Molars of Perissodactyla, with Remarks on Molar Occlusion." *Proceedings of the Zoological Society of London* 121 (4): 777–817. http://dx.doi.org/10.1111/j.1096-3642.1952.tb00784.x.

Calandra, I., U. B. Göhlich, and G. Merceron. 2008. "How Could Sympatric Megaherbivores Coexist? Example of Niche Partitioning within a Proboscidean Mommunity from the Miocene of Europe." *Naturwissenschaften* 95 (9): 831–8. http://dx.doi.org/10.1007/s00114-008-0391-y. Medline:18542904

Calandra, I., U. B. Göhlich, and G. Merceron. 2010. "Feeding Preferences of *Gomphotherium subtapiroideum* (Proboscidea, Mammalia) from the Miocene of Sandelzhausen (Northern Alpine Foreland Basin, Southern Germany) through Life and Geological Time: Evidence from Dental Microwear Analysis." *Paläontologische Zeitschrift* 84 (1): 205–15. http://dx.doi.org/10.1007/s12542-010-0054-0.

Campbell, C. J., A. Fuentes, K. C. MacKinnon, M. Panger, and S. K. Bearder, eds. 2006. *Primates in Perspective*. New York: Oxford University Press.

Covert, H. H., and R. F. Kay. 1981. "Dental Microwear and Diet: Implications for Determining the Feeding Behaviors of Extinct Primates, with a Comment on the Dietary Pattern of *Sivapithecus*." *American Journal of Physical Anthropology* 55 (3): 331–6. http://dx.doi.org/10.1002/ajpa.1330550307. Medline:6267943

Daegling, D. J., and F. E. Grine. 1994. "Bamboo Feeding, Dental Microwear, and Diet of the Pleistocene Ape *Gigantopithecus blacki*." *South African Journal of Science* 90:527–32.

Daegling, D. J., and F. E. Grine. 1999. "Terrestrial Foraging and Dental Microwear in *Papio ursinus*." *Primates* 40 (4): 559–72. http://dx.doi.org/10.1007/BF02574831.

Dahlberg, A. A., and W. G. Kinzey. 1962. "Etude microscopique de l'abrasion et de l'attrition sur la surface des dents." *Bulletin du Groupement International pour la Recherche Scientifique en Stomatologie & Odontologie* 5:242–51.

Danielson, D. R., and K. J. Reinhard. 1998. "Human Dental Microwear Caused by Calcium Oxalate Phytoliths in Prehistoric Diet of the Lower Pecos Region, Texas." *American Journal of Physical Anthropology* 107 (3): 297–304. http://dx.doi.org/10.1002/(SICI)1096-8644(199811)107:3<297::AID-AJPA6>3.0.CO;2-M. Medline:9821494

DeMiguel, D., B. Azanza, and J. Morales. 2010. "Trophic Flexibility within the Oldest Cervidae Lineage to Persist through the Miocene Climatic Optimum." *Palaeogeography, Palaeoclimatology, Palaeoecology* 289 (1-4): 81–92. http://dx.doi.org/10.1016/j.palaeo.2010.02.010.

Dennis, J. C., P. S. Ungar, M. F. Teaford, and K. E. Glander. 2004. "Dental Topography and Molar Wear in *Alouatta palliata* from Costa Rica." *American Journal of Physical Anthropology* 125 (2): 152–61. http://dx.doi.org/10.1002/ajpa.10379. Medline:15365981

Dirks, W., D. J. Reid, C. J. Jolly, J .E. Phillips-Conroy, and F. L. Brett. 2002. "Out of the Mouths of Baboons: Stress, Life History, and Dental Development in the Awash National Park Hybrid Zone, Ethiopia." *American Journal of Physical Anthropology* 118 (3): 239–52. http://dx.doi.org/10.1002/ajpa.10089. Medline:12115280

El Zaatari, S., F. E. Grine, M. F. Teaford, and H. F. Smith. 2005. "Molar Microwear and Dietary Reconstructions of Fossil Cercopithecoidea from the Plio-Pleistocene Deposits of South Africa." *Journal of Human Evolution* 49 (2): 180–205. http://dx.doi.org/10.1016/j.jhevol.2005.03.005. Medline:15964607

Fine, D., and G. T. Craig. 1981. "Buccal Surface Wear of Human Premolar and Molar Teeth: A Potential Indicator of Dietary and Social Differentiation." *Journal of Human Evolution* 10 (4): 335–44. http://dx.doi.org/10.1016/S0047-2484(81)80056-5.

Fiorillo, A. R. 1998. "Dental Microwear Patterns of the Sauropod Dinosaurs *Camarasaurus* and *Diplodocus*: Evidence for Resource Partitioning in the Late Jurassic of North America." *Historical Biology* 13 (1): 1–16. http://dx.doi.org/10.1080/08912969809386568.

Firmat, C., H. Gomes Rodrigues, R. Hutterer, J. C. Rando, J. A. Alcover, and J. Michaux. 2011. "Diet of the Extinct Lava Mouse Malpaisomys insularis from the Canary Islands: Insights from Dental Microwear." *Naturwissenschaften* 98 (1): 33–7. http://dx.doi.org/10.1007/s00114-010-0738-z. Medline:21107517

Galbany, J., J. S. Altmann, and A. Pérez-Pérez. 2009a. "Tooth Wear, Age and Diet in a Living Population of Baboons from Amboseli (Kenya)." *American Journal of Physical Anthropology* 138 (S48): 131.

Galbany, J., F. Estebaranz, L. M. Martínez, and A. Pérez-Pérez. 2009b. "Buccal Dental Microwear Variability in Extant African Hominoidea: Taxonomy versus Ecology." *Primates* 50 (3): 221–30. http://dx.doi.org/10.1007/s10329-009-0139-0. Medline:19296198

Godfrey, L. R., G. M. Semprebon, W. L. Jungers, M. R. Sutherland, E. L. Simons, and N. Solounias. 2004. "Dental Use Wear in Extinct Lemurs: Evidence of Diet and Niche Differentiation." *Journal of Human Evolution* 47 (3): 145–69. http://dx.doi.org/10.1016/j.jhevol.2004.06.003. Medline:15337413

Gordon, K. D. 1982. "A Study of Microwear on Chimpanzee Molars: Implications for Dental Microwear Analysis." *American Journal of Physical Anthropology* 59 (2): 195–215. http://dx.doi.org/10.1002/ajpa.1330590208. Medline:7149017

Gordon, K. D. 1984a. "Taphonomy of Dental Microwear, II." *American Journal of Physical Anthropology* 63:164–5.

Gordon, K. D. 1984b. "Hominoid Dental Microwear: Complications in the Use of Microwear Analysis to Detect Diet." *Journal of Dental Research* 63 (8): 1043–6. http://dx.doi.org/10.1177/00220345840630080601. Medline:6589263

Gordon, K. D. 1984c. "The Assessment of Jaw Movement Direction from Dental Microwear." *American Journal of Physical Anthropology* 63 (1): 77–84. http://dx.doi.org/10.1002/ajpa.1330630110. Medline:6703036

Gordon, K. D. 1988. "A Review of Methodology and Quantification in Dental Microwear Analysis." *Scanning Microscopy* 2 (2): 1139–47. Medline:3041571

Gordon, K. D., and A. C. Walker. 1983. "Playing 'Possum': A Microwear Experiment." *American Journal of Physical Anthropology* 60 (1): 109–12. http://dx.doi.org/10.1002/ajpa.1330600115. Medline:6869498

Grine, F. E. 1981. "Trophic Differences between 'Gracile' and 'Robust' Australopithecines: A Scanning Electron Microscope Analysis of Occlusal Events." *South African Journal of Science* 77:203–30.

Grine, F. E. 1986. "Dental Evidence for Dietary Differences in *Australopithecus* and *Paranthropus*: A Quantitative Analysis of Permanent Molar Microwear." *Journal of Human Evolution* 15 (8): 783–822. http://dx.doi.org/10.1016/S0047-2484(86)80010-0.

Grine, F. E. 1987. "Quantitative Analysis of Occlusal Microwear in *Australopithecus* and *Paranthropus*." *Scanning Microscopy* 1 (2): 647–56. Medline:3112937

Grine, F. E., and R. F. Kay. 1988. "Early Hominid Diets from Quantitative Image Analysis of Dental Microwear." *Nature* 333 (6175): 765–8. http://dx.doi.org/10.1038/333765a0. Medline:3133564

Grine, F. E., M. Sponheimer, P. S. Ungar, J. Lee-Thorp, and M. F. Teaford. 2012. "Dental Microwear and Stable Isotopes Inform the Paleoecology of Extinct Hominins." *American Journal of Physical Anthropology* 148 (2): 285–317. http://dx.doi.org/10.1002/ajpa.22086. Medline:22610903

Grine, F. E., P. S. Ungar, and M. F. Teaford. May-Jun 2002. "Error Rates in Dental Microwear Quantification Using Scanning Electron Microscopy." *Scanning* 24 (3): 144–53. http://dx.doi.org/10.1002/sca.4950240307. Medline:12074496

Grine, F. E., P. S. Ungar, and M. F. Teaford. 2006a. "Molar Microwear and Diet of '*Australopithecus*' *anamensis*." *South African Journal of Science* 102:301–10.

Grine, F. E., P. S. Ungar, M. F. Teaford, and S. El Zaatari. 2006b. "Molar Microwear in *Praeanthropus afarensis*: Evidence for Dietary Stasis through Time and under Diverse Paleoecological Conditions." *Journal of Human Evolution* 51 (3): 297–319. http://dx.doi.org/10.1016/j.jhevol.2006.04.004. Medline:16750841

Gügel, I. L., G. Grupe, and K.H. Kunzelmann. 2001. "Simulation of Dental Microwear: Characteristic Traces by Opal Phytoliths Give Clues to Ancient Human Dietary Behavior." *American Journal of Physical Anthropology* 114 (2): 124–38. http://dx.doi.org/10.1002/1096-8644(200102)114:2<124::AID-AJPA1012>3.0.CO;2-S. Medline:11169902

Hojo, T. 1996. "Quantitative Analyses of Microwear and Honing on the Sloping Crest of the P3 in Female Japanese Monkeys (*Macaca fuscata*)." *Scanning Microscopy* 10 (3): 727–36. Medline:9813635

Janal, M. N., K. G. Raphael, J. Klausner, and M. F. Teaford. 2007. "The Role of Tooth-Grinding in the Maintenance of Myofascial Face Pain: A Test of Alternate Models." *Pain Medicine* 8 (6): 486–96. http://dx.doi.org/10.1111/j.1526-4637.2006.00206.x. Medline:17716322

Kay, R. F. 1987. "Analysis of Primate Dental Microwear Using Image Processing Techniques." *Scanning Microscopy* 1 (2): 657–62. Medline:3616563

Kay, R. F., and H. H. Covert. 1983. "True Grit: A Microwear Experiment." *American Journal of Physical Anthropology* 61 (1): 33–8. http://dx.doi.org/10.1002/ajpa.1330610104. Medline:6869511

Kay, R. F., and F. E. Grine. 1988. "Tooth Morphology, Wear and Diet in *Australopithecus* and *Paranthropus* from Southern Africa." In *The Evolutionary History of the Robust Australopithecines*, ed. F. E. Grine, 427–444. New York: Aldine de Gryter.

Kay, R. F., and P. S. Ungar. 1997. "Dental Evidence for Diet in Some Miocene Catarrhines with Comments on the Effects of Phylogeny on the Interpretation of Adaptation." In

Function, Phylogeny and Fossils: Miocene Hominoids and Great Ape and Human Origins, ed. D. R. Begun, C. Ward, and M. Rose, 131–151. New York: Plenum Press.

Kelley, J. 1986. "Paleobiology of Miocene Hominoids." PhD dissertation, Yale University, New Haven, CT.

Kelley, J. 1990. "Incisor Microwear and Diet in Three Speices of *Colobus*." *Folia Primatologica* 55 (2): 73–84. http://dx.doi.org/10.1159/000156502.

King, T., L. C. Aiello, and P. Andrews. 1999a. "Dental Microwear of *Griphopithecus alpani*." *Journal of Human Evolution* 36 (1): 3–31. http://dx.doi.org/10.1006/jhev.1998.0258. Medline:9924132

King, T., P. Andrews, and B. Boz. 1999b. "Effect of Taphonomic Processes on Dental Microwear." *American Journal of Physical Anthropology* 108 (3): 359–73. http://dx.doi.org/10.1002/(SICI)1096-8644(199903)108:3<359::AID-AJPA10>3.0.CO;2-9. Medline:10096686

Krueger, K. L., and P. S. Ungar. 2010. "Incisor Microwear Textures of Five Bioarcheological Groups." *International Journal of Osteoarchaeology* 20:549–60.

Lalueza Fox, C. 1992. "Dental Striation Pattern in Andamanese and Veddahs from Skulls Collections of the British Museum." *Man in India* 72:377–84.

Lalueza Fox, C., and A. Pérez-Pérez. 1993. "The Diet of the Neanderthal Child Gibralter 2 (Devils' Tower) through the Study of the Vestibular Striation Pattern." *Journal of Human Evolution* 24 (1): 29–41. http://dx.doi.org/10.1006/jhev.1993.1004.

Leakey, M. G., M. F. Teaford, and C. W. Ward. 2003. "Cercopithecidae from Lothagam." In *Lothagam: Dawn of Humanity in Eastern Africa*, ed. M. G. Leakey and J. M. Harris, 201–248. New York: Columbia University Press.

Lewis, P. J., M. Gutierrez, and E. Johnson. 2000. "*Ondatra zibethicus* (Arvicolinae, Rodentia) Dental Microwear Patterns as a Potential Tool for Palaeoenvironmental Reconstruction." *Journal of Archaeological Science* 27 (9): 789–98. http://dx.doi.org/10.1006/jasc.1999.0502.

Lucas, P. W. 1991. "Fundamental Physical Properties of Fruits and Seeds in the Diet of Southeast Asian Primates." In *Primatology Today*, ed. A. Ehara, T. Kimura, O. Takenaka, and M. Iwamoto, 152–158. Amsterdam: Elsevier.

Lucas, P. W., and M. F. Teaford. 1994. "The Functional Morphology of Colobine Teeth." In *Colobine Monkeys: Their Evolutionary Ecology*, ed. J. Oates and A. G. Davies, 173–203. New York: Cambridge University Press.

Lucas, P. W., and M. F. Teaford. 1995. "Significance of Silica in Leaves to Long-Tailed Macaques (*Macaca fascicularis*)." *Folia Primatologica* 64 (1-2): 30–6. http://dx.doi.org/10.1159/000156829. Medline:7665120

Ma, P. H., and M. F. Teaford. 2010. "Dental Reconstruction in Antebellum Baltimore: Insights from Dental Microwear Analysis." *American Journal of Physical Anthropology* 141 (4):571–82. Medline:19927276.

MacFadden, B. J., N. Solounias, and T. E. Cerling. 1999. "Ancient Diets, Ecology, and Extinction of 5-Million-Year-Old Horses from Florida." *Science* 283 (5403): 824–7. http://dx.doi.org/10.1126/science.283.5403.824. Medline:9933161

Mainland, I. L. 1998. "Dental Microwear and Diet in Domestic Sheep (*Ovis aries*) and Goats (*Capra Hircus*): Distinguishing Grazing and Fodder-Fed Ovicaprids Using a Quantitative Analytical Approach." *Journal of Archaeological Science* 25 (12): 1259–71. http://dx.doi.org/10.1006/jasc.1998.0301.

Mainland, I. L. 2000. "A Dental Microwear Study of Seaweed-Eating and Grazing Sheep from Orkney." *International Journal of Osteoarchaeology* 10 (2): 93–107. http://dx.doi.org/10.1002/(SICI)1099-1212(200003/04)10:2<93::AID-OA513>3.0.CO;2-U.

Mainland, I .L. 2003. "Dental Microwear in Grazing and Browsing Gotland Sheep (*Ovis aries*) and Its Implications for Dietary Reconstruction." *Journal of Archaeological Science* 30 (11): 1513–27. http://dx.doi.org/10.1016/S0305-4403(03)00055-4.

Mainland, I. L. 2006. "Pastures Lost? A Dental Microwear Study of Ovicaprine Diet and Management in Norse Greenland." *Journal of Archaeological Science* 33 (2): 238–52. http://dx.doi.org/10.1016/j.jas.2005.07.013.

Mannerberg, F. 1960. "Appearance of Tooth Surface as Observed in Shadowed Replicas." *Odontologisk Revy* 11 (Suppl. 6).

Merceron, G., C. Blondel, M. Brunet, S. Sen, N. Solounias, L. Viriot, and E. Heintz. 2004a. "The Late Miocene Paleoenvironment of Afghanistan as Inferred from Dental Microwear in Artiodactyls." *Palaeogeography, Palaeoclimatology, Palaeoecology* 207 (1-2): 143–63. http://dx.doi.org/10.1016/j.palaeo.2004.02.008.

Merceron, G., C. Blondel, L. De Bonis, G. D. Koufos, and L. Viriot. 2005b. "A New Method of Dental Microwear Analysis: Application to Extant Primates and *Ouranopithecus macedoniensis* (Late Miocene of Greece)." *Palaios* 20 (6): 551–61. http://dx.doi.org/10.2110/palo.2004.p04-17.

Merceron, G., L. De Bonis, L. Viriot, and C. Blondel. 2005a. "Dental Microwear of Fossil Bovids from Northern Greece: Paleoenvironmental Conditions in the Eastern Mediterranean during the Messinian." *Palaeogeography, Palaeoclimatology, Palaeoecology* 217 (3-4): 173–85. http://dx.doi.org/10.1016/j.palaeo.2004.11.019.

Merceron, G., G. Escarguel, J. M. Angibault, and H. Verheyden-Tixier. 2010. "Can Dental Microwear Textures Record Inter-Individual Dietary Variations?" *PLoS ONE* 5 (3): e9542. http://dx.doi.org/10.1371/journal.pone.0009542. Medline:20209051

Merceron, G., and P. S. Ungar. 2005. "Dental Microwear and Palaeoecology of Bovids from the Early Pliocene of Langebaanweg, Western Cape Province, South Africa." *South African Journal of Science* 101:365–70.

Merceron, G., L. Viriot, and C. Blondel. 2004b. "Tooth Microwear Pattern in Roe Deer (*Capreolus capreolus* L.) from Chizé (Western France) and Relation to Food Composition." *Small Ruminant Research* 53 (1-2): 125–32. http://dx.doi.org/10.1016/j.smallrumres.2003.10.002.

Mills, J.R.E. 1955. "Ideal Dental Occlusion in the Primates." *Dental Practice* 6:47–61.

Mills, J.R.E. 1963. "Occlusion and Malocclusion of the Teeth of Primates." In *Dental Anthropology*, ed. D. R. Brothwell, 29–52. New York: Pergamon Press.

Molleson, T., and K. Jones. 1991. "Dental Evidence for Dietary Change at Abu Hureyra." *Journal of Archaeological Science* 18 (5): 525–39. http://dx.doi.org/10.1016/0305-4403(91)90052-Q.

Molleson, T., K. Jones, and S. Jones. 1993. "Dietary Change and the Effects of Food Preparation on Microwear Patterns in the Late Neolithic of Abu Hureyra, Northern Syria." *Journal of Human Evolution* 24 (6): 455–68. http://dx.doi.org/10.1006/jhev.1993.1031.

Morel, A., E. Albuisson, and A. Woda. 1991. "A Study of Human Jaw Movements Deduced from Scratches on Occlusal Wear Facets." *Archives of Oral Biology* 36 (3): 195–202. http://dx.doi.org/10.1016/0003-9969(91)90086-A. Medline:1877893

Nagel, U. 1973. "A Comparison of Anubis Baboons, Hamadryas Baboons and Their Hybrids at a Species Border in Ethiopia." *Folia Primatologica* 19 (2): 104–65. http://dx.doi.org/10.1159/000155536. Medline:4201907.

Noble, V. E., and M. F. Teaford. 1995. "Dental Microwear in Caucasian American *Homo sapiens*: Preliminary Results." *American Journal of Physical Anthropology* Suppl. 20: 162.

Nystrom, P., J. E. Phillips-Conroy, and C. J. Jolly. 2004. "Dental Microwear in Anubis and Hybrid Baboons (*Papio hamadryas*, sensu lato) Living in Awash National Park, Ethiopia." *American Journal of Physical Anthropology* 125 (3): 279–91. http://dx.doi.org/10.1002/ajpa.10274. Medline:15386258

Oliveira, E. V. 2001. "Micro-desgaste dentário em alguns Dasypodidae (Mammalia, Xenarthra)." *Acta Biologica Leopoldensia* 23:83–91.

Organ, J. M., M. F. Teaford, and C. S. Larsen. 2005. "Dietary Inferences from Dental Occlusal Microwear at Mission San Luis de Apalachee." *American Journal of Physical Anthropology* 128 (4): 801–11. http://dx.doi.org/10.1002/ajpa.20277. Medline:16134151

Pastor, R. F. 1992. "Dietary Adaptations and Dental Microwear in Mesolithic and Chalcolithic South Asia." *Journal of Human Ecology (Delhi, India)* 2 (spec issue): 215–28.

Pastor, R. F. 1993. "Dental Microwear among Inhabitants of the Indian Subcontinent: A Quantitative and Comparative Analysis." PhD dissertation, University of Oregon, Eugene.

Pastor, R. F., and T. L. Johnston. 1992. "Dental Microwear and Attrition." In *Human Skeletal Remains from Mahadaha: A Gangetic Mesolithic Site*, ed. K.A.R. Kennedy, 271–304. Ithaca, NY: Cornell University Press.

Pérez-Pérez, A., V. Espurz, J. M. Bermúdez de Castro, M. A. de Lumley, and D. Turbón. 2003. "Non-Occlusal Dental Microwear Variability in a Sample of Middle and Late Pleistocene Human Populations from Europe and the Near East." *Journal of Human Evolution* 44 (4): 497–513. http://dx.doi.org/10.1016/S0047-2484(03)00030-7. Medline:12727465

Pérez-Pérez, A., C. Lalueza, and D. Turbón. 1994. "Intraindividual and Intragroup variability of Buccal Tooth Striation Pattern." *American Journal of Physical Anthropology* 94 (2): 175–87. http://dx.doi.org/10.1002/ajpa.1330940203. Medline:8085610

Peters, C. R. 1982. "Electron-Optical Microscopic Study of Incipient Dental Microdamage from Experimental Seed and Bone Crushing." *American Journal of Physical Anthropology* 57 (3): 283–301. http://dx.doi.org/10.1002/ajpa.1330570306. Medline:7114194

Phillips-Conroy, J. E. 1978. "Dental Variability in Ethiopian Baboons: An Examination of the Anubis-Hamadryas Hybrid Zone in the Awash National Park, Ethiopia." PhD Dissertation, New York University, New York.

Phillips-Conroy, J. E., T. Bergman, and C. J. Jolly. 2000. "Quantitative Assessment of Occlusal Wear and Age Estimation in Ethiopian and Tanzanian Baboons." In *Old World Monkeys*, ed. P. F. Whitehead and C. J. Jolly, 321–340. Cambridge: Cambridge University Press. http://dx.doi.org/10.1017/CBO9780511542589.013

Phillips-Conroy, J. E., and C. J. Jolly. 1986. "Changes in the Structure of the Baboon Hybrid Zone in the Awash National Park, Ethiopia." *American Journal of Physical Anthropology* 71 (3): 337–50. http://dx.doi.org/10.1002/ajpa.1330710309.

Phillips-Conroy, J. E., C. J. Jolly, and F. L. Brett. 1991. "Characteristics of Hamadryas-like Male Baboons Living in Anubis Baboon Troops in the Awash Hybrid Zone, Ethiopia." *American Journal of Physical Anthropology* 86 (3): 353–68. http://dx.doi.org/10.1002/ajpa.1330860304. Medline:1746643

Puech, P.-F. 1977. "Usure dentaire en anthropolgie étude par la technique des répliques." *Revue d'Odonto-Stomatol* 6:51–6.

Puech, P.-F. 1984a. "A la recherche du menu des premiers hommes." *Cahiers Lig. Préhist. et Protohist* 1:46–53.

Puech, P.-F. 1984b. "Acidic Food Choice in *Homo habilis* at Olduvai." *Current Anthropology* 25 (3): 349–50. http://dx.doi.org/10.1086/203146.

Puech, P.-F., F. Cianfarani, and H. Albertini. 1986. "Dental Microwear Features as an Indicator for Plant Food in Early Hominids: A Preliminary Study of Enamel." *Human Evolution* 1 (6): 507–15. http://dx.doi.org/10.1007/BF02437467.

Puech, P.-F., and A. Prone. 1979. "Reproduction experimentale des processes d'usure dentaire par abrasion: Implications paleoecologique chex l'Homme fossile." *Comptes rendus de l'Académie des Sciences Paris* 289:895–8.

Puech, P.-F., A. Prone, and H. Albertini. 1981. "Reproduction expérimentale des processus d-altération de la surface dentaire par friction non abrasive et non adhésive: Application à l'étude de alimentation de L'Homme fossile." *Comptes rendus de l'Académie des Sciences Paris* 293:729–34.

Puech, P.-F., A. Prone, and R. Kraatz. 1980. "Microscopie de l'usure dentaire chez l'homme fossile: Bol alimnetaire et environnement." *Comptes rendus de l'Académie des Sciences Paris* 290:1413–6.

Puech, P.-F., A. Prone, H. Roth, and F. Cianfarani. 1985. "Reproduction expérimentale de processus d'usure des surfaces dentaires des Hominides fossiles: Conséquences morphoscopiques et exoscopiques avec application à l'Hominidé I de Garusi." *Comptes rendus de l'Académie des Sciences Paris* 301:59–64.

Radlanski, R. J., and A. Jäger. 1989. "Micromorphology of the Approximal Contact Surfaces and the Occlusal Abrasion Facets of Human Permanent Teeth." *Deutsche* zahnärztliche *Zeitschrift* 44 (3): 196–7. Medline:2639036.

Rafferty, K. L., M. F. Teaford, and W .L. Jungers. 2002. "Molar Microwear of Subfossil Lemurs: Improving the Resolution of Dietary Inferences." *Journal of Human Evolution* 43 (5): 645–57. http://dx.doi.org/10.1006/jhev.2002.0592. Medline:12457853

Ramdarshan, A., G. Merceron, P. Tafforeau, and L. Marivaux. 2010. "Dietary Reconstruction of the Amphipithecidae (Primates, Anthropoidea) from the Paleogene of South Asia and Paleoecological Implications." *Journal of Human Evolution* 59 (1): 96–108. http://dx.doi.org/10.1016/j.jhevol.2010.04.007. Medline:20510435

Ranjitkar, S., J. A. Kaidonis, G. C. Townsend, A. M. Vu, and L. C. Richards. 2008. "An In Vitro Assessment of the Effect of Load and pH on Wear between Opposing Enamel and Dentine Surfaces." *Archives of Oral Biology* 53 (11): 1011–6. http://dx.doi.org/10.1016/j.archoralbio.2008.05.013. Medline:18603226

Raphael, K. G., J. J. Marbach, J. J. Klausner, M. F. Teaford, and D. K. Fischoff. 2003. "Is Bruxism Severity a Predictor of Oral Splint Efficacy in Patients with Myofascial Face Pain?" *Journal of Oral Rehabilitation* 30 (1): 17–29. http://dx.doi.org/10.1046/j.1365-2842.2003.01117.x. Medline:12485379

Reinhard, K., S.M.F. de Souza, C. Rodrigues, E. Kimmerle, and S. Dorsey-Vinton. 1999. "Microfossils in Dental Calculus: A New Perspective on Diet and Dental Disease." In *Human Remains: Conservation, Retrieval, and Analysis*, ed. E. Williams, 113–118. Oxford: Archaeopress.

Reinhard, K. J., and D. R. Danielson. 2005. "Pervasiveness of Phytoliths in Prehistoric Southwestern Diet and Implications for Regional and Temporal Trends for Dental Microwear." *Journal of Archaeological Science* 32 (7): 981–8. http://dx.doi.org/10.1016/j.jas.2005.01.014.

Rensberger, J. M. 1978. "Scanning Electron Microscopy of Wear and Occlusal Events in Some Small Herbivores." In *Development, Function and Evolution of Teeth*, ed. P. N. Butler and K. A. Joysey, 415–438. New York: Academic Press.

Rensberger, J. M. 1982. "Patterns of Change in Two Locally Persistent Successions of Fossil Aplodontid Rodents." In *Teeth: Form, Function, and Evolution*, ed. B. Kurten, 323–349. New York: Columbia University Press.

Rensberger, J. M. 1986. "Early Chewing Mechanisms in Mammalian Herbivores." *Paleobiology* 12:474–94.

Rensberger, J. M., and H. B. Krentz. 1988. "Microscopic Effects of Predator Digestion on the Surfaces of Bones and Teeth." *Scanning Microscopy* 2 (3): 1541–51. Medline:3201198

Richard, A. F. 1985. *Primates in Nature*. New York: W. H. Freeman.

Rivals, F., and B. Deniaux. 2003. "Dental Microwear Analysis for Investigating the Diet of an Argali Population (*Ovis ammon antiqua*) of Mid-Pleistocene Age, Caune de l'Arago Cave, Eastern Pyrenees, France." *Palaeogeography, Palaeoclimatology, Palaeoecology* 193 (3-4): 443–55. http://dx.doi.org/10.1016/S0031-0182(03)00260-8.

Rivals, F., and B. Deniaux. 2005. "Investigation of Human Hunting Seasonality through Dental Microwear Analysis of Two Caprinae in Late Pleistocene Localities in Southern France." *Journal of Archaeological Science* 32 (11): 1603–12. http://dx.doi.org/10.1016/j.jas.2005.04.014.

Rivals, F., and G. M. Semprebon. 2010. "What Can Incisor Microwear Reveal about the Diet of Ungulates?" *Mammalia* 74 (4): 401–6. http://dx.doi.org/10.1515/mamm.2010.044.

Rodrigues, H. G., G. Merceron, and L. Viriot. 2009. "Dental Microwear Patterns of Extant and Extinct Muridae (Rodentia, Mammalia): Ecological Implications." *Naturwissenschaften* 96 (4): 537–42. http://dx.doi.org/10.1007/s00114-008-0501-x. Medline:19127354

Romero, A., and J. De Juan. 2007. "Intra- and Interpopulation Human Buccal Tooth Surface Microwear Analysis: Inferences about Diet and Formation Processes." *L'Anthropologie* 45:61–70.
Romero, A., J. Galbany, N. Martinez-Ruiz, and J. De Juan. 2009. "In Vivo Turnover Rates in Human Buccal Dental-Microwear." *American Journal of Physical Anthropology* 138 (S48): 223–4.
Rose, K. D., A. Walker, and L. L. Jacobs. 1981. "Function of the Mandibular Tooth Comb in Living and Extinct Mammals." *Nature* 289 (5798): 583–5. http://dx.doi.org/10.1038/289583a0. Medline:7007889
Ryan, A. S. 1979. "Wear Striation Direction on Primate Teeth: A Scanning Electron Microscope Examination." *American Journal of Physical Anthropology* 50 (2): 155–67. http://dx.doi.org/10.1002/ajpa.1330500204. Medline:443353
Ryan, A. S. 1981. "Anterior Dental Microwear and Its Relationship to Diet and Feeding Behavior in Three African Primates (*Pan troglodytes troglodytes, Gorilla gorilla gorilla*, and *Papio hamadryas*)." *Primates* 22 (4): 533–50. http://dx.doi.org/10.1007/BF02381245.
Ryan, A. S., and D. C. Johanson. 1989. "Anterior Dental Microwear in *Australopithecus afarensis*: Comparisons with Human and Nonhuman Primates." *Journal of Human Evolution* 18 (3): 235–68. http://dx.doi.org/10.1016/0047-2484(89)90051-1.
Schmidt, C. W. 2001. "Dental Microwear Evidence for a Dietary Shift between Two Nonmaize-Reliant Prehistoric Human Populations from Indiana." *American Journal of Physical Anthropology* 114 (2): 139–45. http://dx.doi.org/10.1002/1096-8644(200102)114:2<139::AID-AJPA1013>3.0.CO;2-9. Medline:11169903
Schubert, B. W., and P. S. Ungar. 2005. "Wear Facets and Enamel Spalling in Tyrannosaurid Dinosaurs." *Acta Palaeontologica Polonica* 50:93–9.
Schubert, B. W., P. S. Ungar, M. Sponheimer, and K. E. Reed. 2006. "Microwear Evidence for Plio-Pleistocene Bovid Diets from Makapansgat Limeworks Cave, South Africa." *Palaeogeography, Palaeoclimatology, Palaeoecology* 241 (2): 301–19. http://dx.doi.org/10.1016/j.palaeo.2006.04.004.
Scott, R. S., P. S. Ungar, T. S. Bergstrom, C. A. Brown, F. E. Grine, M. F. Teaford, and A. Walker. 2005. "Dental Microwear Texture Analysis Shows Within-Species Diet Variability in Fossil Hominins." *Nature* 436 (7051): 693–5. http://dx.doi.org/10.1038/nature03822. Medline:16079844
Semprebon, G. M., L. R. Godfrey, N. Solounias, M. R. Sutherland, and W. L. Jungers. 2004. "Can Low-Magnification Stereomicroscopy Reveal Diet?" *Journal of Human Evolution* 47 (3): 115–44. http://dx.doi.org/10.1016/j.jhevol.2004.06.004. Medline:15337412
Semprebon, G. M., P. J. Sise, and M. C. Coombs. 2011. "Potential Bark and Fruit Browsing as Revealed by Stereomicrowear Analysis of the Peculiar Clawed Herbivores Known as Chalicotheres (Perissodactyla, Chalicotherioidea)." *Journal of Mammalian Evolution* 18 (1): 33–55. http://dx.doi.org/10.1007/s10914-010-9149-3.
Silcox, M. T., and M. F. Teaford. 2002. "The Diet of Worms: An Analysis of Mole Dental Microwear and Its Relevance to Dietary Inference in Primates and Other Mammals."

Journal of Mammalogy 83 (3): 804–14. http://dx.doi.org/10.1644/1545-1542(2002)083<0804:TDOWAA>2.0.CO;2.

Solounias, N., and B. Dawson-Saunders. 1988. "Dietary Adaptations and Paleoecology of the Late Miocene Ruminants from Pikermi and Samos in Greece." *Palaeogeography, Palaeoclimatology, Palaeoecology* 65 (3-4): 149–72. http://dx.doi.org/10.1016/0031-0182(88)90021-1.

Solounias, N., and L. C. Hayek. 1993. "New Methods of Tooth Microwear Analysis and Application to Dietary Determination of Two Extinct Antelopes." *Journal of Zoology (London, England)* 229 (3): 421–45. http://dx.doi.org/10.1111/j.1469-7998.1993.tb02646.x.

Solounias, N., and S.M.C. Moelleken. 1992a. "Tooth Microwear Analyses of *Eotragus sansaniensis* (Mammalia: Ruminantia), One of the Oldest Known Bovids." *Journal of Vertebrate Paleontology* 12 (1): 113–21. http://dx.doi.org/10.1080/02724634.1992.10011437.

Solounias, N., and S.M.C. Moelleken. 1992b. "Dietary Adaptations of Two Goat Ancestors and Evolutionary Considerations." *Geobios* 25 (6): 797–809. http://dx.doi.org/10.1016/S0016-6995(92)80061-H.

Solounias, N., and S.M.C. Moelleken. 1994. "Differences in Diet between Two Archaic Ruminant Species from Sansan, France." *Historical Biology* 7 (3): 203–20. http://dx.doi.org/10.1080/10292389409380454.

Solounias, N., F. Rivals, and G. M. Semprebon. 2010. "Dietary Interpretation and Paleoecology of Herbivores from Pikermi and Samos (Late Miocene of Greece)." *Paleobiology* 36 (1): 113–36. http://dx.doi.org/10.1666/0094-8373-36.1.113.

Solounias, N., and G. Semprebon. 2002. "Advances in the Reconstruction of Ungulate Ecomorphology with Application to Early Fossil Equids." *American Museum Novitates* 3366:1–49. http://dx.doi.org/10.1206/0003-0082(2002)366<0001:AITROU>2.0.CO;2.

Solounias, N., M. F. Teaford, and A. Walker. 1988. "Interpreting the Diet of Extinct Ruminants: The Case of a Non-Browsing Giraffid." *Paleobiology* 14:287–300.

Strait, S. G. 1993. "Molar Microwear in Extant Small-Bodied Faunivorous Mammals: An Analysis of Feature Density and Pit Frequency." *American Journal of Physical Anthropology* 92 (1): 63–79. http://dx.doi.org/10.1002/ajpa.1330920106. Medline:8238292

Sugawara, K. 1979. "Sociological Study of a Wild Group of Hybrid Baboons between *Papio anubis* and *P. hamadryas* in the Awash Valley, Ethiopia." *Primates* 20 (1): 21–56. http://dx.doi.org/10.1007/BF02373827.

Szmulewicz, M. N., L. M. Andino, E. P. Reategui, T. Woolley-Barker, C. J. Jolly, T. R. Disotell, and R. J. Herrera. 1999. "An Alu Insertion Polymorphism in a Baboon Hybrid Zone." *American Journal of Physical Anthropology* 109 (1): 1–8. http://dx.doi.org/10.1002/(SICI)1096-8644(199905)109:1<1::AID-AJPA1>3.0.CO;2-X. Medline:10342460

Teaford, M. F. 1985. "Molar Microwear and Diet in the Genus Cebus." *American Journal of Physical Anthropology* 66 (4): 363–70. http://dx.doi.org/10.1002/ajpa.1330660403. Medline:3993762

Teaford, M. F. 1988. "A Review of Dental Microwear and Diet in Modern Mammals." *Scanning Microscopy* 2 (2): 1149–66. Medline:3041572

Teaford, M. F. 1991. "Dental Microwear: What Can It Tell Us about Diet and Dental Function?" In *Advances in Dental Anthropology*, ed. M. A. Kelley and C. S. Larsen, 341–356. New York: Alan R. Liss.

Teaford, M. F. 1993. "Dental Microwear and Diet in Extant and Extinct *Theropithecus*: Preliminary Analyses." In *Theropithecus: The Life and Death of a Primate Genus*, ed. N. G. Jablonski, 331–50. Cambridge: Cambridge University Press. http://dx.doi.org/10.1017/CBO9780511565540.013

Teaford, M. F. 1994. "Dental Microwear and Dental Function." *Evolutionary Anthropology* 3 (1): 17–30. http://dx.doi.org/10.1002/evan.1360030107.

Teaford, M. F. 2002. "Dental Enamel Microwear Analysis." In *Foraging, Farming and Coastal Biocultural Adaptation in Late Prehistoric North Carolina*, ed. D. L. Hutchinson, 169–177. Gainesville: University Press of Florida.

Teaford, M. F. 2007. "What Do We Know and Not Know about Dental Microwear and Diet?" In *Evolution of the Human Diet: The Known, the Unknown, and the Unknowable*, ed. P. S. Ungar, 106–131. New York: Oxford University Press.

Teaford, M. F., and K. E. Byrd. 1989. "Differences in Tooth Wear as an Indicator of Changes in Jaw Movement in the Guinea Pig (*Cavia porcellus*)." *Archives of Oral Biology* 34 (12): 929–36. http://dx.doi.org/10.1016/0003-9969(89)90048-4. Medline:2610627

Teaford, M. F., and K. E. Glander. 1991. "Dental Microwear in Live, Wild-Trapped *Alouatta palliata* from Costa Rica." *American Journal of Physical Anthropology* 85 (3): 313–9. http://dx.doi.org/10.1002/ajpa.1330850310. Medline:1897604

Teaford, M. F., and K. E. Glander. 1996. "Dental Microwear and Diet in a Wild Population of Mantled Howlers (*Alouatta palliata*)." In *Adaptive Radiations of Neotropical Primates*, ed. M. Norconk, A. Rosenberger, and P. Garber, 433–449. New York: Plenum Press. http://dx.doi.org/10.1007/978-1-4419-8770-9_25

Teaford, M. F., C. S. Larsen, R. F. Pastor, and V. E. Noble. 2001. "Pits and Scratches: Microscopic Evidence of Tooth Use and Masticatory Behavior in La Florida." In *Bioarchaeology of Spanish Florida: The Impact of Colonialism*, ed. C. S. Larsen, 82–112. Gainesville: University Press of Florida.

Teaford, M. F., P. W. Lucas, P. S. Ungar, and K. E. Glander. 2006. "Mechanical Defenses in Leaves Eaten by Costa Rican Howling Monkeys (*Alouatta palliata*)." *American Journal of Physical Anthropology* 129 (1): 99–104. http://dx.doi.org/10.1002/ajpa.20225. Medline:16136580

Teaford, M. F., and J. D. Lytle. 1996. "Brief Communication: Diet-Induced Changes in Rates of Human Tooth Microwear: A Case Study Involving Stone-Ground Maize." *American Journal of Physical Anthropology* 100 (1): 143–7. http://dx.doi.org/10.1002/(SICI)1096-8644(199605)100:1<143::AID-AJPA13>3.0.CO;2-0. Medline:8859961

Teaford, M. F., and O. J. Oyen. 1989a. "*In Vivo* and *In Vitro* Turnover in Dental Microwear." *American Journal of Physical Anthropology* 80 (4): 447–60. http://dx.doi.org/10.1002/ajpa.1330800405. Medline:2513725

Teaford, M. F., and O. J. Oyen. 1989b. "Differences in the Rate of Molar Wear between Monkeys Raised on Different Diets." *Journal of Dental Research* 68 (11): 1513–8. http://dx.doi.org/10.1177/00220345890680110901. Medline:2584518

Teaford, M. F., and J. G. Robinson. 1989. "Seasonal or Ecological Differences in Diet and Molar Microwear in *Cebus nigrivittatus*." *American Journal of Physical Anthropology* 80 (3): 391–401. http://dx.doi.org/10.1002/ajpa.1330800312. Medline:2686463

Teaford, M. F., and J. A. Runestad. 1992. "Dental Microwear and Diet in Venezuelan Primates." *American Journal of Physical Anthropology* 88 (3): 347–64. http://dx.doi.org/10.1002/ajpa.1330880308. Medline:1642321

Teaford, M. F., and C. A. Tylenda. 1991. "A New Approach to the Study of Tooth Wear." *Journal of Dental Research* 70 (3): 204–7. http://dx.doi.org/10.1177/00220345910700030901. Medline:1999560

Teaford, M. F., and P. S. Ungar. 2000. "Diet and the Evolution of the Earliest Human Ancestors." *Proceedings of the National Academy of Sciences of the United States of America* 97 (25): 13506–11. http://dx.doi.org/10.1073/pnas.260368897. Medline:11095758

Teaford, M. F., P. S. Ungar, and F. E. Grine. 2002. "Fossil Evidence for the Evolution of Human Diet." In *Human Diet: Its Origins and Evolution*, ed. P. S. Ungar and M .F. Teaford, 143–166. Wesport, CT: London, Bergen, and Garvey.

Teaford, M. F., P. S. Ungar, and R. F. Kay. 2008. "Molar Shape and Molar Microwear in the Koobi Fora Monkeys: Ecomorphological Implications." In *Koobi Fora Research Project, Volume 6: The Fossil Monkeys*, ed. N. G. Jablonski and M. G. Leakey, 337–358. Occasional Paper of the California Academy of Sciences, San Francisco.

Teaford, M. F., and A. Walker. 1983. "Dental Microwear in Adult and Still-Born Guinea Pigs (Cavia porcellus)." *Archives of Oral Biology* 28 (11): 1077–81. http://dx.doi.org/10.1016/0003-9969(83)90067-5. Medline:6581764

Teaford, M. F., and A. Walker. 1984. "Quantitative Differences in Dental Microwear between Primate Species with Different Diets and a Comment on the Presumed Diet of *Sivapithecus*." *American Journal of Physical Anthropology* 64 (2): 191–200. http://dx.doi.org/10.1002/ajpa.1330640213. Medline:6380302

Turssi, C. P., J. L. Ferracane, and K. Vogel. 2005. "Filler Features and Their Effects on Wear and Degree of Conversion of Particulate Dental Resin Composites." *Biomaterials* 26 (24): 4932–7. http://dx.doi.org/10.1016/j.biomaterials.2005.01.026. Medline:15769527

Ungar, P. S. 1990. "Incisor Microwear and Feeding Behavior in *Alouatta seniculus* and *Cebus olivaceus*." *American Journal of Primatology* 20 (1): 43–50. http://dx.doi.org/10.1002/ajp.1350200107.

Ungar, P. S. 1992 "Incisor Microwear and Feeding Behavior of Four Sumatran Anthropoids." PhD Dissertation, State University of New York, Stony Brook.

Ungar, P. S. 1994. "Incisor Microwear of Sumatran Anthropoid Primates." *American Journal of Physical Anthropology* 94 (3): 339–63. http://dx.doi.org/10.1002/ajpa.1330940305. Medline:7943190

Ungar, P. S. Jan-Feb 1995. "A Semiautomated Image Analysis Procedure for the Quantification of Dental Microwear II." *Scanning* 17 (1): 57–9. http://dx.doi.org/10.1002/sca.4950170108. Medline:7704317

8

Hominin Ecology from Hard-Tissue Biogeochemistry

Julia A. Lee-Thorp
and Matt Sponheimer

Hominin dietary ecology has been the subject of lively debate for many years, beginning with the discoveries of the first australopiths in Africa, in what seemed to be unlikely habitats for great apes (e.g., Dart 1926, 1957; Robinson 1954). There is good reason for this interest. Large primates spend much of their time searching for or consuming food (e.g., Altmann and Altmann 1970; Teleki 1981; Goodall 1986) and diet is considered one of the most important factors underlying behavioral and ecological differences among extant primates (Ungar 1998; Fleagle 1999). Similarly, the habitats in which they preferred to live are of considerable interest because they essentially form the ecological framework in which various food types may or may not be available at different times of the annual cycle. Thus dietary ecology and environments are closely linked, and both are amenable to investigation by biogeochemical approaches.

Information about past diet and ecology can be gleaned from many sources, but in the case of early hominins they are likely to provide only very partial glimpses, and each approach has distinct constraints. The first potential hominins precede the earliest archaeological traces by millions of years, so dietary information from conventional archaeological evidence—butchered animal bones and stone tools—is unavailable except for the more recent periods. Even then this approach encourages a strong emphasis on animal foods at the expense of plants. A good deal of

DOI: 10.5876/9781607322252:c08

the information about early hominin diet has come from the fossils themselves, and comparisons with extant primates. Comparative morphology and allometry of the dental "equipment" used for processing and consuming foods have received by far the most attention (e.g., Kay 1985). But there are limitations, as some extant primates are, apparently, poorly equipped for their day-to-day diets. For instance, the relatively large incisors and bunodont molars in modern *Papio* are consistent with frugivory (Hylander 1975; Fleagle 1999; Ungar 1998), but many modern populations also consume quantities of grass (Altmann and Altmann 1970; Harding 1976; Dunbar 1983; Strum 1987) for which they are less well-equipped. Attempts to explain such apparent contradictions in the fossil record have argued that the dental morphology of extant apes and early hominins is most informative about their "fallback" or stress-season foods than about their "typical" or preferred foods (Ungar 2004). That may be the case, although on closer inspection some of the evidence cited does not support this argument (see below). The dietary problem can be summed up by the observation that most primates are equipped to eat a wide range of foods, and this is especially likely to have been the case for early hominins.

Two techniques based on the chemical composition of bones and teeth have been developed as alternative, or rather, complementary, approaches to paleoecology. They are both based on the principle that "you are what you eat," or that the chemical composition of food is ultimately traceable in the tissues. In this chapter, we outline how trace elements and stable light isotope biogeochemistry has contributed to our understanding of early hominin diets, and also of the ecology of the associated animal assemblages. We begin with a discussion of how trace element abundances in hard tissues are related to diet, as they were the first biogeochemical data to be published for early hominins (Boaz and Hampel 1978). We then discuss the application of stable-isotope data, mostly carbon isotopes. Last, we discuss the ways that stable isotopes can contribute to the ongoing dialogue about the environments with which early hominins were associated.

TRACE ELEMENTS

Principles

There is a long history of using alkaline earth metal concentrations in ancient hard tissues to investigate diet, which began with the work of Toots and Voorhies in 1965, following on from studies of the pathways of radioactive strontium-90 in the human foodweb (Toots and Voorhies 1965). The approach is based on the well-known principle that mammals discriminate against the alkaline earth metals, strontium (Sr) and barium (Ba), with respect to calcium

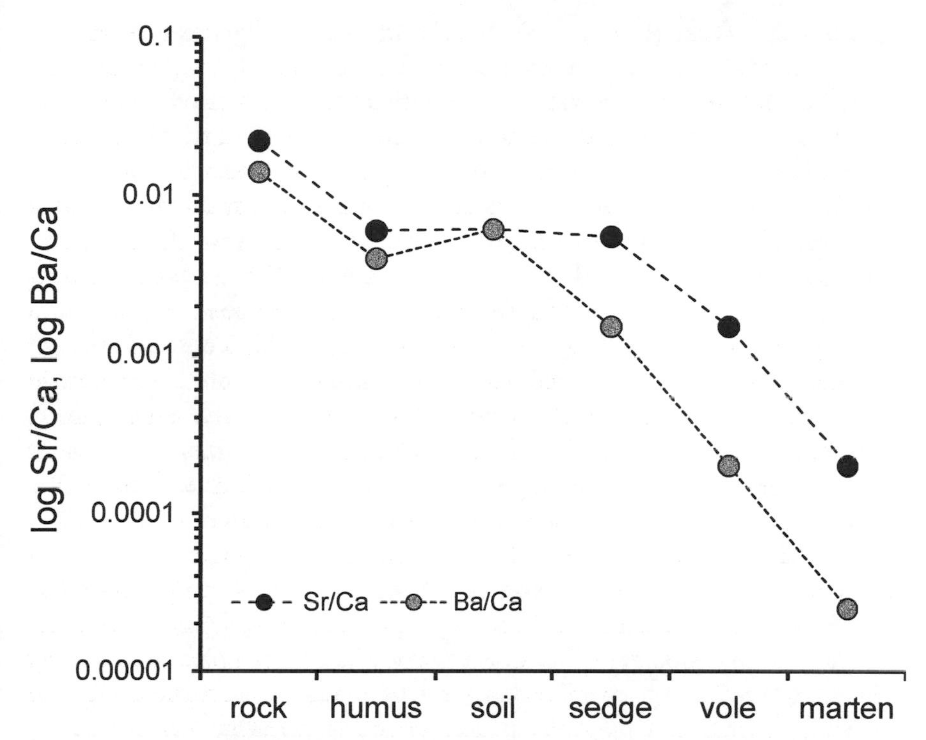

Figure 8.1. *The results of the classic study of trace element distributions in a North American ecosystem, with Sr/Ca and Ba/Ca ratios plotted on a logarithmic scale (Elias et al. 1982). The study was designed to calculate biopurification factors for calcium with respect to strontium and barium uptake. Soil is used as shorthand for soil moisture. The plant:vole:pine marten steps illustrate systematic reduction in Sr/Ca and Ba/Ca in this foodweb, with stronger discrimination against Ba. This study was subsequently taken as representing trophic relations everywhere.*

(Ca) in the digestive tract and kidneys in a process known as biopurification of calcium (Spencer et al. 1973). The result, as famously demonstrated for a North American ecosystem (Elias et al. 1982), is that herbivore tissues have lower Sr/Ca and Ba/Ca ratios than do the plants that they eat, and carnivores in turn have lower Ba/Ca and Sr/Ca than do the herbivores they eat (Figure 8.1). Since strontium and barium are found in bones and teeth where they substitute for calcium in the calcium-phosphate apatite structure, they can in principle be used to investigate trophic behavior of ancient fauna. The approach is site specific because these alkali earth element concentrations differ geographically, largely following geological inputs to the soils. Attempts have been made to apply other trace elements, for instance zinc (Zn), but applications have been unsuccessful because so little is known about their distribution in foodwebs, and their fixation in bone or tooth mineral (Ezzo 1994).

There are two major constraints in application of strontium and barium concentrations to paleodietary reconstruction. One is diagenesis, the alteration of original chemical composition in fossils, and the other is their complex pathways in foodwebs, which can result in significant intrahabitat and intertrophic level variability. Early researchers were largely unaware of the extent of the problem of diagenesis (e.g., Toots and Voorhies 1965; Wyckoff and Doberenz 1968; Brown 1974), but it was subsequently widely recognized (e.g., Boaz and Hampel 1978; Sillen 1981, 1989). Most archaeological and paleontological trace element studies have been carried out on bone. This came about partly because bones are so much more numerous in the fossil record, and also because many teeth are formed early and infants lack the adult capacity for biopurification of calcium (Lough et al. 1963). A major drawback of bone, however, is its susceptibility to postmortem chemical alteration in the form of recrystallization or penetration of a suite of elements from the external environment (e.g., Tuross et al. 1989; Sillen 1989). These processes can quickly obliterate the original, biogenic Sr/Ca and Ba/Ca ratios.

To address the problem, Sillen (1981, 1992) developed a "solubility profiling" technique, based on the premise that altered bone apatite differs in solubility compared to unaltered, biogenic apatite. Following the principles of calcified tissue chemistry, it holds that diagenetic apatite is both more soluble (when it includes more carbonate and other ions that distort the crystal structure), and more resistant (due to penetration of ions such as fluoride that confer greater crystallinity), than biogenic apatite. In this technique these highly soluble and poorly soluble diagenetic apatites, both, are in effect stripped away. The solutes, not the solid materials, are measured for strontium, barium, and calcium concentrations, and the levels of other important components, such as phosphates, are monitored (Sillen 1981, 1992). While ingenious, this method is technically challenging and laborious, greatly limiting wider application. Perhaps more importantly, however, several studies have shown that in many cases, even when it is applied, diagenetic strontium cannot be eradicated from bone (Budd et al. 2000; Hoppe et al. 2003; Trickett et al. 2003; Lee-Thorp and Sponheimer 2003). The most recent attempts have investigated trace element patterns in enamel (Sponheimer et al. 2005a; Sponheimer and Lee-Thorp 2006), which as a denser, more crystalline and ordered apatitic tissue (LeGeros 1991; Elliott 1994) is much more resistant to postmortem elemental alteration than bone. The problem of poor biopurification in infants can be avoided simply by analyzing late-developing teeth.

A more lingering constraint in trace element studies is the requirement for understanding their very complex pathways in ecosystems. We now understand something of the complexity and degree of variability in modern foodwebs, which can result in significant variation between habitats and within a trophic level. The importance of local geology in controlling absolute availability of alkaline earth elements has been known from the early stages of development of the trace element method (Toots and Voorhies 1965), even though sometimes ignored. However, inherent variability *within* trophic levels in ecosystems and indeed within sympatric species has been largely unappreciated. For instance, it has been suggested that there are differences in the Ba/Ca ratios of foregut and hindgut fermenting herbivores (Balter et al. 2002), and sympatric browsing and grazing herbivores can be as readily distinguished by their Sr/Ca and Ba/Ca ratios as can carnivores and insectivores in South African savannas (Sillen 1988; Gilbert et al. 1994; Sponheimer et al. 2005a; Sponheimer and Lee-Thorp 2006).

The mechanisms that lead to such differences are at present poorly understood, but the key to many of them undoubtedly resides in plants—taxa, individuals, and parts (e.g., underground, stem, fruit, leaves) can differ considerably in their strontium distributions due to capillary action in their vascular systems (Runia 1987). This is certainly one of the reasons that Sr/Ca coefficients of variation for a single mammalian species in a single location are typically 30–40 percent (Price et al. 1992; Sillen 1988; Sponheimer et al. 2005a). This large-scale natural variation in mammalian elemental compositions, even in a single ecosystem or region, means that large numbers of samples are always required to adequately characterize dietary ecology. It also implies that obtaining information about trophic behavior at the level of individuals is problematic.

Early hominin diets

The first attempt to investigate the diets of Plio-Pleistocene hominins using chemical composition was made by Boaz and Hampel in 1978. This study sought to build on the work of Toots and Voorhies (1965), who found that, as expected, the bones of Pliocene herbivores from Nebraska had higher strontium concentrations than carnivores from the same site. Boaz and Hampel (1978) determined strontium of bone and/or enamel from *Paranthropus boisei*, *Homo*, and associated herbivores and carnivores from the Omo Group. However, unlike their predecessors, they did not find that carnivores had lower strontium than herbivores or that browsers had higher strontium than grazers, leading them to conclude that strontium "is not preserved in its original life values in the apatite structure of fossilized enamel, dentine and bone." This may in fact have been the case for the fossils they analyzed, but it is impossible to make that judgment from their data. For example, we now know that their assumption that browsing herbivores should have higher strontium than grazing herbivores is untrue, and is in fact the opposite of observations in modern African ecosystems where browsers may have strontium low enough to overlap with those of carnivores (Sillen 1988; Sponheimer and Lee-Thorp 2006). We also do not know whether they sampled only late-developing teeth to avoid problems associated with the inability of infants to discriminate against strontium (Lough et al. 1963). Thus, poorly founded assumptions mean that much of their data set was either misinterpreted or uninterpretable.

The study serves as a sharp reminder of the need to adequately investigate and understand biogeochemical patterns in modern systems before application of techniques to deep time. Without such knowledge it is nearly impossible to determine the potential scope of diagenesis, or even a technique's ability to answer the dietary questions of interest. Yet, for a variety of reasons, such detailed modern studies have been few.

Sillen (1992) developed and applied the solubility-profiling method and a much improved understanding of trace element distributions in mammals to investigate the diet of hominins at Swartkrans. He found that bones of *Paranthropus robustus* had similar Sr/Ca ratios that lay between those of carnivores and primarily herbivorous taxa such as *Papio* and *Procavia* (Figure 8.2a). This, in conjunction with observations from dental microwear, morphology (Grine and Kay 1988), and stable isotopes (Lee-Thorp 1989), led him to conclude that *Paranthropus* was unlikely to be "purely herbivorous." Subsequently, two bone specimens of early *Homo* from Swartkrans were observed to have slightly higher Sr/Ca ratios than *P. robustus* (Sillen et al. 1995), a result that was quite unanticipated, given the generally accepted belief that *Homo* included significant amounts of animal food in its diet (e.g., Aiello and Wheeler 1995). Sillen et al. (1995) argued that the higher Sr/Ca ratio of early *Homo* could have resulted from the consumption of strontium-rich underground plant-storage organs (roots, rhizomes, and bulbs). This argument had precedents. Hatley and Kappelman (1980) had argued that ursids, suids, and early hominins occupied similar adaptive zones, in that all relied upon highly abundant underground resources. Brain (1981) showed that wear on Swartkrans bone tools was consistent with their use as digging "sticks," although the makers of the tools could have been *Paranthropus* or *Homo*, and Backwell and d'Errico (2001) later argued that termites rather than geophytes were targeted. O'Connell et al. (1999), Wrangham et al., (1999), and Laden and Wrangham (2005) further developed the underground storage-organ hypothesis, noting that with the expedient of a

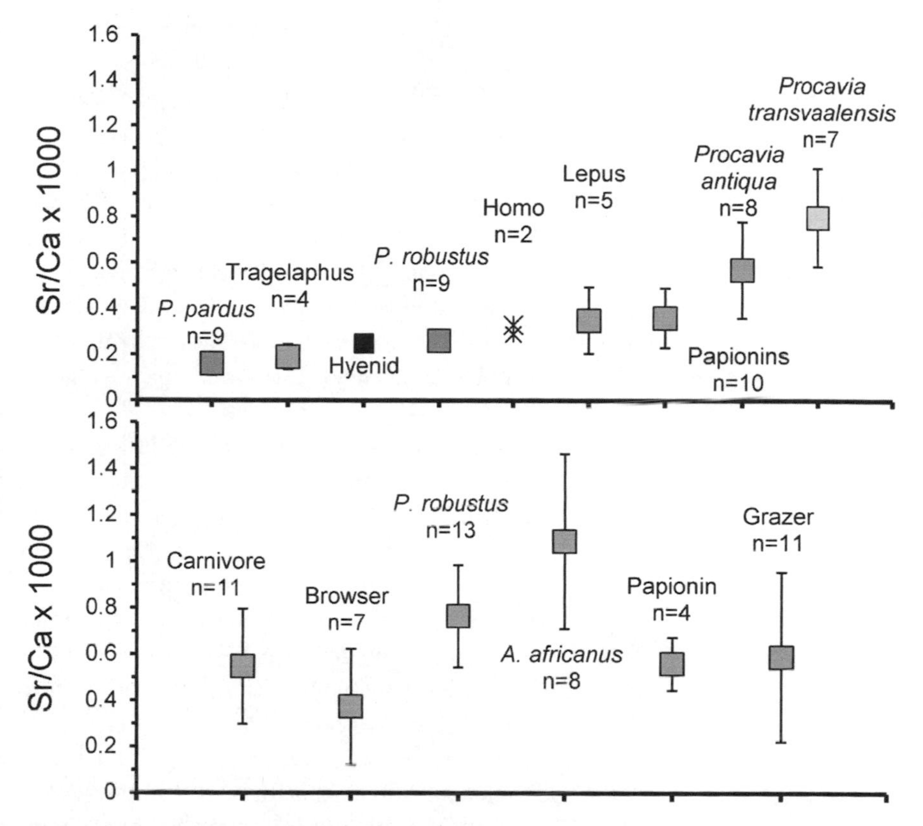

FIGURE 8.2. *The upper panel shows Sr/Ca distribution data for* Paranthropus, Homo, *and a suite of fauna from Swartkrans based on bone analysis, shown as mean Sr/Ca x 1000 and standard deviations (from Sillen 1992). The lower panel shows enamel data for* A. africanus *and* P. robustus *and associated fauna from Makapansgat, Sterkfontein, and Swartkans, shown as mean Sr/Ca x 1000 and standard deviations (data from Sponheimer et al. 2005a; Sponheimer and Lee-Thorp 2006). The data from the three sites were combined because of the similarity in geology and in Sr/Ca ratios for modern fauna from the Sterkfontein and Makapansgat valleys.*

crude digging stick, early *Homo* could have had access to reliable underground resources for which there was relatively little competition.

Important though these trace element studies were for injecting new life into the debate on early hominin diets, Sr/Ca ratios have limitations as a simple trophic-level indicator. As discussed above, Sr/Ca ratios are highly variable in plants at the base of the food web (Bowen and Dymond 1955; Runia 1987) and this fact seriously complicates interpretation of Sr/Ca data (Burton et al. 1999).

For instance, we cannot rule out the possibility that low Sr/Ca ratios could result from fruit consumption, as fruit has been shown to have a relatively low Sr/Ca ratio in some instances (Haghiri 1964). Furthermore, most of the trace element studies were carried out on bone, which is particularly vulnerable to diagenetic overprinting (Sillen 1981; Budd et al. 2000; Hoppe et al. 2003; Lee-Thorp and Sponheimer 2003). A recent study of *Paranthropus* and *Australopithecus* enamel from Swartkrans and Sterkfontein, respectively, showed that neither has lower Sr/Ca ratios than contemporaneous baboons and most herbivores (Sponheimer et al. 2005a) (Figure 8.2b). The enamel-based Sr/Ca results offer little evidence for australopith omnivory. They also demonstrate that, as shown in modern foodwebs (Sillen 1988), there is a high coefficient of variation, which suggests that interpretations at the level of individual are best avoided.

Strontium and calcium may, however, offer another type of information if used in combination with barium. A bivariate plot of two physiologically related trace element ratios (Ba/Ca and Sr/Ba) for early hominins and associated nonhominin fauna (Figure 8.3) shows first that clear ecological patterning is retained in fossil enamel, since grazers, browsers, and carnivores are very nicely separated in much the same way we see with modern African fauna (Sponheimer and Lee-Thorp 2006). Second, the hominins do overlap with the carnivores. But as this is also the case for the papionins, which are demonstrably not carnivorous, little should be made of this resemblance. Most strikingly, *Australopithecus africanus* fossil enamel is characterized by high Sr/Ba ratios, a pattern distinct from all other fossil specimens that have been analyzed. It suggests that they consumed different foods than all of these groups—foods with unusually high strontium and relatively low barium concentrations. Foods that could meet this requirement include grass seeds and underground storage organs. The evidence for the latter is indirect, being based partly on observations that three specimens of African mole rat (*Cryptomys hottentotus*), a species that is known to consume only underground roots and bulbs, had the highest Sr/Ba ratio of any animal we have studied (Sponheimer and Lee-Thorp 2006). The possibilities of both grass seed and underground storage-organ consumption, both of which have been suggested as possible early hominin foods based on other methods (see below, require further consideration.

In summary, although there is clearly ecological patterning to be found in the trace element ratios of early hominins and associated fauna when diagenesis is circumvented, interpretation of these data remains problematic. The difficulty stems from the lack of work on trace element distributions in modern African ecosystems. No detailed studies have been published that demonstrate the elemental distributions in African plants and animals, although some promising

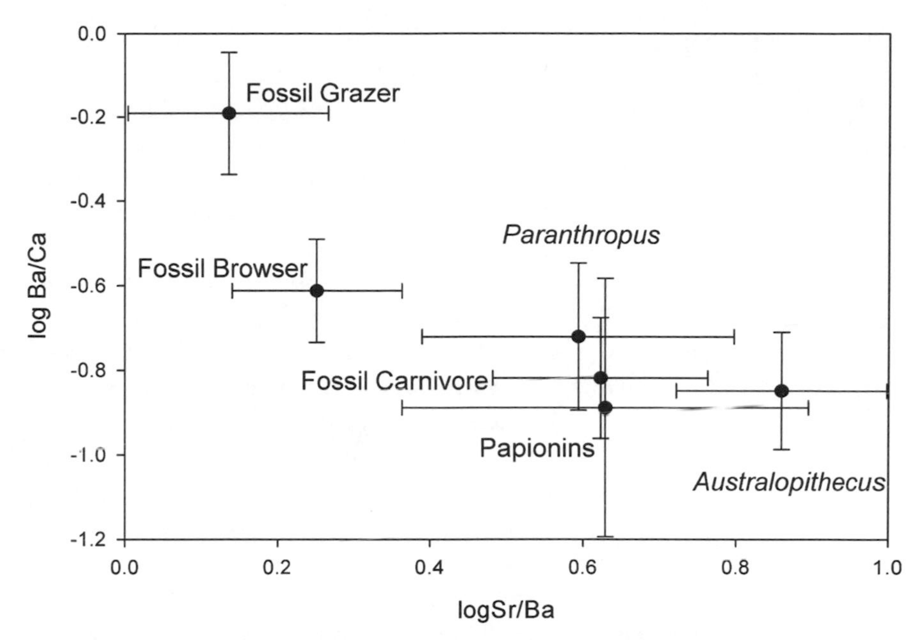

Figure 8.3. *A bivariate plot of log Ba/Ca versus Sr/Ba ratios for fauna and hominins from Makapansgat, Sterkfontein, and Swartkrans separates* A. africanus *from* P. robustus, *although they overlap in Sr/Ca ratios (Figure 8.2). Data are from Sponheimer and Lee-Thorp 2006.*

work has been carried out in North America (Burton et al. 1999). The reason is twofold. In the early days of trace element studies, there was insufficient appreciation for the variation that existed in plants and animals, and therefore it was assumed that trace element ratios simply reflected trophic level. Later, even while researchers began to appreciate the complexity (e.g. Sillen 1988), concerns about diagenesis greatly reduced the time and effort put into trace element studies. Thus, soon after trace element analysis was first applied to early hominins in 1992, it lapsed into neglect except for a few specialized applications. Now that it has been demonstrated that trace element compositions retain much of their fidelity in enamel, studies investigating elemental distribution in modern foodwebs are urgently required.

STABLE CARBON ISOTOPE ANALYSIS

Principles

The basis for stable carbon isotope applications to diet rests on the primary isotopic distinctions that occur in the two main plant photosynthetic pathways. The

RuBisCO (Ribulose Bisphosphate Carboxylase Oxygenase) enzyme involved in catalyzing the conversion of CO_2 to plant sugars discriminates strongly against $^{13}CO_2$. Thus, plants using this cycle alone—known as the Calvin-Benson or C_3 pathway—have low $\delta^{13}C$[1] values compared to atmospheric CO_2 (O'Leary 1981; Farquhar et al. 1982). Some families have evolved a mechanism to concentrate and fix CO_2 by means of another enzyme, PEPCase (Phosphoenolpyruvate Carboxylase), before entering the RuBisCO cycle, so the isotope effect of the latter enzyme is not expressed (Sage 2004). This is known as the C_4 pathway. As a result the $\delta^{13}C$ values of C_3 and C_4 plants are distinct (Smith and Epstein 1971). Almost all trees, shrubs, and herbs, and temperate or shade-adapted grasses follow the C_3 pathway, while C_4 photosynthesis is common among grasses and sedge families in warm, often dry, environments with strong solar radiation in the growing season.

Thus in African savannas, the trees, shrubs, herbs, and their edible parts are distinctly lower in $\delta^{13}C$ than the grasses and many of the sedges. C_4 plants exhibit a relatively narrow $\delta^{13}C$ range (from about −9 to −14%, global mean of ca. −12%), whereas C_3 plants are subject to environmental influences and have much broader ranges (from about −22 to −38‰ at the extremes, with a global mean ca. −26‰) (Smith and Epstein 1971). Environmental influences acting on C_3 plants include low light and recycling of CO_2 in dense forests (leading to very low $\delta^{13}C$ values) (Vogel 1978; Farquhar et al. 1982; van der Merwe and Medina 1989), and humidity or temperature effects (leading to higher $\delta^{13}C$ values under more arid and/or warm conditions and vice versa) (see Tieszen 1991 for a review). A third photosynthetic pathway, Crassulacean Acid Metabolism (CAM), fixes CO_2 by night and utilizes RuBisCO by day, so that $\delta^{13}C$ values can vary depending on whether they are "obligate" CAM or not, and upon environmental conditions (Winter and Smith 1996). They are primarily succulents, such as euphorbias that are rare outside of arid environments and sparsely used by most mammals (but see Codron et al. 2006 for use by baboons), and are thus not considered important components of environments inhabited by Plio-Pleistocene hominins (Peters and Vogel 2005).

The carbon isotopes in these plants are ultimately incorporated into the tissues of animals that consume them. Thus, zebra, which consume C_4 grasses, are

---

1. By international convention stable light-isotope ratios of hydrogen, carbon, nitrogen, oxygen and sulphur are expressed in the delta (δ) notation relative to an international standard, in parts per thousand. So for $^{13}C/^{12}C$ the expression is $\delta^{13}C$ (‰) = $(R_{UNKNOWN}/R_{STANDARD} - 1) \times 1000$, where $R = {}^{13}C/{}^{12}C$, and the international standard is VPDB.

isotopically distinct from giraffe, which browse the leaves of trees. Most forest- or woodland-dwelling primates that consume the edible parts of C_3 plants—such as leaves, fruit, nuts, resins, pith, bark, and bulbs—are distinct from, for instance, gelada baboons that include C_4 grasses in their diets. Secondary consumers of animal tissues, in turn, also reflect the primary basis of the diets of their prey animals. For example, the isotopic composition of insectivorous bat-eared foxes reflects the mix of C_3 and C_4 insects they consume in a particular environment (Lee-Thorp and Sponheimer 2005). Faunivores can provide an integrated reflection on the overall isotopic profile of the local faunal communities. Omnivorous animals, such as baboons, brown hyenas, suids, and of course hominins, are extracting foods from both primary and secondary levels in the local foodweb.

The first studies to apply stable carbon isotopes to investigate paleodiets sought to document the consumption of introduced maize among Native American populations in northeastern North America, where almost all of the local plant foods are C_3, while maize is a C_4 cereal (Vogel and van der Merwe 1977; van der Merwe and Vogel 1978). This study paved the way for a plethora of applications in the following decades, most of which relied on stable isotope abundances in bone collagen. However, since collagen is rarely preserved for very long on geological timescales, most of these studies were confined to the fairly recent past (but see Jones et al. 2001) showing collagen survival for up to 200 kyr (thousand years) years under optimal preservation conditions.

The carbon isotopes in the mineral phase of bone, a bioapatite, can also be used as dietary proxies (Sullivan and Krueger 1981, 1983; Lee-Thorp and van der Merwe 1987). But even though bone mineral clearly can persist for a very long time, it is susceptible to alteration, which may result in the loss of the biogenic $\delta^{13}C$ composition. This susceptibility is due to bone's high organic content, porosity, and minute mineral-crystal size (LeGeros 1991), which make it susceptible to dissolution and recrystallization phenomena that facilitate the incorporation of external carbonate ions, or indeed other ions such as external strontium or rare-earth elements (Wang and Cerling 1994; Lee-Thorp 2000; Lee-Thorp and Sponheimer 2003). Thus, paleodietary studies based on bioapatites were forestalled until it was demonstrated that dental enamel from ancient fauna with well-understood diets retained original biogenic isotope abundances (Lee-Thorp and van der Merwe 1987; Lee-Thorp et al. 1989). Since then, numerous empirical and theoretical studies have substantiated this finding (e.g., Wang and Cerling 1994; Sponheimer and Lee-Thorp 1999b; Hoppe et al. 2003; Lee-Thorp and Sponheimer 2003). Enamel's resistance to diagenetic phenomena is conferred by its virtually organic-free and relatively crystalline state (LeGeros

1991), which promotes stability. Thus, tooth enamel affords the opportunity to investigate the dietary ecology of early Pleistocene and Pliocene hominins and of their associated faunal assemblages.

There is a further advantage to analyzing bioapatite as sample material. Bioapatite carbonate forms directly from blood bicarbonate, and the relationship between breath CO_2 (in equilibrium with blood bicarbonate), diet, and enamel apatite $\delta^{13}C$ has been well documented (Passey et al. 2005). Overall, the diet to enamel shift[2] averages about 13‰ for most large mammals (Lee-Thorp et al. 1989; Passey et al. 2005). Some variability has been documented; for instance measurements on small rodents on controlled diets suggest a diet-apatite $\delta^{13}C$ difference of just under +10‰ (Ambrose and Norr 1993; Tieszen and Fagre 1993), while studies of some large ruminants and hind-gut feeders indicate shifts of +13 to +14‰ (Cerling and Harris 1999; Passey et al. 2005). This variation likely reflects differences in metabolism and/or dietary physiology, such as whether or not an animal produces large quantities of methane (see Passey et al. 2005). Most importantly, bioapatite demonstrably reflects the $\delta^{13}C$ of the bulk diet (all energy sources and proteins), and not largely the protein component as is the case for collagen (Krueger and Sullivan 1984; Lee-Thorp et al. 1989; Ambrose and Norr 1993; Tieszen and Fagre 1993). Thus, apatite and bone collagen $\delta^{13}C$ provide different perspectives on an individual's diet. Analysis of both components would provide the most complete picture, but that has not proven possible with early hominins so far. For the reasons outlined above, only tooth enamel is sampled for stable-isotope analysis of hominin and nonhominin specimens that are millions of years old.

Initially, relatively large samples of 200–400 milligrams were needed, but advances in mass spectrometry have reduced the sample size to just a few milligrams. As a result, it has become possible to sample teeth while producing little to no readily observable damage by removing a little enamel powder from the length of the crown with a diamond-tipped burr and low-speed rotary drill. If a high-resolution sequence along the growth axis is required, powder is extracted sequentially away from the enamel–dentine junction in small increments (e.g., Balasse et al. 2003), or, by means of laser-ablation mass spectrometry (Passey and Cerling 2006; Sponheimer et al. 2006b; Lee-Thorp et al. 2010). Usually, a sequence of pretreatment protocols on the powders is followed in order to

2. The diet-apatite carbonate shift is often reported directly as a difference in isotopic (δ) values (Δ), or more correctly expressed as an enrichment factor, ε ($\varepsilon_{DIET\text{-}APATITE\ CO3} = [(\delta_{DIET} + 1000)/(\delta_{APATITE\ CO3} + 1000) - 1] \times 1000$).

remove organic contaminants (including glue, if present) and diagenetic carbonates. It is worth noting that these protocols can vary between laboratories, and as they can lead to small but significant differences in oxygen isotope composition in particular, they should be applied, and the results compared, with caution.

Diets of australopiths and early *Homo*

Most isotope research has been directed at the dietary ecology of hominins from the South African sites, and the situation has changed only recently with the publication of several East African studies. Among the South African hominins, it was generally believed until the early 1990s that *A. africanus* was more omnivorous compared to *P. robustus,* which was considered to be a plant-food specialist. Based on the latter's dental morphology with huge molars and small anterior teeth, and occlusal molar microwear showing relatively high proportions of pitting, these plant foods were thought to be small, hard objects such as hard fruits and nuts (Grine 1981, 1986; Grine and Kay 1988; Ungar and Grine 1991). As these are all C_3 foods, one would then expect that *P. robustus* would have $\delta^{13}C$ values indistinguishable from those of C_3 consumers such as browsing and/or frugivorous bovids. Similarly, if *A. africanus* consumed a chimpanzee-like diet, their $\delta^{13}C$ values should also be relatively depleted in ^{13}C. Several studies have demonstrated that this is not the case; the $\delta^{13}C$ values of both australopiths are distinct from that of their coeval C_3-consuming mammals (Lee-Thorp et al. 1994, 2000; Sponheimer and Lee-Thorp 1999a; van der Merwe et al. 2003; Sponheimer et al. 2005a; Figure 8.4). In addition, $\delta^{13}C$ of the few *Homo* specimens analyzed, from Swartkrans, overlaps with the australopith values (Lee-Thorp et al. 2000).

Analysis of variance and Fisher's PLSD test show that both *Australopithecus* ($\bar{x}$ = −7.1 ± 1.8‰, N = 19) and *Paranthropus* ($\bar{x}$ = −7.6 ± 1.1, N = 18) differ strongly from contemporaneous C_3 ($\bar{x}$ = −11.5 ± 1.3‰, N = 61) and C_4 consumers ($\bar{x}$ = −0.6 ± 1.8‰, N = 60) ($P < 0.0001$), but cannot be distinguished from each other (P = 0.18). This distinction between the hominins and other fauna cannot be ascribed to diagenesis, as there is no evidence for alteration of browser or grazer $\delta^{13}C$ values. If we take the mean $\delta^{13}C$ values of C_4- and C_3-consuming herbivores at these sites as indicative of "pure" C_4 and C_3 diets respectively, then we calculate that, on average, about 35 percent of their diets are from C_4 resources. This is a large quantity of C_4 resources, and they must necessarily consist of grasses or sedges, or animals that ate these plants, or a combination of these. None of these possibilities was expected, as extant apes are not known to consume

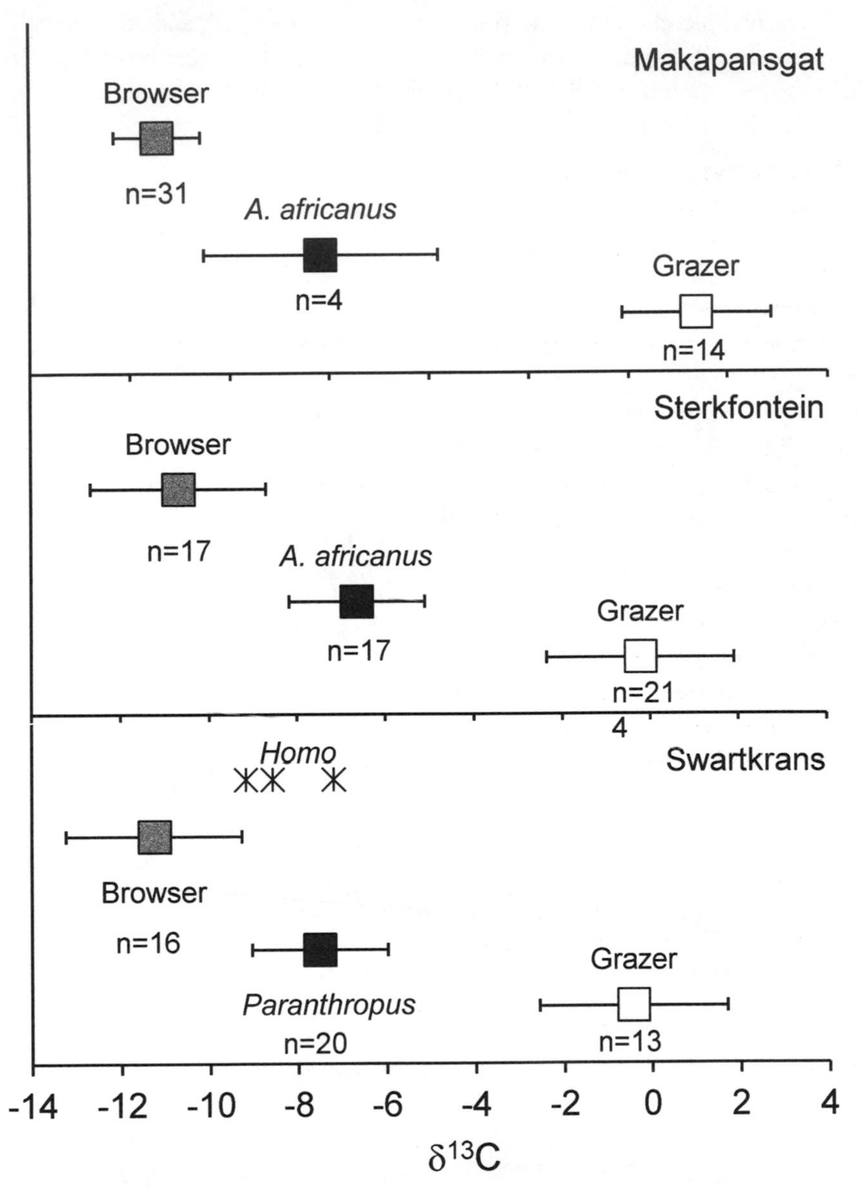

FIGURE 8.4. *Enamel* $\delta^{13}C$ *data for* A. africanus *and* P. robustus *specimens from the sites of Makapansgat, Sterkfontein, and Swartkrans, compared with* C_3 *plant consumers and* C_4 *plant consumers; all data are shown as means (squares) and standard deviations. Numbers (N) of individuals are given. Data are from Lee-Thorp et al. (1994, 2000) for Swartkrans; Sponheimer and Lee-Thorp (1999a) and Sponheimer (1999) for Makapansgat, van der Merwe et al. (2003) and Sponheimer et al. (2005b) for Sterkfontein.*

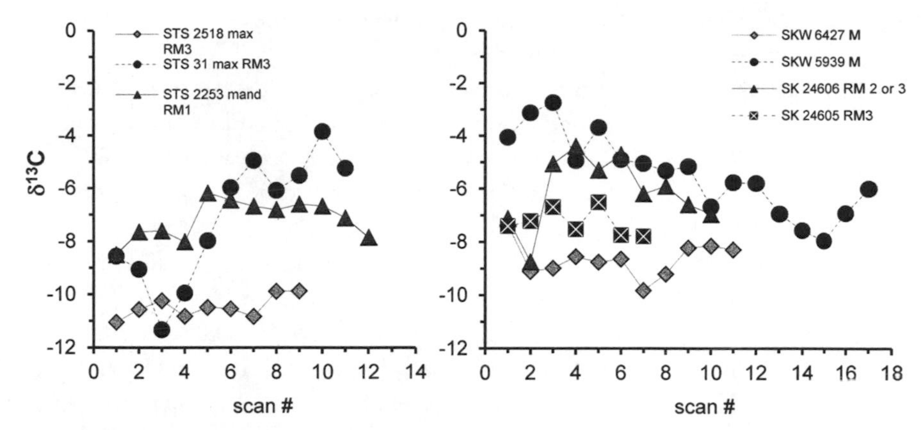

Figure 8.5. *High-resolution laser-ablation* $\delta^{13}C$ *sequences for* A. africanus *(N = 3, left) and* P. robustus *(N = 4, right) plotted against scan number. Sample increments were approximately 0.3 mm. The* A. africanus *data are from Lee-Thorp et al. (2010) and the* Paranthropus *data are from Sponheimer et al. (2006b), where the data were originally plotted according to a time-sequence model based on perikymata counts. Here we avoided using an age model because of uncertainty about the length of time involved in enamel maturation.*

these foods (Goodall 1986; McGrew et al. 1982). Even in environments where C_4 foods are readily available, it is hard to detect any signs of C_4 consumption from chimpanzee $\delta^{13}C$ values (Schoeninger et al. 1999; Carter 2001; Sponheimer et al. 2006a). We have argued that these data suggest a fundamental niche difference between the South African australopiths and extant apes (Lee-Thorp et al. 2003; Lee-Thorp and Sponheimer 2006; Lee-Thorp et al. 2010; Sponheimer et al. 2005b) although recent, but limited, data hint that *Australopithecus sediba* might have been a C_3 specialist (Henry et al. 2012).

Hominin $\delta^{13}C$ values were also more variable than almost all other modern and fossil taxa that have been analyzed in South Africa (Lee-Thorp et al. 1994, 2000; Sponheimer 1999; Sponheimer and Lee-Thorp 1999a, 2001; Codron et al. 2005, 2006; van der Merwe et al. 2003). There are no significant differences in hominin $\delta^{13}C$ values between Makapansgat Member 3, Sterkfontein Member 4, or Swartkrans Member 1 (ANOVA, P = 0.14) so it cannot be argued that environmental change drove the australopiths to modify their diets over time, leading to this variable $\delta^{13}C$. In fact, hominin $\delta^{13}C$ values are highly variable within any given unit (Member) or time interval. This could be interpreted as the result of more rapid environmental shifts within units that drove the dietary diversity

(Hopley and Maslin 2010), or it could simply mean that they were opportunistic primates with wide habitat tolerances.

Moreover, the variability is not only found between the teeth of different individuals, but also *within* teeth, demonstrating annual and/or seasonal-scale dietary shifts by an individual. Recent studies using laser-ablation stable isotope ratio mass spectrometry to obtain multiple analyses along the growth axis of individual teeth showed that $\delta^{13}C$ values for an individual *P. robustus* from Swartkrans may vary by up to 5‰ (Sponheimer et al. 2006b) and almost as much for two out of three *A. africanus* teeth from Sterkfontein (Lee-Thorp et al. 2010) (Figure 8.5). In contrast, there was no evidence for such isotopic shifts in three browsing herbivores (*Raphicerus campestris*) from the same assemblage. It is not possible to extract completely discrete information about shifts using intratooth sampling, even using high-resolution sampling, because enamel maturation continues for several months after initial mineral deposition, resulting in isotopic overprinting that dampens dietary shifts (Passey and Cerling 2002; Balasse 2002). Thus a shift of approximately 5‰ indicates a much more extreme shift from a C_3- to a C_4-dominated diet during the period of crown formation.

In East Africa, recent stable light isotope studies of hominin diets have yielded startling results, which have caused us to rethink their dietary ecology. Analyses of *P. boisei* from Tanzania (van der Merwe et al. 2008) and from Kenya (Cerling et al. 2011) showed consistently high enamel $\delta^{13}C$ values. The data, now firmly established with the latest study of a reasonable number of individuals (N = 22 individuals) from different sites and covering a period of almost half a million years (Cerling et al. 2011), indicate very large C_4 dietary contributions indeed, of up to 80 percent or more. The mean $\delta^{13}C$ value for *P. boisei* ($\bar{x}$ = −1.3 ± 0.9‰) is slightly but not significantly higher (indicating more C_4) even than the mean of all published specimens of fossil *Theropithecus oswaldi*, the grass-eating baboon ($\bar{x}$ = −2.3 ± 1.5‰). It differs significantly from that of contemporaneous *P. robustus* ($\bar{x}$ = −7.6 ± 1.1‰) in South Africa and early *Homo* throughout Africa ($\bar{x}$ = −7.8 ± 1.5‰) ($P < 0.0001$) (Figure 8.6).

The results are consistent with *P. boisei* having had a specialist adaptation for the consumption of C_4 grass, or C_4 sedges such as *Cyperus papyrus*, or both. While introduction of a small amount of C_4 carbon indirectly via animal foods cannot be ruled out, a plant staple is the most plausible explanation. Whatever the C_4 food source was exactly, the results are inconsistent with consumption of anything more than perhaps 20 percent C_3 fruits and nuts. In fact, they are inconsistent with much of what has been written about their presumed "nut-cracker" diet over the last fifty years (but see Jolly 1970). They are, furthermore, irreconcilable with the idea that these individuals ate diets even broadly similar

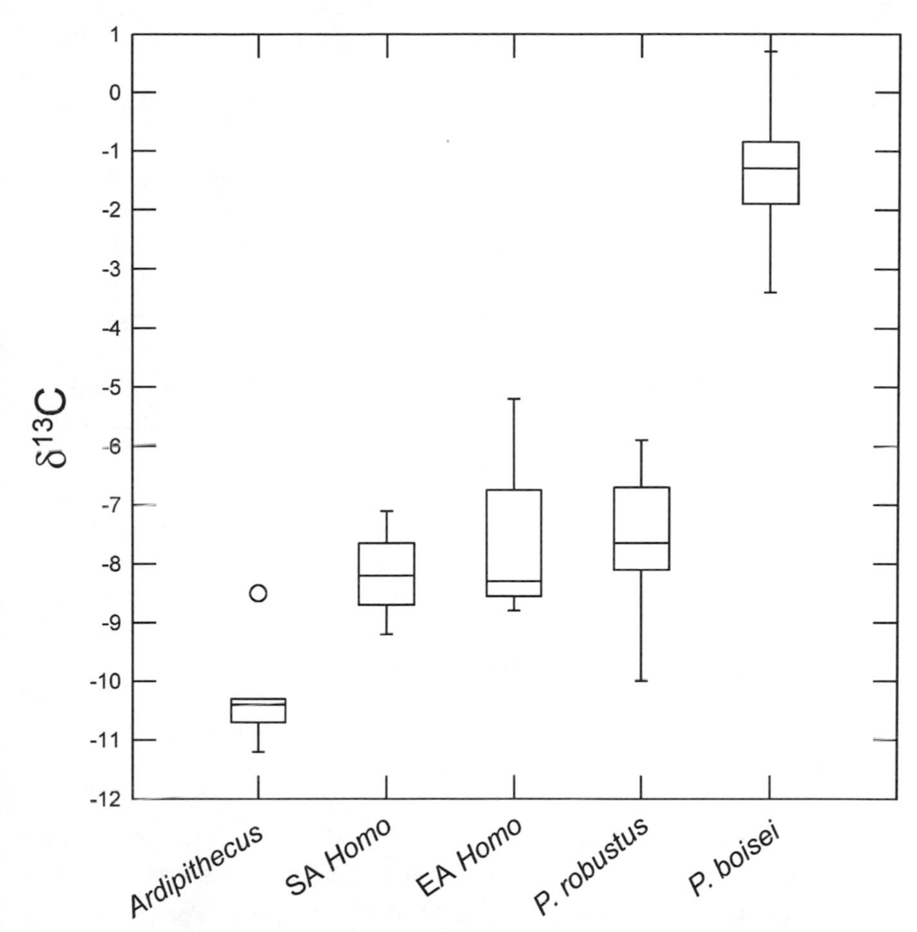

Figure 8.6. *A comparison of hominin data as box-and-whisker plots for* Ardipithecus ramidus, Homo *from South Africa (SA* Homo*) and East Africa (EA* Homo*),* P. robustus, *and* P. boisei. *Data are from Lee-Thorp et al. (2000), Sponheimer et al. (2005b), van der Merwe et al. (2008), Cerling et al. (2011) and White et al. (2009).*

to those of African apes. The outcome is singularly important for our understanding of early hominin diets in general and of australopiths in particular. One consideration must be the nutritional quality of the diet, since both grasses and sedges are low in protein and high in cellulose and carbohydrates.

In contrast, specimens of *Homo* from Olduvai yielded values similar to the Swartkrans specimens analyzed previously ($\delta^{13}C = -7.4 \pm 2.0$, and -8.2 ± 1.0‰ respectively, $N = 3$ in both cases) (van der Merwe et al. 2008; Lee-Thorp et al. 2000). These values indicate C_3 and C_4 dietary contributions of roughly 75 percent and 25 percent, respectively, and it is striking that they are so similar in such widely separated locations.

Analyses of an earlier 4.4-myr-old putative hominin, *Ardipithecus ramidus* from Aramis, Ethiopia, revealed none of this close engagement with C_4 resources (White et al. 2009). Rather, its isotopic composition most resembles that of "savanna" chimpanzees, which tend to avoid C_4 resources (Schoeninger et al. 1999; Sponheimer et al. 2006a). Although White et al. (2009) stress the woody nature of the Aramis environment, when taken together, the faunal $\delta^{13}C$ values show that a good deal of C_4 grassy vegetation was present in the environs. Therefore, *Ardipithecus* was apparently not participating in the broader opportunities offered beyond the woodland/riparian components of its habitat. In contrast, the next-oldest hominins for which we have published $\delta^{13}C$ data—*Australopithecus afarensis* and *Australopithecus bahrelghazali* from about about 3.5 mya (million years ago), and *A. africanus* from about 2.8 mya—had clearly shifted their dietary niche to exploit C_4 resources (Lee-Thorp et al. 2012; Sponheimer et al. 2013). This occurred in spite of the relatively closed nature of the Makapansgat environment (Reed 1997, and see below). If an engagement with C_4 foods marked a fundamental shift in hominin evolution as we have argued elsewhere, then such a shift occurred *sans Ardipithecus*.

C_4 FOODS AND DIETARY BREADTH VERSUS FALLBACK HYPOTHESES

So what were these C_4 resources that early hominins (discounting *Ar. ramidus* for the moment) were consuming? And were these foods that were used in times of stress—fallback foods—as suggested by Ungar (2004)? The answer to these questions is particularly significant for the australopiths, because the outcome has a variety of physiological, social, and behavioral implications. For instance, if australopiths had a grass-based (graminivorous) diet similar to the modern gelada baboon (*Theropithecus gelada*), such a relatively low-nutrient diet would place limitations on brain expansion and sociality as argued by Aiello and Wheeler (1995) and Milton (1999). Conversely, diets rich in animal foods would indicate a leap in dietary quality over modern apes.

Carbon isotope analysis alone cannot fully answer these questions, but it can provide some useful constraints. We cannot be specific about C_3 dietary contributions, rather we can say with more certainty that C_4 contributions must include (directly) tropical grasses and/or sedges, or (indirectly) animals that ate these plants, or some combination of these possibilities. How do we, or can we, distinguish among them? Consideration of potential food items suggests that for the most part, none on its own can account for the observed mixed $\delta^{13}C$ values in the South African australopiths and *Homo* in both eastern and South Africa. Grasses were discounted earlier (Lee-Thorp et al. 2000), because hominin teeth are unsuited to processing abrasive, phytolith-rich grass blades, microwear studies showed little evidence for characteristic scratches, and moreover grasses offer poor returns as they are small packages and their more nutritious period is limited. Some C_4 sedges provide larger starch-rich food packages (van der Merwe et al. 2008) and could be easier to collect. However, they are not very abundant in southern Africa, where recent surveys have shown that two-thirds of sedge species use C_3 photosynthesis (Stock et al. 2004; Sponheimer et al. 2005b). Yet, C_4 sedges are more abundant in East African wetlands (Hesla et al. 1982). One other pertinent observation is their toughness and dubious nutritional value (depending on the species and part of the plant consumed), presumably explaining why few mammals are known to exploit them to any significant extent (but see van der Merwe et al. 2008 for edibility and energy values for *Cyperus papyrus*, and use by humans in the Okavango Delta).

It can be expected that a significant proportion of animal foods will be enriched in ^{13}C in biomes where a large proportion of the primary biomass is C_4. However, the easily foraged items, such as termites, do not show such a pattern. Termites have been suggested as likely animal foods partly because the wear patterns on Swartkrans bone tools suggest their use as tools used for digging termite mounds (Backwell and d'Errico 2001). However, a survey of termites showed that most have mixed C_3/C_4 compositions in South African woodlands, meaning that a huge number would need to be eaten to fulfill the approximately 35-percent C_4 "quota." Thus, the most reasonable argument for the South African hominins, and East African *Homo*, is that they consumed a wide variety of C_4 foods such as grass seeds, sedges, arthropods, large and small vertebrates, and birds' eggs (e.g., Peters and Vogel 2005; Sponheimer and Lee-Thorp 2003).

This solution, however falls short when we consider that C_4 contributed as much as 80 percent to *P. boisei* diets, and indeed when we consider that some South African australopiths seasonally consumed very large amounts of C_4 resources. It is hard to merit any explanation for such positive $\delta^{13}C$ values that does not invoke the large-scale consumption of a reliable plant C_4 resource such as sedges in wetland environments or grasses. Van der Merwe et al. (2008) favored the C_4 sedge explanation as most plausible since C_4 sedges are likely to have been abundant in the wetlands at Olduvai (Bamford et al. 2006; Hesla et al. 1982), and they are available throughout the year. Palatable grasses, on the other hand, are restricted to the wet season, as they become desiccated in the dry season. For some habitual grass consumers such as warthogs or Gelada baboons, grass rhizomes and herbs fill the seasonal gap (e.g., Iwamoto 1993).

Other biogeochemical proxies may help constrain possibilities further, but we know less about them. Trace elements are currently of little help since so little is known about their distributions in African plants; for instance trace element distributions in sedges or even fruits are unknown. It is not yet clear whether the unusual trace element pattern in *A. africanus*, which suggests consumption of foods high in strontium and low in barium (perhaps grass seeds and/or underground storage organs), reflects the C_4 or C_3 diet contributions. Enamel apatite $\delta^{18}O$ may provide some hints. Oxygen isotopes are usually considered as climate indicators, since the primary influence in ecosystems is from environmental drinking water, which is subject to a range of strong climate influences (e.g., vapor source, storm paths, temperature, and altitude) (Dansgaard 1964). However, mammalian $\delta^{18}O$ is also influenced by dietary ecology (Bocherens et al. 1996; Kohn et al. 1996; Sponheimer and Lee-Thorp 2001, Bowen et al. 2009). In herbivores the oxygen from plant water and carbohydrates in leaves that are enriched in ^{18}O as a result of evapotranspiration represents a strong influence in nonobligate drinkers (Bocherens et al. 1996; Levin et al. 2006). The distribution of $\delta^{18}O$ in bioapatites also reflects trophic behavior. The faunivores *Otocyon megalotis*, *Crocuta crocuta,* and *Orycteropus afer* were observed to be depleted in ^{18}O compared to herbivores in two modern ecosystems (Lee-Thorp and Sponheimer 2005). Low faunivore $\delta^{18}O$ values could be linked to their high-lipid, high-protein diets. Many suids and some primates also have relatively low $\delta^{18}O$, but the causes are uncertain—it may be simply related to high water requirements, or, a component from the diet (Sponheimer and Lee-Thorp 1999b; Carter 2001).

We do not have very exact comparative data from those hominin sites with published $\delta^{18}O$ data, but even so, gross comparisons of the Aramis, Makapansgat, and Koobi Fora fauna and hominins can be revealing (Figure 8.7). In contrast with the australopiths from Makapansgat and Koobi, the mean $\delta^{18}O$ for *Ar. ramidus* is high compared to the suids and hyenids. At Makapansgat, *A. africanus* is depleted in ^{18}O compared to suids, and similar to the hyenids. Interestingly, even lower values are observed for *P. boisei*; their mean $\delta^{18}O$ is lower than all other fauna excepting the hippopotamus (Figure 8.7). Although these patterns are tantalizing, their interpretation is not simple because there are multiple possible causes for the low $\delta^{18}O$ values in primates and suids. The first possibility is simply a strong water dependency, while dietary factors may be linked to foods that have low $\delta^{18}O$ values, including underground storage organs, sedges, and fruits. Given our present limited understanding of $\delta^{18}O$ patterning in foodwebs, we simply cannot distinguish among the options.

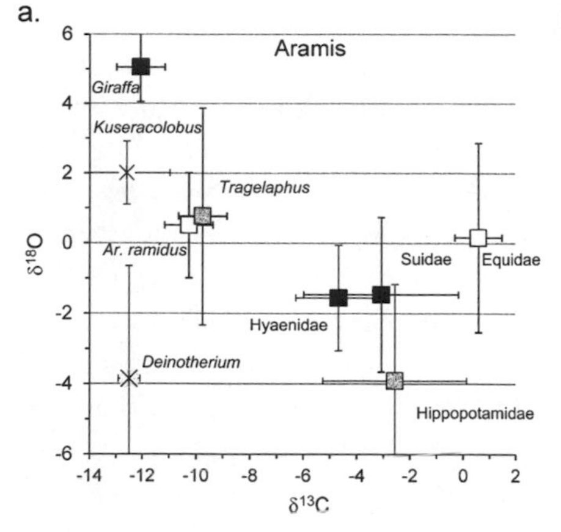

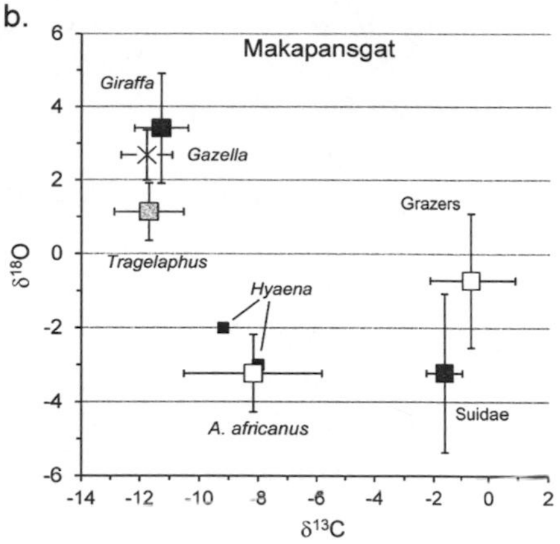

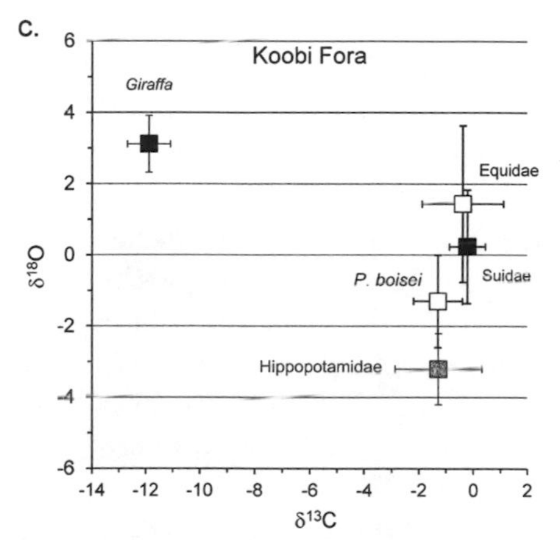

FIGURE 8.7. *Bivariate* $\delta^{18}O$ *and* $\delta^{13}C$ *plots for (a)* Ardipithecus ramidus *and selected fauna from Aramis, (b)* Australopithecus africanus *and selected fauna from Makapansgat Member 3, and (c)* Paranthropus boisei *and selected fauna from Koobi Fora, shown as means and standard deviations. Data are from White et al. (2009), Sponheimer (1999), and Cerling et al. (2011).*

Despite these uncertainties, we should not lose sight of a significant finding from the isotope data, namely that at some point australopiths began to consume novel C_4 resources. Thus, a fundamental difference between australopiths and extant apes might be that, when confronted with increasingly open areas, apes continued to use the foods that are most abundant in forest environments, whereas australopiths recognized and began to exploit new resources. Should we regard this development as reflecting an increase in dietary breadth, or as an adaptive response to increasing seasonal stress when lower quality, mechanically demanding foods had to be consumed (i.e., the Fallback Hypothesis)? Or is there some other explanation?

The Fallback Hypothesis has been extensively explored in recent literature, stimulated by an apparent discrepancy between australopith morphology that appears to emphasize the importance of increasingly harder and/or more-abrasive foods from the mid-Pliocene onwards, and the complete lack of dental microwear evidence for mastication of such foods in the East African australopiths, *A. anamensis*, *A afarensis*, or *P. boisei* (Grine et al. 2006a, 2006b; Ungar et al. 2008). One interpretation of the discrepancy, based on the microwear data for South African *P. robustus* (that *do* suggest hard-object feeding) and coupled with isotope data that show seasonally variable C_4 contributions, argues that the latter were fallback foods, while for the remaining time their diets were more C_3 and resembled those of chimpanzees. Following this argument, the dentognathic adaptations of the australopiths are for fallback, rather than preferred dietary resources (Ungar 2004; Laden and Wrangham 2005; Scott et al. 2005; Grine et al. 2006a, 2006b; Ungar et al. 2008). The principle that morphology can be governed by fallback rather than by preferred foods, also known as Liem's Paradox, is grounded in the broader ecological literature. For instance, Lambert et al. (2004) found that despite large differences in the enamel thickness of *Lophocebus albigena* and *Cercopithecus ascanius*, their diets only differed in hardness during times of fruit scarcity, suggesting that fallback foods were driving these dental differences. Among African apes, chimpanzees and gorillas both prefer fruits over herbaceous vegetation (Tutin and Fernandez 1985; Stanford and Nkurunungi 2003), but during periods of scarcity, gorillas can rely on terrestrial herbaceous and pithy vegetation, and this can be argued to be reflected in their greater occlusal relief. It would follow then, that differences in the dentition of gorillas and chimpanzees are largely a function of their fallback foods, and not a function of their preferred or even typical diets (Ungar 2004).

Application of the Fallback Hypothesis to the hominin record presents contradictions, however. The most intractable problem is the question about why not one of twenty-nine East African australopith teeth analyzed (Grine et al. 2006a, 2006b; Ungar et al. 2008) shows any evidence for the effects of hard foods, which should be expected given that these australopiths are all megadont and possess a number of features consistent with the consumption of hard foods (McHenry and Coffing 2000; Teaford et al. 2002). If they switched to harder fallback foods than those consumed by chimpanzees, why don't we see it? Since dental microwear best preserves dietary information about the period immediately preceding death (Grine 1986), and since primate mortality is often highest during resource stress (Cheney et al. 1981; Milton 1990; Gould et al. 1999; Richard et al. 2002), it could be argued that fallback foods will be *overrepresented* in the dental microwear record (although it might also be expected that this would vary by habitat and taxon). But that is not the case for the eastern

African australopiths. Even for the South African australopiths the argument is difficult to sustain because seasonal switching to C_4 foods as demonstrated (Sponheimer et al. 2006b; Lee-Thorp et al. 2010) is more likely to have occurred in the rainy, summer season (i.e., less stressful times in most savannas).

How then do we explain the morphology? Consideration of the very high $\delta^{13}C$ values for *P. boisei* requires a different sort of explanation that goes beyond dietary broadening and begins to resemble a new specialization entirely. The *P. boisei* values almost certainly reflect a diet that was dominated by either edible C_4 sedge parts (e.g., their starchy tubers) or edible grass parts (e.g., fresh blades, seeds, corms, and rhizomes). It follows then that since *P. boisei* is the quintessence of the australopith masticatory package, the package itself is an adaptation for such C_4 foods. In other words, it is possible that we have been misreading the morphology, or that the morphology is consistent with more than one type of dietary specialization, including one that requires a great deal of repetitive biomechanical loading (Daegling et al. 2011; Ungar and Sponheimer 2011).

Regardless, new isotopic data from eastern and central Africa will allow broad taxonomic, temporal, and regional comparisons among hominins (Sponheimer et al. 2013). It now appears that before 4 Ma, hominins had diets that were dominated by C_3 vegetation and were in that sense similar to those of extant chimpanzees. By about 3.5 Ma, multiple hominin taxa (e.g., *A. afarensis*, *A. bahrelghazali*) incorporated C_4 or CAM foods in their diets, and by about 2.5 Ma, *Paranthropus* in eastern Africa diverged toward C_4 specialization and occupied an isotopic niche unknown in catarrhine primates except in the fossil relations of grass-eating gelada baboons. At the same time, other taxa (e.g., *A. africanus*) continued to have highly mixed and varied C_3/C_4 diets or possibly C_3 diets (*A. sediba*). Overall, there is also a trend toward greater C_4 or CAM consumption in early hominins over time, although this varies by region. Hominin C_4 or CAM consumption also increases with postcanine tooth area and mandibular robusticity, which could indicate that these foods played a role in furthering australopith masticatory robusticity. Nevertheless, the C_4 or CAM resources that hominins ate remain unknown (Sponheimer et al. 2013).

HABITAT RECONSTRUCTION

As evident in the discussion above, the local environment provides the opportunities and constraints for obtaining the basic resources of life—food, water, and shelter. Further, it is widely understood that environmental and climate shifts have played a major role in shaping the course of faunal and hominin evolution (e.g., Vrba 1985; Potts 1996; Feibel 1997). Establishing the environmental

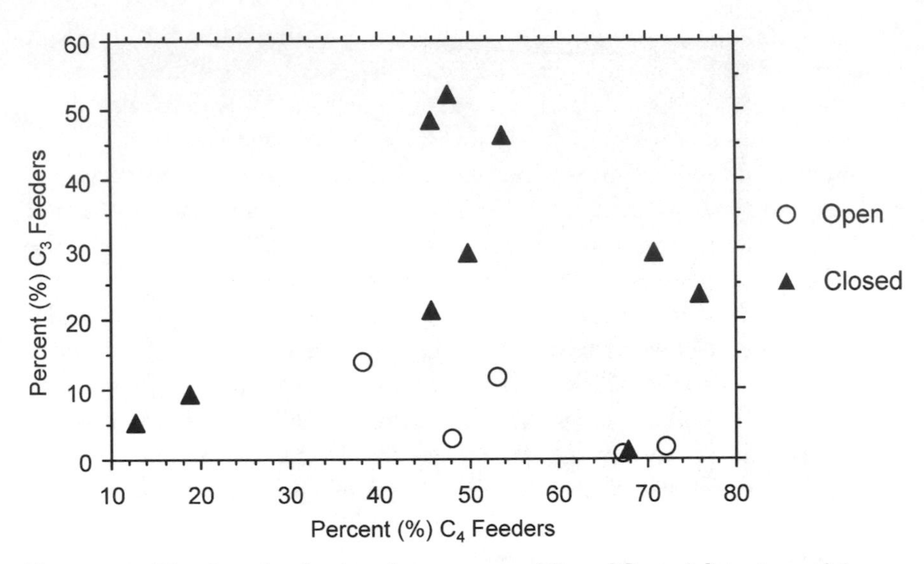

FIGURE 8.8. *Bivariate plot showing the percentages of C_3 and C_4 specialists at a variety of African game reserves. Open habitats are represented by open circles and closed by filled triangles. The relatively "open" environments (as defined by Vrba 1980) are confined to the lower-right corner of the graph.*

context requires multiple lines of evidence gleaned from geological, floral, and faunal studies. Following the same principles outlined above, isotope ratios in fossil-tooth enamel have become an increasingly important component of environment and habitat reconstruction. Isotopic approaches provide information of relevance to the diets and habitat preferences of animals in the fossil-bone assemblages that in turn help to address fundamental questions about habitat structure and ecology.

The rationale for using carbon isotope data from herbivores to investigate paleoecology has followed several strands. First, isotope analysis has been used to track the emerging importance of C_4 grasses in tropical African ecosystems from the late Miocene onwards by assessing shifts in the dietary habits of various families of herbivores. Thus it has been shown by carbon isotope analysis of herbivores that although C_4 grasses began to spread at about 8 mya in the tropics, they seem to have reached the mid-latitudes somewhat later, in the early Pliocene (e.g., Cerling et al. 1997; Hopley et al. 2006; Ségalen et al. 2007). This important shift has in turn raised questions about the evolutionary implications for mammalian lineages and communities, and thus the different rates and times of adoption of C_4 grazing by the Bovidae, Elephantidae, Equidae,

Deinotheridae, Giraffidae, Hippopotamidae, and Suidae families, can also be tracked across sub-Saharan Africa. Finally, at site level, the general relationship between the proportions of C_4 grass consumers present and the availability of palatable grasses in the local environment, and vice versa, the proportion of browsers and the availability of browse can be employed to yield information about the nature of the local habitat. For instance, if virtually every animal at a site is found to have been a C_4 grass-consumer, it is reasonable to conclude that the area was open and dominated by grasses. This approach of course, has immediate parallels with ecomorphologically sensitive faunal abundance approaches.

The idea is illustrated in Figure 8.8 where the percentage of C_3 and C_4 consumers at fifteen African game reserves generally distinguishes between "closed" and "open" habitats (as defined by Vrba 1980), with the latter being isolated in the bottom-right corner (few C_3 specialists and many C_4 specialists). The only "closed" area that clusters incorrectly with "open" habitats is Lake Manyara, where buffaloes comprise 66 percent of the total bovid population. However, Lake Manyara also contains many "open" areas, so this apparent contradiction is at least partly a matter of definition. These data strongly suggest that when an area has fewer than 20 percent C_3 consumers and more than 35 percent C_4 consumers, it is likely to be an "open" environment. One advantage of this approach, when applied to the fossil record, is that it requires no *a priori* assumptions about the behaviors of fossil taxa. This is particularly useful as many extinct species have no modern congeners.

It is also worth pointing out the difference between carbon isotope data from herbivore enamel and pedogenic carbonates (discussed by Quade and Levin, Chapter 3, this volume). Put simply, the former provides a perspective through the filter of an herbivore's dietary preferences. It is an averaging mechanism in which the animals do the sampling of the vegetation. If we have a good understanding of the herbivores' dietary preferences (i.e., whether browsers or grazers), it allows us to distinguish not just the proportions of C_4 grasses, but also the contributions of C_3 grasses from C_3 shrubs and trees, which is not visible from pedogenic-carbonate analyses.

For the most part, researchers have followed an approach in which they interpret the patterns of $\delta^{13}C$ data from the various taxa in a site or sites in terms of vegetation coverage and habitat type. Before turning to the data that we now have from across sub-Saharan Africa, we discuss an example of the isotopic version of Vrba's approach applied to the South African hominin sites, shown in Figure 8.8. We divided the $\delta^{13}C$ data for large herbivore specimens from the sites of Makapansgat, Sterkfontein, and Swartkrans into three broad

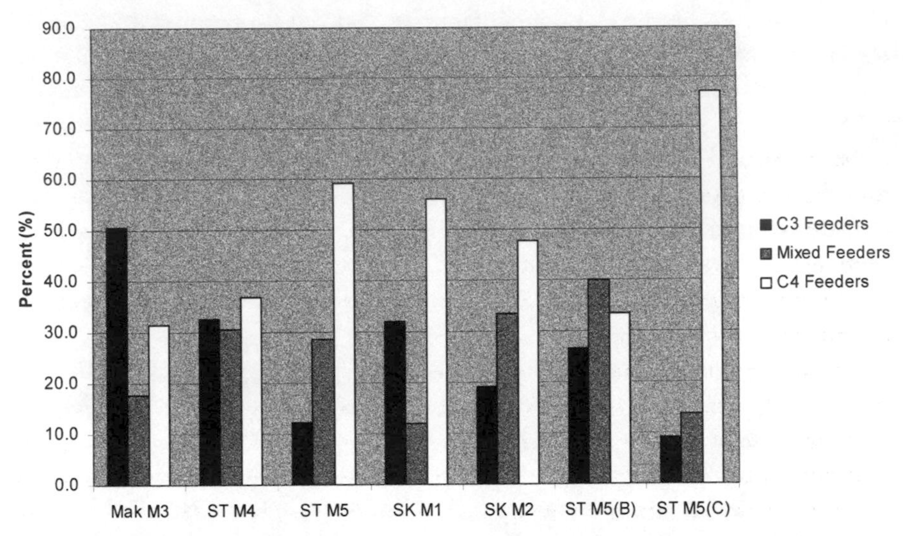

Figure 8.9. *Distribution of herbivores with predominantly browsing, mixed feeding and grazing ecology at Makapansgat Member 3 (ca. 2.6–2.8 mya), Sterkfontein Member 4 (ca. 2.0–2.6 mya), Sterkfontein Member 5 (ca. 1.6–2 mya), and Swartkrans Members 1 and 2 (ca. 1.7–2 mya and < 1.6 mya, respectively). Data are also shown for Sterkfontein Member 5 divided into two of its units: Unit B is associated with Oldowan tools and Unit C is associated with Acheulean tools (see text). Data are from Lee-Thorp et al. (2007).*

categories: predominantly C_3 consumers ($\delta^{13}C < -9.0‰$), predominantly C_4 consumers ($\delta^{13}C > -3.0‰$), and mixed feeders ($\delta^{13}C$ values between these two) (Lee-Thorp et al. 2007; Figure 8.9), and compared the patterns of these three categories through time. Conventionally, these deposits have been divided into a series of members believed to form a sequence from older to younger on the basis of lithostratigraphy and biostratigraphy, but new uranium–lead (U–Pb) absolute dates for flowstones show more overlap than previously entertained (Pickering et al. 2011). They do nevertheless form a sequence sufficient for our purposes here. Figure 8.9 shows a general but uneven decline in the proportions of C_3 consumers (browsers) through this sequence, and a concomitant rise in the proportions of C_4 consumers (grazers). The two *A. africanus*–bearing members, Makapansgat Member 3 and Sterkfontein Member 4, show more than 30 percent C_3 consumers and fewer than 40 percent C_4 consumers. In contrast, all of the members that contain *Homo* (Swartkrans Members 1 and 2, and Sterkfontein Member 5) have more than 70 percent C_4 consumers and mixed feeders combined. Thus, these data suggest that *A. africanus* inhabited more "closed," woody environments than *Homo* and its contemporary, *P. robus-*

tus. The result is broadly consistent with results from faunal studies (e.g., Vrba 1985; Reed 1997).

One contrast with the faunal studies, however, is that the $\delta^{13}C$ data suggest a stronger dominance by C_3 consumers in Makapansgat Member 3, which appears to have a rather dense wooded environment. Another distinction can be found in Sterkfontein Member 5 when the data are subdivided according to Unit B (containing sparse Oldowan tools) and Unit C (yielding Acheulean technology; suggesting a more recent age) (Kuman and Clarke 2000; Clarke and Partridge 2002). The $\delta^{13}C$ data suggest that about 30 percent of the Unit B herbivores sampled were C_3 consumers and just over 30 percent were C_4 consumers, whereas fewer than 10 percent of the Unit C herbivores were C_3 consumers and nearly 80 percent were C_4 consumers. This suggests that the area was dominated by grassy vegetation when Unit C accumulated, and that a major environmental change occurred between the deposits associated with Oldowan and Acheulean tools respectively.

We can also compare continuous, rather than categorical, $\delta^{13}C$ data from each member, which reveals a similar pattern (Figure 8.10). In aggregate, the herbivores from Makapansgat Member 3 have the most negative $\delta^{13}C$ values ($\bar{x}$ = –6.8‰), followed by Sterkfontein Member 4 ($\bar{x}$ = –5.4‰), Swartkrans Members 1 and 2 ($\bar{x}$ = –4.4‰), Sterkfontein Member 5 Unit B ($\bar{x}$ = –3.5‰), and lastly Sterkfontein Member 5 Unit C ($\bar{x}$ = –2.1‰). Again, $\delta^{13}C$ data for the *Australopithecus*-bearing members suggest environments more heavily wooded than those bearing early *Homo* and/or *Paranthropus*. Makapansgat Member 3 always emerges as the most closed environment, with Sterkfontein Member 5 Unit C the grassiest.

One potential problem with using central-tendency measures of such isotope-based techniques is that, ideally, one should produce $\delta^{13}C$ values for all herbivore specimens, which is impractical as well as expensive. Thus, one must either (1) sample a random subset of the fauna preserved in the deposit, or (2) establish mean values for all taxa, and then produce a site mean adjusted for the relative abundance of each taxon (as the site may be dominated by one or a few taxa). The latter will only be possible with well-studied faunal assemblages in which specimens have been classified to genus or species and for which relative-abundance data are available. In contrast, selecting a random sample of herbivores should be quite easy, although it is rarely if ever done in practice.

We now have comprehensive published isotope data for a number of important sites in South, eastern, and central Africa, but detailed comparisons can present difficulties because of the different sampling strategies applied. For the

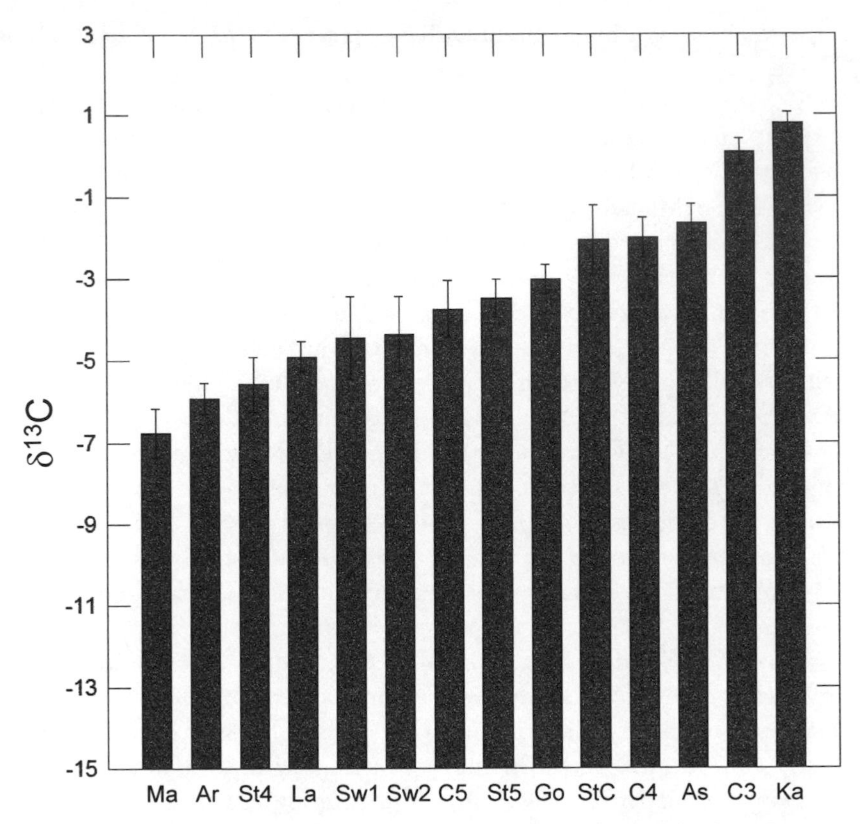

Figure 8.10. *A simple way of representing the distribution of C_3- and C_4-dependent animals at a site is to average all the herbivore $\delta^{13}C$ data. Here we show mean $\delta^{13}C$ values (with standard-error bars) for central, eastern, and southern African herbivore assemblages. Abbreviations are as follows: C5 = Kossom Bougoudi, Chad (~5.3 mya), C4 = Kolle, Chad (~5–4 mya), C3 is Koro Toro, Chad (3.5 mya), Ar is Aramis, Ethiopia (4.4 mya), Go = Gona, Ethiopia (~4.3 mya), As = Asbole, Ethiopia (0.6–0.8 mya), La = Upper Laetoli Beds, Tanzania (~3.6–3.8 mya), Ka = Kanjera South, Kenya (~2.0 mya), Ma = Makapansgat Member 3, South Africa (~2.8 mya), St4 = Sterkfontein Member 4, South Africa (~2–2.5 mya), Sw1 = Swartkrans Member 1, South Africa (~1.7–2 mya), Sw2 = Swartkrans Member 2, South Africa (~1.8–1.5 mya), St5 = Sterkfontein Member 5 Unit B, South Africa (~1.7–1.9 mya), StC = Sterkfontein Member 5 Unit C, South Africa (~1.6–1.7 mya). Data from Zazzo et al. (2000), Levin et al. (2004), Lee-Thorp et al. (2007), Plummer et al. (2009), White et al. (2009), Bedaso et al., (2010), and Kingston (2011).*

most part, researchers have used a site- or locality-specific approach in order to ascertain the structure of the vegetation from composite $\delta^{13}C$ data for several species or tribes. They have also examined the nature of changes when different periods are represented, as for example, for the Lower and Upper Laetoli and Ndolanya Beds at Laetoli (Kingston and Harrison 2007; Kingston 2011). These approaches generally produce a rather complex reflection of prevailing environments, which can be difficult to compare across sites. But the advantage is that the ecological behavior of particular taxa can be examined and compared in detail, and the information is more valuable still when compared with species-abundance data, as for instance for the Aramis (White et al. 2009) and Asbole faunas (Bedaso et al. 2010).

Here we have attempted to use a simple, continuous, rather than categorical, approach, by calculating the $\delta^{13}C$ means for many of these sites from the published data for herbivores (Figure 8.10). This is admittedly a rather crude approach but nevertheless useful for broad comparisons. Two patterns emerge in Figure 8.10. First, the eastern and central African sites, taken together, show generally higher $\delta^{13}C$ means than do the South African sites, suggesting that the former have always tended to more C_4 grass consumers. That is not unexpected. However, the overview does help to contextualize the animated discussions about whether closed or rather more open woodlands represented the preferred habitats of the early putative hominins in East Africa, such as at Aramis (White et al. 2009; Cerling et al. 2010). Based on these simple averages, Makapansgat Member 3 represents the most closed habitat associated with australopiths. In eastern Africa the Upper Laetoli Beds and Aramis show the lowest average $\delta^{13}C$ value (Figure 8.10), and the Early Pleistocene Kanjera South fauna shows the highest. The Mid-Pliocene Chad sites appear to be the most open and grassy overall.

Next, where different periods are represented, as in the Chad and Sterkfontein Valley sites, the dominant trend from the Pliocene to the Pleistocene is toward increasingly positive $\delta^{13}C$ averages, mirroring trends also seen in pedogenic carbonates, faunal abundances and ecomorphology. The Chad sites show a trend in which mean $\delta^{13}C$ suggests that at about 5 mya (Kossom Bougoudi; $\bar{x}$ = −3.8‰), the area was already as open as the areas inhabited by *Homo* in South Africa about 1.8 mya, and that by about 3.5 mya (Koro Toro; $\bar{x}$ = +0.3‰) the area was largely dominated by open C_4 grassland (Figure 8.10; data from Zazzo et al. 2000). This might indicate a real paleoenvironmental difference between the two regions, with Chad having more open environments, but it is also likely to reflect differences in contingent sampling. If, for instance, the giraffid components of the ecosystem could not be analyzed because only postcranial material is represented (as noted by Zazzo et al. 2000), this would represent a strong bias. Indeed, ecomorphological attributes of the Chadian assemblages suggest that community structure differed far less than the isotope data imply (Fara et al. 2005). The same can be said for the approximately 4.3-myr-old Gona fauna, with a mean herbivore $\delta^{13}C$ value (–3.0‰) (data from Levin et al. 2004), which is higher (more C_4) than that of all the South African data sets, save for Sterkfontein Member 5 Unit C. This site contrasts with the data from the Aramis sites, which averages –5.8‰ (data from White et al. 2009), but here the range including low, forestlike values for some primates suggests differences in various sectors of the landscape. Again, differences in sampling strategy may have emphasized distinctions.

This is not to say that any of the isotopic studies of herbivore enamel in South, central, or eastern Africa are flawed, but that they were not necessarily concerned with determining the dietary and habitat preferences of the herbivore faunas *as a whole*. If herbivore $\delta^{13}C$ data are to be used in this way for establishing paleoenvironmental patterns and trends in future, the results may be more useful when obtained from collections for which the alpha taxonomy and relative-abundance studies are well advanced, or, from large, randomly selected sample suites.

There are other difficulties. One is related to altitude and whether any C_3 grasses are present, as they would invalidate the assumption that percentage of C_3 feeders equates with the abundance of browse/trees and vice versa for grass. For instance, phytolith analyses suggest that a moderate proportion of local grasses at Laetoli are C_3 (Rossouw and Scott 2011). Another problem, particularly relevant in the South African hominin sites, is the unknown length of time during which a fossil assemblage accumulated (Hopley and Maslin 2010). Therefore, the assemblages may or may not fairly represent the fauna found in the vicinity of the site at any given time. Furthermore, the relative percentages of taxa found in each deposit may not accurately reflect the living community at any given moment due to collection biases (Brain 1981; Behrensmeyer 1991; Lyman 1994), but such problems plague all faunal analyses. On a positive note, a recent study of the Sterkfontein Valley sites revealed that while there are taphonomic biases among particular assemblages (e.g., more craniodental remains relative to postcranial material in calcified compared to decalcified/uncalcified sediments), there was no evidence that they influenced the taxonomic composition of the faunas (de Ruiter et al. 2008).

Although the focus of this section has been the use of carbon isotopes for paleohabitat studies, we would be remiss in not mentioning that oxygen isotope ratios (expressed as $\delta^{18}O$) in herbivore enamel carry information about climate and environment, although interpretation of such data is complex. In principle, the $\delta^{18}O$ composition of locally available environmental water (from rainfall and open-water sources) exerts a strong influence on an animal's body water from which bioapatites are precipitated. Studies based on $\delta^{18}O$ from PO_4 in mammalian bioapatites showed that information about paleowaters and paleotemperatures can be extracted (e.g., Longinelli 1984) based on well-known hydrological principles—that latitude and temperature exert major effects on rainfall $\delta^{18}O$ values (Dansgaard 1964). In practice these effects work well at cool, high latitudes, but in the low to mid-latitude regions of Africa, other factors, such as low relative humidity, become increasingly dominant influences on environmental waters. Moreover, it soon emerged that where significant amounts of body water come from leaf water, aridity caused further enrichment in $\delta^{18}O$ via leaf-water evapotranspiration isotope effects (Luz and Kolodny 1989). It is now known that mammalian $\delta^{18}O$ compositions can vary strongly among different species in the same place depending on parameters related to diet, drinking behavior, and thermophysiology (see, e.g., Kohn et al. 1996). For instance, several studies have found that browsers, such as giraffes, have higher $\delta^{18}O$ values than many grazing animals such as zebra, while highly water-dependent taxa such as hippos have the lowest $\delta^{18}O$ values (Bocherens et al. 1996; Cerling et al. 1997; Kohn et al. 1996; Sponheimer and Lee-Thorp 2001). Levin et al. (2006) argued that differences in the oxygen isotope compositions of more and less water-dependent taxa could be used as a "paleoaridity" indicator. The principle behind this paleoaridity index is that the bioapatite $\delta^{18}O$ compositions of mammalian taxa that obtain most of their water by drinking copiously are primarily records of meteoric water $\delta^{18}O$. Some mammals, however, obtain their water primarily from plants. Since strong evaporative enrichment in $\delta^{18}O$ occurs in leaves under conditions of low humidity (or aridity), these mammals too are sensitive to such differences and are termed "evaporation sensitive" (ES), while the former are "evaporation insensitive" (EI). In mesic environments with high humidity or lower water deficits, there should be little difference between ES and EI taxa but the difference should increase with higher water deficit/lower humidity and thus greater aridity. So far this index has been seldom applied (but see White et al. 2009; Bedaso et al. 2010), but it should prove more reliable than simple inter-site comparisons of $\delta^{18}O$ values, as they are influenced by many factors including altitude, distance inland, changes in storm tracks, and taxonomic composition.

CONCLUDING REMARKS

Biogeochemical approaches have helped to transform the landscape of hominin ecology studies over the last decade or so. They have contributed new proxies with which to investigate old questions, and in some cases, the results have challenged long-held beliefs about the ecology of our forebears. Of course, these methods cannot, and should not, operate in a vacuum. Biogeochemical data can be used to test hypotheses generated by other methods, and in turn, interpretations of biogeochemical studies must be subject to the scrutiny of complementary methods. We should bear in mind that different paleoecological methods rarely, if ever, address exactly the same questions. For instance, many techniques address paleodiet, but are they really asking the same questions about diet or extracting information that is comparable? The answer must be a resounding no. For example, dental microwear-texture analysis appears to be telling us about the mechanical properties of an individual's diet, carbon isotope analysis about the carbon isotope composition of an individual's diet, and ecomorphology about the foods that challenged the masticatory apparatus of an individual's ancestors. Thus, all of these techniques are in fact addressing a constellation of questions related to diet, *but not the diet itself.* It is important to be mindful of this distinction, for, when different methods seem to be telling us different things, it likely results from the fact that they are asking different questions and that the constraints of each approach differ. Only when we are able to integrate these sundry answers into a cohesive whole will we have really begun to understand hominin paleodiet.

It is also worth stressing, again, as we tried to convey earlier in the chapter, that much more research is required if we are to be in a position to interpret these data in a robust manner. Trace element abundances, in particular, are poorly understood in African ecosystems, and this severely limits our ability to interpret the data even though there are glimpses of opportunity. Carbon isotope abundances, in contrast, have been more broadly studied from both empirical and theoretical perspectives and are correspondingly much better understood. Thus we can be more confident about interpretation of the results. This is not to say that we do not have much to learn. For instance, use of stable carbon isotopes to provide environmental information of relevance to the study of early hominins must be considered to be in its infancy. Clearly there is much work to be done.

ACKNOWLEDGMENTS

We are very grateful to Nikolaas van der Merwe and Andrew Sillen for their leadership and inspiration, to Bob Brain, Francis Thackeray, Mike Raath, and Stephanie Potze for allowing us access to fossil collections, and to our colleagues and former students for their contributions in the field and laboratory, and in discussions. This research has received support from the NRF (South

Africa), the NSF, the Leakey Foundation, and the Universities of Cape Town, Bradford, Utah, Colorado at Boulder, and Oxford.

REFERENCES

Aiello, L. C., and P. Wheeler. 1995. "The Expensive Tissue Hypothesis: The Brain and the Digestive System in Human and Primate Evolution." *Current Anthropology* 36 (2): 199–221. http://dx.doi.org/10.1086/204350.

Altmann, S. A., and J. Altmann. 1970. *Baboon Ecology*. Chicago: University of Chicago Press.

Ambrose, S. H., and L. Norr. 1993. "Experimental Evidence for the Relationship of the Carbon Isotope Ratios of Whole Diet and Dietary Protein to Those of Bone Collagen and Carbonate." In *Prehistoric Human Bone: Archaeology at the Molecular Level*, ed. J. B. Lambert and G. Grupe, 1–37. Berlin: Springer-Verlag.

Backwell, L. R., and F. d'Errico. 2001. "Evidence of Termite Foraging by Swartkrans Early Hominids." *Proceedings of the National Academy of Sciences of the United States of America* 98 (4): 1358–63. http://dx.doi.org/10.1073/pnas.98.4.1358. Medline:11171955

Balasse, M. 2002. "Reconstructing Dietary and Environmental History from Enamel Isotopic Analysis: Time Resolution of Intra-Tooth Sequential Sampling." *International Journal of Osteoarchaeology* 12 (3): 155–65. http://dx.doi.org/10.1002/oa.601.

Balasse, M., A. B. Smith, S. H. Ambrose, and S. R. Leigh. 2003. "Determining Sheep Birth Seasonality by Analysis of Tooth Enamel Oxygen Isotope Ratios: The Late Stone Age Site of Kasteelberg (South Africa)." *Journal of Archaeological Science* 30 (2): 205–15. http://dx.doi.org/10.1006/jasc.2002.0833.

Balter, V., H. Bocherens, A. Person, N. Labourdette, M. Renard, and B. Vandermeersch. 2002. "Ecological and Physiological Variability of Sr/Ca and Ba/Ca in Mammals of West European Mid-Wurmian Food Webs." *Palaeogeography, Palaeoclimatology, Palaeoecology* 186 (1-2): 127–43. http://dx.doi.org/10.1016/S0031-0182(02)00448-0.

Bamford, M. K., R. M. Albert, and D. Cabanes. 2006. "Plio-Pleistocene Macroplant Fossil Remains and Phytoliths from Lowermost Bed II in the Eastern Palaeolake Margin of Olduvai Gorge, Tanzania." *Quaternary International* 148 (1): 95–112. http://dx.doi.org/10.1016/j.quaint.2005.11.027.

Bedaso, Z., J. G. Wynn, Z. Alemseged, and D. Geraads. 2010. "Paleoenvironmental Reconstruction of the Asbole Fauna (Busidima Formation, Afar, Ethiopia) Using Stable Isotopes." *Geobios* 43 (2): 165–77. http://dx.doi.org/10.1016/j.geobios.2009.09.008.

Behrensmeyer, A. K. 1991. "Terrestrial Vertebrate Accumulations." In *Taphonomy: Releasing the Data Locked in the Fossil Record*, ed. P. A. Allison and D.E.G. Briggs, 229–235. New York: Plenum.

Boaz, N. T., and J. Hampel. 1978. "Strontium Content of Fossil Tooth Enamel and Diet of Early Hominids." *Journal of Paleontology* 52:928–33.

Bocherens, H., P. L. Koch, A. Mariotti, D. Geraads, and J J. Jaeger. 1996. "Isotopic Biogeochemistry (^{13}C, ^{18}O) of Mammalian Enamel from African Pleistocene Hominid Sites." *Palaios* 11 (4): 306–18. http://dx.doi.org/10.2307/3515241.

Bowen, G. J., J. R. Ehleringer, L. A. Chesson, A. H. Thompson, D. W. Podlesak, and T. E. Cerling. 2009. "Dietary and Physiological Controls on the Hydrogen and Oxygen Isotope Ratios of Hair from Mid-20th Century Indigenous Populations." *American Journal of Physical Anthropology* 139 (4): 494–504. http://dx.doi.org/10.1002/ajpa.21008. Medline:19235792

Bowen, H.J.M., and J.A. Dymond. 1955. "Strontium and Barium in Plants and Soils." *Philosophical Transactions of the Royal Society of London. B* 144:355–68.

Brain, C. K. 1981. *The Hunters or the Hunted?* Chicago: University of Chicago Press.

Brown, A. B. 1974. "Bone Strontium as a Dietary Indicator in Human Skeletal Populations." *Contributions to Geology (Copenhagen)* 13:47–8.

Budd, P., J. Montgomery, B. Barreiro, and R. G. Thomas. 2000. "Differential Diagenesis of Strontium in Archeological Human Dental Tissues." *Applied Geochemistry* 15 (5): 687–94. http://dx.doi.org/10.1016/S0883-2927(99)00069-4.

Burton, J. H., T. D. Price, and W. D. Middleton. 1999. "Correlation of Bone Ba/Ca and Sr/Ca Due to Biological Purification of Calcium." *Journal of Archaeological Science* 26 (6): 609–16. http://dx.doi.org/10.1006/jasc.1998.0378.

Carter, M. L. (2001). "Sensitivity of Stable Isotopes (13C, 15N, and 18O) in Bone to Dietary Specialization and Niche Separation among Sympatric Primates in Kibale National Park,Uganda." PhD Dissertation, University of Chicago.

Cerling, T. E., and J. M. Harris. 1999. "Carbon Isotope Fractionation between Diet and Bioapatite in Ungulate Mammals and Implications for Ecological and Paleoecological Studies." *Oecologia* 120 (3): 347–63. http://dx.doi.org/10.1007/s004420050868.

Cerling, T. E., J. M. Harris, S. H. Ambrose, M. G. Leakey, and N. Solounias. 1997. "Dietary and Environmental Reconstruction with Stable Isotope Analyses of Herbivore Tooth Enamel from the Miocene Locality of Fort Ternan, Kenya." *Journal of Human Evolution* 33 (6): 635–50. http://dx.doi.org/10.1006/jhev.1997.0151. Medline:9467773

Cerling, T. E., N. E. Levin, J. Quade, et al. 2010. "Comment on the Paleoenvironment of *Ardipithecus ramidus*." *Science* 328: 1105–d.

Cerling, T. E., E. Mbua, F. M. Kirera, F. K. Manthi, F. E. Grine, M. G. Leakey, M. Sponheimer, and K. T. Uno. 2011. "Diet of *Paranthropus boisei* in the Early Pleistocene of East Africa." *Proceedings of the National Academy of Sciences of the United States of America* 108 (23): 9337–41. http://dx.doi.org/10.1073/pnas.1104627108. Medline:21536914

Cheney, D. L., P. C. Lee, and R. M. Seyfarth. 1981. "Behavioral Correlates of Non-Random Mortality among Free-Ranging Female Vervet Monkeys." *Behavioral Ecology and Sociobiology* 9 (2): 153–61. http://dx.doi.org/10.1007/BF00293587.

Clarke, R. J., and T. C. Partridge. 2002. "On the Unrealistic 'Revised Age Estimates' for Sterkfontein." *South African Journal of Science* 98:415–9.

Codron, D., J. A. Lee-Thorp, M. Sponheimer, D. de Ruiter, and J. Codron. 2006. "Inter- and Intra-Habitat Dietary Variability of Chacma Baboons (*Papio ursinus*) in South African Savannas Based on Fecal $\delta^{13}C$, $\delta^{15}N$ and %N." *American Journal of Physical Anthropology* 129 (2): 204–214. http://dx.doi.org/10.1002/ajpa.20253.

Codron, D., J. Luyt, J. A. Lee-Thorp, M. Sponheimer, D. de Ruiter, and J. Codron. 2005. "Utilization of Savanna-Based Resources by Plio-Pleistocene Baboons." *South African Journal of Science* 101:245–8.

Daegling, D. J., W. S. McGraw, P. S. Ungar, J. D. Pampush, A. E. Vick, and E. A. Bitty. 2011. "Hard-Object Feeding in Sooty Mangabeys (*Cercocebus atys*) and Interpretation of Early Hominin Feeding Ecology." *PLoS ONE* 6 (8): e23095. http://dx.doi.org/10.1371/journal.pone.0023095. Medline:21887229.

Dansgaard, W. 1964. "Stable Isotopes in Precipitation." *Tellus* 16 (4): 436–68. http://dx.doi.org/10.1111/j.2153-3490.1964.tb00181.x.

Dart, R. A. 1926. "Taungs and Its Significance." *Natural History* 26:315–27.

Dart, R. A. 1957. "The Osteodontokeratic Culture of *Australopithecus prometheus*." *Transvaal Museum Memoir* 10:1–105.

de Ruiter, D. J., M. Sponheimer, and J. A. Lee-Thorp. 2008. "Indications of Habitat Association of *Australopithecus robustus* in the Bloubank Valley, South Africa." *Journal of Human Evolution* 55 (6): 1015–30. http://dx.doi.org/10.1016/j.jhevol.2008.06.003. Medline:18824254

Dunbar, R.I.M. 1983. "Theropithecines and Hominids: Contrasting Solutions to the Same Ecological Problem." *Journal of Human Evolution* 12 (7): 647–58. http://dx.doi.org/10.1016/S0047-2484(83)80004-9.

Elias, R. W., Y. Hirao, and C. C. Patterson. 1982. "The Circumvention of the Natural Biopurification of Calcium along Nutrient Pathways by Atmospheric Inputs of Industrial Lead." *Geochimica et Cosmochimica Acta* 46 (12): 2561–80. http://dx.doi.org/10.1016/0016-7037(82)90378-7.

Elliott, J. C. 1994. *Structure and Chemistry of the Apatites and Other Calcium Orthophosphates*. Amsterdam: Elsevier.

Ezzo, J. A. 1994. "Putting the Chemistry Back into Archaeological Bone Chemistry Analysis: Modeling Potential Paleodietary Indicators." *Journal of Anthropological Archaeology* 13: 1–34.

Fara, E., A. Likius, H. T. Mackaye, P. Vignaud, and M. Brunet. 2005. "Pliocene Large-Mammal Assemblages from Northern Chad: Sampling and Ecological Structure." *Naturwissenschaften* 92 (11): 537–41. http://dx.doi.org/10.1007/s00114-005-0041-6. Medline:16220286.

Farquhar, G. D., M. H. O'Leary, and J. A. Berry. 1982. "The Relation between Carbon Isotope Discrimination and the Intercellular Carbon Dioxide Concentration in Leaves." *Australian Journal of Plant Physiology* 9 (2): 121–38. http://dx.doi.org/10.1071/PP9820121.

Feibel, C. 1997. "Debating the Environmental Factors in Hominid Evolution." *GSA Today* 7 (3): 1–7.

Fleagle, J. G. 1999. *Primate Adaptation and Evolution*. 2nd ed. New York: Academic Press.

Gilbert, C., J. Sealy, and A. Sillen. 1994. "An Investigation of Barium, Calcium and Strontium as Paleodietary Indicators in the Southwestern Cape, South Africa." *Journal of Archaeological Science* 21 (2): 173–84. http://dx.doi.org/10.1006/jasc.1994.1020.

Goodall, J. 1986. *The Chimpanzees of Gombe*. Cambridge: Cambridge University Press.

Gould, L., R. W. Sussman, and M. L. Sauther. 1999. "Natural Disasters and Primate Populations: The Effects of a 2-Year Drought on a Naturally Occurring Population of Ring-Tailed Lemurs (*Lemur catta*) in Southwestern Madagascar." *International Journal of Primatology* 20 (1): 69–84. http://dx.doi.org/10.1023/A:1020584200807.

Grine, F. E. 1981. "Trophic Differences between Gracile and Robust Australopithecines." *South African Journal of Science* 77:203–30.

Grine, F. E. 1986. "Dental Evidence for Dietary Differences in *Australopithecus* and *Paranthropus*: A Quantitative Analysis of Permanent Molar Microwear." *Journal of Human Evolution* 15 (8): 783–822. http://dx.doi.org/10.1016/S0047-2484(86)80010-0.

Grine, F. E., and R. F. Kay. 1988. "Early Hominid Diets from Quantitative Image Analysis of Dental Microwear." *Nature* 333 (6175): 765–8. http://dx.doi.org/10.1038/333765a0. Medline:3133564

Grine, F. E., P. S. Ungar, and M. F. Teaford. 2006a. "Was the Early Pliocene Hominin *'Australopithecus' anamensis* a Hard Object Feeder?" *South African Journal of Science* 102:301–10.

Grine, F. E., P. S. Ungar, M. F. Teaford, and S. El Zaatari. 2006b. "Molar Microwear in *Praeanthropus afarensis*: Evidence for Dietary Stasis through Time and under Diverse Paleoecological Conditions." *Journal of Human Evolution* 51 (3): 297–319. http://dx.doi.org/10.1016/j.jhevol.2006.04.004. Medline:16750841

Haghiri, F. 1964. "Strontium-90 Accumulation in Some Vegetable Crops." *Ohio Journal of Science* 64:371–5.

Harding, R.S.O. 1976. "Ranging Patterns of a Troop of Baboons (*Papio anubis*) in Kenya." *Folia Primatologica* 25 (2-3): 143–85. http://dx.doi.org/10.1159/000155711. Medline:817989

Hatley, T., and J. Kappelman. 1980. "Bears, Pigs, and Plio-Pleistocene Hominids: Case for Exploitation of Belowground Food Resources." *Human Ecology* 8 (4): 371–87. http://dx.doi.org/10.1007/BF01561000.

Henry, A.G., P. S. Ungar, B. H. Passey, M. Sponheimer, L. Rossouw, M. Bamford, P. Sandberg, D. de Ruiter, and L. Berger. 2012. "The Diet of *Australopithecus sediba*." *Nature* 487: 90–93. http://dx.doi.org/10.1038/nature11185.

Hesla, A.B.I., L. L. Tieszen, and S. K. Imbamba. 1982. "A Systematic Survey of C_3 and C_4 Photosynthesis in the Cyperaceae of Kenya, East Africa." *Photosynthetica* 16:196–205.

Hopley, P. J., A. G. Latham, and J. D. Marshall. 2006. "Palaeoenvironments and Palaeodiets of Mid-Pliocene Micromammals from Makapansgat Limeworks, South Africa: A Stable Isotope and Dental Microwear Approach." *Palaeogeography, Palaeoclimatology, Palaeoecology* 233 (3-4): 235–51. http://dx.doi.org/10.1016/j.palaeo.2005.09.011.

Hopley, P. J., and M. Maslin. 2010. "Climate-Averaging of Terrestrial Faunas: An Example from the Plio-Pleistocene of South Africa." *Paleobiology* 36 (1): 32–50. http://dx.doi.org/10.1666/0094-8373-36.1.32.

Hoppe, K. A., P. L. Koch, and T. T. Furutani. 2003. "Assessing the Preservation of Biogenic Strontium in Fossil Bones and Tooth Enamel." *International Journal of Osteoarchaeology* 13 (1-2): 20–8. http://dx.doi.org/10.1002/oa.663.

Hylander, W. L. 1975. "Incisor Size and Diet in Anthropoids with Special Reference to Cercopithecidae." *Science* 189 (4208): 1095–8. http://dx.doi.org/10.1126/science.808855. Medline:808855

Iwamoto, T. 1993. "The Ecology of *Theropithecus gelada*." In *Theropithecus: The Rise and Fall of a Primate Genus*, ed. N. G. Jablonski, 441–452. Cambridge: Cambridge University Press.

Jolly, C. J. 1970. "The Seed-Eaters: A New Model of Hominid Differentiation Based on a Baboon Analogy." *Man* 5 (1): 5–26. http://dx.doi.org/10.2307/2798801.

Jones, A. M., T. C. O'Connell, E. D. Young, K. Scott, C. M. Buckingham, P. Iacumin, and M. D. Brasier. 2001. "Biogeochemical Data from Well Preserved 200 ka Collagen and Skeletal Remains." *Earth and Planetary Science Letters* 193 (1-2): 143–9. http://dx.doi.org/10.1016/S0012-821X(01)00474-5.

Kay, R. 1985. "Dental Evidence for the Diet of *Australopithecus*." *Annual Review of Anthropology* 14 (1): 315–41. http://dx.doi.org/10.1146/annurev.an.14.100185.001531.

Kingston, J. D. 2011. "Stable Isotopic Analyses of Laetoli Fossil Herbivores." In *Paleontology and Geology of Laetoli: Human Evolution in Context*, vol. 1., ed. T. Harrison, 293–328. Dordrecht: Springer. http://dx.doi.org/10.1007/978-90-481-9956-3_15

Kingston, J. D., and T. Harrison. 2007. "Isotopic Dietary Reconstructions of Pliocene Herbivores at Laetoli: Implications for Early Hominin Paleoecology." *Palaeogeography, Palaeoclimatology, Palaeoecology* 243 (3-4): 272–306. http://dx.doi.org/10.1016/j.palaeo.2006.08.002.

Kohn, M. J., M. J. Schoeninger, and J. W. Valley. 1996. "Herbivore Tooth Oxygen Isotope Compositions: Effects of Diet and Physiology." *Geochimica et Cosmochimica Acta* 60 (20): 3889–96. http://dx.doi.org/10.1016/0016-7037(96)00248-7.

Krueger, H. W., and C. H. Sullivan. 1984. "Models for Carbon Isotope Fractionation between Diet and Bone." In *Stable Isotopes in Nutrition*, ed. J. F. Turnlund and P. E. Johnson, 205–222. ACS Symposium Series 258. Washington, DC: American Chemical Society.

Kuman, K., and R. J. Clarke. 2000. "Stratigraphy, Artefact Industries and Hominid Associations for Sterkfontein, Member 5." *Journal of Human Evolution* 38 (6): 827–47. http://dx.doi.org/10.1006/jhev.1999.0392. Medline:10835264

Laden, G., and R. W. Wrangham. 2005. "The Rise of the Hominids as an Adaptive Shift in Fallback Foods: Plant Underground Storage Organs (USOs) and Australopith Origins." *Journal of Human Evolution* 49 (4): 482–98. http://dx.doi.org/10.1016/j.jhevol.2005.05.007. Medline:16085279

Lambert, J. E., C. A. Chapman, R. W. Wrangham, and N. L. Conklin-Brittain. 2004. "Hardness of Cercopithecine Foods: Implications for the Critical Function of Enamel Thickness in Exploiting Fallback Foods." *American Journal of Physical Anthropology* 125 (4): 363–8. http://dx.doi.org/10.1002/ajpa.10403. Medline:15386250

Lee-Thorp, J. A. 1989. "Stable Carbon Isotopes in Deep Time: The Diets of Fossil Fauna and Hominids." Ph.D. Dissertation. University of Cape Town.

Lee-Thorp, J. A. 2000. "Preservation of Biogenic Carbon Isotope Signals in Plio-Pleistocene Bone and Tooth Mineral." In *Biogeochemical Approaches to Paleodietary*

Analysis, ed. S. Ambrose and K. A. Katzenberg, 89–115. New York: Plenum Press. http://dx.doi.org/10.1007/0-306-47194-9_5

Lee-Thorp, J., A. Likius, T. H. Mackaye, P. Vignaud, M. Sponheimer, M. Brunet. 2012. "Isotopic evidence for an early shift to C4 resources by Pliocene hominins in Chad." *Proceedings of the National Academy of Science USA* 109, 20369–20372.

Lee-Thorp, J. A., J. C. Sealy, and N. J. van der Merwe. 1989. "Stable Carbon Isotope Ratio Differences between Bone Collagen and Bone Apatite, and Their Relationship to Diet." *Journal of Archaeological Science* 16 (6): 585–99. http://dx.doi.org/10.1016/0305-4403(89)90024-1.

Lee-Thorp, J. A., and M. Sponheimer. 2003. "Three Case Studies Used to Reassess the Reliability of Fossil Bone and Enamel Isotope Signals for Paleodietary Studies." *Journal of Anthropological Archaeology* 22 (3): 208–16. http://dx.doi.org/10.1016/S0278-4165(03)00035-7.

Lee-Thorp, J. A., and M. Sponheimer. 2005. "Opportunities and Constraints for Reconstructing Palaeoenvironments from Stable Light Isotope Ratios in Fossils." *Geological Quarterly* 49:195–203.

Lee-Thorp, J. A., and M. Sponheimer. 2006. "Biogeochemical Approaches to Investigating Hominin Diets." *Yearbook of Physical Anthropology* 49:131–48.

Lee-Thorp, J. A., M. Sponheimer, and J. C. Luyt. 2007. "Tracking Changing Environments Using Stable Carbon Isotopes in Fossil Tooth Enamel: An Example from the South African Hominin Sites." *Journal of Human Evolution* 53 (5): 595–601. http://dx.doi.org/10.1016/j.jhevol.2006.11.020. Medline:17920103

Lee-Thorp, J. A., M. Sponheimer, B. H. Passey, D. J. de Ruiter, and T. E. Cerling. 2010. "Stable Isotopes in Fossil Hominin Tooth Enamel Suggest a Fundamental Dietary Shift in the Pliocene." *Philosophical Transactions of the Royal Society of London. Series B* 365 (1556): 3389–96. http://dx.doi.org/10.1098/rstb.2010.0059. Medline:20855312

Lee-Thorp, J. A., M. Sponheimer, and N. J. van der Merwe. 2003. "What Do Stable Isotopes Tell Us about Hominid Dietary and Ecological Niches in the Pliocene?" *International Journal of Osteoarchaeology* 13:104–13. http://dx.doi.org/10.1002/oa.659.

Lee-Thorp, J. A., J. F. Thackeray, and N. J. van der Merwe. 2000. "The Hunters and the Hunted Revisited." *Journal of Human Evolution* 39 (6): 565–76. http://dx.doi.org/10.1006/jhev.2000.0436. Medline:11102267

Lee-Thorp, J. A., and N. J. van der Merwe. 1987. "Carbon Isotope Analysis of Fossil Bone Apatite." *South African Journal of Science* 83:712–5.

Lee-Thorp, J. A., N. J. van der Merwe, and C. K. Brain. 1994. "Diet of *Australopithecus robustus* at Swartkrans from Stable Carbon Isotopic Analysis." *Journal of Human Evolution* 27 (4): 361–72. http://dx.doi.org/10.1006/jhev.1994.1050.

LeGeros, R. Z. 1991. *Calcium Phosphates in Oral Biology and Medicine*. Paris: Karger.

Levin, N. E., T. E. Cerling, B. H. Passey, J. M. Harris, and J. R. Ehleringer. 2006. "A Stable Isotope Aridity Index for Terrestrial Environments." *Proceedings of the National Academy of Sciences of the United States of America* 103 (30): 11201–5. http://dx.doi.org/10.1073/pnas.0604719103. Medline:16840554

Levin, N. E., J. Quade, S. W. Simpson, S. Semaw, and M. Rogers. 2004. "Isotopic Evidence for Plio-Pleistocene Environmental Change at Gona, Ethiopia." *Earth and Planetary Science Letters* 219 (1-2): 93–110. http://dx.doi.org/10.1016/S0012-821X(03)00707-6.

Longinelli, A. 1984. "Oxygen isotopes in mammal bone phosphate: a new tool for paleohydrological and paleoclimatological research?" *Geochimica et Cosmochimica Acta* 48, 385–390.

Lough, S. A., J. Rivera, and C. L. Comar. 1963. "Retention of Strontium, Calcium and Phosphorous in Human Infants." *Proceedings of the Society for Experimental Biology and Medicine* 112:631–6.

Luz, B. and Y. Kolodny. 1985. "Oxygen isotopes variations in phosphate of biogenic apatites, IV. Mammal teeth and bones." *Earth and Planetary Science Letters* 75, 29-36.

Lyman, R. L. 1994. *Vertebrate Taphonomy*. Cambridge: Cambridge University Press.

McGrew, W. C., M. J. Sharman, P. J. Baldwin, and C.E.G. Tutin. 1982. "On Early Hominid Plant-Food Niches." *Current Anthropology* 23:213–4.

McHenry, H. M., and K. Coffing. 2000. "*Australopithecus* to *Homo*: Transformations in Body and Mind." *Annual Review of Anthropology* 29 (1): 125–46. http://dx.doi.org/10.1146/annurev.anthro.29.1.125.

Milton, K. 1990. "Annual Mortality Patterns of a Mammal Community in Central Panama." *Journal of Tropical Ecology* 6 (04): 493–9. http://dx.doi.org/10.1017/S0266467400004909.

Milton, K. 1999. "A Hypothesis to Explain the Role of Meat-Eating in Human Evolution." *Evolutionary Anthropology* 8 (1): 11–21. http://dx.doi.org/10.1002/(SICI)1520-6505(1999)8:1<11::AID-EVAN6>3.0.CO;2-M.

O'Connell, J. F., K. Hawkes, and N. G. Blurton Jones. 1999. "Grandmothering and the Evolution of *Homo erectus*." *Journal of Human Evolution* 36 (5): 461–85. http://dx.doi.org/10.1006/jhev.1998.0285. Medline:10222165

O'Leary, M. 1981. "Carbon Isotope Fractionation in Plants." *Phytochemistry* 20 (4): 553–67. http://dx.doi.org/10.1016/0031-9422(81)85134-5.

Passey, B. H., and T. E. Cerling. 2002. "Tooth Enamel Mineralization in Ungulates Implications for Recovering a Primary Isotopic Time-Series." *Geochimica et Cosmochimica Acta* 66 (18): 3225–34. http://dx.doi.org/10.1016/S0016-7037(02)00933-X.

Passey, B. H., and T. E. Cerling. 2006. "In situ Stable Isotope Analysis ($\delta^{13}C$, $\delta^{18}O$) of Very Small Teeth Using Laser Ablation GC/IRMS." *Chemical Geology* 235: 238–49.

Passey, B. H., T. F. Robinson, L. K. Ayliffe, T. E. Cerling, M. Sponheimer, M. D. Dearing, B. L. Roeder, and J. R. Ehleringer. 2005. "Carbon Isotope Fractionation between Diet Breadth, CO_2, and Bioapatite in Different Mammals." *Journal of Archaeological Science* 32 (10): 1459–70. http://dx.doi.org/10.1016/j.jas.2005.03.015.

Peters, C. R., and J. C. Vogel. 2005. "Africa's Wild C4 Plant Foods and Possible Early Hominid Diets." *Journal of Human Evolution* 48 (3): 219–36. http://dx.doi.org/10.1016/j.jhevol.2004.11.003. Medline:15737391

Pickering, R., J. D. Kramers, P. J. Hancox, D. J. de Ruiter, and J. D. Woodhead. 2011. "Contemporary Flowstone Development Links Early Hominin Bearing Cave Deposits in South Africa." *Earth and Planetary Science Letters* 306 (1–2): 23–32. http://dx.doi.org/10.1016/j.epsl.2011.03.019.

Plummer, T. W., P. W. Ditchfield, L. C. Bishop, J. D. Kingston, J. V. Ferraro, D. R. Braun, F. Hertel, and R. Potts. 2009. "Oldest Evidence of Tool Making Hominins in a Grassland-Dominated Ecosystem." *PLoS ONE* 4 (9): e7199. http://dx.doi.org/10.1371/journal.pone.0007199. Medline:19844568

Potts, R. 1996. "Evolution and Climate Variability." *Science* 273 (5277): 922–3. http://dx.doi.org/10.1126/science.273.5277.922.

Price, T. D., J. Blitz, J. H. Burton, and J. Ezzo. 1992. "Diagenesis in Prehistoric Bone: Problems and Solutions." *Journal of Archaeological Science* 19 (5): 513–29. http://dx.doi.org/10.1016/0305-4403(92)90026-Y.

Reed, K. E. Feb-Mar 1997. "Early Hominid Evolution and Ecological Change through the African Plio-Pleistocene." *Journal of Human Evolution* 32 (2-3): 289–322. http://dx.doi.org/10.1006/jhev.1996.0106. Medline:9061560

Richard, A. F., R. E. Dewar, M. Schwartz, and J. Ratsirarson. 2002. "Life in the Slow Lane? Demography and Life Histories of Male and Female Sifaka (*Propithecus verreauxi verreauxi*)." *Journal of Zoology (London, England)* 256 (4): 421–36. http://dx.doi.org/10.1017/S0952836902000468.

Robinson, J. T. 1954. "Prehominid Dentition and Hominid Evolution." *Evolution; International Journal of Organic Evolution* 8 (4): 324–34. http://dx.doi.org/10.2307/2405779.

Rossouw, L., and L. Scott. 2011. "Phytoliths and Pollen, the Microscopic Plant Remains in Pliocene Volcanic Sediments around Laetoli, Tanzania." In *Paleontology and Geology of Laetoli: Human Evolution in Context*, vol. 1, ed. T. Harrison, 201–215. Dordrecht: Springer. http://dx.doi.org/10.1007/978-90-481-9956-3_9

Runia, L. J. 1987. "Strontium and Calcium Distribution in Plants: Effect on Paleodietary Studies." *Journal of Archaeological Science* 14 (6): 599–608. http://dx.doi.org/10.1016/0305-4403(87)90078-1.

Sage, R. F. 2004. "The Evolution of C_4 Photosynthesis." *New Phytologist* 161 (2): 341–70. http://dx.doi.org/10.1111/j.1469-8137.2004.00974.x.

Schoeninger, M. J., J. Moore, and J. M. Sept. 1999. "Subsistence Strategies of Two 'Savanna' Chimpanzee Populations: The Stable Isotope Evidence." *American Journal of Primatology* 49 (4): 297–314. http://dx.doi.org/10.1002/(SICI)1098-2345(199912)49:4<297::AID-AJP2>3.0.CO;2-N. Medline:10553959

Scott, R. S., P. S. Ungar, T. S. Bergstrom, C. A. Brown, F. E. Grine, M. F. Teaford, and A. Walker. 2005. "Dental Microwear Texture Analysis Shows Within-Species Diet Variability in Fossil Hominins." *Nature* 436 (7051): 693–5. http://dx.doi.org/10.1038/nature03822. Medline:16079844

Ségalen, L., J. A. Lee-Thorp, and T. C. Cerling. 2007. "Timing of C4 Grass Expansion across Sub-Saharan Africa." *Journal of Human Evolution* 53 (5): 549–59. http://dx.doi.org/10.1016/j.jhevol.2006.12.010. Medline:17905413

Sillen, A. 1981. "Strontium and Diet at Hayonim Cave." *American Journal of Physical Anthropology* 56 (2): 131–7. http://dx.doi.org/10.1002/ajpa.1330560204. Medline:7325216

Sillen, A. 1988. "Elemental and Isotopic Analysis of Mammalian Fauna from Southern Africa and Their Implications for Paleodietary Research." *American Journal of Physical Anthropology* 76 (1): 49–60. http://dx.doi.org/10.1002/ajpa.1330760106.
Sillen, A. 1989. "Diagenesis of the Inorganic Phase of Cortical Bone." In *The Chemistry of Prehistoric Human Bone*, ed. T. D. Price, 211–299. Cambridge: Cambridge University Press.
Sillen, A. 1992. "Strontium-Calcium Ratios (Sr/Ca) of *Australopithecus robustus* and Associated Fauna from Swartkrans." *Journal of Human Evolution* 23 (6): 495–516. http://dx.doi.org/10.1016/0047-2484(92)90049-F.
Sillen, A., G. Hall, and R. Armstrong. 1995. "Strontium Calcium Ratios (Sr/Ca) and Strontium Isotope Ratios (87Sr/86Sr) of *Australopithecus robustus* and *Homo sp.* from Swartkrans." *Journal of Human Evolution* 28 (3): 277–85. http://dx.doi.org/10.1006/jhev.1995.1020.
Smith, B. N., and S. Epstein. 1971. "Two Categories of $^{13}C/^{12}C$ Ratios for Higher Plants." *Plant Physiology* 47 (3): 380–4. http://dx.doi.org/10.1104/pp.47.3.380. Medline:16657626
Spencer, H., J. M. Warren, L. Kramer, and J. Samachson. Mar-Apr 1973. "Passage of Calcium and Strontium across the Intestine in Man." *Clinical Orthopaedics and Related Research* 91 (91): 225–34. http://dx.doi.org/10.1097/00003086-197303000-00031. Medline:4703151
Sponheimer, M. 1999. "Isotopic Ecology of the Makapansgat Limeworks Fauna." Ph.D. Dissertation, Rutgers University.
Sponheimer, M., Z. Alemseged, T. E. Cerling, F. E. Grine, W. H. Kimbel, M. G. Leakey, J. A. Lee-Thorp, F. K. Manthi, K. Reed, B. A. Wood, J. G. Wynn. Pre-Review Version of "Isotopic Evidence of Early Hominin Diets: Past, Present, and Future." Figshare. http://dx.doi.org/10.6084/m9.figshare.628068. Retrieved 19:19, February 21, 2013 (GMT).
Sponheimer, M., D. de Ruiter, J. A. Lee-Thorp, and A. Späth. 2005a. "Sr/Ca and Early Hominin Diets Revisited: New Data from Modern and Fossil Tooth Enamel." *Journal of Human Evolution* 48 (2): 147–56. http://dx.doi.org/10.1016/j.jhevol.2004.09.003. Medline:15701528
Sponheimer, M., and J. A. Lee-Thorp. 1999a. "Isotopic Evidence for the Diet of an Early Hominid, *Australopithecus africanus*." *Science* 283 (5400): 368–70. http://dx.doi.org/10.1126/science.283.5400.368. Medline:9888848
Sponheimer, M., and J. A. Lee-Thorp. 1999b. "Oxygen Isotopes in Enamel Carbonate and Their Ecological Significance." *Journal of Archaeological Science* 26 (6):723–8. http://dx.doi.org/10.1006/jasc.1998.0388.
Sponheimer, M., and J. A. Lee-Thorp. 2001. "The Oxygen Isotope Composition of Mammalian Enamel Carbonate from Morea Estate, South Africa." *Oecologia* 126:153–7. http://dx.doi.org/10.1007/s004420000498.
Sponheimer, M., and J. A. Lee-Thorp. 2003. "Differential Resource Utilization by Extant Great Apes and Australopithecines: Towards Solving the C4 Conundrum." *Comparative Biochemistry and Physiology: Part A, Molecular and Integrative Physiology* 136 (1): 27–34. http://dx.doi.org/10.1016/S1095-6433(03)00065-5. Medline:14527627

Sponheimer, M., and J. A. Lee-Thorp. 2006. "Enamel Diagenesis at South African Australopith Sites: Implications for Paleoecological Reconstruction with Trace Elements." *Geochimica et Cosmochimica Acta* 70 (7): 1644–54. http://dx.doi.org/10.1016/j.gca.2005.12.022.
Sponheimer, M., J. A. Lee-Thorp, D. de Ruiter, D. Codron, J. Codron, A. T. Baugh, and F. Thackeray. 2005b. "Hominins, Sedges, and Termites: New Carbon Isotope Data from the Sterkfontein Valley and Kruger National Park." *Journal of Human Evolution* 48 (3): 301–12. http://dx.doi.org/10.1016/j.jhevol.2004.11.008. Medline:15737395
Sponheimer, M., J. E. Loudon, D. Codron, M. E. Howells, J. D. Pruetz, J. Codron, D. J. de Ruiter, and J. A. Lee-Thorp. 2006a. "Do 'Savanna' Chimpanzees Consume C4 Resources?" *Journal of Human Evolution* 51 (2): 128–33. http://dx.doi.org/10.1016/j.jhevol.2006.02.002. Medline:16630647
Sponheimer, M., B. H. Passey, D. J. de Ruiter, D. Guatelli-Steinberg, T. E. Cerling, and J. A. Lee-Thorp. 2006b. "Isotopic Evidence for Dietary Variability in the Early Hominin *Paranthropus robustus*." *Science* 314 (5801): 980–2. http://dx.doi.org/10.1126/science.1133827. Medline:17095699
Stanford, C. B., and J. B. Nkurunungi. 2003. "Sympatric Ecology of Chimpanzees and Gorillas in Bwindi Impenetrable National Park, Uganda: Diet." *International Journal of Primatology* 24 (4): 901–18. http://dx.doi.org/10.1023/A:1024689008159.
Stock, W. D., D. K. Chuba, and G. A. Verboom. 2004. "Distribution of South African C–3 and C–4 Species of Cyperaceae in Relation to Climate and Phylogeny." *Austral Ecology* 29 (3): 313–9. http://dx.doi.org/10.1111/j.1442-9993.2004.01368.x.
Strum, S. C. 1987. *Almost Human: A Journey into the World of Baboons*. New York: Random House.
Sullivan, C. H., and H. W. Krueger. 1981. "Carbon Isotope Analysis of Separate Chemical Phases in Modern and Fossil Bone." *Nature* 292 (5821): 333–5. http://dx.doi.org/10.1038/292333a0. Medline:7019719
Sullivan, C. H., and H. W. Krueger. 1983. "Carbon Isotope Ratios of Bone Apatite and Animal Diet Reconstruction." *Nature* 301 (5896): 177–8. http://dx.doi.org/10.1038/301177a0. Medline:6337340
Teaford, M. F., P. S. Ungar, and F. E. Grine. 2002. "Paleontological Evidence for the Diets of African Plio–Pleistocene Hominins with Special Reference to Early *Homo*." In *Human Diet: Its Origin and Evolution*, ed. P. S. Ungar and M. F. Teaford, 143–166. Westport, CT: Bergin and Garvey.
Teleki, G. 1981. "The Omnivorous Diet and Eclectic Feeding Habits of Chimpanzees in Gombe National Park, Tanzania." In *Omnivorous Primates*, ed. R.S.O. Harding and G. Teleki, 303–343. New York: Columbia University Press.
Tieszen, L. L. 1991. "Natural Variations in the Carbon Isotope Values of Plants: Implications for Archaeology, Ecology, and Paleoecology." *Journal of Archaeological Science* 18 (3): 227–48. http://dx.doi.org/10.1016/0305-4403(91)90063-U.
Tieszen, L. L., and T. Fagre. 1993. "Effect of Diet Quality and Composition on the Isotopic Composition of Respiratory CO2, Bone Collagen, Bioapatite, and Soft Tissues." In

Prehistoric Human Bone: Archaeology at the Molecular Level, ed. J. B. Lambert and G. Grupe, 121–155. Berlin: Springer.
Toots, H., and M. R. Voorhies. 1965. "Strontium in Fossil Bones and the Reconstruction of Food Chains." *Science* 149 (3686): 854–5. http://dx.doi.org/10.1126/science.149.3686.854. Medline:17737382
Trickett, M. A., P. Budd, J. Montgomery, and J. Evans. 2003. "An Assessment of Solubility Profiling as a Decontamination Procedure for the $^{87}Sr/^{86}Sr$ Analysis of Archaeological Human Skeletal Tissue." *Applied Geochemistry* 18 (5): 653–8. http://dx.doi.org/10.1016/S0883-2927(02)00181-6.
Tuross, N., A. K. Behrensmeyer, and E. D. Eanes. 1989. "Strontium Increases and Crystallinity Changes in Taphonomic and Archaeological Bone." *Journal of Archaeological Science* 16 (6): 661–72. http://dx.doi.org/10.1016/0305-4403(89)90030-7.
Tutin, C.E.G., and M. Fernandez. 1985. "Food Consumed by Sympatric Populations of *Gorilla g. gorilla* and *Pan. t. troglodytes* in Gabon: Some Preliminary Data." *International Journal of Primatology* 6 (1): 27–43. http://dx.doi.org/10.1007/BF02693695.
Ungar, P. 2004. "Dental Topography and Diets of *Australopithecus afarensis* and Early Homo." *Journal of Human Evolution* 46 (5): 605–22. http://dx.doi.org/10.1016/j.jhevol.2004.03.004. Medline:15120268
Ungar, P. S. 1998. "Dental Allometry, Morphology, and Wear as Evidence for Diet in Fossil Primates." *Evolutionary Anthropology* 6 (6): 205–17. http://dx.doi.org/10.1002/(SICI)1520-6505(1998)6:6<205::AID-EVAN3>3.0.CO;2-9.
Ungar, P. S., and F. E. Grine. 1991. "Incisor Size and Wear in *Australopithecus africanus* and *Paranthropus robustus*." *Journal of Human Evolution* 20 (4): 313–40. http://dx.doi.org/10.1016/0047-2484(91)90013-L.
Ungar, P. S., F. E. Grine, and M. F. Teaford. 2008. "Dental Microwear and Diet of the Plio-Pleistocene Hominin *Paranthropus boisei*." *PLoS ONE* 3 (4): e2044. http://dx.doi.org/10.1371/journal.pone.0002044. Medline:18446200
Ungar, P. S., and M. Sponheimer. 2011. "The Diets of Early Hominins." *Science* 334 (6053): 190–3. http://dx.doi.org/10.1126/science.1207701. Medline:21998380
van der Merwe, N. J., F. T. Masao, and R. J. Bamford. 2008. "Isotopic Evidence for Contrasting Diets of Early Hominins *Homo habilis* and *Australopithecus boisei* of Tanzania." *South African Journal of Science* 104:153–5.
van der Merwe, N. J., and E. Medina. 1989. "Photosynthesis and $^{13}C/^{12}C$ Ratios in Amazonian Rain Forests." *Geochimica et Cosmochimica Acta* 53 (5): 1091–4. http://dx.doi.org/10.1016/0016-7037(89)90213-5.
van der Merwe, N. J., J. F. Thackeray, J. A. Lee-Thorp, and J. Luyt. 2003. "The Carbon Isotope Ecology and Diet of *Australopithecus africanus* at Sterkfontein, South Africa." *Journal of Human Evolution* 44 (5): 581–97. http://dx.doi.org/10.1016/S0047-2484(03)00050-2. Medline:12765619
van der Merwe, N. J., and J. C. Vogel. 1978. "13C Content of Human Collagen as a Measure of Prehistoric Diet in Woodland North America." *Nature* 276 (5690): 815–6. http://dx.doi.org/10.1038/276815a0. Medline:364321.

Vogel, J. C. 1978. "Recycling of Carbon in a Forest Environment." *Oecologia Plantarum* 13:89–94.

Vogel, J. C., and N. J. van der Merwe. 1977. "Isotopic Evidence for Early Maize Cultivation in New York State." *American Antiquity* 42 (2): 238–42. http://dx.doi.org/10.2307/278984.

Vrba, E. S. 1980. "The Significance of Bovid Remains as Indicators of Environment and Predation Patterns." In *Fossils in the Making*, ed. A .K. Behrensmeyer and A. P. Hill, 247–272. Chicago: University of Chicago Press.

Vrba, E. 1985. "Ecological and Adaptive Changes Associated with Early Hominid Evolution." In *Ancestors: The Hard Evidence*, ed. E. Delson, 63–71. New York: Alan R. Liss.

Wang, Y., and T. E. Cerling. 1994. "A Model of Fossil Tooth and Bone Diagenesis: Implications for Paleodiet Reconstruction from Stable Isotopes." *Palaeogeography, Palaeoclimatology, Palaeoecology* 107 (3-4): 281–9. http://dx.doi.org/10.1016/0031-0182(94)90100-7.

White, T. D., S. H. Ambrose, G. Suwa, D. F. Su, D. DeGusta, R. L. Bernor, J. R. Boisserie, M. Brunet, E. Delson, S. Frost, et al. 2009. "Macrovertebrate Paleontology and the Pliocene Habitat of *Ardipithecus ramidus*." *Science* 326 (5949): 87–93. http://dx.doi.org/10.1126/science.1175822 www.sciencemag.org/cgi/content/full/326/5949/67/DC1. Medline:19810193

Winter, K., and J.A.C. Smith. 1996. *Crassulacean Acid Metabolism: Biochemistry, Ecophysiology and Evolution*. Berlin: Springer-Verlag.

Wrangham, R. W., J. H. Jones, G. Laden, D. Pilbeam, and N. L. Conklin-Brittain. 1999. "The Raw and Stolen. Cooking and the Ecology of Human Origins." *Current Anthropology* 40 (5): 567–94. http://dx.doi.org/10.1086/300083. Medline:10539941

Wyckoff, R.W.G., and A. R. Doberenz. 1968. "The Strontium Content of Fossil Teeth and Bones." *Geochimica et Cosmochimica Acta* 32 (1): 109–15. http://dx.doi.org/10.1016/0016-7037(68)90090-2.

Zazzo, A., H. Bocherens, M. Brunet, A. Beauvilain, D. Billiou, H. T. Mackaye, P. Vignaud, and A. Mariotti. 2000. "Herbivore Paleodiet and Paleoenvironment Changes in Chad during the Pliocene Using Stable Isotope Ratios in Tooth Enamel Carbonate." *Paleobiology* 26 (2): 294–309. http://dx.doi.org/10.1666/0094-8373(2000)026<0294:HPAPCI>2.0.CO;2.

9

The Behavior of Plio-Pleistocene Hominins: Archaeological Perspectives

David R. Braun

The ecology of early Pleistocene hominins (members of the human clade; Wood and Richmond 2000) is a complicated relationship between cultural mechanisms and biological adaptations. Although skeletal remains of hominins represent the most concrete evidence of human evolution, the archaeological record is the most abundant record of the ecology of our ancestors. The relatively large amount of stone artifacts and associated animal bones allows archaeologists to view hominin behavior through time and across ancient landscapes. The combination of time transgressive (across time periods at one location) and synchronic (across one time horizon at many locations) approaches is unique to an archaeological perspective. It also allows us to ask questions about behavior that can act as independent lines of evidence to compare and contrast with skeletal evidence of hominin evolution.

The period between 2.6 to 1.6 mya (million years ago) was a crucial time in the history of human evolution. The archaeological assemblages associated within this time horizon are often termed the *Oldowan* after the site of Olduvai Gorge in Tanzaniza (Leakey 1967, 1971). This period hosted the first appearance of chipped-stone artifacts (Semaw 2000; Semaw et al. 1997). A recent report suggests that hominins may have accessed animal resources (meat and marrow) even earlier than this (~3.4 mya; McPherron et al. 2010), possibly with naturally sharp stones (but see Dominguez-Rodrigo et al. 2010). Evidence of hominins butchering

DOI: 10.5876/9781607322252:c09

large mammals prior to 2.0 mya exists in the Afar, yet this evidence is infrequent and currently poorly documented (de Heinzelin et al. 1999; Domínguez-Rodrigo et al. 2005; Domínguez-Rodrigo 2009). These major shifts in hominin behavior have implications for the nature and timing of hominin dietary changes (Blumenschine 1987; Blumenschine et al. 1991; Dominguez-Rodrigo and Pickering 2003), the interrelationship of behavioral and environmental variability through time (Potts 1998), and even the eventual global distribution of the genus *Homo* (Antón et al. 2002). The archaeological record represents a snapshot of decisions made by hominins to acquire resources necessary for their survival. We may even be able to investigate hominin cognitive development independent of biological changes in the hominin brain (Stout et al. 2000; Stout 2002; Stout and Chaminade 2009).

However, all high-order models of hominin behavior are based on very meager evidence. Despite the volume of material that represents the early Pleistocene archaeological record, it is after all just broken rocks and bones. We extrapolate hypotheses about hominin behavior from what we think are the processes that produced the archaeological record. Continued examination of every inferential step between the battered cobbles and hypotheses about hominin behavioral variation is a vital aspect of the archaeology of human origins (Braun et al. 2006).

The aim of this chapter is to investigate how we "know" what we think we know about early Pleistocene hominin behavior by reviewing studies that focus on archaeological collections from 2.6 to 1.9 mya. The last three decades of research have provided archaeologists with large collections of shattered rocks and fossilized bones that were collected in a controlled manner from deposits of known geological age. Thus archaeologists now have the collections needed to begin to test hypotheses about how hominins made a living. Almost all of the evidence derives from two major proxies of human behavior and their associated specialized fields of analysis: stone tools (lithic analysis) and fossil animal bones (zooarchaeological analysis). These two fields are unfortunately rarely integrated (although see Brantingham 1998 for one attempt), despite the clearly complementary nature of these two proxies of behavior.

HOMININ TRACE FOSSILS

Stone artifacts and bones with evidence of hominin butchery represent traces of hominins that can answer particular questions about the behavior of our ancestors. Lithic and zooarchaeological analyses can be distilled down to a few main questions. Lithic analysts study stone artifacts to determine (1) how and why hominins collected stones, (2) why they broke stones to make sharp edges in particular patterns, (3) why they used certain stones for certain tasks (if they did), and (4) how and why they decided to drop broken stones in particular places. In a similar fashion zooarchaeologists study bones to understand (1) which parts of mammal carcasses were accessed by hominins, (2) where and under what ecological conditions hominins gained access to carcasses, (3) how hominins interacted with other large carnivores, and (4) why hominins discarded evidence of prehistoric meals at certain points on the landscape.

Answers to these questions are almost universally disputed among those who study the behavior of early Pleistocene hominins. Archaeologists do agree about a few things concerning the origin of tool use:

1. It all began in Africa. All of the earliest evidence of hominin stone-tool use and carnivory is in Africa. Three main areas represent the major foci of research: the Turkana Basin in northern Kenya, the Afar Depression in Ethiopia, and Olduvai Gorge in Tanzania. Some exceptions to this East African–centric view include sites in western Kenya (Kanjera) (Plummer et al. 2009), Algeria (Ain Hanech) (Sahnouni et al. 2002), and South Africa (Sterkfontein) (Kuman and Clarke 2000).

2. It happened relatively quickly. There are theoretical reasons and some empirical evidence to believe that tool use may have been an ancestral trait of all hominins (Panger et al. 2002). However, the relatively contemporaneous appearance of dense accumulations of sharp-edged stone tools and cutmarks (macroscopic striations on fossil bones indicating the use of a stone edge against the bone surface) is probably not coincidental. Observations from Dikika (McPherron et al. 2010) may suggest precursors to tool-use behavior. Oldowan archaeologists are still divided as to whether or not hominin populations prior to 2.5 mya likely used stone tools regularly (Roche et al. 1999; Semaw 2000; Delagnes and Roche 2005). Yet shortly after the appearance of chipped-stone artifacts (Semaw et al. 1997), hominins begin to produce large concentrations of stone artifacts throughout Africa and the frequency of cutmarked bones increases dramatically after 2.0 mya (Kimbel et al. 1996; Sahnouni and de Heinzelin 1998; Plummer et al. 1999; Roche et al. 1999; Braun et al. 2010).

ARCHAEOLOGICAL SITE INTEGRITY

Almost all archaeological assemblages are collected from "sites." However, the reality of an isolated location where hominin behavior was concentrated is difficult to substantiate. We tend to assume that early Pleistocene hominins conducted many of their activities at one spot on a landscape because there is good evidence to believe this is the case later in time (especially in sites that are constrained by cave walls). However, the impetus for accumulations of bones and stones on a landscape is still a matter of substantial dispute (Plummer 2004).

One of the most influential debates in the archaeology of human origins stems from Glynn Isaac's initial discussion of site formation based on sites from the Koobi Fora Formation in northern Kenya. Isaac (1976, 1978, 1983) suggested that dense concentrations of artifacts and bones represented a pattern of behavior that separated early Pleistocene hominins from their more apelike ancestors. The "home-base hypothesis" suggested that hominins occupied a central location that was the center of social activities. This hypothesis carried with it a series of social implications, including the sharing of resources among conspecifics and sexual division of labor (males hunted, females gathered). In a series of rebuttals to the home-base hypothesis, Lewis Binford (1981, 1985; Binford et al. 1988) challenged not only the social implications of this hypothesis but also the behavioral connection between stone artifacts and fossil bones at many Oldowan localities. Binford posited that multiple possible scenarios could explain the association of bones and stones at one locality on the ancient landscape. A river depositing its bedload on a river bank could have concentrated bones and stones that had no behavioral association (i.e., palimpsests). Several studies rose to the challenge of differentiating between sites that were evidence of hominin behavior as opposed to fluvial activity or even coincidental accumulations of carnivore food refuse and hominin artifacts (Behrensmeyer 1978; Behrensmeyer and Hill, 1980; Schick, 1986, 1987a, 1987b, 1997). Using variables such as bone-weathering stages (the longer it takes for a bone to get buried the more degraded the surface of the resultant fossil), artifact-size distributions (smaller fragments tend to be washed away when water washes over a site), and the orientation of specimens when they are excavated (rivers tend to line up bones and artifacts when they deposit them), it was possible to distinguish between foci of hominin activity and accidental associations of bones and stones.

However, archaeologists rarely agree on the reasons why hominins created dense accumulations of artifacts and bones. Models that try to explain stone and bone accumulations include a modified home-base hypothesis that focuses more on central-place foraging (Isaac 1983) and modified versions of the central-place model that focuses on the defense of resources (Rose and Marshall 1996). The "stone cache" model suggested that accumulations represent hominin attempts to redistribute stone resources to increase the likelihood of finding stone for making artifacts and carcasses for butchery in similar places on the landscape (Potts 1984, 1988, 1991). Another model suggested sites represented favored places where hominins would return several times. Eventually enough discarded stone would make the area a source of stone for hominins (Schick 1987b). Studies of the ecology of modern African ecosystems suggest that woodlands along the edges of rivers would have offered refuges from large carnivores for hominins. Hence, these preferred locations may have accumulated artifacts and bones through repeated use (Blumenschine 1986; 1995). In this scenario hominins scavenged carcass parts from the kills of lions, cheetahs, or leopards (Cavallo and Blumenschine 1989), or sabertooth cats (Marean 1989). The most recent model proposed is based on the similarities between Hadza hunter–gatherer kill localities and early Pleistocene archaeological collections (O'Connell et al. 1988; 2002). It suggests that male hominins actively scavenged from carnivores in an attempt to display their genetic fitness. This interpretation implies that although faunal acquisition was important for social reasons, it was relatively unimportant for dietary needs. O'Connell and colleagues (1988, 2002) argue rather that underground storage organs (USOs; i.e., roots and tubers) found in savanna environments, were the main source of calories for early Pleistocene hominins. They believe that the combination of USOs and social mechanisms supported the increases in brain and body size that characterize early Pleistocene hominin evolution.

These models are all based on attempts to understand why hominins made concentrations of artifacts and bones. As debates surrounding home bases continued, Isaac turned his interests to those areas between concentrations. In the early 1980s Isaac suggested shifting our focus away from areas that are "unusually crammed with material" (Isaac 1983, 258) to those areas where artifacts are sparse. The idea initiated an interest in how early hominins used the ancient landscape. This perspective on the archaeological record was seen as a transition from a study of artifacts to a study of the "ecological systems with which hominids interacted" (Potts 1991: 173). Landscape approaches to Oldowan behavior at Olduvai (Blumenschine and Masao 1991; Peters and Blumenschine 1996; Blumenschine and Peters 1998) and Koobi Fora (Rogers 1997; Rogers et al. 1994; Braun et al. 2008b) are examples of this change in Oldowan studies.

STONE ARTIFACTS

The study of Oldowan artifacts has a long history although real understanding of how and why hominins made stone artifacts is still in its infancy (Dibble and Rezek 2009; Nonaka et al. 2010). Studies of Lower Paleolithic industries in Europe heavily influenced initial studies of Oldowan industries. Research on "pebble tool" industries were rife with various systems wherein types of artifacts were distinguished by their two-dimensional shape (Biberson 1967; Chavaillon 1976). This approach is often termed *typology* for its focus on "types" of artifacts. Often typological studies used terms laden with functional implications (e.g., *awls*, *burins*). Unfortunately, every locality appeared to have types that were so specific to a particular site that meaningful comparisons were difficult. In contrast, Mary Leakey's description of the enormous collections of artifacts at Olduvai Gorge provided a concise typological description that could be applied elsewhere (Leakey 1971). It was a major step because Leakey showed how different percentages of these types documented industrial changes through time. For example, increased percentages of certain core forms like burins or handaxe-like forms indicated a change from the Oldowan to the Developed Oldowan.

Aspects of Leakey's typology are still in use today, however, Isaac, Harris, and Kroll (1997) modified the typology to remove the functionally laden terms in Leakey's system (Table 9.1). Three main categories of artifacts are relatively universal to studies of Oldowan industries:

Flaked Pieces: Pieces of battered stone with chips of rock removed from them. Evidence of this chipping is in the form of negative impressions of conchoidal fracture. Flaked pieces are often called a *core* or *nucleus*.

Detached Pieces: Fragments of stone that have been removed from a flaked piece. These pieces are usually smaller than flaked pieces and they usually have many features of conchoidal fracture that are distinct to humanly controlled fracture. Detached pieces are often called *flakes*, *angular fragments*, or *debitage*.

Pounded Pieces: Pieces of battered rock with evidence of battering in the form of pits and furrows. These pieces are usually very difficult to distinguish from cobbles battered by river action. Pounded pieces are sometimes called *hammerstones* and are usually distinguished by battering clustered in specific areas of the stone.

Since Leakey's original work on the Oldowan there have been many advances in the study of the earliest technology. Experimental studies and new collections are the major driving forces behind new advances in Oldowan research (Sahnouni et al. 1997; Stout et al. 2000). The last twenty years of research on the Oldowan can really be divided into two main types of research focused on (1) artifact form and function at one locality and (2) the distribution of artifacts at many localities.

Table 9.1. Different typologies used by Oldowan archaeologists. Leakey's typology allowed the formal recognition of distinct forms in the archaeological record. Isaac, Harris, and Kroll's typology parallels Leakey's system but removes functional interpretations. Notice the lack of detail Leakey associated with detached pieces relative to flaked pieces. Manuports featured heavily in Olduvai assemblages, yet were rare at Koobi Fora, which is why Isaac and colleagues seldom used this category.

Typology	*Leakey (1971)*	*Isaac et al. (1997)*
Tools/Cores/Flaked Pieces	Choppers	Choppers
	Polyhedrons	Polyhedrons
	Discoids	Discoids, regular
	Protobifaces	Discoids, partial
	Scrapers, heavy duty	Discoids, elongate
	Scrapers, light duty	Scrapers, core
	Burins	Scrapers, flake
	Awls	Miscellaneous forms
	Outils ecailles	Acheulean forms
	Laterally trimmed flakes	
	Sundry Tools	
	Bifaces	
Pounded Pieces	Spheroids/Subspheroids	Hammerstones
	Modified (battered)	Battered cobbles
	Hammerstones	Anvils
	Anvils	
	Utilized flakes	
Detached Pieces/Debitage	Unmodified flakes	Whole flakes
	Other fragments	Broken flakes
		Angular fragments
		Core fragments
Unmodified		Manuports

Artifact form and function

Initial research on artifact form was an attempt to describe and understand the variation within types of artifacts. This research focused mostly on flaked pieces (Clark 1970). Variation was often thought to be the result of different functional requirements. However, it was not until Toth's landmark experimental studies (Toth 1982, 1985, 1987) that researchers began to quantify what tool types were actually efficient at conducting specific activities. Remarkably Toth's studies showed that core forms were not very efficient for butchery or many other tasks. The most efficient tool type was detached pieces, specifically whole flakes. Further, Toth (1985, 1987) realized that the best use of core forms was for producing detached pieces. Experimental replication of Oldowan artifacts suggested that core forms were probably flaked over and over again, passing

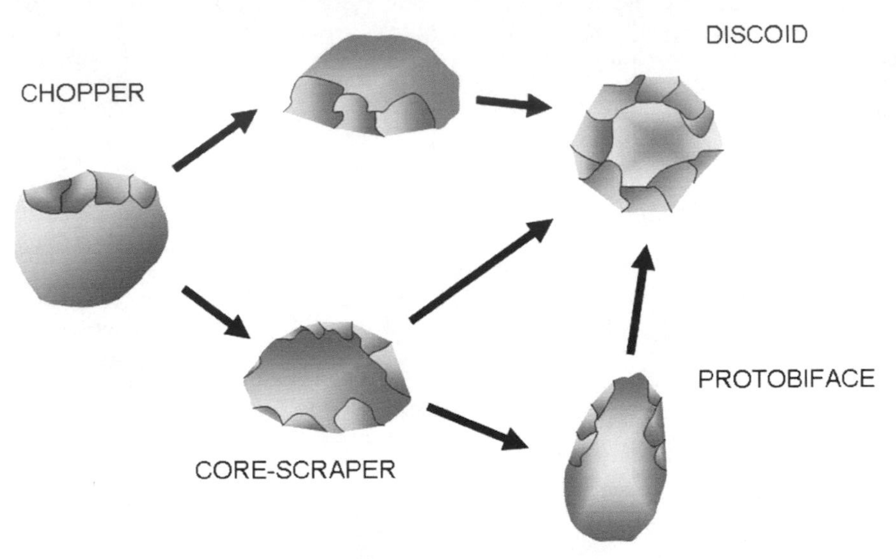

Figure 9.1. *Modification can transform a tool from one form to another through the process of reduction. This is often termed the life history of the tool. Experiments by Toth and others showed that types are not static but rather change as the extent of reduction continues. (Adapted from Potts, 1991.)*

from one core form to another as they were reduced (Figure 9.1) (Potts 1991; Sahnouni et al. 1997). The most important feature of this realization is that core forms do not represent what hominins wanted to produce, but rather, what hominins decided to discard. The principle is called the *final artifact fallacy* (Davidson and Noble 1993; McPherron 2000), and it is a pervasive concept in stone-tool analysis throughout the Paleolithic. This concept has major implications for the use of tool form for inferring intelligence in hominins. If artifacts represent those parts of their cultural repertoire that they no longer wanted, can they really represent the intellectual capabilities of hominins?

Reduction Sequences and the *Chaîne Opératoire*

Another major change in the last twenty years has been a shift away from descriptions focused on artifact shape. Instead, Oldowan archaeologists focus on how an artifact was made. Much of this research involves replicating tool forms and determining the different techniques required to make certain artifact forms (Toth 1982, 1985; Sahnouni et al. 1997; Ludwig and Harris 1998). Toth's experiments determined the different types of detached pieces expected from the production of specific flaked-piece forms. Using flaked-piece assemblages from Koobi Fora, Toth determined what the expected assemblage of detached pieces should be. When he compared expected versus actual assemblages, Toth determined the extent of transport of materials in and out of a site.

A technique called *chaîne opératoire* has become very influential in stone-artifact analysis in recent years. Originally this methodology was applied to much younger collections in western Europe to determine the connection between technological techniques and prehistoric societies (Leroi-Gourhan 1964). *Chaîne opératoire* attempted to connect ethnography and archaeology through an understanding of material culture (Lemmonnier 1990). Assemblages were defined by the techniques of manufacture and (sometimes ambiguous) links were made between artifacts and "technical systems" (Soressi and Geneste 2006). However, a basic underlying concept that drew attention from Oldowan archaeologists is that *chaîne opératoire* focused on the continuous transformation of artifacts and the ability to reconstruct artifact-production techniques. Oldowan archaeologists saw great promise in the *chaîne opératoire* approach for understanding the cognitive aspects of tool manufacture (de la Torre and Mora 2005, 2009).

Most recently, using a remarkably well-preserved collection from the site of Lokalalei 2C on the western shores of Lake Turkana, Delagnes and Roche (2005) reconstructed specific rules that hominins followed when making artifacts. By refitting tools together, Delagnes and Roche (2005) could literally reconstruct the patterns of detached piece removal. Although refitting analysis is incredibly time intensive, the benefits are astounding. It is as close as we can come to watching Oldowan hominins make stone artifacts. Delagnes and Roche (2005) determined that hominins consistently used the largest possible flaking surface. Core surfaces are maintained throughout the reduction of the core. Cortical surfaces appear to have been favored for reduction, and the extent of core reduction was determined by the natural angles of the cobble being reduced.

Stone tools and cognition

The interest in artifact-production techniques came at a critical time in Oldowan research. When the earliest sites in West Turkana were dated to 2.3 mya, they were some of the oldest archaeological localities in the world at that time (Kibunjia 1994; Kibunjia et al. 1992). The analysis of these assemblages appeared to show that these earliest hominins lacked proficiency in tool manufacture

(Kibunjia et al. 1992). Data from these West Turkana sites suggested a gradual increase in hominin ability to manufacture stone artifacts, prompting some to suggest these earliest sites were part of an industry termed the *pre-Oldowan* (Roche 1988, 2000). Shortly afterwards researchers working in Ethiopia confirmed that the earliest evidence of stone-artifact manufacture from the site of Gona, dated to 2.6 mya (Semaw et al. 1997; Semaw 2000). The research team working at Gona argued that the earliest toolmakers understood all aspects of stone manufacture (Semaw 2000). Those who studied the earliest evidence of artifacts were at an impasse. Did hominin material culture appear fully developed with the cognition required for a complete understanding of stone-tool manufacture? Or was there a gradual increase in the proficiency of hominins, as suggested by the gradual increase in number of Oldowan sites and the density of material at these localities through time (Harris 1983; Harris and Capaldo 1993)? The appearance of modified bones at 3.4 mya makes this debate even more complex (McPherron et al. 2010). Hominins may have had the ability to use tools long before they started making them. Although there have been some theoretical models of what these pre-Oldowan industries might look like (Carbonell et al. 2006), there are currently no means for identifying these earliest industries. Our present understanding suggests that hominins were proficient at producing tools by 2.6 mya and that prior to that they appear to have not made chipped-stone artifacts.

One of the major impediments in this debate was that understanding skill and cognition is very difficult in an archaeological collection. As described above, most artifacts are what hominins did *not* want. How can we judge their competence based on the discarded elements of their material culture? Oldowan archaeologists needed a comparative framework for the production of the earliest assemblages. Fortunately, research was already ongoing to study modern nonhuman-primate tool use. This research tried to determine if our closest living relative had the mental capacity to learn how to make stone tools (Toth et al. 1993; Schick et al. 1999). Toth, Schick, and colleagues trained a bonobo (*Pan paniscus*) named Kanzi to produce sharp-edged flakes using freehand hard-hammer percussion. Thus, they determined the baseline of nonhuman-primate stone-tool manufacture. Although this research shows the cognitive capabilities for tool use, this behavior had to be learned and has not been observed in the wild. This comparative framework supported the argument that the last common ancestor between chimpanzees and humans may have possessed the cognitive capabilities for stone-tool production.

More recently some researchers have suggested that modern chimpanzee nutcracking behavior displays the incipient knowledge required for tool manufacture (Mercader et al. 2002). Many similarities are intriguing. Modern chimpanzees in the Tai Forest of Cote d'Ivoire appear to select appropriate stones for the specific task of cracking nuts, and they transport these stones over moderate distances. In particular, comparisons between industries with a high percentage of bipolar technique and the refuse of nutcracking behavior had piqued the interest of those interested in the origins of stone technology. Bipolar percussion is similar to regular hard-hammer percussion except the flaked piece is placed on an anvil. In this type of percussion, fracture initiation occurs at two opposed points and platforms display extensive crushing (Carvalho et al. 2008; Haslam et al. 2009). The high percentage of shatter usually associated with bipolar percussion seemed to mimic the products of nonhuman-primate tool use (Mercader et al. 2002). However, the distinctions between ape tools and Oldowan artifacts are vast. Even the earliest Oldowan artifacts show an understanding of platform maintenance and angle selection unparalleled even in trained, captive, nonhuman primates (Schick et al. 1999; de la Torre 2004).

So while nonhuman-primate studies provide the lowest common denominator in tool use, they are still far from even the earliest Oldowan tools. So how do we judge skill and cognition in more advanced tool use? Replication studies by modern flintknappers prove useful, but even the most skilled flintknapper is not the best analogy for Oldowan hominins. Few modern human groups still depend on stone-tool manufacture on a regular basis. Possible exceptions include adze production in Irian Jaya, Indonesia (Stout 2002). At the village of Langda, tool production is a deeply ingrained part of the culture of the local people. Stout's ethnoarchaeological work provides quantitative and qualitative evidence to suggest that skill in tool production can be identified in detached-piece assemblages. Stout combined this evidence with extensive work using brain imaging techniques on modern western flintknappers (Stout et al. 2000, 2008). The combination of these data sets represents the closest archaeologists have ever gotten to understanding Oldowan cognition and skill and represents one of the most exciting avenues of Oldowan research to date. More recent work has tried to focus the understanding of how hominins learned to flintknap. These details have been largely elusive but some experiments suggest that specific aspects of angle selection and the emplacement of specific amounts of force on tool edges during the flintknapping process are key (Nonaka et al. 2010).

One problem with modern replication studies is that our current understanding of fracture mechanics of stone is oddly understudied (Dibble and Rezek 2009). Flintknapping is useful for developing hypotheses as to why certain attributes are related. Yet the lack of control in these experiments makes it difficult to isolate cause-and-effect relationships (Dibble 1998). Initial work provided archaeologists with the majority of our understanding of various techniques in the Paleolithic (Dibble and Pelcin 1995; Dibble 1997; Pelcin 1997; Brantingham and Kuhn 2001). More controlled experiments are badly needed before an understanding of hominin cognition and skill can be developed.

Artifacts and landscapes

The previously discussed, approaches to Oldowan technology focused on describing technology at the level of an individual artifact. However, obviously hominin tool-manufacture behavior was not confined by the limits of archaeologists' excavations. On the contrary, lithic production is remarkable because it requires hominins to select certain types of stones and transport them to different places on an ancient landscape. This feature of the Oldowan technical system has been capitalized on by studies that focus on the types of stones selected for artifact manufacture

Initial ideas on how and why Oldowan hominins select stones for artifact manufacture suggested functional differences between raw material types (Leakey 1971). Later Schick (1987b) astutely observed differences in types of raw materials present at localities at Olduvai and suggested differences in transport as the primary impetus for these patterns. Further investigations of land use among Oldowan hominins showed that raw material availability seemed to affect hominins differently through time (Rogers et al. 1994). Rogers and colleagues noticed that sites older than 1.6 myr (million years) in the Turkana Basin show evidence that hominins used raw materials from only the most proximal sources of stone. Hominins also appear to have selected the largest possible cores for transport to areas that are the most distal from raw-material sources (Braun et al. 2008b). Although the earliest hominins did not appear to have transported stone great distances (at least not in the Turkana Basin), they did have an acute understanding of the appropriate types of stone for the manufacture of artifacts. At Gona, Stout and colleagues (Stout et al. 2005) showed that hominins selected stones that were easily flaked, even when other raw material was abundantly available. Raw materials, such as trachyte, are consistently overrepresented in artifact assemblages compared to the local conglomerates. Assemblages from West Turkana display similar patterns (Harmand 2009). Extensive raw-material sourcing and studies of raw-material engineering properties suggest hominins at the site of Kanjera South transported materials great distances and selected specific materials based on the properties of these stones (Braun et al. 2008a,).

Unfortunately the picture of hominin raw-material transport and selection is likely more complex than previously expected. Hominin lithic-transport decisions recorded in a landscape study of Olduvai Gorge suggest transport decisions were made based on a variety of factors (Blumenschine et al. 2008). Although transport decisions at Olduvai were related to the relative availability of different raw materials, this can only explain a fraction of the variation in industries at Olduvai. Blumenschine and colleagues (2008) suggested that landscape-scale variation in carcass availability also has a major influence on hominin lithic-discard patterns. To further complicate patterns, a study of raw-material selectivity from sites at Hadar found none of the patterns expressed in the Gona assemblages (Goldman and Hovers 2009). This is especially disturbing considering that the two localities are directly adjacent to each other. Understanding the synchronic variability in raw-material selection and transport behaviors across landscapes clearly needs to be addressed.

The dawn of technological evolution

Archaeologists studying Oldowan assemblages determined that by the time that early Pleistocene hominins started producing large concentrations of stone artifacts, they understood some basics aspects of fracture mechanics and maintenance of platforms during flaked-piece reduction. Hominins also appear to be selecting and transporting certain types of stone to make artifacts. However, expression of these behaviors appears variably through time and across landscapes. Clearly archaeologists interested in Oldowan tool use need to begin applying the detailed analyses that provide understanding of aspects of technical competence to landscape scale assemblages. We need to understand why some behaviors are expressed in certain ecological conditions and not others. Recent approaches have begun to apply these types of models with surprising results. Stout and colleagues (2010) documented significant variation throughout the sequence of sites at Gona. They proposed a number of possible models to explain this variability and suggested ways of testing the explanations of this variability. Unfortunately, it is clear that these types of studies require large collections from many different localities. Continued fieldwork combined with hypothesis-driven experimental research may be the future of Oldowan lithic analysis.

FAUNAL REMAINS

Bones from archaeological sites are often referred to as archaeofaunas. The patterns displayed in the frequencies of different types of bones and the condition of those bones represent patterns of hominin diet, if accumulations of bones can be reliably attributed to hominins. Zooarchaeological data are informative because they allow archaeologists to characterize the trophic level and niche of our early Pleistocene ancestors. However, converting bits of broken bone into an ecological understanding of hominin lifeways is not without serious complications. The most pervasive problem is that bones, unlike stones, are easily destroyed. Bones attract the attention of carnivores and microorganisms that break bones down for their nutrients. Rapid burial is vital to the survival of bones in the archaeological record. Even after burial diagenetic processes (such as soil formation and the acidity of the surrounding sediment) can destroy them. However, through the course of carnivore ravaging and subsequent burial, bones record signatures of these different processes that zooarchaeologists can use to determine the history of a bone from living animal to hominin meal to fossil.

Some studies of Oldowan bone assemblages focus on the different carcass parts that are found in association with stone tools. These are often referred to as *skeletal-part profile* studies. Other investigations focus on marks on the surfaces of bones. These are termed *bone-surface modification* studies. The history of how these studies began to be applied to Oldowan archaeofaunas tracks prevailing thoughts about hominin behavior over the last thirty years (Dominguez-Rodrigo and Pickering 2003).

Skeletal-part profiles

Shortly after Isaac posited his home-base hypothesis, zooarchaeologists began to investigate what an accumulation of bones at a home base might look like. At the time, the most pervasive reference was that developed by archaeologists working in younger time periods. These studies focused on how and why humans transported different parts of a carcass (White 1953) and on modeling on which carcass parts were consistently transported away from kill sites. Subsequently Binford (1981, 1985; Binford et al. 1988) tested models of carcass-part transport. He confirmed that transport decisions were based on the energetic nutrients associated with specific body parts. It was clear that transported assemblages were usually dominated by limb bones. At first this seemed an easy way to distinguish accumulations created by hominins at central places. However, Binford also noticed that when carnivores ravage carcasses they, too, often leave assemblages dominated by limb elements. In fact, limb-dominated assemblages could even be the result of hominin scavenging from carnivore kills. Further, studies of the Hadza hunter–gatherers show that they often deflesh carcasses and transport the axial skeleton (i.e., the shoulder and pelvic girdles plus the vertebral column) (O'Connell et al. 1990). This means that hunter–gatherer campsites would have no limb elements. If early hominins followed a similar pattern, accumulations of limb bones found at archaeological sites could not have been the result of hominin behavior. The relative representation of skeletal parts was plagued by equifinalities (when many different processes produce the same pattern).

Experimental work determined patterns that separated hominin accumulations from carnivore kills. Robert Blumenschine's studies of the Serengeti ecosystem were a major step in solving this confusion surrounding skeletal-part profiles. Through a series of articles that described how predators consume prey on the modern savanna, Blumenschine determined that many of the Plio-Pleistocene assemblages had distinct similarities with carnivore kills (Blumenschine 1986; Blumenschine 1987; Blumenschine 1988a; Blumenschine and Marean 1993; Blumenschine 1995). In fact he concluded that hominins' role in these assemblages was relegated to breaking open limb bones that had already been defleshed by carnivores.

However, skeletal-part profiles encountered another difficulty. In order to reconstruct the profile of bones that had been transported, it was necessary for all bones to fossilize in similar frequencies. Unfortunately, not all bones have the same density, and therefore those bones more likely to survive diagenetic processes are overrepresented in fossil assemblages. Another problem was that many bones, once broken open by hominins or carnivores to gain access to marrow cavities, are only represented by small fragments of the shaft of limb bones (Binford and Bertram 1977; Brain 1967; Grayson 1989; Lyman et al. 1992; Lyman 1994; Lam et al. 2003). If recovery and proper identification of these small fragments does not occur, these limb bones are underrepresented in Oldowan archaeofaunas. Zooarchaeologists needed a new methodology for determining the ecological position of hominins.

Bone-surface modifications

Archaeological assemblages from the Koobi Fora Formation were some of the first to be studied using an approach that focused directly on bone-surface modifications (Bunn et al. 1980). Henry Bunn's research focused on conspicuous marks that showed evidence that stone tools had been scraped across the surfaces of bones (Bunn 1981, 1983). Bunn believed that the placement of these marks on upper-limb bones that are associated with large muscle masses testified that hominins repeatedly gained access to carcasses before carnivores. At a similar time others were identifying cut marks on bone surfaces at Olduvai

(Potts and Shipman 1981). These marks on bone surfaces seemed to be the evidence needed to determine that hominins were accessing significant portions of meat in their diet.

However, surfaces of bones also exhibit evidence of carnivore activity in the form of pits and furrows (Blumenschine 1988b; Selvaggio 1994; Capaldo 1997). Assemblages with abundant tooth marks also displayed abundant cutmarks. In order to understand if hominins or carnivores gained access to carcasses first, it was necessary to model these interactions in a modern setting. Observation of modern ecological processes for comparison with the archaeological record, sometimes known as *actualistic studies*, allowed archaeologists to link different scenarios of carcass access with specific frequencies of marks on bone surfaces. A series of experimental studies addressed this problem (Cavallo and Blumenschine 1989; Blumenschine and Marean 1993; Selvaggio 1994; Capaldo 1997). Several different scenarios were put forth that described frequencies of tooth marks and cutmarks expected for different patterns of carnivore and hominin involvement. Research into the size of tooth marks (specifically tooth pits) has allowed some researchers to determine the size of the carnivores with which the hominins were in competition (Dominguez-Rodrigo and Piqueras 2003). Further, zooarchaeologists began to identify marks on bone surfaces that resulted when hominins broke open bones for marrow (percussion marks) (Blumenschine and Selvaggio 1988). Based on this research, Blumenschine, Marean, and their students (Blumenschine 1988a, 1988b, 1995; Blumenschine and Marean 1993; Selvaggio 1994; Capaldo 1997, 1998) developed a model in which hominins gained access to carcasses only after large felids stripped off the majority of meat. Hominins accessed defleshed limbs and broke open bones for marrow. Then hyenas, with their powerful bone-crunching jaws, removed the ends of limb bones (epiphyses), which are rich in fat but inaccessible to hominins and felids. This model, sometimes referred to as the *multiple-stage model*, was developed using experimental collections. When these experiments were compared to archaeological collections, the high number of tooth marks matched the carnivore-first model. This suggested that hominins must have gained access to carcasses only after carnivores consumed most of the flesh. In this scenario hominins must have been passive scavengers.

But there was still the matter of all those cutmarks. If hominins were really only interested in within-bone nutrients (i.e., marrow), then why were hominins cutting at bones with no flesh on them? It is possible that hominins were removing periosteum (a connective-tissue casing on bones beneath muscle masses) before breaking bones open for marrow (Binford 1981; Blumenschine 1986). Yet removing periosteum often leaves marks characteristic of scraping, which are not found very often in the archaeological record of this time period. Further, periosteum removal does not explain why there are cutmarks on portions of bones where muscles are attached. Subsequent experimental work showed that when hominins get access to carcasses before carnivores, they tend to make cutmarks on bones with the highest meat yield (i.e., associated with the greatest muscle mass, such as with the humerus and femur). In Bunn's initial description of Olduvai assemblages, he noted that the majority of cutmarks appear on these bones (Bunn and Kroll 1986). Further, Domínguez-Rodrigo and Barba (2006) reanalyzed the FLK 22 (Zinj) horizon from Olduvai and noted that many of the tooth marks were actually the product of biochemical destruction of bone surfaces. Further, cutmark locations matched patterns from experimental butcheries of fully fleshed carcasses. From this perspective it appears that Oldowan hominins might have been less of a passive scavenger and more of a power scavenger or even marginal hunter.

However, almost all of these models are based entirely on one site from Olduvai (FLK 22 Zinj). This assemblage substantially postdates the first appearance of stone tools. Faunal remains from sites older than 2 myr are poorly known. Small assemblages from the Ethiopian site of Bouri in the Middle Awash research area suggest some butchery and breaking of bones for marrow removal (de Heinzelin et al. 1999). Another small assemblage from the Gona research area also in Ethiopia suggests that hominins gained access to fleshy limb elements (Domínguez-Rodrigo et al. 2005). New collections of modified bone from the Koobi Fora Formation and the Kanjera South Formation in Western Kenya may provide the first large collections of modified bone in the early Pleistocene (Braun et al. 2010).

OLDOWAN HOMININ BEHAVIOR

Despite the limited evidence that we have of early Pleistocene hominins, the extensive research regarding their stone artifacts and the remains of their meals allows us to speculate about hominin behavior. We can be reasonably sure that sometime between 2.5 and 2.0 mya selective pressures forced hominins to begin extracting resources that required the use of stone tools. At this point we don't know if those resources included animal tissue but the synchronicity of the first appearance of cutmarked bone and stone artifacts seems to suggest it. Stone tools possibly also allowed for a more efficient extraction of resources other than animal tissue (e.g., USOs).

Whatever the resources were that drove hominins to make stone tools, they must have been important enough to force hominins to become proficient at selecting the right types of stones for artifact manufacture and to consistently transport these specific types of stones. Further, stone tools must have been important enough to necessitate strict rules of artifact manufacture. By 2.3 mya (and maybe earlier) hominins understood which angles on stones facilitated conchoidal fracture.

Also by this time hominins were accessing large-mammal carcasses to some extent. The earliest evidence of this behavior seems to be some variation between carcasses that were possibly fully fleshed (Gona) and those that appear to have been scavenged from a primary consumer (Bouri). Within the first 600 kyr (thousand years) of stone-tool manufacture, the incorporation of some large-mammal tissue (either meat or marrow) likely increased. There is the distinct possibility that the FLK Zinj collection represents an anomaly in the archaeological record. Localities with evidence of mammal-tissue acquisition may reflect times of stress that forced hominins to fall back on secondary resources.

Although these models of early Pleistocene hominin behavior are elaborate, they are based on interpretations from only a handful of sites. The vast majority of archaeological data that provide insights into hominin diet are based on interpretations from sites at Olduvai Gorge. If archaeology is to contribute to an understanding of hominin behavior, new sites must be incorporated into our understanding. One troubling aspect of the Oldowan archaeological record is that inferences of hominin behavior are based on very little evidence. Some early Pleistocene assemblages contain fewer than ten cores and fifty whole flakes (e.g., the Omo collections); inferring behavior from such small assemblages is very difficult. A continued focus on experimentation to elucidate what archaeological patterns mean is vital.

Variability also appears as a major feature of the Oldowan archaeological record. This may be the result of the numerous hominin species that existed throughout the Oldowan period (Delagnes and Roche 2005). Yet a more testable explanation is that Oldowan behavior is very sensitive to its ecological context (Plummer 2004). If this is the case, paleoecological analyses conducted in concert with Oldowan archaeological projects is crucial. If Oldowan archaeologists can understand behavioral patterns relative to varied environmental contexts, it may be possible to really understand the ecology of early Pleistocene hominins.

REFERENCES

Antón, S. C., W. R. Leonard, and M. L. Robertson. 2002. "An Ecomorphological Model of the Initial Hominid Dispersal from Africa." *Journal of Human Evolution* 43 (6): 773–85. http://dx.doi.org/10.1006/jhev.2002.0602. Medline:12473483

Behrensmeyer, A. K. 1978. "Taphonomic and Ecologic Information from Bone Weathering." *Paleobiology* 4:150–62.
Behrensmeyer, A. K., and A. P. Hill. 1980. *Fossils in the Making: Vertebrate Taphonomy and Paleoecology*. Chicago: University of Chicago Press.
Biberson, P. J. 1967. "Some Aspects of the Lower Paleolithic of Northwest Africa." In *Background to Evolution in Africa*, ed. W. W. Bishop and J. D. Clark, 447–475. Chicago: University of Chicago Press.
Binford, L. R. 1981. *Bones: Ancient Men and Modern Myths*. New York: Academic Press.
Binford, L. R. 1985. "Human Ancestors: Changing Views of Their Behavior." *Journal of Anthropological Archaeology* 4 (4): 292–327. http://dx.doi.org/10.1016/0278-4165(85)90009-1.
Binford, L. R., and J. B. Bertram. 1977. "Bone Frequencies and Attritional Processes." In *For Theory Building in Archaeology*, ed. L. R. Binford, 77–153. New York: Academic Press.
Binford, L. R., H. T. Bunn, and E. M. Kroll. 1988. "Fact and Fiction about the *Zinjanthropus* Floor: Data, Arguments and Interpretation." *Current Anthropology* 29 (1): 123–35. http://dx.doi.org/10.1086/203618.
Blumenschine, R. J. 1986. "Early Hominid Scavenging Opportunities: Implications of Carcass Availability in the Serengeti and Ngorongoro Ecosystems." BAR International Series 283. Oxford: BAR.
Blumenschine, R. J. 1987. "Characteristics of an Early Hominid Scavenging Niche." *Current Anthropology* 28 (4): 383–94. http://dx.doi.org/10.1086/203544.
Blumenschine, R. J. 1988a. "An Experimental Model of the Timing of Hominid and Carnivore Influence on Archaeological Bone Assemblages." *Journal of Archaeological Science* 15 (5): 483–502. http://dx.doi.org/10.1016/0305-4403(88)90078-7.
Blumenschine, R. J. 1988b. "A Taphonomic Trio for Archaeology: Mechanics, Energetics and Ecology." Society for American Archaeology 53rd Annual Meeting. Society for American Archaeology, Phoenix.
Blumenschine, R. J. 1995. "Percussion Marks, Tooth Marks, and Experimental Determinations of the Timing of Hominid and Carnivore Access to Long Bones at Flk Zinjanthropus, Olduvai Gorge, Tanzania." *Journal of Human Evolution* 29 (1): 21–51. http://dx.doi.org/10.1006/jhev.1995.1046.
Blumenschine, R. J., and C. W. Marean. 1993. "A Carnivore's View of Archaeological Bone Assemblages." In *From Bones to Behavior: Ethnoarchaeological and Experimental Contributions to the Interpretations of Faunal Remains*, ed. J. Hudson, 273–300. Carbondale: University of Southern Illinois Press.
Blumenschine, R. J., and F. T. Masao. 1991. "Living Sites at Olduvai Gorge, Tanzania?: Preliminary Landscape Archaeology Results in the Basal Bed II Lake Margin Zone." *Journal of Human Evolution* 21 (6): 451–62. http://dx.doi.org/10.1016/0047-2484(91)90095-D.
Blumenschine, R. J., F. T. Masao, J. Tactikos, and J. Ebert. 2008. "Effects of Distance from Stone Source on Landscape-Scale Variation in Oldowan Artifact Assemblages in the Paleo-Olduvai Basin, Tanzania." *Journal of Archaeological Science* 35 (1): 76–86. http://dx.doi.org/10.1016/j.jas.2007.02.009.

Blumenschine, R. J., and C. R. Peters. 1998. "Archaeological Predictions for Hominid Land Use in the Paleo-Olduvai Basin, Tanzania, during Lowermost Bed II Times." *Journal of Human Evolution* 34 (6): 565–607. http://dx.doi.org/10.1006/jhev.1998.0216. Medline:9650101

Blumenschine, R. J., and M. M. Selvaggio. 1988. "Percussion Marks on Bone Surfaces as a New Diagnostic of Hominid Behavior." *Nature* 333 (6175): 763–5. http://dx.doi.org/10.1038/333763a0.

Blumenschine, R. J., A. Whiten, and K. Hawkes. 1991. "Hominid Carnivory and Foraging Strategies, and the Socio-economic Function of Early Archaeological Sites." *Philosophical Transactions of the Royal Society of London. Series B, Biological Sciences* 334 (1270): 211–21. http://dx.doi.org/10.1098/rstb.1991.0110. Medline:1685579.

Brain, C. K. 1967. "Hottentot Food Remains and Their Bearing on the Interpretation of Fossil Bone Assemblages." *Scientific Papers of the Namib Desert Research Station* 32:1–7.

Brantingham, P. J. 1998. "Hominid-Carnivore Coevolution and Invasion of the Predatory Guild." *Journal of Anthropological Archaeology* 17 (4): 327–53. http://dx.doi.org/10.1006/jaar.1998.0326.

Brantingham, P. J., and S. L. Kuhn. 2001. "Constraints on Levallois Core Technology: A Mathematical Model." *Journal of Archaeological Science* 28 (7): 747–61. http://dx.doi.org/10.1006/jasc.2000.0594.

Braun, D. R., J.W.K. Harris, N. E. Levin, J. T. McCoy, A.I.R. Herries, M. K. Bamford, L. C. Bishop, B. G. Richmond, and M. Kibunjia. 2010. "Early Hominin Diet Included Diverse Terrestrial and Aquatic Animals 1.95 Ma in East Turkana, Kenya." *Proceedings of the National Academy of Sciences of the United States of America* 107 (22): 10002–10007. http://dx.doi.org/10.1073/pnas.1002181107. Medline:20534571

Braun, D. R., T. Plummer, P. Ditchfield, J. V. Ferraro, D. Maina, L. C. Bishop, and R. Potts. 2008a. "Oldowan Behavior and Raw Material Transport: Perspectives from the Kanjera Formation." *Journal of Archaeological Science* 35 (8): 2329–45. http://dx.doi.org/10.1016/j.jas.2008.03.004.

Braun, D. R., M. J. Rogers, J.W.K. Harris, and S. J. Walker. 2008b. "Landscape-Scale Variation in Hominin Tool Use: Evidence from the Developed Oldowan." *Journal of Human Evolution* 55 (6): 1053–63. http://dx.doi.org/10.1016/j.jhevol.2008.05.020. Medline:18845314

Braun, D. R., J. C. Tactikos, J. V. Ferraro, and J.W.K. Harris. 2006. "Archaeological Inference and Oldowan Behavior." *Journal of Human Evolution* 51 (1): 106–8. http://dx.doi.org/10.1016/j.jhevol.2006.04.002. Medline:16780924

Bunn, H. T. 1981. "Archaeological Evidence for Meat-eating by Plio-Pleistocene Hominids from Koobi Fora and Olduvai Gorge." *Nature* 291 (5816): 574–7. http://dx.doi.org/10.1038/291574a0.

Bunn, H. T. 1983. "Evidence on the Diet and Subsistence Patterns of Plio-Pleistocene Hominids at Koobi Fora, Kenya, and Olduvai Gorge, Tanzania." In *Animals and Archaeology*, ed. J. Clutton-Brock and C. Grigson, 21–30. Oxford: BAR.

Bunn, H. T., J.W.K. Harris, G. L. Isaac, Z. Kaufulu, E. M. Kroll, K. D. Schick, N. Toth, and A. K. Behrensmeyer. 1980. "FxJj 50: An Early Pleistocene Site in Northern Kenya." *World Archaeology* 12 (2): 109–36. http://dx.doi.org/10.1080/00438243.1980.9979787.

Bunn, H. T., and E. M. Kroll. 1986. "Systematic Butchery by Plio-Pleistocene Hominids at Olduvai Gorge, Tanzania." *Current Anthropology* 27 (5): 431–52. http://dx.doi.org/10.1086/203467.

Capaldo, S. D. 1995. "Inferring Hominid and Carnivore Behavior from Dual-Patterned Archaeofaunal Assemblages." PhD dissertation, Rutgers University, New Brunswick, NJ.

Capaldo, S. D. 1997. "Experimental Determinations of Carcass Processing by Plio-Pleistocene Hominids and Carnivores at FLK 22 (*Zinjanthropus*). Olduvai Gorge, Tanzania." *Journal of Human Evolution* 33 (5): 555–97. http://dx.doi.org/10.1006/jhev.1997.0150. Medline:9403079

Capaldo, S. D. 1998. "Simulating the Formation of Dual-Patterned Archaeofaunal Assemblages with Experimental Control Samples." *Journal of Archaeological Science* 25: 311–30.

Carbonell, E., R. Sala, D. Barsky, and V. Celiberti. 2006. "From Homogeneity to Multiplicity: A New Approach to the Study of Archaic Stone Tools." In *Interdisciplinary Approaches to the Oldowan*, ed. E. Hovers and D. R. Braun, 25–37. Dordrecht: Springer. http://dx.doi.org/10.1007/978-1-4020-9060-8_3

Carvalho, S., E. Cunha, C. Sousa, and T. Matsuzawa. 2008. "Chaînes opératoires and Resource-Exploitation Strategies in Chimpanzee (*Pan troglodytes*) Nut Cracking." *Journal of Human Evolution* 55 (1): 148–63. http://dx.doi.org/10.1016/j.jhevol.2008.02.005. Medline:18359504

Cavallo, J. A., and R. J. Blumenschine. 1989. "Tree-Stored Leopard Kills: Expanding the Hominid Scavenging Niche." *Journal of Human Evolution* 18 (4): 393–9. http://dx.doi.org/10.1016/0047-2484(89)90038-9.

Chavaillon, J. 1976. "Evidence for the Technical Practices of Early Pleistocene Hominids, Shungura Formation, Lower Omo Valley, Ethiopia." In *Earliest Man and Environments in the Lake Rudolf Basin*, ed. Y. Coppens, F. C. Howell, G. L. Isaac, and R.E.F. Leakey, 565–573. Chicago: University of Chicago Press.

Clark, D. J. 1970. *The Prehistory of Africa*. London: Thames and Hudson.

Davidson, I., and W. Noble. 1993. "Tools, Language and Cognition in Human Evolution." In *Tools, Language and Cognition in Human Evolution*, ed. K. Gibson and T. Ingold, 363–388. Cambridge: Cambridge University Press.

de Heinzelin, J., J. D. Clark, T. D. White, W. S. Hart, P. R. Renne, G. WoldeGabriel, Y. Beyene, and E.S. Vrba. 1999. "Environment and Behavior of 2.5-million-year-old Bouri Hominids." *Science* 284 (5414): 625–9. http://dx.doi.org/10.1126/science.284.5414.625. Medline:10213682

Delagnes, A., and H. Roche. 2005. "Late Pliocene Hominid Knapping Skills: The Case of Lokalalei 2C, West Turkana, Kenya." *Journal of Human Evolution* 48 (5): 435–72. http://dx.doi.org/10.1016/j.jhevol.2004.12.005. Medline:15857650

de la Torre, I. 2004. "Omo Revisited: Evaluating the Technological Skills of Pliocene Hominids." *Current Anthropology* 45 (4): 439–65. http://dx.doi.org/10.1086/422079.
de la Torre, I., and R. Mora. 2005. *Technological Strategies in the Lower Pleistocene at Olduvai Beds I and II.* Liege: Service de Prehistoire, Universite de Liege.
de la Torre, I., and R. Mora. 2009. "Remarks on the Current Theoretical and Methodological Approaches to the Study of Early Technological Strategies in Eastern Africa." In *Interdisciplinary Approaches to the Oldowan*, ed. E. Hovers and D. R. Braun, 15–24. Dodrecht: Springer Science and Business Media B.V. http://dx.doi.org/10.1007/978-1-4020-9060-8_2
Dibble, H. L. 1997. "Platform Variability and Flake Morphology: Comparison of Experimental and Archaeological Data and Implications for Interpreting Prehistoric Lithic Technological Strategies." *Lithic Technology* 22:150–70.
Dibble, H. L. 1998. "Comment on "Quantifying Lithic Curation, an Experimental Test of Dibble and Pelcin's Original Flake-Tool Mass Predictor," by Zachary J. Davis and John J. Shea." *Journal of Archaeological Science* 25 (7): 611–3. http://dx.doi.org/10.1006/jasc.1997.0254.
Dibble, H. L., and A. Pelcin. 1995. "The Effect of Hammer Mass and Velocity on Flake Mass." *Journal of Archaeological Science* 22 (3): 429–39. http://dx.doi.org/10.1006/jasc.1995.0042.
Dibble, H. L., and Z. Rezek. 2009. "Introducing a New Experimental Design for Controlled Studies of Flake Formation: Results for Exterior Platform Angle, Platform Depth, Angle of Blow, Velocity, and Force." *Journal of Archaeological Science* 36 (9): 1945–54. http://dx.doi.org/10.1016/j.jas.2009.05.004.
Dominguez-Rodrigo, M. 2009. "Are All Oldowan Sites Palimpsests? If So, What Can They Tell Us about Hominid Carnivory?" In *Interdisciplinary Approaches to the Oldowan*, ed. E. Hovers and D. R. Braun, 129–147. Dordrecht: Springer. http://dx.doi.org/10.1007/978-1-4020-9060-8_11
Domínguez-Rodrigo, M., and R. Barba. 2006. "New Estimates of Tooth Mark and Percussion Mark Frequencies at the FLK Zinj Site: The Carnivore-Hominid-Carnivore Hypothesis Falsified." *Journal of Human Evolution* 50 (2): 170–94. http://dx.doi.org/10.1016/j.jhevol.2005.09.005. Medline:16413934
Dominguez-Rodrigo, M., and T. R. Pickering. 2003. "Early Hominid Hunting and Scavenging: A Zooarcheological Review." *Evolutionary Anthropology* 12 (6): 275–82. http://dx.doi.org/10.1002/evan.10119.
Dominguez-Rodrigo, M., R. Pickering, and H. T. Bunn. 2010. "Configurational Approach to Identifying Earliest Hominin Butchers." *Proceedings of the National Academy of Sciences* 107: 20929–20934.
Domínguez-Rodrigo, M., T. R. Pickering, S. Semaw, and M. J. Rogers. 2005. "Cutmarked Bones from Pliocene Archaeological Sites at Gona, Afar, Ethiopia: Implications for the Function of the World's Oldest Stone Tools." *Journal of Human Evolution* 48 (2): 109–21. http://dx.doi.org/10.1016/j.jhevol.2004.09.004. Medline:15701526

Dominguez-Rodrigo, M., and A. Piqueras. 2003. "The Use of Tooth Pits to Identify Carnivore Taxa in Tooth-Marked Archaeofaunas and Their Relevance to Reconstruct Hominid Carcass Processing Behaviours." *Journal of Archaeological Science* 30 (11): 1385–91. http://dx.doi.org/10.1016/S0305-4403(03)00027-X.

Goldman, T., and E. Hovers. 2009. "Methodological Considerations in the Study of Oldowan Raw Material Selectivity: Insights from A.L. 894 (Hadar, Ethiopia)." In *Interdisciplinary Approaches to the Oldowan*, ed. E. Hovers and D. R. Braun, 71–84. Dordrecht: Springer. http://dx.doi.org/10.1007/978-1-4020-9060-8_7

Grayson, D. K. 1989. "Bone Transport, Bone Destruction, and Reverse Utility Curves." *Journal of Archaeological Science* 16 (6): 643–52. http://dx.doi.org/10.1016/0305-4403(89)90028-9.

Harmand, S. 2009. "Variability in Raw Material Selectivity at the Late Pliocene Sites of Lokalalei, West Turkana, Kenya." In *Interdisciplinary Approaches to the Oldowan*, ed. E. Hovers and D. R. Braun, 85–97. Dordrecht: Springer. http://dx.doi.org/10.1007/978-1-4020-9060-8_8

Harris, J.W.K. 1983. "Cultural Beginnings: Plio-Pleistocene Archaeological Occurrences from the Afar, Ethiopia." *African Archaeological Review* 1 (1): 3–31. http://dx.doi.org/10.1007/BF01116770.

Harris, J.W.K., and S. D. Capaldo. 1993. "The Earliest Stone Tools." In *The Use of Tools by Human and Non-Human Primates*, ed. A. Berthelet and J. Chavaillon, 196–220. Oxford: Clarendon Press.

Haslam, M., A. Hernandez-Aguilar, V. Ling, S. Carvalho, I. de la Torre, A. DeStefano, A. Du, B. Hardy, J.W.K. Harris, L. Marchant, et al. 2009. "Primate Archaeology." *Nature* 460 (7253): 339–44. http://dx.doi.org/10.1038/nature08188. Medline:19606139

Isaac, G. L. 1976. "The Activities of Early African Hominids: A Review of Archaeological Evidence from the Time Span Two and a Half to One Million Years Ago." In *Human Origins: Louis Leakey and the East African Evidence*, ed. G. L. Isaac and E. McCown, 483–514. Menlo Park, CA: W. A. Benjamin.

Isaac, G. L. 1978. "Food Sharing and Human Evolution: Archaeological Evidence from the Plio-Pleistocene of East Africa." *Journal of Anthropological Research* 34:311–25.

Isaac, G. L. 1983. "Early Stages in the Evolution of Human Behavior: The Adaptive Significance of Stone Tools." *Zesdekroon-voordraft gehouden voor die Stichting Nederlands Museum voor anthropologie en praehistorie* 5–43.

Isaac, G. L., J.W.K. Harris, and E. M. Kroll. 1997. "The Stone Artefact Assemblages: A Comparative Study." *Koobi Fora Research Project* 262–362.

Kibunjia, M. 1994. "Pliocene Archaeological Occurrences in the Lake Turkana Basin." *Journal of Human Evolution* 27 (1-3): 159–71. http://dx.doi.org/10.1006/jhev.1994.1040.

Kibunjia, M., H. Roche, F. H. Brown, and R. E. Leakey. 1992. "Pliocene and Pleistocene Archaeological Sites West of Lake Turkana, Kenya." *Journal of Human Evolution* 23 (5): 431–38. http://dx.doi.org/10.1016/0047-2484(92)90091-M.

Kimbel, W. H., R. C. Walter, D. C. Johanson, K. E. Reed, J. L. Aronson, Z. Assefa, C. W. Marean, G. G. Eck, R. Bobe, E. Hovers, Y. Rak, C. Vondra, T. Yemane, D. York, Y. Chen,

N. M. Evensen, and P. E. Smith. 1996. "Late Pliocene Homo and Oldowan Tools from the Hadar Formation (Kada Hadar Member), Ethiopia." *Journal of Human Evolution* 31 (6): 549–61. http://dx.doi.org/10.1006/jhev.1996.0079.

Kuman, K., and R. J. Clarke. 2000. "Stratigraphy, Artefact Industries and Hominid Associations for Sterkfontein, Member 5." *Journal of Human Evolution* 38 (6): 827–47. http://dx.doi.org/10.1006/jhev.1999.0392. Medline:10835264

Lam, Y. M., O. M. Pearson, C. W. Marean, and X. B. Chen. 2003. "Bone Density Studies in Zooarchaeology." *Journal of Archaeological Science* 30 (12): 1701–8. http://dx.doi.org/10.1016/S0305-4403(03)00065-7.

Leakey, M. D. 1967. "Preliminary Survey of the Cultural Material from Beds I and II, Olduvai Gorge, Tanzania." In *Background to Evolution in Africa*, ed. W. W. Bishop and J. D. Clark, 417–446. Chicago: University of Chicago Press.

Leakey, M. D. 1971. *Olduvai Gorge*. Cambridge: Cambridge University Press.

Lemmonnier, P. 1990. "Topsy Turvy Techniques: Remarks on Social Representation of Techniques." *Cambridge Archaeological Review* 9:27–37.

Leroi-Gourhan, A. 1964. *Le geste et la parole: I- Technique et language, II- La memoire et les rythmes*. Paris: Albin Michel.

Ludwig, B. V., and J.W.K. Harris. 1998. "Towards a Technological Reassessment of East African Plio-Pleistocene Lithic Assemblages." In *The Rise and Diversity of the Lower Paleolithic*, ed. M. Petraglia and K. Paddaya, 84–106. New York: Academic Press.

Lyman, R. L. 1994. *Vertebrate Taphonomy*. Cambridge: Cambridge University Press.

Lyman, R. L., L.E. Houghton, and A. L. Chambers. 1992. "The Effect of Structural Density on Marmot Skeletal Part Representation in Archaeological Sites." *Journal of Archaeological Science* 19 (5): 557–73. http://dx.doi.org/10.1016/0305-4403(92)90028-2.

Marean, C. W. 1989. "Sabertooth Cats and Their Relevance for Early Hominid Diet and Evolution." *Journal of Human Evolution* 18 (6): 559–82. http://dx.doi.org/10.1016/0047-2484(89)90018-3.

McPherron, S. P. 2000. "Handaxes as a Measure of the Mental Capabilities of Early Hominids." *Journal of Archaeological Science* 27 (8): 655–63. http://dx.doi.org/10.1006/jasc.1999.0467.

McPherron, S. P., Z. Alemseged, C. W. Marean, J. G. Wynn, D. Reed, D. Geraads, R. Bobe, and H. A. Béarat. 2010. "Evidence for Stone-Tool-Assisted Consumption of Animal Tissues before 3.39 Million Years Ago at Dikika, Ethiopia." *Nature* 466 (7308): 857–60. http://dx.doi.org/10.1038/nature09248. Medline:20703305

Mercader, J., M. Panger, and C. Boesch. 2002. "Excavation of a Chimpanzee Stone Tool Site in the African Rainforest." *Science* 296 (5572): 1452–5. http://dx.doi.org/10.1126/science.1070268. Medline:12029130

Nonaka, T., B. Bril, and R. Rein. 2010. "How Do Stone Knappers Predict and Control the Outcome of Flaking? Implications for Understanding Early Stone Tool Technology." *Journal of Human Evolution* 59 (2): 155–67. http://dx.doi.org/10.1016/j.jhevol.2010.04.006. Medline:20594585

O'Connell, J. F., K. Hawkes, and N. Blurton Jones. 1988. "Hadza Scavenging: Implications for Plio/Pleistocene Hominid Subsistence." *Current Anthropology* 29 (2): 356–63. http://dx.doi.org/10.1086/203648.

O'Connell, J. F., K. Hawkes, and N. Blurton Jones. 1990. "Reanalysis of Large Mammal Body Part Transport among the Hadza." *Journal of Archaeological Science* 17 (3): 301–16. http://dx.doi.org/10.1016/0305-4403(90)90025-Z.

O'Connell, J. F., K. Hawkes, K. D. Lupo, and N. G. Blurton Jones. 2002. "Male Strategies and Plio-Pleistocene Archaeology." *Journal of Human Evolution* 43 (6): 831–72. http://dx.doi.org/10.1006/jhev.2002.0604. Medline:12473486

Panger, M. A., A. S. Brooks, B. G. Richmond, and B. Wood. 2002. "Older than the Oldowan? Rethinking the Emergence of Hominin Tool Use." *Evolutionary Anthropology* 11 (6): 235–45. http://dx.doi.org/10.1002/evan.10094.

Pelcin, A. W. 1997. "The Formation of Flakes: The Role of Platform Thickness and Exterior Platform Angle in the Production of Flake Initiations and Terminations." *Journal of Archaeological Science* 24 (12): 1107–13. http://dx.doi.org/10.1006/jasc.1996.0190.

Peters, C. R., and R. J. Blumenschine. 1996. "Landscape Perspectives on Possible Land Use Patterns for Early Hominids in the Olduvai Basin, Tanzania: Part II, Expanding the Landscape Models." *Kaupia* 6:175–221.

Plummer, T. 2004. "Flaked Stones and Old Bones: Biological and Cultural Evolution at the Dawn of Technology." *American Journal of Physical Anthropology* 125 (Suppl. 39): 118–64. http://dx.doi.org/10.1002/ajpa.20157. Medline:15605391

Plummer, T., L. C. Bishop, P. Ditchfield, and J. Hicks. 1999. "Research on Late Pliocene Oldowan Sites at Kanjera South, Kenya." *Journal of Human Evolution* 36 (2): 151–70. http://dx.doi.org/10.1006/jhev.1998.0256. Medline:10068064

Plummer, T. W., P. W. Ditchfield, L. C. Bishop, J. D. Kingston, J. V. Ferraro, D. R. Braun, F. Hertel, and R. Potts. 2009. "Oldest Evidence of Tool Making Hominins in a Grassland-Dominated Ecosystem." *PLoS ONE* 4 (9): e7199. http://dx.doi.org/10.1371/journal.pone.0007199. Medline:19844568

Potts, R. 1984. "Home Bases and Early Hominids." *American Scientist* 72:338–47.

Potts, R. 1988. *Early Hominid Activities at Olduvai.* Hawthorne: Aldine de Gruyter.

Potts, R. 1991. "Why the Oldowan? Plio-Pleistocene Toolmaking and the Transport of Resources." *Journal of Anthropological Research* 47:153–76.

Potts, R. 1998. "Variability Selection in Hominid Evolution." *Evolutionary Anthropology* 7 (3): 81–96. http://dx.doi.org/10.1002/(SICI)1520-6505(1998)7:3<81::AID-EVAN3>3.0.CO;2-A.

Potts, R., and P. Shipman. 1981. "Cutmarks Made by Stone Tools on Bones from Olduvai Gorge, Tanzania." *Nature* 291 (5816): 577–80. http://dx.doi.org/10.1038/291577a0.

Roche, H. 1988. "Technological Evolution in Early Hominids." *OSSA: International Journal of Skeletal Research* 14:97–8.

Roche, H. 2000. "Variability of Pliocene Lithic Productions in East Africa." *ACTA Anthropological Sinica* 19:98–103.

Roche, H., A. Delagnes, J.-P. Brugal, C. S. Feibel, M. Kibunjia, V. Mourre, and P. J. Texier. 1999. "Early Hominid Stone Tool Production and Technical Skill 2.34 Myr ago in West Turkana, Kenya." *Nature* 399 (6731): 57–60. http://dx.doi.org/10.1038/19959. Medline:10331389

Rogers, M. J. 1997. *A Landscape Archaeological Study at East Turkana, Kenya, Anthropology*. New Brunswick, NJ: Rutgers University.

Rogers, M. J., J.W.K. Harris, and C. S. Feibel. 1994. "Changing Patterns of Land Use by Plio-Pleistocene Hominids in the Lake Turkana Basin." *Journal of Human Evolution* 27 (1-3): 139–58. http://dx.doi.org/10.1006/jhev.1994.1039.

Rose, L., and F. Marshall. 1996. "Meat Eating, Hominid Sociality, and Home Bases Revisited." *Current Anthropology* 37 (2): 307–38. http://dx.doi.org/10.1086/204494.

Sahnouni, M., and J. de Heinzelin. 1998. "The Site of Ain Hanech Revisited: New Investigations at this Lower Pleistocene Site in Northern Algeria." *Journal of Archaeological Science* 25 (11): 1083–101. http://dx.doi.org/10.1006/jasc.1998.0278.

Sahnouni, M., D. Hadjouis, J. van der Made, A. el-K. Derradji, A. Canals, M. Medig, H. Belahrech, Z. Harichane, and M. Rabhi. 2002. "Further Research at the Oldowan Site of Ain Hanech, North-eastern Algeria." *Journal of Human Evolution* 43 (6): 925–37. http://dx.doi.org/10.1006/jhev.2002.0608. Medline:12473489

Sahnouni, M., K. D. Schick, and N. Toth. 1997. "An Experimental Investigation into the Nature of Faceted Limestone "Spheroids" in the Early Palaeolithic." *Journal of Archaeological Science* 24 (8): 701–13. http://dx.doi.org/10.1006/jasc.1996.0152.

Schick, K. D. 1986. *Stone Age Sites in the Making: Experiments in the Formation and Transformation of Archaeological Occurrences*. Oxford: BAR.

Schick, K. D. 1987a. "Experimentally-Derived Criteria for Assessing Hydrologic Disturbance of Archaeological Sites." In *Natural Formation Processes and the Archaeological Record*, ed. D. T. Nash and M. D. Petraglia, 86–107. Oxford: BAR Series.

Schick, K. D. 1987b. "Modeling the Formation of Early Stone Age Artifact Concentrations." *Journal of Human Evolution* 16 (7-8): 789–807. http://dx.doi.org/10.1016/0047-2484(87)90024-8.

Schick, K. D. 1997. "Experimental Studies of Site-Formation Processes." In *Koobi Fora Research Project*, ed. G. L. Isaac, 244–256. Oxford: Clarendon Press.

Schick, K. D., N. Toth, G. Garufi, E. Savage-Rumbaugh, D. Rumbaugh, and R. Sevcik. 1999. "Continuing Investigations into the Stone Tool-Making and Tool-Using Capabilities of a Bonobo (*Pan paniscus*)." *Journal of Archaeological Science* 26 (7): 821–32. http://dx.doi.org/10.1006/jasc.1998.0350.

Selvaggio, M. M. 1994. "Carnivore Tooth Marks and Stone Tool Butchery Marks on Scavenged Bones: Archaeological Implications." *Journal of Human Evolution* 27 (1-3): 215–28. http://dx.doi.org/10.1006/jhev.1994.1043.

Semaw, S. 2000. "The World's Oldest Stone Artefacts from Gona, Ethiopia: Their Implications for Understanding Stone Technology and Patterns of Human Evolution between 2.6–1.5 Million Years Ago." *Journal of Archaeological Science* 27 (12): 1197–214. http://dx.doi.org/10.1006/jasc.1999.0592.

Semaw, S., P. Renne, J.W.K. Harris, C. S. Feibel, R. L. Bernor, N. Fesseha, and K. M. Mowbray. 1997. "2.5-million-year-old Stone Tools from Gona, Ethiopia." *Nature* 385 (6614): 333–6. http://dx.doi.org/10.1038/385333a0. Medline:9002516
Soressi, M., and M. Geneste. 2006. "Discussing the History and Efficacy of the Chaine Operatoire Approach to Lithic Analysis." 71st Annual Meeting of the Society for American Archaeology, San Juan, Puerto Rico.
Stout, D. 2002. "Skill and Cognition in Stone Tool Production: An Ethnographic Case Study from Irian Jaya." *Current Anthropology* 43 (5): 693–722. http://dx.doi.org/10.1086/342638.
Stout, D., and T. Chaminade. 2009. "Making Tools and Making Sense: Complex, Intentional Behaviour in Human Evolution." *Cambridge Archaeological Journal* 19 (01): 85–96. http://dx.doi.org/10.1017/S0959774309000055.
Stout, D., J. Quade, S. Semaw, M. J. Rogers, and N. E. Levin. 2005. "Raw Material Selectivity of the Earliest Stone Toolmakers at Gona, Afar, Ethiopia." *Journal of Human Evolution* 48 (4): 365–80. http://dx.doi.org/10.1016/j.jhevol.2004.10.006. Medline:15788183
Stout, D., S. Semaw, M. J. Rogers, and D. Cauche. 2010. "Technological Variation in the Earliest Oldowan from Gona, Afar, Ethiopia." *Journal of Human Evolution* 58 (6): 474–91. http://dx.doi.org/10.1016/j.jhevol.2010.02.005. Medline:20430417
Stout, D., N. Toth, K. Schick, and T. Chaminade. 2008. "Neural Correlates of Early Stone Age Toolmaking: Technology, Language and Cognition in Human Evolution." *Philosophical Transactions of the Royal Society of London. Series B, Biological Sciences* 363 (1499): 1939–49. http://dx.doi.org/10.1098/rstb.2008.0001. Medline:18292067
Stout, D., N. Toth, K. D. Schick, J. Stout, and G. Hutchins. 2000. "Stone Tool-Making and Brain Activation: Position Emission Tomography (PET) Studies." *Journal of Archaeological Science* 27 (12): 1215–23. http://dx.doi.org/10.1006/jasc.2000.0595.
Toth, N. 1982. *The Stone Age Technology of Early Hominids at Koobi Fora, Kenya: An Experimental Approach, Anthropology*. Berkeley: University of California.
Toth, N. 1985. "Oldowan Reassessed: A Close Look at Early Stone Artifacts." *Journal of Archaeological Science* 12 (2): 101–20. http://dx.doi.org/10.1016/0305-4403(85)90056-1.
Toth, N. 1987. "Behavioral Inferences from Early Stone Artifact Assemblages: An Experimental Model." *Journal of Human Evolution* 16 (7-8): 763–87. http://dx.doi.org/10.1016/0047-2484(87)90023-6.
Toth, N., K. D. Schick, E. S. Savage-Rumbaugh, R. A. Sevcik, and D. M. Rumbaugh. 1993. "*Pan* the Tool-Maker: Investigations into the Stone Tool-Making and Tool-Using Capabilities of a Bonobo (*Pan paniscus*)." *Journal of Archaeological Science* 20 (1): 81–91. http://dx.doi.org/10.1006/jasc.1993.1006.
White, T. E. 1953. "A Method of Calculating the Dietary Percentage of Various Food Animals Utilized by Aboriginal Peoples." *American Antiquity* 18 (4): 396–8. http://dx.doi.org/10.2307/277116.
Wood, B., and B. G. Richmond. 2000. "Human Evolution: Taxonomy and Paleobiology." *Journal of Anatomy* 197 (1): 19–60. http://dx.doi.org/10.1046/j.1469-7580.2000.19710019.x. Medline:10999270.

PART 3

Analogies and Models

10

Plants and Protopeople: Paleobotanical Reconstruction and Early Hominin Ecology

Jeanne Sept

Vegetation was a core component of the early hominin landscape, whether providing hominins with staple plant foods, shade, and arboreal refuge, or creating habitats and hiding places for their predators. Trying to reconstruct and understand the paleoecological relationships between hominin species and the plant communities in which they lived is a fascinating scientific challenge because it is dependent on the integration of so many types of modern and ancient data. It is also challenging because direct fossil evidence of ancient plants and vegetation patterns is relatively rare, compared to the vertebrate or invertebrate fossil records; thus many of our reconstructions of ancient vegetation patterns are based on indirect analogies with modern contexts.

While the fossil record suggests that plant taxa have evolved significantly through the Cenozoic, relatively few morphological changes are apparent in individual plant taxa during the last 7 myr (million years) of hominin evolution in Africa. This allows paleobotanists to use morphological comparisons to identify ancient macro- and microbotanical remains with relative precision. It also allows paleobotanists to make uniformitarian assumptions that the basic physiological processes and ecological principles fundamental to plants today can be applied to interpret the paleobotanical record. It thus forms the basis for our ability to understand plant variety, abundance, and distribution in the past.

DOI: 10.5876/9781607322252:c10

For example, when considering evidence for the plant-food diet of early hominins, we can reasonably assume that the fruit of an ancient fig would have had similar nutritional properties to the fruit of closely related modern figs, and would have been eaten when available by hominins, just as living primates and people commonly feed on raw figs today (Peters and O'Brien 1981). We can make comparable inferences about other nutritional attributes of plant taxa that hominins would have encountered. Considering questions of habitat on a larger scale, while it is clear that modern vegetation patterns cannot be used as direct analogs for ancient environments, an understanding of the ecological structure and diversity of modern vegetation types in relation to local climate and soils allows us to develop hypotheses about the attributes of vegetation that would have grown under comparable conditions in the past. Such models of vegetation composition and physiognomy then can be tested against different types of direct and indirect paleobotanical evidence.

BACKGROUND TO PLANT STUDIES IN AFRICA

Vegetation Patterns

Before considering the different methods available for reconstructing ancient plant communities and paleoecological relationships, it is useful to quickly review the current vegetation patterns in Africa. Africa is such a large continent, straddling the equator and stretching to temperate regions in both the northern and southern hemispheres, that its vegetation includes a number of distinct vegetation types and floral zones (White 1983). Vegetation is shaped by the intersection of climatic conditions with the regional and local landforms, particularly the topography and soil chemistry and structure.

The bulk of the African continent is characterized by low-altitude, gentle topography. The only significant highlands occur as ancient, metamorphic folded mountain belts on the northern and southern tips of the continent, and the eastern volcanic highlands that flank the Rift Valley. Significant altitude creates temperature gradients in the highlands, resulting in strong vertical zonation in montane vegetation types, especially in East Africa. These altitudinal vegetation zones range from moist evergreen and deciduous forests on the flanks of the mountains, through floristically distinctive belts of montane forest, including bamboo, to zones of ericaceous (plants requiring acidic soil) heath and moorland on the highest peaks above the treeline, such as the Ruwenzoris, Mount Kenya, and Mount Kilimanjaro.

Climate patterns produce dramatic differences in rainfall across Africa. The tropical air masses track from west to east, bringing moist equatorial air from the Atlantic and producing considerable rainfall across the center of the continent, up to the margins of the eastern highlands. As the westerlies rise over the highlands, they release most of their remaining moisture on the western flanks of the mountains, creating a rain shadow in the Rift Valley to the east. Shifting in latitude either north or south from the equator, the climate is drier. In the mid-latitudes the equatorial air converges with easterly prevailing winds, creating an Intertropical Convergence Zone (ITCZ). Storms are generated along the ITCZ in both hemispheres, and the path of the ITCZ shifts north and south annually, creating the seasonal rainfall patterns. Much of the continent experiences a single rainy season but in East Africa monsoons off the Indian Ocean contribute to a biseasonal rainfall pattern. In combination with high rates of seasonal evapotransipiration, these broad rainfall patterns limit plant growth, and result in latitudinal belts of vegetation—humid equatorial forests flanked by belts of drier woodlands, savanna grasslands, and then deserts—regional distribution patterns that are roughly symmetrical in both hemispheres. The northern and southern edges of the continent extend into temperate latitudes and have distinctive dry-summer/winter-rainfall regimes that produce a zone of Mediterranean vegetation in north Africa and a vegetation with similar structure but unique flora in the southern Cape. While the composition and structure of local plant communities have been significantly affected by patterns of human land-use, the large-scale biogeographic patterning of the vegetation is still evident in global land-cover analyses, as illustrated in Figure 10.1 (Mayaux et al. 2004).

Some plants have adapted specialized photosynthetic pathways to cope with the ecological challenges of such tropical climates (Dawson et al. 2002; Smith and Epstein 1971). While most dicots use the Calvin cycle for photosynthesis, which initially fixes atmospheric CO_2 in three-carbon molecules, this C_3 photosynthetic process is inefficient under the intense solar radiation typical of tropical conditions. A number of plant taxa have evolved alternative photosynthetic pathways that allow them to continue growth even during the hottest tropical conditions. These C_4 plants, which include many tropical grasses, sedges, and some forbs, use a different enzyme and a four-carbon compound as a precursor to glucose synthesis; this alternative photosynthetic pathway functions faster under very hot or dry conditions, limiting water loss and photorespiration. Crassulacean acid metabolism (CAM) plants, including many succulents, limit evaporative water loss by opening their stomata at night, using an acid to initially store the CO_2, and then closing their stomata during the heat of the day while photosynthesizing the stored carbon. Because of the strong zonation in climate in Africa, the relative abundance and distribution of C_3, C_4, and CAM

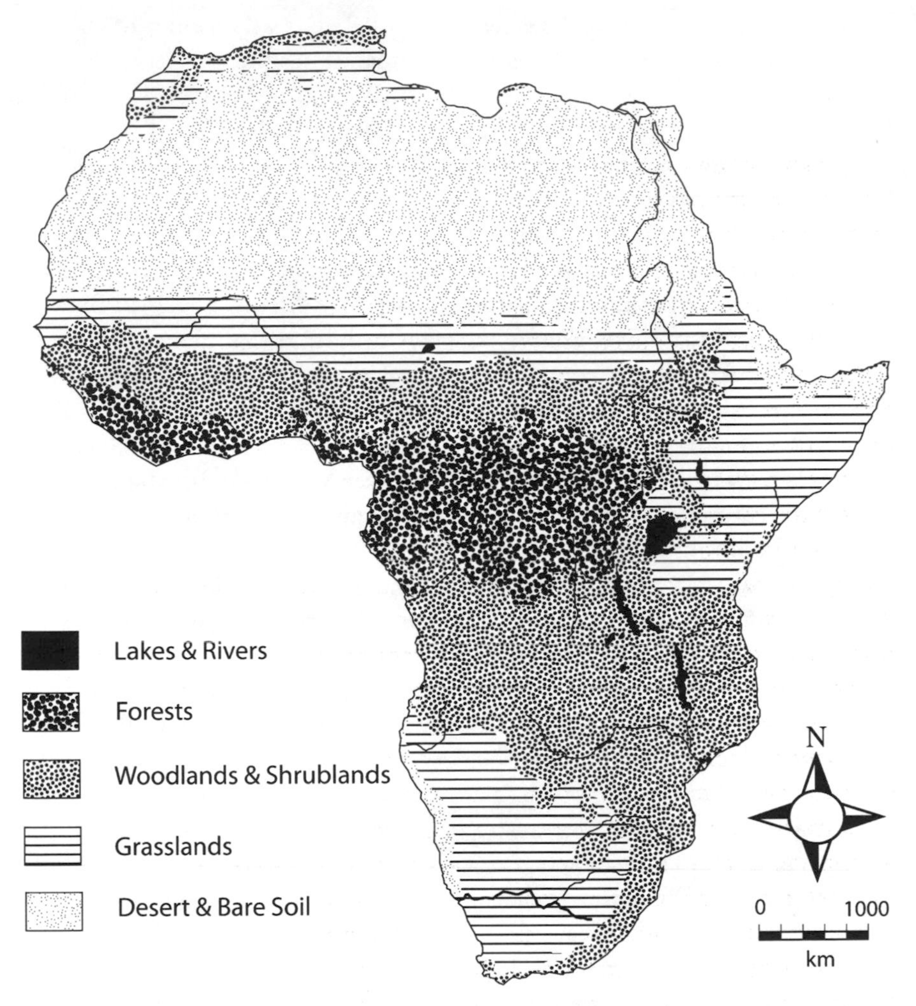

Figure 10.1. *Land-cover zones of Africa (after Mayaux et al. 2004) with land-cover classifications grouped into broad categories of Forest (including montane and submontane forest, closed evergreen and deciduous forests in lowlands, swamp and mangrove forests, and mosaic forest/croplands); Woodlands (including deciduous woodlands such as miombo, both open and closed shrublands, and croplands with open woody vegetation and tree crops); Grasslands (including open and woody grasslands and croplands); and Bare Soil (including deserts, bare rock, and salt pans).*

plants in different African habitats is also strongly patterned with vegetation type. For example, in East Africa, most grasses below 2500 m of elevation are C_4, with the exception of grasses in moist or understory habitats (Tieszen et al.

1979; Livingstone and Clayton 1980), while over 65 percent of the sedge species (Cyperaceae) in Kenya are C_4 (Hesla et al. 1982). In contrast, in South Africa, where the temperate environments support a wider distribution of C_3 grasses, C_3 sedges are dominant (Stock et al. 2004) and sometimes locally very dense (Codron et al. 2005; Sponheimer et al. 2005).

Within these broad climatic zones, the porosity and chemistry of the soils also influence the vegetation structure; this is particularly evident in the semi-arid tropics, which have highly seasonal rainfall patterns and high evaporation rates. In sandy soils, rain can penetrate deeply, and such moisture conditions can favor the growth of large trees and other perennials with deep roots. Fine-grained or compacted soils, on the other hand, cannot absorb rainfall as quickly, often resulting in surface runoff or rapid evaporation loss and the development of hardpans or relatively impermeable layers of carbonate (*caliche*); such edaphic conditions can favor plants with shallow, extensive root systems and/or salt-tolerant taxa. Edaphic variation is apparent both at local and regional scales.

Local, microhabitat variation in response to soil conditions exists in any region, mediated by local topography and groundwater sources. For example, the deep root systems of many tree species prefer well-drained soils, so riparian woodlands often develop along the high banks and sandy overbank deposits along rivers and streams, and on slopes around lakes (Haslam 1978). Poorly drained soils in the fine-grained floodplains around lakes and rivers, however, are generally dominated by either herbaceous cover, or small, perennial dicots commonly called *forbs*. Low-lying silty and clay-rich floodplains near large rivers or along lake margins can be quite barren, in part because the unstable sedimentary conditions make it difficult for perennial plants to become established and favor annuals with a "weedy" or colonizing habit (Carr 1976). Seasonally inundated lowlands, or perennial swamps, also maintain edaphic grasslands, or patches of sedges or other monocots.

Edaphic conditions can also influence vegetation structure and floristic patterns at a regional scale. For example, large areas of southeastern Africa have relatively porous soil catenas that have developed from ancient acidic bedrock, such as granites and sandstone (Lind and Morrison 1974). Such soil conditions encourage woody growth, and when these regions receive intermediate rainfall (500–1200 mm) the characteristic vegetation is a deciduous woodland known as *miombo* (derived from the local name for *Brachystegia*, one of the genera of trees characteristic of these woodlands). However, regions of East Africa such as the Serengeti plains, which have comparable climate but richer, volcanic soils, are dominated by edaphic grasslands with relatively few trees. In the eastern Serengeti, for example, grasslands dominate on shallow, alkaline soils that

have evolved a specialized stomach that enables microbial activity both to ferment cellulose and detoxify some secondary compounds; living African apes have guts adapted to processing relatively large volumes of fiber, and often use behavioral strategies, such as tool use, to work around "antinutrient" properties of plants to acquire high-quality foods (Milton 1984; Milton 1987; Wrangham et al. 1992; Milton 2006; Lambert 2007). Stahl noted that the phytochemical burdens of a plant-food diet would have also constrained early hominins until fire was controlled and used for cooking (Stahl 1984). Despite debates about the adaptive value of cooking in human evolution, evidence for cooking by Early Stone Age hominins is still limited (Wrangham et al. 1999; Wrangham and Conklin-Brittain 2003; Wrangham 2007). In addition to cooking, people have traditionally used a variety of technical strategies such such as pounding, soaking, or grinding to improve palatability of wild plant foods. Some Oldowan and Acheulian tools could have been used in this fashion (Goren-Inbar et al. 2002a).

Primate studies have documented how the abundance of food in time and space can affect primate subsistence behaviors. Important variables to consider include both the density and patchiness of the plant distribution in different habitats, the relative productivity of the plants across different habitats, and the seasonal availability and predictability of the foods. Documenting such variation in primate habitats has proven time consuming. However, long-term field studies such as Altmann's research on baboon populations in Amboseli Park, Kenya, or the work by Wrangham's team on chimpanzee feeding in Kibale Forest, Uganda, have provided a basis for understanding variables such as energetics, nutrient balance, and social value as alternative currencies that constrain food choice. Detailed field studies such as these provide baseline data for evaluating the predictions of different models of ranging and foraging behavior, applied to primates (Norton et al. 1987; Altmann 1988; Wrangham et al. 1992; Barton et al. 1992; Chapman and Chapman 2000).

METHODOLOGICAL HISTORY

Vegetation Survey Research Methods

Over the years plant ecologists have debated the value of different approaches to vegetation description and classification, ranging from the theoretical perspective that distinct assemblages of plants recur, and can be identified as plant communities, to the more individualistic view that emphasizes dynamic variation—that populations of plant species are distributed across environmental gradients and that classification schemes impose artificial boundaries on meaningful spatiotemporal ecological continua (Kent and Coker 1992). Many phytosociological methods have been developed to recognize and define plant communities including floristic composition (focusing on species composition and relative abundance), physiognomic classifications (focused on structural and functional attributes), and various ordination methods. Early approaches, such as the Braun–Blanquet method of the Zurich–Montpelier school, relied on subjective classifications of species absence or abundance, defining vegetation associations based on samples (*relevés*) thought to be typical or representative of each plant community. Subsequent computational techniques facilitated more objective classifications of sampled species abundance (e.g., collected from measured plots or transects), using a wide variety of statistical and quantitative clustering techniques (Greig-Smith 1983; Gibson 2002). The design of the sampling methodology (e.g., plot size, shape, and distribution) varies with the scale and scope of the study; quantifying herbaceous groundcover requires a different sampling strategy than measuring forest canopy, for example. Analytical methods chosen depend upon the research goals.

Advances in remote-sensing technology have created new tools for the evaluation, classification, and mapping of large-scale vegetation zones across the continent. A new generation of sensors on satellites have allowed improved spatial and temporal resolution, compared to earlier sensors, and allowed ecologists to develop a variety of measures of land cover as well as comparative vegetation indices. One of the simplest and most widely used is the Normalized Difference Vegetation Index (NDVI), or "greenness index," which is particularly useful for monitoring land cover and land-use changes, differences in primary productivity, and seasonal growth patterns (Box et al. 1989; Mayaux et al. 2004). Callibrating remotely sensed data with "ground truth"—nested, geo-referenced vegetation sampling—is proving to be an effective strategy for monitoring vegetation patterning at a variety of scales useful to human and nonhuman-primate ecology (Moran, Skole, and Turner 2004; Lu 2005a, 2005b; Vogt et al. 2006). Even medium-resolution sensors, such as the MODIS (Hansen et al. 2005), can yield valuable land-cover data for calibrating the interpretation of microbotanical samples. Recent methodological advances have allowed the integration of multidisciplinary, high-resolution satellite data from several different sensors to produce a comprehensive, spatially detailed data set and map of land cover across Africa in the year 2000 (Mayaux et al. 2004).

Whether collecting comparative ethnobotanical information about the uses of wild plants today, or doing an on-the-ground vegetation survey, a fundamental first step is learning how to systematically collect plant samples for taxonomic identification, and the creation of botanical voucher specimens for subsequent curation in an herbarium. This step not only adds scientific credibility to the work, but is also a professional courtesy that facilitates future botanical research by others. Specimens are collected by location and date, with as much information as possible recorded about the morphology of the plant, along with additional field notes about the ecological context and local names and ethnographic uses noted by informants. The goal for a good-quality herbarium specimen should be to collect a representative sample of the plant, including flowers, fruits, stems, leaves (and even roots of smaller plants). In the field, plant samples are flattened and dried between sheets of newspaper, using a plant press, although drying can be a challenge in the humid tropics. Subsequently, if the specimen is curated by a herbarium, it will be mounted on a stiff card with a label containing all relevant information. To collect samples for phytochemical analysis (of nutrients, or secondary compounds, for example), it is important first to record the fresh weight of the plant part you are sampling, and then to preserve it in a way to facilitate subsequent laboratory analysis. For some analyses, quickly drying the samples is sufficient. For other analyses freezing the specimens, or preserving them in alcohol is preferred. Simple phytochemical screening tests on fresh specimens can be done in the field with portable kits. *Ethnobotany: A Methods Manual* (Martin 1995) is a good, practical guide to details about sampling and collecting methodology.

Paleobotanical research methods

As in all branches of paleontology, paleobotanical studies have shifted over the last fifty years from a focus on the basic recovery and description of direct evidence of ancient flora to interpreting the available data within the context of broader, taphonomic analyses. Jacobs provides a very useful overview of the long-term record of palaeobotanical studies in Africa (Jacobs 2004). Technological developments have paced the recovery techniques and quantitative analyses of more varied and detailed forms of evidence, but the basic analytical foci have remained the same: microbotanical remains, macrobotanical remains, and indirect or contextual evidence of the impact of plants on other aspects of the ancient environment. Still, much basic taphonomic and actualistic work needs to be done to strengthen the theoretical framework for interpreting different types of evidence.

Microbotanical evidence: pollen

Pollen grains, which contain the male genetic material of plants, are housed within a tough, carbon-based microscopic shell, or exine. Unlike the organic components of pollen grains, which decompose quickly, the exines are preserved under a wide variety of sedimentary conditions, and are frequently abundant in marine, lacustrine, and riverine sediments, dry cave sediments, or other nonalkaline sediments that have avoided extremely oxidizing or reducing conditions. Indeed, pollen grains are so durable that they are extracted from ancient sediment samples by washing the fine fraction with a series of acid baths and other solutions, which ultimately dissolve most of the other particles in the samples, leaving pollen and spores as a residue for analysis. Since pollen are produced in abundance by many plant species, and most taxonomic groups of plants have distinctive pollen-grain morphologies, fossil pollen have great potential as records of ancient flora. While the interpretation of the relative abundance of different pollen types recovered in samples is complicated by a number of factors related to the pollen-dispersal strategies of the plants, and other taphonomic variables, analytical progress has allowed the study of fossil pollen (*palynology*) to play an increasingly important role both in reconstructing changing vegetation patterns and in investigating patterns of climate change.

Pollen analysis has a history that stretches back over two centuries; however, modern paleoenvironmental pollen analysis did not begin until the 1940s and 1950s (Boyd and Hall 1998). Pollen studies for the reconstruction of vegetation history in sub-Saharan Africa began in the 1950s, with pioneering work on late Pleistocene and Holocene sediments (van Zinderen Bakker 1957; van Zinderen Bakker 1962; van Zinderen Bakker 1967). These studies used broad biogeographical patterns and uniformitarian principles of "ecological actuality" (van Zinderen Bakker 1967) to interpret the composition of the recovered pollen samples. Several of these early studies focused on reconstructing the vegetation around particular paleoanthropological sites, while others aimed to document shifting vegetation zones and climate change.

For example, van Zinderen Bakker (1957) sampled layers of peaty clay from the early *Homo sapiens* site of Florisbad, in South Africa, finding pollen from taxa that suggested that during Middle Stone Age times the site was surrounded by a scrub vegetation, comparable to southern Africa's semiarid interior "Karoo," rather than the lush grasslands present in the region today. The first paleobotanical research focused on an Early Stone Age site was done at Kalambo Falls, Zambia (Clark and van Zinderen Bakker 1964), where Acheulean assemblages in waterlogged sediments were associated with both pollen and macrobotanical fossils. While determining the age of the assemblages proved problematic because they were beyond the range of radiocarbon dating, the pollen demonstrated that the vegetation at the site during the Acheulean occupations

understanding of the sedimentological context of the ancient samples. The development and application of these methodological advances in support of studies of early hominin evolution was pioneered by Raymonde Bonnefille, whose initial collaborations with Maurice Taieb, Jean Chavaillon, Clark Howell, and others in multidisciplinary explorations of Plio-Pleistocene deposits in Ethiopia (Bonnefille 1976), has founded much of our contemporary understanding of the vegetation context of early hominin evolution in Africa (see below). The power of Bonnefille's approach has made it clear that pollen analysis is not merely a technical skill and should not be decontextualized in a lab—it is best done by fieldworkers who not only understand the ecological relationships of contemporary plant communities as a context for the formation of modern pollen assemblages, but who also understand the geomorphology of ancient sediments, so that pollen samples can be collected with an eye to the variables important for their interpretation. This is particularly important for interpreting the palynological record associated with early hominin sites in East Africa, where researchers are challenged to distinguish the relative influence of global climatic changes compared with the local or regional effects of tectonic activity on ancient vegetation patterns (Vincens et al. 2006).

Microbotanical evidence: phytoliths

Both silica and calcium accumulate within and between the cells of many plants, from parts of the roots, stems, wood, and leaves, to the fruits and seeds. In some cases these biogenic minerals form amorphous clumps, but in others they assume the microscopic shape of the tissues where they precipitate, creating a variety of forms of phytolith within a single plant that persist after the plant tissue decomposes. Opal (silica-based) phytoliths are resistant to oxidation, and thus are preserved in a wide range of sedimentary conditions, although they can suffer from diagenesis in acidic conditions. Their potential for paleoecological research has been recognized since the 1960s (Rovner 1971), and now researchers commonly isolate phytoliths from sediment samples with heavy-liquid extraction techniques. Opal phytoliths are particularly abundant, diverse, and distinctive in grasses and other monocots, but are produced by a number of tropical dicot species as well. However, phytoliths are polymorphic and show considerable variation within and between individual plants because multiple phytolith forms are produced within an individual plant and redundant phytolith forms produced across many taxa. Their morphotypes thus have limited taxonomic use (cf. Mulholland and Rapp 1992). So far phytolith assemblages have been most effective at providing evidence for broad structural groupings of plants (e.g., tropical forest versus savanna vegetation; woody-vegetation cover versus more-open cover). Taxonomically they can be used to distinguish the presence of various monocot families and subfamilies in fossil assemblages (e.g., grasses, sedges, and palms), and can sometimes be used as an indicator of the relative abundance of C_3 and C_4 plants as well.

Compared to pollen, phytoliths are preserved in a wider variety of sedimentary conditions but have more-limited taxonomic value; they best complement pollen studies by providing data on subfamilies of grass, which cannot be distinguished through pollen analysis. However, because phytolith research has developed much more recently than palynology, a wide range of both morphological and taphonomic studies remain to be done.

Patricia Palmer began paleoecological phytolith research in East Africa with studies of grass cuticles recovered from lake sediments (Palmer 1976), later developing an extensive reference collection of the micromorphology of the epidermis of modern East African grasses to aid further research. However, while phytolith systematics and paleoecological analysis have developed through the work of Piperno and others in the Americas (Piperno 1988; Fredlund and Tieszen 1994; Strömberg 2004), only recently have researchers begun to focus on the application of such research in Africa (Barboni and Bremond 2009; Mercader at al. 2009).

Some studies have evaluated the phytolithic signatures of different African biomes. Bremond, Alexandre, and others have examined how phytolith assemblages can help trace the shifting forest–grassland boundary (Alexandre et al. 1997), using indices of different morphotypes associated with broad vegetation groupings today. In particular, they propose that an index of tree-cover density can be established by calculating a D/P ratio of phytoliths typical of woody dicots (D) compared to Poaceae (P) grass phytoliths (Bremond et al. 2005). Runge researched phytolith-assemblage composition in the soils of lowland equatorial forests, trying to understand taphonomic processes, since phytoliths can be transported by both wind and water over long distances, and are also subject to mixing in soils through bioturbation (Runge 1999). Runge's foundational work on phytolith classification and modern sampling has been followed by work with Mercader on Holocene archaeological sites in the tropical forest (Mercader et al. 2000). Other Holocene archaeological applications include phytolith evidence for the timing of the introduction of bananas into Africa (Lejju et al. 2006), and analyses of the spatial distribution of phytolith across living floors to identify dung concentrations and aspects of site structure at early pastoralist sites in East Africa (Shahack-Gross et al. 2004). The application of phytolith research to early hominin sites, particularly those where pollen has not been preserved, is only just beginning (Barboni et al. 1999; Albert et al. 2006).

Phytolith analyses are also playing a growing role in paleoclimate reconstructions. The long sedimentary sequences recovered from deep-sea cores contain a variety of proxy indicators of climate, including fossils of both marine organisms, such as plankton and algae, that have settled to the bottom of the ocean, and organisms from terrestrial environments, including freshwater diatoms, pollen, and phytoliths, that have been blown or washed in with the dust from nearby continents. For example, phytoliths from a core off the west coast of equatorial Africa have been analyzed to track climate changes over the last 340 kyr (thousand years) (Abrantes 2003). The total volume of phytoliths being transported into the ocean is interpreted as an indicator of changing wind strength, a proxy for variations in the moisture budget of the Sahel region, as the winter tradewinds blow plumes of dust from the dry interior southwest into the Gulf of Guinea. The composition of the phytolith assemblages can also be used to distinguish varying proportions of C_3 and C_4 grasses in the region. Today C_4 grasses dominate the hot, dry savannas and Sahel vegetation zones in Africa, and these have distinctive phytolith morphotypes. Because the abundance of different grasses varies with the strong latitudinal zonation in climate and vegetation across west Africa, phytolith analysis has been used to track shifting boundaries between drier and more humid grasslands in the region (Alexandre et al. 1997), or as a proxy of tree-cover density (Bremond et al. 2005).

Barboni et al. (2007) reviewed 149 comparative modern phytolith assemblages from terrestrial samples in Africa, to test the potential of phytolith assemblages to discriminate between vegetation types. They examined associations between different phytolith types in different phytogeographical zones and estimates of land cover derived from MODIS satellite NDVI data (Hansen et al. 2005). While they recommend refinement in the methods currently used to collect and analyze phytolith assemblages, as well as additional research on the formation processes of such assemblages, they were able to identify a number of promising multivariate associations between average land cover and the incidence of various phytolith types. They found that the relative abundance of various types of globular phytoliths may be more reliable indicators of tree cover in many zones than are frequencies of arboreal pollen.

Macrobotanical evidence

Plant macrofossils are less common in African sediments than either pollen or phytoliths. However, when they have been recovered they provide a valuable glimpse of the plant communities that grew near an ancient site, complementing the microbotanical record. Impressions or casts of wood, fruits, seeds, leaves, roots, and other structural parts of plants have been occasionally preserved. In general, macrobotanical remains are assumed to form autochthonous assemblages, derived from local plant communities and unlikely to have been transported far before fossilization. This common taphonomic assumption can be explicitly evaluated through the analysis of the precise spatial relationship and orientation of individual macrobotanical fossils within an assemblage (Jacobs and Winkler 1992).

Some of the macrofossils have taxonomically diagnostic morphology. Genera of plants have been identified from fossil flowers and fruit at a number of sites in Africa, particularly at sites from the Miocene or more recent deposits (Chesters 1957; Hamilton 1968; Bonnefille and Letouzey 1976; Dechamps et al. 1992). Fossil wood can be identified to particular families or genera, identifying associations of riverine forest, for example (Dechamps and Maes 1985), in addition to preserving evidence of fluctuating conditions of growth and occasional trauma (e.g., lesions probably caused by bush fires). Leaf morphology, on the other hand, can be largely independent of taxonomy. Assemblages of leaves, or undiagnostic fragments of wood, can be compared in size and morphology to contemporary taxa, and help diagnose the presence of different types of vegetation, differentiating moist forest from dry woodland or savanna, for example (Jacobs and Kabuye 1987; Kingston et al. 2002), or documenting the presence of grasses (Retallack et al. 1990).

Jacobs (2004) has summarized the contributions macrobotanical evidence has made to arguments about the development of the current, angiosperm-dominated forest, woodland, and savanna biomes in Africa during the Cenozoic. But macrobotanical evidence is often so rare or isolated that it is used mainly to test site-specific paleobotanical hypotheses. The identification of fossil *Antrocaryon* fruit found in Pliocene deposits at the Omo site complex is an example (Bonnefille and Letouzey 1976); the likelihood that this fruit represents a riverine forest tree is supported by evidence from sedimentological and vertebrate microfossils, as well as pollen data. Similarly, out of 300 specimens of fossil wood collected from Member 4 breccia at Sterkfontein (Bamford 1999), two genera were identified: a liana typical of Central African gallery forests today (*Dichapetalum*), and a moist forest-margin shrub (*Anastrabe*). These woody taxa, coupled with the presence of colobus monkeys, suggested to Bamford that gallery forest was present near the site during Member 4 times, although other micro- and macrofaunal records from the site have been interpreted as evidence of the proximity of everything from grasslands to bushland thicket and open woodlands. But beyond noting that gallery forests can be found in a variety of phytogeographic zones today, it is difficult to evaluate the paleoecological meaning of the juxtaposition of the fossil wood and open country bovids and rodents associated with australopithecines in this fairly diachronic cave deposit.

Indirect evidence of ancient vegetation patterns

Where direct paleobotanical evidence is missing from a paleontological or archaeological site, there are a number of proxy indicators that can be used to support vegetation reconstruction. For example, deep-sea cores recovered off the African coast can contain not only stratified evidence of marine environments but also a record of neighboring terrestrial environments. In addition to pollen and phytoliths, transported off the land by wind and water currents, terrestrial dust and other particulates (including freshwater diatoms) can be transported great distances to become incorporated into the ocean-floor sedimentary record. Paleoclimatologists generally assume that periods of continental aridity—which expose shallow lakebeds, reduce vegetation cover, and are characterized by more vigorous winds—result in larger volumes of dust entering the atmosphere to contribute to deep-sea sediments. In some regions dust volume correlates with other climatic proxies, such as oxygen isotopes, and can be interpreted as a record of changing climate (deMenocal 1995; deMenocal 2004). In West Africa, for example, the seasonal dust plumes over the eastern Atlantic are legendary, and taphonomic studies have shown that seasonal and annual variations in the moisture budget across the Sahel, as far inland as the Chad Basin, correlate with this dust supply. Though variable in detail, overall the dust record seems to correlate well with a microbotanical record that tracks the north–south shifts in the boundaries of the Saharan, Sahelian, and savanna woodland biomes during the last 6 myr (Leroy and Dupont 1994; Abrantes 2003).

Another indirect link between vegetation and the sedimentary record can be traced through the stable-isotopic signature of sediments and paleosols. The key is the differential photosynthetic pathways of plants, described earlier (Dawson et al. 2002). Atmospheric carbon dioxide includes carbon isotopes of different weights, most of which are the stable isotopes ^{12}C and ^{13}C. Further, C_4 and CAM plants are more likely to fix carbon dioxide composed of heavier isotopes of carbon ($^{13}CO_2$ rather than $^{12}CO_2$) than are C_3 plants. The tissues of the living plants reflect this chemical difference in their $^{13}C/^{12}C$ ratios; when compared with the relative isotopic concentrations in a standard, C_3 plant tissues have an average $\delta^{13}C$ value of –27‰ (a difference of 27 parts per thousand or "permil" less than the standard) and C_4 plants have an average $\delta^{13}C$ value of –12‰. When plants decompose, these differences become incorporated into the isotopic signatures of the organic-carbon and soil-carbonate profiles. Although some fractionation occurs at each chemical step in the process, ultimately soils where more C_4 plants grew will have a relatively heavier (e.g., less negative) carbon isotopic signature. This method was applied to reconstructions of African vegetation from fossil soils first by Thure Cerling and colleagues, who analyzed first the carbonates from hominin sites at Olduvai Gorge and Koobi Fora (Cerling et al. 1988; Cerling and Hay 1986) and then from earlier Miocene sediments in eastern Africa. He found a positive shift in the isotopic signature occurring around 8–9 mya (million years ago) (Cerling 1992), and argued that this indicated the spread of tropical C_4 grasslands across eastern Africa at this time. Since these formative studies, the method has undergone considerable development and a number of applications to paleoenvironmental reconstruction at early hominin sites in eastern and southen Africa, as described in selected case studies, below, and in other contributions to this volume. Similarly, because animals "are what they eat," stable-isotope analysis of fossil-tooth enamel from hominins and other vertebrates has proven to be a very useful way to study the paleoenvironmental context of the evolution of hominin diet (Lee-Thorp and Sponheimer 2006), as described by Julia Lee-Thorp, Matt Sponheimer, and others in chapters in this volume. Methods to analyze the stable isotopes of carbon, nitrogen, and oxygen, have all been developed and are beginning to yield very consistent and replicable results, in part because of the taphonomic studies documenting the variation in isotopic composition between individual plants and the animals that eat them.

Actualistic studies and analogies

Taphonomic studies have become critical to the interpretation of a variety of paleocological data. As described earlier, to calibrate the relationship between the taxonomic composition of a plant community and the pollen assemblages that accumulate in nearby sediments, palynologists must do the research on modern pollen rain (Gajewski et al. 2002), and comparable taphonomic studies are beginning for phytoliths as well (Albert et al. 2006). Similarly, before it is possible to interpret carbon isotopes in paleosols as indicators of paleovegetation, one must first establish baseline data on how the mix of C_3 and C_4 plants on contemporary landscapes contribute to the variable chemical signatures of the soils they grow in (Codron et al. 2005; Sponheimer et al. 2005).

A related challenge faced by paleoanthropologists is that the questions we ask about modern vegetation and its relationship to ancient vegetation are generally not the same questions that interest botanists and plant ecologists. While many paleoanthropologists have developed hypotheses about the relative abundance of different types of plant foods in different paleohabitats, most vegetation description and analysis done by plant ecologists has focused on vegetation classification, phytosociology, or quantitative plant ecology, as described earlier. A search of the botanical literature yields few studies in Africa of the habitat-specific availability of different plant foods. Ecological studies have been done on forage quality for ungulates (both wild and domestic), but these have little direct bearing on hominin dietary choices. Primatologists have collected some very useful data on the availability and nutritional quality of foods eaten by primates at selected study sites. However, while ethnographic studies of modern foragers have commonly created inventories of the wide range of food types eaten by hunter–gatherers (Lee 1979; Kelly 1995), few studies have collected quantitative field data on the availability (frequency, density, or distribution) or variable quality of these foods across different habitats. Thus, paleoanthropologists interested in studying the relationships between ancient vegetation type and likely plant food availability need to establish such relationships empirically themselves. Such actualistic studies focused on plant-food analogs, initially undertaken by Peters in South Africa (Peters and Maguire 1981), and Sept (Sept 1984, 1986, 1990, 1992) and Vincent (Vincent 1985a; Vincent 1985b) in East Africa, have formed the foundation for subsequent analyses and paleoecological modeling (Sept 2001, 2007; Copeland 2007; Griffith et al. 2010).

DISCUSSION: CURRENT PALEOBOTANICAL APPLICATIONS IN HOMININ PALEOECOLOGY

Efforts to evaluate the relationship between early hominins and the plant communities they lived in have been undertaken at different temporal and geographical resolutions, ranging from generalized reconstructions to site-specific habitat variations. For many years the only direct evidence for vegetation at early hominin sites was that provided by pollen analysis, but pollen samples are rarely recovered from the same stratigraphic horizons and localities as hominin fossils or archaeological sites, and thus are most commonly interpreted as evidence of regional vegetation patterns, rather than as precise indicators of early hominin habitats. These efforts have been led by Raymonde Bonnefille in East Africa, who recovered pollen samples from both *Australopithecus* and early *Homo* sites at Olduvai and Laetoli in Tanzania, from sites in the Omo–Turkana Basin in northern Kenya and southern Ethiopia, and from Melka Kunture, Gadeb, and Hadar in Ethiopia (Bonnefille 1995). While each of these pollen spectra preserved both local and regional signals of vegetation at the sampled horizons, her main analytical focus has been to reconstruct the generalized vegetation setting and biome of the hominins—to investigate macroevolutionary trends in hominin–vegetation–climate relationships rather than microhabitat variation and paleoecological questions. This type of palynological research has benefited both from increasingly high-resolution fossil records and from the application of more quantitative analysis and formal modeling. Overall she concluded that the East Africa pollen record between 4 and 1 mya reflected vegetation changes in response to global shifts in paleoclimate documented by other proxy indicators. Before 3.2 mya, East African vegetation included more widespread and diverse woodlands. Increasingly, after 2.4 mya, mosaics of wooded and open grasslands would have been prominent in the lowland landscapes, particularly in response to increased aridity after 1.8 mya, but the pollen data document notable diversity in fluctuating vegetation patterns, rather than clear trends.

The following sites provide brief case studies of the interpretive challenges that exist at the other end of the temporal and geographical spectrum when attempting to integrate different sources of direct and indirect paleobotanical information for more localized or site-specific interpretations of hominin habitats. They have been chosen to illustrate hominin fossil or Early Stone Age archaeological sites where several types of paleobotanical data have been compared to provide a framework for arguments about early hominin ecology.

Case study: Hadar

Palynologists face challenges when trying to determine whether a regional record of vegetation change is due to global (climatic) changes or to variability in local conditions (e.g., tectonic activity). This is illustrated by the pollen research at the site of Hadar in the Afar region of Ethiopia. Numerous fossil specimens of *Australopithecus afarensis* have been recovered from a well-calibrated stratigraphic sequence at Hadar, dating to between 3.4 and 2.9 mya.

At Hadar, Bonnefille initially collected over 206 sediment samples throughout the stratigraphic sequence at Hadar, but only 27 produced pollen spectra (Bonnefille et al. 1987). This is not unusual for Plio-Pleistocene sediments, where pollen preservation and recovery is the exception rather than the rule. The best pollen spectra from Hadar were recovered from fine-grained, somewhat acidic lacustrine sediments, but they were unevenly distributed, chronologically and paleogeographically. Overall, 116 different taxa were initially identified among the fossil pollen, including arboreal taxa commonly associated with with East African biomes today, such as montane forest (*Podocarpus*, *Juniperus*,

Olea, and *Hagenia*) and semiarid bushland and grassy woodlands (*Acacia*, *Euclea*, *Commiphora*), and several taxa that are not present in the Ethiopian flora today. Some of the samples have high representations of insect-pollinated trees, such as relatively high frequencies of *Garcinia* and *Alangium*, suggesting local presence of forest or woodland. While all the samples were dominated either by grasses or by sedge and reed pollen in the lake-margin samples, few taxa typical of semiarid steppe conditions (such as "saltbush" Chenopodiaceae/Amaranthaceae) were represented in the assemblage. Although significant fluctuations in land cover occurred during the sampled interval, overall Bonnefille originally suggested that the Pliocene vegetation at Hadar had more in common with upland semi-evergreen bushlands than it did with the lowland semidesert plant communities around the site today. One possible explanation for this difference could be tectonic activity—extensive down-faulting in the region during the subsequent Pleistocene could have lowered the local basin elevation by as much as 1000 meters. Alternatively, more-humid climatic conditions might have caused the vegetation changes documented at the site.

Since this initial work at Hadar, the relationship between modern pollen rain and climatic parameters has been studied more extensively, and a new, multivariate methodology is being developed that helps infer biome reconstructions from the association of pollen taxa, grouped by plant functional types (Jolly et al. 1998; Gajewski et al. 2002). Applied to ancient samples, similarity indices can be developed between a given fossil sample and modern pollen sample analogs, drawn from a sample taphonomic database of over 1,100 modern pollen samples. A reevaluation of the Hadar pollen data using these new comparative methods has allowed Bonnefille and her colleagues (Bonnefille et al. 2004) to refine their reconstruction of the vegetation. They characterize the site as having been located in a transition zone between relatively warm mixed forest and xerophytic cool steppe along an escarpment slope, and that a diversity of biomes would have been available to the australopithecines. Their analysis supports the original interpretation of a Pliocene Hadar with twice the precipitation of the modern region and cooler temperatures. While they note that some of this may have been due to altitudinal shifts, they maintain that the patterns of vegetation variability at the site through time also corresponds well with global records.

While fossil wood was also recovered from Hadar, it has not yet been fully described and analyzed. However, in the last few years, good paleobotanical records have been established at several early hominin archaeological sites. These Early Stone Age sites provide opportunities to compare and integrate different types of paleoecological evidence for early hominin behavior in specific paleolandscape contexts.

Case study: Olduvai Gorge

Paleoanthropology has benefited from intensive paleontological and archaeological research at Olduvai Gorge for almost half a century. Following the discovery of the famous *Zinjanthropus* skull in 1959, Louis and Mary Leakey devoted decades of their lives to fossil discovery and archaeological excavation at the sites in the gorge, working closely with Richard Hay to document the geological context of their discoveries, with particular focus on the rich record of Early Stone Age sites (Leakey 1971, 1994; Hay 1976). More recently, a multidisciplinary team led by Blumenschine (Peters and Blumenschine 1995; Blumenschine and Peters 1998) has renewed focus on reconstructing the paleolandscape context of the archaeological sites, and has developed conceptual models of hominin land use in the region to be tested against fossil and archaeological evidence from the sites. The paleobotanical record at Olduvai provides an interesting case study in the challenges of integrating different types of evidence to reconstruct ancient vegetation patterns at a site complex, and in the limitations of interpreting early hominin behavior in paleoenvironmental context. Pollen, phytoliths, and a variety of macrobotanical fossils have been recovered from different sites at Olduvai, with indirect evidence of plant communities also derived from paleosol chemistry, complemented by analogies with current vegetation patterns.

The nine pollen spectra from Olduvai (Bonnefille 1984, 1994; Bonnefille et al. 1982) are drawn from sample paleosols and lake-margin sediments from Bed I and lowermost Bed II, a time span of roughly 200 kyr (Bonnefille 1995). Compared taphonomically to the contemporary pollen rain in the region, it is clear that the Oldowan vegetation has long been part of the Sudano-Zambezian flora, comprising a mix of woodlands and grasslands in the lowlands, near the ancient lake, and fluctuating forests in the regional highlands. For analysis, Bonnefille subdivides the pollen assemblages from Olduvai into different groups. For example, pollen from water-dependent plants, such as sedges, is interpreted as locally derived, and grass pollen, which dominate all the assemblages, are assumed to be largely regional indicators. Pollen from two families of small, arid-loving plants, the Chenopodiaceae and the Amaranthaceae, are interpreted as indicators of local conditions of aridity and relatively saline soils. Rather than trying to distinguish between montane forest, woodland, or riparian forest, all arboreal pollen are treated as an index of general tree cover in the basin in order to track climatic change. This avoids the dangerous assumption that floristic associations were the same in the past as today. In general, the pollen demonstrate that the Plio-Pleistocene Olduvai Basin was characterized by fluctuating wooded grassland habitats, with evidence of trees and shrubs (such as *Acacia*, *Boscia*, *Grewia*, and *Ximenia*) characteristic of semiarid wooded savanna grasslands, as well as sedges and palms characteristic of lake-margin vegetation.

Localized phytolith assemblages have been recovered from several carefully excavated sites at Olduvai, particularly lake-margin sediments in upper Bed I and lower Bed II (Albert et al. 2006; Bamford et al. 2006). These phytolith samples offer a fine-grained spatial and temporal resolution to vegetation reconstruction that complements the regional record of the pollen assemblages. In the context of new taphonomic studies of phytolith accumulation in several East African contexts, these have been interpreted as autochthonous (locally derived) assemblages, documenting the presence of palms, and fluctuating communities of sedges, various grasses, and grasslands along the eastern margins of the large, shallow, saline–alkaline paleolake Olduvai, as well as dense patches of ground water palm forest close to freshwater springs. Fine-scale vegetation fluctuations over time and space should be expected in such unstable sedimentary environments (Western and Van Praet 1973; Carr 1976; Sept 1984), and these data seem consistent with the record of localized macrobotanical fossils recovered from associated sediments.

Fossil wood, parts of stems and twigs, rootcasts, and a variety of other fragmentary macrobotanical fossils have been recovered from different sites in lowermost Bed II at Olduvai, and some of these fossils have retained diagnostic morphology (Bamford 2005). Two fragments of fossil wood have been identified as a tree from the Caesalpiniaceae, a leguminous family characterized by trees and shrubs from a variety of tropical woodland, wooded grassland, and riparian habitats. Caesalpiniaceae are only represented by a few pollen grains in overlying sediments and cannot be identified through phytoliths, so the silicified wood contributes new floristic detail to the vegetation reconstruction of the site. The particular morphology of the specimens most closely matches the current species *Guibourtia coleosperma*, a tree with edible seed arils, common today in the sandy soils of the mixed woodlands and bushlands of southern Africa. These fossils provide a tantalizing glimpse into a moment in time—a shady tree growing near the lakeshore—and they also remind us that even as we can have confidence in how the wooded grasslands near this lake would have been generally structured, the phytogeography of particular taxa has undoubtedly changed significantly through time.

This paleobotanical record from Olduvai can be interpreted as consistent with the istotopic signatures of paleosol horizons from middle Bed I and lower Bed II. Sikes collected paleosol samples from rootmarked claystones exposed in seventeen trenches near the lake-margin localities of HWK and FLK (Sikes 1994), analyzing the isotopic composition of organic matter and pedogenic carbonate from the horizon. The Bed II paleosol samples span a depositional episode of approximately 10 kyr, and Sikes argues that they document floral microhabitats near the eastern margin of the ancient lakeshore. Organic carbon $\delta^{13}C$ values ranged from –23.7 to –19.5%, averaging –21.1%, and pedogenic carbonates averaged –4.8%. Compared with modern analogs, these chemical signatures suggest that the local plant biomass was composed of only one-fourth to one-half C_4 plants. Using the isotopic composition of soil samples from modern analogs as an interpretive guide, Sikes suggests the Olduvai paleosols formed under a relatively closed canopy plant community, ranging from patches of riparian forest to grassy woodland.

Copeland (2007) studied the availability of wild plant foods in several modern habitats in northern Tanzania chosen as partial analogs to the ancient riparian sites in lowermost Bed II at Olduvai. Combined with the implications of the paleobotanical evidence preserved at the sites, as well as associated faunal remains, her work underscores the likelihood that fluctuating lake levels at paleo-Olduvai would have affected the plant-food-foraging options near the paleolake. High-lake stands would have narrowed zones of fruit- and legume-rich riparian woodlands near the lake and flooded marshes formerly abundant with edible rootstocks. Thus it is likely that hominins would have been more dependent upon stream margins and higher ground along the basin margins for plant-food foraging during times when lake levels were high.

Case studies: Acheulean sites

Three Acheulean sites from different regions provide tantalizing samples of not only evidence of local hominin habitats, but also some direct evidence of hominin interaction with the plant communities during the early and middle Pleistocene.

Peninj. The site of Peninj, in northern Tanzania, is an early Acheulean/Developed Oldowan site complex, which preserves some palynological evidence for early hominin habitat, as well as phytoliths found within site sediments and adhering to several artifacts from the site. Pollen were recovered from a sandy horizon with interstratified clays in the upper Humbu Formation, and the ancient assemblage resembled the composition of modern assemblages in the region, dominated by grasses and sedges, with occasional pollen from savanna trees such (as *Acacia*), or woody taxa characteristic of local riparian habitats (*Rhus*, *Moraceae*, *Salvadora*) and a few allocthonous Afromontane forest taxa represented (Domínguez-Rodrigo et al. 2001a). The riparian trees and shrubs documented in the Peninj pollen assemblage commonly bear edible fruits today, suggesting that plant-food-foraging opportunities would have been available near the site. Phytoliths were recovered from the sediments and adhering to three stone tools at the site (Dominguez-Rodrigo et al. 2001b). The morphotypes of the soil phytoliths are most probably from grasses, likely C_4 grasses, and reeds, while the phytoliths extracted off the artifacts were significantly different; they were likely from dichot morphotypes commonly produced by woody taxa in the Leguminosae family, such as *Acacia* and Salvadoraceae, both of which are present in the pollen assemblage. Dominguez-Rodrigo argues that the wear patterns on the handaxes and the growth habits of *Acacia* and *Salvadora* plants suggest that the handaxes were likely used to chop or process hard *Acacia* wood. However, this type of taxonomic specificity in reconstructing ancient tool-use is difficult to test; after all, in riparian environments *Salvadora* shrubs can grow into very large trees. Yet the larger point, that hominins were using Early Stone Age technology to process or work wood found in savanna habitats, is significant and is dependent upon the direct paleobotanical evidence. This microbotanical evidence from Peninj for hominin use of plants is indirectly supported by remarkable macrobotanical assemblages preserved at two other Acheulean sites.

Gesher Benot Ya'aqov. Farther north in the Jordan Rift Valley, the 780-kyr-old site of Gesher Benot Ya'aqov preserves an artifact assemblage with distinct morphological and technological similarities to African assemblages of comparable age, associated with a remarkable macrobotanical assemblage preserved in waterlogged sediments at the edge of ancient Lake Hula (Goren-Inbar et al. 2002b). Despite recovering approximately 1,500 fragments of logs, branches, wood, and bark from the site, detailed morphological study suggests that none was worked with stone artifacts. Careful taphonomic analysis suggests that the fragments of wood had likely washed into the site from surrounding hills with seasonal floods. However, 13 of the wood specimens over 20 mm long had been burned, and a large number of small paleobotanical fragments from the site had also been charred, including over 50,000 fragments of wood and over 23,000 fragments of fruit (Goren-Inbar et al. 2002b, 2004). The authors conclude that the spatial distribution of the charred remains, coincident with patterns of burnt microdebitage, point to the Acheulean toolmakers as agents with controlled use of fire (Alperson-Afil and Goren-Inbar, 2010). Notably, the remains of a variety of plants with edible seeds, including wild almond, pistachio, and ash, were found at the site, associated with pitted stones, strongly suggesting that nut-cracking occurred at the site (Goren-Inbar et al. 2002b). Most of the plant species identified from these paleobotanical samples are still locally common in the region today, indicating continuity in the natural woodland associations in the Levant during the last million years, facilitating actualistic studies of modern vegetation which helped with the interpretation of the site. Goren-Inbar and her coauthors have demonstrated how critical detailed taphonomic and morphological analysis is to the interpretation of such macrobotanical remains. In the years ahead, ethnoarchaeological and experimental research will be key to defining detectable signatures of plant food processing, such as peeling or roasting tubers (Mallol et al. 2007; Dominy et al. 2008; Langejans 2010).

Kalambo Falls. Another Acheulean site noted for both its macro- and microbotanical remains is Kalambo Falls, a site in northern Zambia on the banks of Kalambo River, close to Lake Tanganyika. Kalambo archaeological assemblages range from recent Iron Age levels, back to Acheulean assemblages that are undoubtedly of middle Pleistocene age but that are difficult to date. Excavations in waterlogged deposits at the site recovered remarkable assemblages of macrobotanical remains from different localities and Acheulean and Sangoan horizons, including leaves, seeds and fruits, bark, wood and charcoal, carbonized tree trunks, and smaller pieces of possibly worked wood. Pollen recovered from the site sediments were originally analyzed by van Zinderen Baker and published in the first site monograph (van Zinderen Bakker 1969), with additional comments on macrobotanical remains (Chalk 1969; Clark 1969; White 1969; Whitmore 1969). Subsequently, J. D. Clark's completion of the third Kalambo monograph at the very end of his distinguished career (Clark 2001) included a reanalysis of thirty pollen samples from the site (Taylor et al. 2001) and a detailed report on possible wooden artifacts recovered from the site.

The vegetation in the Kalambo region today is dominated by a mosaic of miombo woodland and bushlands, with edaphic grasslands in seasonally flooded dambos. Van Zinderen Baker's original analysis of pollen from six horizons at the site distinguished proportions of grasses and other monocots he associated with edaphic wetlands from pollen taxa he felt represented shifting bushland, woodland, and riparian forest communities, as well as regional fluctuations in montane vegetation zones. As discussed earlier, van Zinderen Baker sought to interpret the pollen samples in the context of global climate fluctuations, and his understanding of European evidence of glacial and interglacial cycles. His correlations of climatic fluctuations proved to be inappropriate, in part because of difficulties in interpreting the radiocarbon chronology from the site. In their subsequent reanalysis, Taylor and colleagues (2001) have corrected some analytical

hominin–plant interaction, it is difficult to evaluate the relative importance of associations between hominin fossils or archaeological localities and particular plant resources. As White reminded us when evaluating the context of robust australopithecines (White 1988), the sedimentary resting places where hominin remains became fossilized may be quite removed from the habitats in which the hominins lived. Similarly, while primary-context archaeological sites mark the localities where hominin activities occurred, and sometimes reoccurred, the taphonomic biases in the samples of sites that have been preserved, discovered, and described make it difficult to generalize about the behavioral meaning of the local vegetation contexts of the sites.

Second, the temporal sampling can be problematic as well. Rarely are hominin fossils recovered from exactly the same horizons as botanical samples, and studies of the dynamic history of African vegetation patterns demonstrate that the structure and composition of plant communities can shift dramatically within the hundreds and thousands of years that separate ancient paleobotanical samples from the horizons of archaeological or hominin fossil sites. So paleoanthropologists are always left to extrapolate, or connect the dots, between samples that are not, generally, comparable.

This is one of the reasons why formal modeling has been gaining in popularity as a complement to paleohabitat investigations of early hominins—it helps make the dimensions and assumptions inherent in such extrapolations quite explicit. Models can be both heuristic devices (as in the case of broad, conceptual models), and essential analytical tools. All models are simplifications of a complex world, and make trade-offs between their balance of generality, precision, and realism. Whether evaluating the dynamic, taphonomic relationship between pollen and phytolith accumulation, the NDVI, and vegetation type (Jolly et al. 1998; Gajewski et al. 2002) or trying to simulate how the distribution of particular plant foods on a paleolandscape would influence foraging decisions of different hominin species (Sept 2001, 2007; Griffith et al. 2010), models are becoming an important analytical tool for interpreting how evidence of ancient and modern vegetation can be integrated to develop nuanced understandings of early hominin ecological relationships. At the same time, the computational power of computers has given paleobotanists and paleoanthropologists alike powerful tools to model complex taphonomic and ecological systems; these analyses are vital because the interpretation of ancient evidence is heavily dependent upon statistical inference. They also allow the integration of multivariate evidence at different scales of resolution—comparisons of local samples acquired in different contexts across the entire African continent, for example.

Even as scientific fields become more technically specialized and computationally complex, however, we are reminded how important it is that the specialists also maintain a broad contextual knowledge of their disciplines. For example, it is advantageous for those who identify ancient pollen grains to also have firsthand field knowledge. Palynologists benefit, on the one hand, by knowing how plant communities are structured and how individual species grow in varying conditions in the wild. On the other hand, when palynologists help design and implement sediment-sampling strategies in the field in concert with paleoanthropologists, such contextual information can be used to develop thoughtful interpretations of palynological results. Similarly, it is vital for those who are interpreting the presence and distribution of macrobotanical remains to be familiar with details of the depositional context of the samples and to have experience with taphonomic analysis of contemporary macrobotanical assemblages in analogous sedimentary situations, whether for sites with close floristic and structural affinities to modern settings or for sites from much earlier periods.

Overall, when paleoanthropologists and paleobotanists collaborate closely and pay careful attention to the taphonomic context of their data, we benefit from more nuanced and holistic interpretations of the vegetation context of hominin sites. And whether plants provided shade and refuge, were eaten as staple foods, or provided sources of raw materials for tools, we can be certain that they were a vital part of the habitat of all early hominin species and a foundation for paleoecological adaptations.

ACKNOWLEDGMENTS

I thank Matt Sponheimer and the other organizers for their invitation to participate in a fascinating conference and contribute to this volume. It is also a privilege to acknowledge how the generous, pioneering research of Raymonde Bonnefille has rooted so much of our current understanding of early hominin paleoecology.

REFERENCES

Abrantes, F. 2003. "A 340,000 Year Continental Climate Record from Tropical Africa—News from Opal Phytoliths from the Equatorial Atlantic." *Earth and Planetary Science Letters* 209 (1-2): 165–79. http://dx.doi.org/10.1016/S0012-821X(03)00039-6.

Albert, R. M., M. K. Bamford, and D. Cabanes. 2006. "Taphonomy of Phytoliths and Macroplants in Different Soils from Olduvai Gorge (Tanzania) and the Application to

Plio-Pleistocene Palaeoanthropological Samples." *Quaternary International* 148 (1): 78–94. http://dx.doi.org/10.1016/j.quaint.2005.11.026.

Alexandre, A., J. D. Meunier, A. M. Lezine, A. Vincens, and D. Schwartz. 1997. "Phytoliths: Indicators of Grassland Dynamics during the Late Holocene in Intertropical Africa." *Palaeogeography, Palaeoclimatology, Palaeoecology* 136 (1-4): 213–29. http://dx.doi.org/10.1016/S0031-0182(97)00089-8.

Alperson-Afil, N., Goren-Inbar, N., 2010. *The Acheulian Site of Gesher Benot Ya'aqov: Ancient Flames and Controlled Use of Fire*. Springer, Dordrecht.

Altmann, S. A. 1998. *Foraging for Survival: Yearling Baboons in Africa*. Chicago: University of Chicago Press.

Backéus, I., B. Pettersson, L. Stromquist, and C. Ruffo. 2006. "Tree Communities and Structural Dynamics in Miombo (Brachystegia-Julbernardia) Woodland, Tanzania." *Forest Ecology and Management* 230 (1–3): 171–8. http://dx.doi.org/10.1016/j.foreco.2006.04.033.

Bamford, M. K. 1999. "Pliocene Fossil Woods from an Early Hominid Cave Deposit, Sterkfontein, South Africa." *South African Journal of Science* 95:231–7.

Bamford, M. K. 2005. "Early Pleistocene Fossil Wood from Olduvai Gorge, Tanzania." *Quaternary International* 129 (1): 15–22. http://dx.doi.org/10.1016/j.quaint.2004.04.003.

Bamford, M. K., R. M. Albert, and D. Cabanes. 2006. "Plio-Pleistocene Macroplant Fossil Remains and Phytoliths from Lowermost Bed II in the Eastern Paleolake Margin of Olduvai Gorge, Tanzania." *Quaternary International* 148 (1): 95–112. http://dx.doi.org/10.1016/j.quaint.2005.11.027.

Barboni, D., R. Bonnefille, A. Alexandre, and J. D. Meunier. 1999. "Phytoliths as Paleoenvironmental Indicators, West Side Middle Awash Valley, Ethiopia." *Palaeogeography, Palaeoclimatology, Palaeoecology* 152 (1-2): 87–100. http://dx.doi.org/10.1016/S0031-0182(99)00045-0.

Barboni, D., and L. Bremond. 2009. "Phytoliths of East African grasses: an assessment of their environmental and taxonomic significance based on floristic data." *Review of Palaeobotany and Palynology* 158: 29–41.

Barboni, D., L. Bremond, and R. Bonnefille. 2007. "Comparative Study of Modern Phytolith Assemblages from Inter-Tropical Africa." *Palaeogeography, Palaeoclimatology, Palaeoecology* 246 (2–4): 457–70.

Barton, R. A., A. Whiten, S. C. Strum, R. W. Byrne, and A. J. Simpson. 1992. "Habitat Use and Resource Availability in Baboons." *Animal Behaviour* 43 (5): 831–44. http://dx.doi.org/10.1016/S0003-3472(05)80206-4.

Belsky, A. J. 1990. "Tree/Grass Ratios in East African Savannas: A Comparison of Existing Models." *Journal of Biogeography* 17 (4/5): 483–9. http://dx.doi.org/10.2307/2845380.

Blumenschine, R. J., and C. R. Peters. 1998. "Archaeological Predictions for Hominid Land Use in the Paleo-Olduvai Basin, Tanzania, during Lowermost Bed II Times." *Journal of Human Evolution* 34 (6): 565–607. http://dx.doi.org/10.1006/jhev.1998.0216. Medline:9650101

Bonnefille, R. 1971. "Atlas des pollens d'Ethiopie: Pollens actuels de la basse vallée de l'Omo (Ethiopie); Récoltes botaniques 1968." *Adansonia* 2:463–518.

Bonnefille, R. 1976. "Palynological Evidence for an Important Change in the Vegetation of the Omo Basin between 2.5 and 2 Million Years Ago." In *Earliest Man and Environments in the Lake Rudolf Basin*, ed. Y. Coppens, F. C. Howell, G. L. Isaac, and R.E.F. Leakey, 421–431. Chicago: University of Chicago Press.
Bonnefille, R. 1984. "Palynological Research at Olduvai Gorge." *Research Reports: 1976 Projects* 17: 227–243.
Bonnefille, R. 1994. "Palynology and Palaeoenvironment of East African Hominid Sites." In *Integrative Paths to the Past: Palaeoanthropological Advances in Honor of F. Clark Howell*, ed. R. S. Corruccini and R. L. Ciochon, 415–27. Advances in Human Evolution Series, 2. Englewood Cliffs, NJ: Prentice Hall.
Bonnefille, R. 1995. "A Reassessment of the Plio-Pleistocene Pollen Recored of East Africa." In *Paleoclimate and Evolution, with Emphasis on Human Origins*, ed. E. Vrba, G. Denton, T. Partridge, and L. Burckle, 299–310. New Haven, CT: Yale University Press.
Bonnefille, R., and R. Letouzey. 1976. "Fruits fossiles d'Antrocaryon dans la vallée de l'Omo (Ethiopie)." *Adansonia* 16 (2): 65–82.
Bonnefille, R., D. Lobreau, and G. Riollet. 1982. "Fossil Pollen of Ximenia (Olacaceae) in the Lower Pleistocene of Olduvai, Tanzania: Palaeoecological Implications." *Journal of Biogeography* 9: 469–86. http://dx.doi.org/10.2307/2844614.
Bonnefille, R., R. Potts, F. Chalié, D. Jolly, and O. Peyron. 2004. "High-Resolution Vegetation and Climate Change Associated with Pliocene *Australopithecus afarensis*." *Proceedings of the National Academy of Sciences of the United States of America* 101 (33): 12125–9. http://dx.doi.org/10.1073/pnas.0401709101. Medline:15304655
Bonnefille, R., A. Vincens, and G. Buchet. 1987. "Palynology, Stratigraphy and Palaeoenvironment of a Pliocene Hominid Site (2.9–3.3 M.Y.) at Hadar, Ethiopia." *Palaeogeography, Palaeoclimatology, Palaeoecology* 60:249–81. http://dx.doi.org/10.1016/0031-0182(87)90035-6.
Booysen, P.D.V. 1984. *Ecological Effects of Fire in South African Ecosystems*. Berlin: Springer-Verlag.
Box, E. O., B. N. Holben, and V. Kalb. 1989. "Accuracy of the AVHRR Vegetation Index as a Predictor of Biomass, Primary Productivity and Net CO_2 Flux." *Plant Ecology* 80 (2): 71–89. http://dx.doi.org/10.1007/BF00048034.
Boyd, W. E., and V. A. Hall. 1998. "Landmarks on the Frontiers of Palynology: An Introduction to the IX International Palynological Congress Special Issue on New Frontiers and Applications." *Palynology Review of Palaeobotany and Palynology* 103 (1–2): 1–10. http://dx.doi.org/10.1016/S0034-6667(98)00020-7.
Bremond, L., A. Alexandre, C. Hely, and J. Guiot. 2005. "A Phytolith Index as a Proxy of Tree Cover Density in Tropical Areas: Calibration with Leaf Area Index along a Forest-Savanna Transect in Southeastern Cameroon." *Global and Planetary Change* 45 (4): 277–93. http://dx.doi.org/10.1016/j.gloplacha.2004.09.002.
Carr, C. J. 1976. "Plant Ecological Variation and Pattern in the Lower Omo Basin." In *Earliest Man and Environment in the Lake Rudolf Basin*, ed. Y. Coppens, F. C. Howell, G. L. Isaac, and R. E. Leakey, 432–470. Chicago: Chicago University Press.

Lu, D. 2005a. "Integration of Vegetation Inventory Data and Landsat TM Image for Vegetation Classification in the Western Brazilian Amazon." *Forest Ecology and Management* 213 (1-3): 369–83. http://dx.doi.org/10.1016/j.foreco.2005.04.004.

Lu, D. 2005b. "Satellite Estimation of Aboveground Biomass and Impacts of Forest Stand Structure." *Photogrammetric Engineering and Remote Sensing* 71 (8): 967–74.

Mallol, C., F. W. Marlowe, B. M. Wood, and C. C. Porter. 2007. "Earth, wind, and fire: ethnoarchaeological signals of Hadza fires." *Journal of Archaeological Science* 34: 2035–2052.

Martin, G. 1995. *Ethnobotany: A Methods Manual*. London: Chapman & Hall.

Mayaux, P., E. Bartholome, S. Fritz, and A. Belward. 2004. "A New Land-Cover Map of Africa for the Year 2000." *Journal of Biogeography* 31 (6): 861–77. http://dx.doi.org/10.1111/j.1365-2699.2004.01073.x.

Mercader, J., T. Bennett, C. Esselmont, S. Simpson, and D. Walde. 2009. "Phytoliths in woody plants from the miombo woodlands of Mozambique." *Annals of Botany* 104: 91–113.

Mercader, J., F. Runge, L. Vrydaghs, H. Doutrelepont, E. Corneille, and J. Juan-Tresseras. 2000. "Phytoliths from Archaeological Sites in the Tropical Forest of Ituri, Democratic Republic of Congo." *Quaternary Research* 54 (1): 102–12. http://dx.doi.org/10.1006/qres.2000.2150.

Milton, K. 1984. "The Role of Food-Processing Factors in Primate Food Choice." In *Adaptations for Foraging in Nonhuman Primates*, ed. P. S. Rodman and J.G.H. Cant, 249–279. New York: Columbia University Press.

Milton, K. 1987. "Primate Diets and Gut Morphology: Implications for Hominid Evolution." In *Food and Evolution: Toward a Theory of Human Food Habits*, ed. M. Harris and E. Ross, 93–115. Philadelphia: Temple University Press.

Milton, K. 2006. "Diet and Primate Evolution." *Scientific American* 16 (2): 22–9. http://dx.doi.org/10.1038/scientificamerican0606-22sp.

Moran, E. F., D. L. Skole, and B. L. Turner, II. 2004. "The Development of the International Land-Use and Land-Cover Change (LUCC) Research Program and Its Links to NASA's Land-Cover and Land-Use Change (LCLUC) Initiative." In Land Change Science, ed. G. Gutman, et al., 1–15. Kluwer Academic Publishers.

Mulholland, S. C., and G. Rapp, Jr., eds. 1992. *Phytolith Systematics: Emerging Issues*. New York: Plenum Press.

Napier-Bax, P., and D. Sheldrick. 1963. "Some Preliminary Observations on the Food of Elephant in the Tsavo Royal National Park (East) in Kenya." *East African Wildlife Journal* 1:40–53.

Norton, G. W., R. J. Rhine, G. W. Wynn, and R. D. Wynn. 1987. "Baboon Diet: A Five-Year Study of Stability and Variability in the Plant Feeding and Habitat of the Yellow Baboons (*Papio cynocephalus*) of Mikumi National Park, Tanzania." *Folia Primatologica* 48 (1-2): 78–120. http://dx.doi.org/10.1159/000156287. Medline:3653819

Palmer, P. G. 1976. "Grass Cuticles: A New Paleoecological Tool for East African Lake Sediments." *Canadian Journal of Botany* 54 (15): 1725–34. http://dx.doi.org/10.1139/b76-186.

Peters, C. R., and R. J. Blumenschine. 1995. "Landscape Perspectives on Possible Land Use Patterns for Early Pleistocene Hominids in the Olduvai Basin, Tanzania." *Journal of Human Evolution* 29 (4): 321–62. http://dx.doi.org/10.1006/jhev.1995.1062.

Peters, C. R., and B. Maguire. 1981. "Wild Plant Foods in the Makapansgat Area: A Modern Ecosystem Analogue for *Australopithecus africanus* Adaptations." *Journal of Human Evolution* 10 (7): 565–83. http://dx.doi.org/10.1016/S0047-2484(81)80048-6.

Peters, C. R., and E. M. O'Brien. 1981. "The Early Hominid Plant-Food Niche: Insights from an Analysis of Plant Exploitation by *Homo, Pan*, and *Papio* in Eastern and Southern Africa." *Current Anthropology* 22 (2): 127–40. http://dx.doi.org/10.1086/202631.

Piperno, D. R. 1988. *Phytolith Analysis: An Archaeological and Geological Perspective*. London: Academic Press.

Retallack, G. J., D. P. Dugas, and E. A. Bestland. 1990. "Fossil Soils and Grasses of a Middle Miocene East African Grassland." *Science* 247 (4948): 1325–8. http://dx.doi.org/10.1126/science.247.4948.1325. Medline:17843796

Rovner, I. 1971. "Potential of Opal Phytoliths for Use in Paleoecological Reconstruction." *Quaternary Research* 1 (3): 343–59. http://dx.doi.org/10.1016/0033-5894(71)90070-6.

Runge, F. 1999. "The Opal Phytolith Inventory of Soils in Central Africa: Quantities, Shapes, Classification and Spectra." *Review of Palaeobotany and Palynology* 107 (1-2): 23–53. http://dx.doi.org/10.1016/S0034-6667(99)00018-4.

Sept, J. M. 1984. "Plants and Early Hominids in East Africa: A Study of Vegetation in Situations Comparable to Early Archaeological Site Locations." PhD dissertation, University of California, Berkeley.

Sept, J. M. 1986. "Plant Foods and Early Hominids at Site FxJj50, Koobi Fora, Kenya." *Journal of Human Evolution* 15 (8): 751–70. http://dx.doi.org/10.1016/S0047-2484(86)80008-2.

Sept, J. M. 1990. "Vegetation Studies in the Semliki Valley, Zaire as a Guide to Paleoanthropological Research." *Virginia Museum of Natural History Memoire* 1:95–121.

Sept, J. M. 1992. "Archaeological Evidence and Ecological Perspectives for Reconstructing Early Hominid Subsistence Behavior." In *Archaeological Method and Theory*, ed. M. B. Schiffer, 1–56. Tucson: University of Arizona Press.

Sept, J. M. 2001. "Modeling the Edible Landscape." In *Meat Eating and Human Evolution*, ed. C. B. Stanford and H. T. Bunn, 73–98. Oxford: Oxford University Press.

Sept, J. M. 2007. "Modeling the Significance of Paleoenvironmental Context for Early Hominin Diets." In *Evolution of Human Diet: The Known, the Unknown, and the Unknowable*, ed. P. S. Ungar, 289–307. Oxford: Oxford University Press.

Shahack-Gross, R., F. Marshall, K. Ryan, and S. Weiner. 2004. "Reconstruction of Spatial Organization in Abandoned Maasai Settlements: Implications for Site Structure in the Pastoral Neolithic of East Africa." *Journal of Archaeological Science* 31 (10): 1395–411. http://dx.doi.org/10.1016/j.jas.2004.03.003.

Sikes, N. 1994. "Early Hominid Habitat Preferences in East Africa: Paleosol Carbon Isotope Evidence." *Journal of Human Evolution* 27 (1-3): 25–45. http://dx.doi.org/10.1006/jhev.1994.1034.

Sinclair, A.R.E. 1979. "Dynamics of the Serengeti Ecosystem: Process and Pattern." In *Serengeti: Dynamics of an Ecosystem*, ed. A.R.E. Sinclair and M. Norton-Griffiths, 1–30. Chicago: University of Chicago Press.

Sinclair, A.R.E. 1995. "Serengeti Past and Present." In *Serengeti II: Dynamics, Management, and Conservation of an Ecosystem*, ed. A.R.E. Sinclair and P. Arcese, 3–30. Chicago: University of Chicago Press.

Sinclair, A.R.E., and M. Norton-Griffiths, eds. 1979. *Serengeti: Dynamics of an Ecosystem.*, 389. Chicago: University of Chicago Press.

Smith, B. N., and S. Epstein. 1971. "Two Categories of C/C Ratios for Higher Plants." *Plant Physiology* 47 (3): 380–4. http://dx.doi.org/10.1104/pp.47.3.380. Medline:16657626

Sponheimer, M., J. Lee-Thorp, D. de Ruiter, D. Codron, J. Codron, A. T. Baugh, and F. Thackeray. 2005. "Hominins, Sedges, and Termites: New Carbon Isotope Data from the Sterkfontein Valley and Kruger National Park." *Journal of Human Evolution* 48 (3): 301–12. http://dx.doi.org/10.1016/j.jhevol.2004.11.008. Medline:15737395

Stahl, A. 1984. "Hominid Dietary Selection before Fire." *Current Anthropology* 25 (2): 151–68. http://dx.doi.org/10.1086/203106.

Stock, W. D., D. K. Chuba, and G. A. Verboom. 2004. "Distribution of South African C-3 and C-4 species of Cyperaceae in Relation to Climate and Phylogeny." *Austral Ecology* 29 (3): 313–9. http://dx.doi.org/10.1111/j.1442-9993.2004.01368.x.

Strömberg, C.A.E. 2004. "Using Phytolith Assemblages to Reconstruct the Origin and Spread of Grass-Dominated Habitats in the Great Plains of North America during the Late Eocene to Early Miocene." *Palaeogeography, Palaeoclimatology, Palaeoecology* 207 (3-4): 239–75. http://dx.doi.org/10.1016/j.palaeo.2003.09.028.

Taylor, D., R. Marchant, and A. C. Hamilton. 2001. "A Reanalysis and Interpretation of Palynological Data from the Kalambo Falls Prehistoric Site." In *Kalambo Falls Prehistoric Site III: The Earlier Cultures; Middle and Earlier Stone Age*, ed. J. D. Clark, J. Cormack, and S. Chin, 66–81. Cambridge: Cambridge University Press.

Tieszen, L. T., M. M. Senyimba, and S. K. Imbamba. 1979. "The Distribution of C3 and C4 Grasses and Carbon Isotope Discrimination along an Altitudinal and Moisture Gradient in Kenya." *Oecologia* 37:337–50.

van Zinderen Bakker, E. M. 1957. *A Pollen Analytical Investigation of the Florisbad Deposit.*, 56–67. South Africa: Livingstone.

van Zinderen Bakker, E. M. 1962. "A Late-Glacial and Post-Glacial Climatic Correlation between East Africa and Europe." *Nature* 194 (4824): 201–203. http://dx.doi.org/10.1038/194201a0.

van Zinderen Bakker, E. M. 1967. "Upper Pleistocene and Holocene Stratigraphy and Ecology on the Basis of Vegetation Changes in Sub-Saharan Africa." In *Background to Evolution in Africa*, ed. W. W. Bishop and J. D. Clark, 125–147. Chicago: University of Chicago Press.

van Zinderen Bakker, E. M. 1969. "The Pleistocene Vegetation and Climate of the Basin." In *Kalambo Falls Prehistoric Site I: The Geology, Palaeoecology and Detailed Stratigraphy of the Excavations*, ed. J. D. Clark, 57–84. Cambridge: Cambridge University Press.

Vincens, A. 1984. "Environnement végétal et sédimentation pollinique lacustre actuelle dans le Bassin du Lac Turkana (Kenya)." *Revue de Paléobiologie volume spécial* 235–242.

Vincens, A., J. J. Tiercelin, and G. Buchet. 2006. "New Oligocene–Early Miocene Microflora from the Southwestern Turkana Basin: Palaeoenvironmental Implications in the Northern Kenya Rift." *Palaeogeography, Palaeoclimatology, Palaeoecology* 239 (3–4): 470–86. http://dx.doi.org/10.1016/j.palaeo.2006.02.007.

Vincent, A. S. 1985a. "Plant Foods in Savanna Environments: A Preliminary Report of Tubers Eaten by the Hadza of Northern Tanzania." *World Archaeology* 17 (2): 131–48. http://dx.doi.org/10.1080/00438243.1985.9979958. Medline:16470992

Vincent, A. S. 1985b. "Underground Plant Foods and Subsistence in Human Evolution." PhD dissertation, University of California, Berkeley.

Vogt, N., J. Bahati, J. Unruh, G. Green, A. Banana, W. Gombya-Ssembajjwe, and S. Sweeney. 2006. "Integrating Remote Sensing Data and Rapid Appraisals for Land-Cover Change Analyses in Uganda." *Land Degradation and Development* 17 (1): 31–43. http://dx.doi.org
/10.1002/ldr.692.

Watrin, J., A.-M. Lezine, K. Gajewski, and A. Vincens. 2007. "Pollen-plant-climate relationships in sub-Saharan Africa." *Journal of Biogeography* 34: 489–499.

Western, D., and D. Maitumo. 2004. "Woodland Loss and Restoration in a Savanna Park: A 20-Year Experiment." *African Journal of Ecology* 42 (2): 111–21. http://dx.doi.org/10.1111
/j.1365-2028.2004.00506.x.

Western, D., and C. Van Praet. 1973. "Cyclical Changes in the Habitat and Climate of an East African Ecosystem." *Nature* 241 (5385): 104–6. http://dx.doi.org/10.1038/241104a0.

White, F. 1969. "Identification of Fruits and Seeds from Site B, Kalambo Falls." In *Kalambo Falls Prehistoric Site I: The Geology, Palaeoecology and Detailed Stratigraphy of the Excavations*, ed. J. D. Clark, 216–217. Cambridge: Cambridge University Press.

White, F. 1983. *The Vegetation of Africa: A Descriptive Memoir to Accompany the UNESCO/AETFAT/UNSO Vegetation Map of Africa*. La Chaux-de-Fonds: UNESCO.

White, T. D. 1988. "The Comparative Biology of 'Robust' *Australopithecus*: Clues from Context." In *Evolutionary History of the "Robust" Australopithecines*, ed. F. E. Grine, 449–484. New York: Aldine de Gruyter.

Whitmore, T. C. 1969. "Report on Bark and Other Specimens from Site B, Kalambo Falls, Depositional Hase F2, White Sands Beds, Mkamba Member: Kalambo Falls Prehistoric Site I." *The Geology, Palaeoecology and Detailed Stratigraphy of the Excavations*, 221–224. Cambridge: Cambridge University Press.

Wrangham, R. W. 2007. "The Cooking Enigma." In *Evolution of the Human Diet: The Known, the Unknown and the Unknowable*, ed. P. S. Ungar, 308–323. Oxford: Oxford University Press.

Wrangham, R. W., N. L. Conklin, C. A. Chapman, and K. D. Hunt. 1991. "The Significance of Fibrous Foods for Kibale Forest Chimpanzees." In *Foraging Strategies and Natural Diet of Monkeys, Apes and Humans*, ed. A. Whiten and E. M. Widdowson, 171–78. Oxford: Clarendon Press. http://dx.doi.org/10.1098/rstb.1991.0106.

Wrangham, R. W., and N. L. Conklin-Brittain. 2003. "Cooking as a Biological Trait." *Comparative Biochemistry and Physiology: A, Molecular and Integrative Physiology* 136 (1): 35–46. http://dx.doi.org/10.1016/S1095-6433(03)00020-5. Medline:14527628.
Wrangham, R. W., J. H. Jones, G. Laden, D. Pilbeam, and N. Conklin-Brittain. 1999. "The Raw and the Stolen: Cooking and the Ecology of Human Origins." *Current Anthropology* 40 (5): 567–94. http://dx.doi.org/10.1086/300083. Medline:10539941.

11

Chimpanzee Models of Human Behavioral Evolution

John C. Mitani

What behavioral changes occurred during human evolution and why? Seeking answers to these questions lies at the heart of anthropological inquiry, but the task is fraught with difficulty. Part of the problem resides in the fact that behavior, unlike skeletal anatomy, does not fossilize, and understanding its evolution requires sources of data beyond the direct evidence furnished by material human remains. The behavior of our closest living relatives, the nonhuman primates, has been recognized for more than fifty years as one potential source of information for reconstructing human behavior (Washburn and DeVore 1961; Washburn 1963). The Order Primates comprises several species that differ in their behavior (Smuts et al. 1987), and we can use this diversity in productive ways to understand our own evolution.

Seeking clues to the evolution of human behavior provided the impetus for some of the earliest studies of primates in the wild. In the late 1950s Sherwood Washburn, an anatomist turned anthropologist, became interested in using primates as models to understand human behavioral evolution. He reasoned that savanna-dwelling baboons and early humans, though distantly related evolutionarily, inhabited similar habitats and thus faced the same types of ecological problems (Washburn and Devore 1961; Washburn 1963). If baboons adapted to their environments in the same manner as early humans, then the behavior of modern baboons would reflect the behavior of our ancestors in

DOI: 10.5876/9781607322252:c11

a very real way. As shown in Chapter 12, baboon models continue to furnish important insights into human evolution.

In Japan, a distinctly different approach to primate field studies developed under the leadership of Kinji Imanishi and Junichiro Itani. Here, too, the underlying motivation was to understand our own evolution, but unlike Washburn and his American students, Imanishi and Itani turned to the study of our closest living relatives, the African apes (Nishida 1990). When political unrest forced them to halt their fieldwork on gorillas in the Belgian Congo, they initiated a series of field studies on chimpanzees in Western Tanzania in 1961. One of these continues to this day at the Mahale Mountains National Park (Nakamura and Nishida 2006). The Gombe Stream National Park lies about 150 km north of the Mahale Mountains along the eastern shore of Lake Tanganyika. It was here that Jane Goodall, inspired by the well-known paleoanthropologist Louis Leakey, began her fieldwork on wild chimpanzees (Goodall 1963, 1986). Leakey also had a keen interest in the evolution of human behavior and believed that a study of chimpanzees inhabiting a lakeshore, woodland habitat similar to that occupied by early humans, would prove illuminating.

From these beginnings, chimpanzees have continued to serve as models of human behavioral evolution. In a provocative paper, Sayers and Lovejoy (2008) have recently argued that knowledge about chimpanzees has frequently been misapplied for such purposes. The lively comments by others along with Sayers and Lovejoy's response to them reveal that disagreement exists regarding why and how we use chimpanzees to model the evolution of human behavior. In this chapter, I attempt to clarify some points surrounding this controversy by reviewing four issues. I start by outlining the conceptual basis for employing chimpanzees to reconstruct the behavior of early humans. Homologous similarities often exist between closely related species; because chimpanzees share a common ancestor with us and represent our closest living relatives, they also share several behavioral traits with us. Second, I summarize our knowledge of the behavior of wild chimpanzees. Here I focus on three aspects—tool use, hunting, and intergroup aggression—that have featured prominently in discussions of human behavioral evolution. Third, I review the literature that has used chimpanzees as models to reconstruct the behavior of ancestral humans. Fourth, I conclude with a discussion of some of the problems and challenges surrounding the use of chimpanzees to understand human behavioral evolution.

HOMOLOGY AS THE BASIS OF BEHAVIORAL RECONSTRUCTION

Biological organisms display similarities in their anatomy, physiology, genetics, and behavior. Some similarities are the product of convergent evolution. The wings of bats, birds, and insects are a familiar example. Wings represent a common solution to the problem of flight and have evolved independently in these distantly related creatures. Products of convergent evolution result from the process of natural selection and furnish important insights into adaptations, but they reveal little about the evolutionary relationships between organisms.

In contrast, some similarities between organisms have nothing to do with the function they perform but instead reflect common ancestry. For example, amphibians, reptiles, birds, and mammals possess pendactyl (i.e., five digit) limbs. These animals occupy a variety of habitats and use their limbs in different ways. Thus, there is no obvious ecological or functional reason why they should have the same number of digits. Evolution provides a solution to this apparent paradox. The pendactyl limb represents a homologous structure that has been inherited by a diverse group of organisms via a common ancestor.

Identifying homologous traits shared by organisms furnishes the principal means to reconstruct their evolutionary relationships (Wiley 1981). Anatomical and genetic data are typically employed for such purposes. Homologies clearly exist in behavior, yet behavior is seldom used, with only about 5 percent of phylogenetic studies including behavioral characters (Sanderson et al. 1993; Proctor 1996). The reasons for this are unclear (Rendall and Di Fiore 2007). The founders of the modern study of ethology emphasized that behavior could be used to determine the evolutionary relationships of animals and frequently asked explicit questions regarding behavioral evolution (e.g., Lorenz 1941; Tinbergen 1959, 1963). More recent research affirms the utility of employing behavioral traits to address phylogenetic questions (Prum 1990; Di Fiore and Rendall 1994; Macedonia and Stanger 1994; Irwin 1996; Rendall and Di Fiore 2007).

If behavior evolves like any other trait, then it should be possible to map behavioral changes onto a phylogeny of a group of organisms. Homologous behaviors that exist in extant members of closely related species are likely to have been present in their common ancestor. By extension, it is logical to assume that the earliest of humans may have retained similar behavior. In contrast, behavioral differences between these same species may have resulted from adaptive responses to changing ecological conditions over time. For purposes of understanding the course of human evolution, the task turns to identifying our closest living relatives and the behavioral similarities and differences that exist between them and us.

EVOLUTIONARY RELATIONSHIPS BETWEEN CHIMPANZEES AND HUMANS

Africa is the home to three species of living apes: the chimpanzee, the bonobo, and the gorilla. Huxley (1863) was one of the first to recognize that humans shared a common ancestry with the African apes, but the precise evolutionary relationships between these species have been elucidated only recently. Untangling these relationships has important implications for reconstructing the behavior of early humans.

Anatomical studies generally support the view that chimpanzees are more closely related to gorillas than they are to humans (e.g., Andrews and Martin 1987; Figure 11.1a). Chimpanzees and gorillas share several characteristics of the postcranial skeleton that appear to link them together to the exclusion of humans. Early genetic data were unable to determine the evolutionary relationships between chimpanzees, gorillas, and humans, but instead painted a picture of an unresolved trichotomy (e.g., Sarich and Wilson 1967; Figure 11.1b). More recent genetic studies, however, now clearly reveal that humans and chimpanzees form a monophyletic clade relative to gorillas (e.g,. Sibley et al. 1990; Ruvolo 1997; Chen and Li 2001; Figure 11.1c). The publication of the entire chimpanzee genome underscores the close genetic relationship between chimpanzees and humans, with single nucleotide substitutions occurring at a rate of about 1 percent between individuals of each species (Chimpanzee Sequencing and Analysis Consortium 2005).

While most genetic studies support the chimpanzee–human relationship displayed in Figure 11.1c, some genetic data indicate that chimpanzees are more closely related to gorillas or that gorillas might be our closest living relatives. These conflicting results arise because the speciation event leading to gorillas occurred only a very short time before chimpanzees diverged from humans. This has resulted in an intermingling of the genomes of chimpanzees, gorillas, and humans, with different parts telling different stories with respect to their evolutionary relationships (Pääbo 2003). Despite these ambiguities, a clear consensus confirms chimpanzees and their sister taxon, the bonobo, as humankind's closest living relatives. Because of this, chimpanzees represent appropriate models to reconstruct the behavior of early humans. To do so requires information about the behavior of living chimpanzees.

CHIMPANZEE BEHAVIOR

Chimpanzees have been studied extensively for more than fifty years in the wild. Most of our understanding of their behavior derives from six long-term

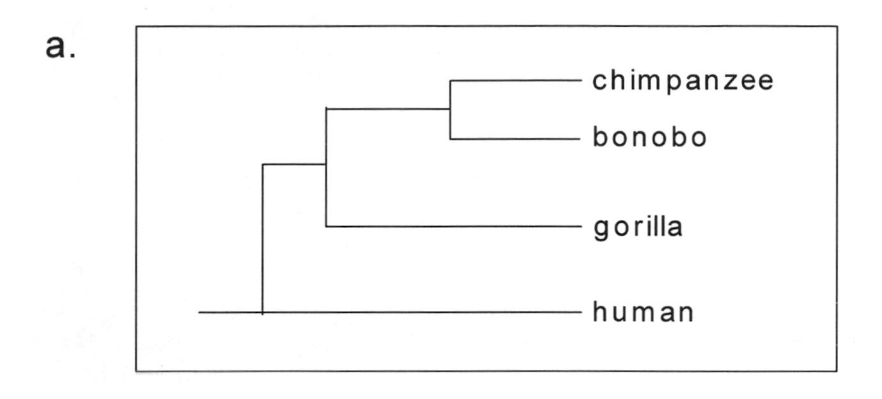

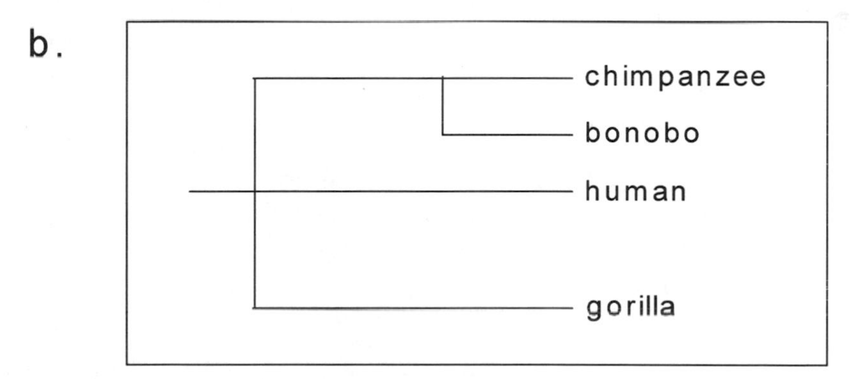

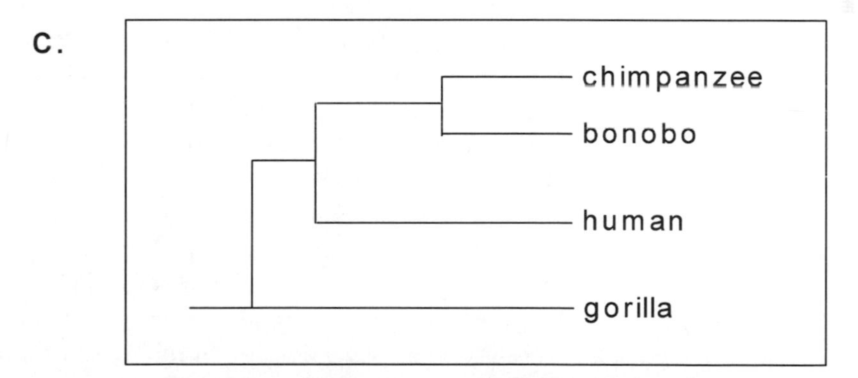

FIGURE 11.1. *Alternative hominoid phylogenies. (a) Anatomical studies suggest that chimpanzees and gorillas share a more recent common ancestry with each other than either does with humans. (b) Early genetic studies were unable to resolve the evolutionary relationships between chimpanzees, gorillas, and humans. (c) The current consensus based on recent genetic research indicates that chimpanzees and humans share a more recent common ancestry with each other than either does with gorillas.*

TABLE 11.1. Long-term chimpanzee field studies

Location	*Community*	*Duration of study*	*Reference*
Budongo Forest Reserve, Uganda	Sonso	1990–present	Reynolds 2005
Bossou, Guinea	Bossou	1976–present	Sugiyama 2004
Gombe National Park, Tanzania	Kasakela and Kahama	1960–present	Goodall 1986
Kibale National Park, Uganda	Kanyawara	1987–present	Wrangham 2000
Kibale National Park, Uganda	Ngogo	1995–present	Mitani et al. 2002b
Mahale Mountains National Park, Tanzania	Kajabala and Mimikiri	1965–present	Nishida 1990
Taï National Park, Ivory Coast	North, Central, and South	1979–present	Boesch and Boesch-Achermann 2000

field studies (Table 11.1). In the following, I review several aspects of chimpanzee behavior with special emphasis given to three topics that have featured significantly in discussions of human behavioral evolution: tool use, hunting and meat sharing, and male dominance and territoriality.

Social organization

Like most other primates, chimpanzees are social. Chimpanzees live in groups, called unit-groups or communities, which vary in size between 20 and 150 individuals (Nishida 1968; Goodall 1986; Boesch and Boesch-Achermann 2000; Wrangham 2000; Nishida et al. 2003; Sugiyama 2004; Reynolds 2005; Mitani 2006a). Unlike the relatively cohesive social groups found in most primates, chimpanzee communities are labile; community members fission and fuse forming temporary subgroups or "parties" that change in size and composition. Parties include between 4 and 10 individuals on average (Sakura 1994; Chapman et al. 1995; Boesch 1996; Matsumoto-Oda et al. 1998; Newton-Fisher et al. 2000; Mitani et al. 2002a). Despite the labile nature of chimpanzee parties, they do not form randomly. Male chimpanzees are typically much more gregarious than are females (Nishida 1968; Halperin 1979; Sakura 1994; Boesch 1996; Pepper et al. 1999; Wrangham 2000).

Demography and dispersal

The composition of chimpanzee communities is not static but changes due to the demographic processes of immigration, emigration, birth, and death.

Female chimpanzees, typically, but not always, disperse from their natal groups after reaching sexual maturity at an age of about 11 years old (Boesch and Boesch-Achermann 2000; Pusey 2001; Nishida et al. 2003; Sugiyama 2004). While virtually all female chimpanzees have been observed to disperse at some study sites (e.g., Mahale: Nishida et al. 2003 and Tai: Boesch and Boesch-Achermann 2000), a substantial proportion of females do not transfer at others (e.g., Gombe: Pusey et al. 1997; Pusey 2001; Bossou: Sugiyama 2004). The failure of some females to disperse is likely due to the limited number of dispersal options available to individuals in some areas, such as Gombe and Bossou. In these cases, habitat degradation has completely eliminated or severely reduced the number of neighboring chimpanzee communities into which females can move. In contrast to females, male chimpanzees are philopatric, and with very rare exceptions (e.g., Nishida and Hiraiwa-Hasegawa 1985), remain in their natal territories for life.

After dispersing and following a 2–3-year period of adolescent sterility, female chimpanzees begin to reproduce at an average age of 14 years old (Wallis 1997; Boesch and Boesch-Achermann 2000; Nishida et al. 2003; Sugiyama 2004). Females subsequently give birth once every 5–6 years to successive offspring who survive to weaning (Wallis 1997; Boesch and Boesch-Achermann 2000; Nishida et al. 2003; Sugiyama 2004). Chimpanzees are extremely long-lived, sometimes reaching 40–50 years old in the wild (Hill et al. 2001; Emery Thompson et al. 2007). Mortality is nonetheless high, often approaching and sometimes exceeding 50 percent during the first 5 years of life (Hill et al. 2001).

Feeding behavior and ecology

Chimpanzees are highly adaptable and live in several different kinds of habitats, including forests, woodlands, and savannas. In the wild, chimpanzees consume a variety of foods, including the reproductive and nonreproductive parts of plants, insects, and vertebrate prey, but they feed principally on ripe fruit (Wrangham 1977; Nishida and Uehara 1983; Wrangham et al. 1998; Yamakoshi 1998; Newton-Fisher 1999). Chimpanzees possess large bodies with adult females weighing over 30 kilograms (Pusey et al. 2005). To satisfy their correspondingly high nutritional demands, chimpanzees must range over large territories in search of seasonally scarce fruit trees that are often widely scattered. Territory sizes average between 5 and 30 square kilometers, depending on habitat type and quality (Hasegawa 1990; Chapman and Wrangham 1993; Herbinger et al. 2001; Williams et al. 2002; Newton-Fisher 2003; Mitani et al. 2010).

Toolmaking and tool use

Termites of the genus *Macrotermes* build mounds of hard soil to ventilate their underground nests. When they gather seasonally near the surface of these mounds prior to the departure of adult reproductives, chimpanzees fashion probes made of sticks, vines, plant stalks, or grass stems and insert them into the holes of mounds to obtain termites. Such termite "fishing" behavior was first observed by Goodall (1963, 1964) over fifty years ago at the Gombe National Park, Tanzania. Prior to that time, conventional wisdom held that humans were the only species to make and use tools. Goodall's startling observations fundamentally altered our conception of what it is to be human. After learning of Goodall's observations, Louis Leakey remarked, "Now we must redefine tool, redefine man, or accept chimpanzees as humans" (Goodall 1998).

Subsequent field research has shown that chimpanzees use tools extensively to feed, communicate, direct aggression toward others, inspect the environment, and clean their bodies (McGrew 1992). Despite these varied functions, the majority of chimpanzee tool use takes place in a feeding context (McGrew 1992). Chimpanzees typically use tools to acquire foods that are difficult to obtain without their aid (Yamakoshi 2001). Examples in addition to termite fishing include the use of wood and stone hammers to crack hard nuts, leaf "sponges" to extract water from tree holes, and sticks to fish for arboreal ants (Goodall 1963, 1964; Nishida 1973; Sugiyama and Koman 1979; Boesch and Boesch 1981). There is a distinct sex difference in the use of tools. Female chimpanzees use probes to fish for termites and ants and hammers to crack nuts much more frequently than males do (McGrew 1979; Boesch and Boesch 1981; Nishida and Hiraiwa 1982). Because resources such as reproductive termites and nuts are available only during certain times of the year, the frequency with which tools are used to acquire them varies seasonally. Observations of chimpanzees in Central Africa, however, reveal that chimpanzees employ "toolkits" to extract termites from their hard, aboveground mounds throughout the year (Suzuki et al. 1995; Bermejo and Illera 1999; Sanz et al. 2004). Chimpanzees use large sticks to perforate these mounds, and then fashion smaller fishing probes to extract termites.

Chimpanzees spend considerable time using tools to feed. For example, chimpanzees in the Taï National Park, Ivory Coast, use wooden and stone hammers over two hours each day to crack open the nuts of five species during their seasonal period of availability between November and March (Boesch and Boesch 1983; Günter and Boesch 1993). Chimpanzees derive three feeding benefits through their seasonal use of tools. First, the energetic payoff from tool use can be substantial. Taï chimpanzees have been estimated to obtain over 3,700 kilocalories per day through their nut-cracking efforts (Günter and Boesch 1993). Second, tool use can facilitate chimpanzees to cope with the seasonal effects of fruit scarcity. At Bossou, chimpanzees use stone hammers to crack the nuts of oil palms and palm fronds to pound and extract the pith of their hearts (Sugiyama and Koman 1979; Sugiyama 1994). Oil-palm nuts for cracking and pith for pounding are available throughout the year, but these activities elevate dramatically during the fruit-poor months of January–May at Bossou (Yamakoshi 1998). Finally, chimpanzees are likely to acquire essential nutrients through tool use. Female chimpanzees fish for arboreal *Camponotus* ants more frequently than do males at the Mahale Mountains National Park, Tanzania (Nishida and Hiraiwa 1982). The frequency of ant fishing decreases during the termiting season (Nishida and Hiraiwa 1982). In contrast, male chimpanzees at Mahale hunt and consume vertebrate prey more often than females do there (Uehara 1997). These complementary relationships suggest that tool use by female chimpanzees at Mahale facilitates the acquisition of protein.

Although tool use is a universal feature of chimpanzee behavior, tool repertoires differ significantly between populations. While West African chimpanzees use wood and stone hammers to crack open nuts, East African chimpanzees do not display this behavior, even though some of the same species of nuts are available to them. In East Africa chimpanzees fish for termites, but West African chimpanzees fail to do so in the presence of the same kinds of termites. These regional variations in tool use are not easily explained by differences in the physical and biotic environments. Some observations suggest that local variations in tool use are learned through social transmission, leading to the proposition that they reflect incipient forms of culture as practiced by humans (Whiten et al. 1999, Whiten 2005). The best evidence for social learning in the wild comes from a study that documents the development of termite fishing and points to some fascinating sex differences in behavior (Lonsdorf et al. 2004; Lonsdorf 2005). Female chimpanzees fish for termites at an earlier age and significantly more than males. After acquiring the habit, females are also more proficient fishers than males. The significance of these sex differences is currently unclear, but they mirror the proclivity of females to fish for termites more than males in adulthood. More recently, Whiten and colleagues (2005) have shown that captive chimpanzees learn different tool-use techniques to acquire food by observing others. These data furnish the strongest evidence thus far for observational learning in chimpanzees.

Hunting and meat sharing

Very few primates other than humans habitually hunt and consume vertebrate prey. Chimpanzees are a well-known exception. Goodall (1963) was the first to observe chimpanzees hunting and eating meat in the wild at Gombe, and these behaviors have been subsequently investigated there and at three other long-term study sites: Mahale, Ngogo, and Taï. As a result, considerable information now exists regarding several aspects of chimpanzee predation, including prey choice, the identity of hunters, hunting frequency, and hunting success (Boesch and Boesch 1989; Uehara et al. 1992; Boesch 1994a, 1994b, 1994c; Stanford et al. 1994a; Stanford 1998; Mitani and Watts 1999, 2001; Hosaka et al. 2001; Watts and Mitani 2002; Gilby et al. 2006).

Red colobus monkeys (*Procolobus badius*) are the preferred prey of chimpanzees wherever they live sympatrically (Boesch and Boesch 1989; Uehara et al. 1992; Stanford et al. 1994a; Mitani and Watts 1999). These monkeys represent between 80 and 90 percent of all prey across study sites (Table 11.2). In some areas, such as Gombe and Mahale, chimpanzees hunt red colobus prey opportunistically whenever they meet them during the course of their normal foraging movements (Nishida et al. 1979; Stanford 1998). In contrast, chimpanzees at Ngogo and Taï frequently actively seek red colobus prey during "patrols" (Boesch and Boesch 1989; Mitani and Watts 1999; Watts and Mitani 2002). In these hunting patrols, chimpanzees search for red colobus prey silently while moving in a directed and single-file fashion. Patrolling chimpanzees scan the environment for signs of monkeys, with searches sometimes lasting several hours.

Chimpanzees prey selectively on individuals in the youngest age classes. Juvenile and infant red colobus monkeys form over 50 percent of all prey (Table 11.2). In addition to red colobus monkeys, chimpanzees also hunt several other species of primates, ungulates, and birds (Table 11.3). Hunting is primarily a male activity, with adult male chimpanzees making the majority of all red colobus kills (Table 11.2). Chimpanzees are extraordinarily successful predators. Hunting success, assayed by the percentage of hunting attempts that result in kills, averages over 50 percent across study sites (Table 11.2). Hunting success is positively correlated with hunting-party size and the number of male hunters; large parties with more male hunters are generally more successful than small parties with fewer male hunters (Boesch and Boesch 1989; Stanford et al. 1994b; Mitani and Watts 1999; Hosaka et al. 2001; Watts and Mitani 2002).

While adult male chimpanzees are the most successful hunters, members of other age–sex classes show a keen interest in hunting and acquire meat as well. Estimates of consumption are difficult to make but suggest that indi-

Table 11.2. Characteristics of chimpanzee hunting behavior (Stanford 1998; Mitani and Watts 1999; Boesch and Boesch-Achermann 2000; Hosaka et al. 2001; Watts and Mitani 2002)

	Gombe	*Mahale*	*Ngogo*	*Taï*
Prey selection				
% of red colobus	85%	84%	88%	81%
n (kills)		(294)	(292)	(267)
Age class of prey				
% immature red colobus	85%	56%	58%	59%
n (kills)	(241)	(155)	(213)	(258)
Hunters				
% of kills by adult males	89%	51%	90%	71%
n (kills)		(117)	(261)	(38)
Hunting success				
% of all red colobus hunts	55%	60%	81%	74%
n (hunts)		(74)	(80)	(162)

vidual males at Taï obtain more than 60 kilograms of meat per year (Boesch and Boesch-Achermann 2000). Considerably lower estimates of meat eating, ranging between 9 and 26 kilograms per individual per year, have been made for female chimpanzees at Taï and for chimpanzees at other study sites (Stanford 1998; Boesch and Boesch-Achermann 2000; Watts and Mitani 2002). It bears emphasizing that all estimates of meat acquisition should be viewed very cautiously since they make several simplifying assumptions and are based on incomplete observations of animals. Estimates do not account for interindividual differences in behavior nor considerable seasonal variation in consumption (see below). It is also important to recognize that despite the seemingly high levels of consumption reflected by these estimates, meat remains a scarce and valuable resource, representing less than 5 percent of a chimpanzee's total annual diet (Goodall 1986; McGrew 1992).

Hunting Seasonality. Hunting frequency varies substantially over time (Mitani and Watts 2005a). During some months, chimpanzees do not hunt at all. Alternatively, chimpanzees occasionally engage in periodic "binges" or "crazes" during which they hunt red colobus prey nearly every day for periods of up to ten weeks (Stanford et al. 1994a; Mitani and Watts 1999; Hosaka et al. 2001; Watts and Mitani 2002). Several factors are likely to affect this seasonal variation in hunting activity (Mitani and Watts 2005a). One prominent hypothesis suggests that chimpanzees hunt because they are hungry. According to this hypothesis, chimpanzees increase their hunting effort to compensate for nutritional shortfalls created by seasonal periods of low fruit availability

Table 11.3. Vertebrate prey of wild chimpanzees (Goodall 1986; Boesch and Boesch 1989; Uehara et al. 1992; Stanford et al. 1994a; Mitani and Watts 1999; Boesch and Boesch-Achermann 2000; Hosaka et al. 2001; Watts and Mitani 2002)

Artiodactyla
Cephalophus callipygus (red duiker)
Cephalophus monticola (blue duiker)
Phacochoerus aethiopicus (warthog)
Potamochoerus porcus (bushpig)
Tragelaphus scriptus (bushbuck)
Carnivora
Viverra civetta (African civet)
Galliformes
Guttera edouardi (guinea fowl)
Primates
Cercopithecus aethiops (vervet monkey)
Cercopithecus ascanius (redtail monkey)
Cercopithecus diana (diana monkey)
Cercopithecus mitis (blue monkey)
Cercopithecus mona (mona monkey)
Cercopithecus petaurista (lesser spot-nosed monkey)
Cercocebus atys (sooty mangabey)
Colobus guereza (guereza monkey)
Colobus polykomos (black and white colobus)
Colobus verus (olive colobus)
Galago spp. (unidentified galago)
Lophocebus albigena (gray-cheeked mangabey)
Papio anubis (olive baboon)
Perodicticus potto (potto)
Procolobus badius (red colobus)
Rodentia
Rat (unidentified species)
Squirrel (unidentified species)
Thryonomys swinderianus (cane rat)

(Teleki 1973, Stanford 1996 1998; Boesch and Boesch-Achermann 2000). The nutritional-shortfall hypothesis is appealing in its simplicity, but the only direct test of it has produced results contrary to prediction. Chimpanzees at Ngogo increase, rather than decrease, their hunting activity during seasonal periods of high fruit availability (Mitani and Watts 2001, 2005a). These observations suggest that instead of providing a fallback energy source, hunting may represent a "leisure" activity that occurs only during permissible ecological conditions. When fruit is plentiful, chimpanzees form large parties (Chapman et al. 1995; Wrangham 2000; Mitani et al. 2002a), which make it easier for them to capture red colobus prey (see above). Chimpanzees forgo hunting attempts when they gather in smaller parties during lean periods of fruit scarcity because the odds of success are reduced considerably.

Hunting Tactics. Wild chimpanzees are frequently portrayed as cooperative hunters in the scientific and lay literature (de Waal 2001a, 2005). It may come as a surprise then to learn that a clear consensus on this issue does not exist among those who actually study the hunting behavior of chimpanzees (review in Muller and Mitani 2005). Boesch and Boesch, (1989) reported that chimpanzees at Taï show a high level of behavioral coordination during group hunts, with individuals performing complementary roles as they pursue arboreal red colobus prey. While acknowledging that limited coordination may take place during hunts, other researchers note that observation conditions make it hard to ascertain with certainty whether chimpanzees cooperate during hunts (Stanford 1998; Hosaka et al. 2001; Watts and Mitani 2002). Because it is difficult to observe the behavior of individual chimpanzees during hunts, others have analyzed the outcomes of hunts to evaluate the fitness effects of cooperation. Cooperative hunting will evolve if individuals who hunt in groups obtain a net fitness payoff relative to solitary hunters (Creel 1997). In the only test of this hypothesis, Boesch (1994b) suggested that at Taï chimpanzees who hunt in groups do better than solitary hunters, but this conclusion is contradicted by a reanalysis of the data that reveals solitary hunters actually acquire *more* meat for their hunting efforts than do individuals who hunt in groups (Muller and Mitani 2005). In sum, whether chimpanzees hunt cooperatively remains an open question. A more convincing case for cooperation can be found in the context of meat sharing.

Meat Sharing. No chimpanzee behavior is as seemingly paradoxical as meat sharing. Meat is costly to obtain and clearly valued by all chimpanzees, yet possessors will readily share with conspecifics. Why? Some observations suggest that chimpanzees share meat with others to ensure their cooperation during hunts (Boesch 1994b) or that males swap meat with estrous females to obtain matings (Stanford 1996, 1998; Stanford et al. 1994b), but neither hypothesis has received widespread empirical support (Muller and Mitani 2005; Gilby 2006). Current evidence is largely consistent with a third hypothesis that proposes male chimpanzees share meat strategically with others to build and strengthen social bonds between them (Nishida et al. 1992; but see Gilby 2006). At Mahale, Ntologi, a particularly clever alpha male, shared meat nonrandomly and selectively with other males, who in turn supported him in long-term alliances. These alliances helped Ntologi maintain his position at the top of dominance hierarchy for an astounding 16-year period (Uehara et al. 1994). Observations at Ngogo furnish additional support for the male social-bonding hypothesis (Mitani and Watts 2001). Male chimpanzees there are the most frequent participants in meat sharing episodes. Males swap meat nonrandomly with specific individuals, and sharing is evenly balanced within dyads (Mitani 2006b). Meat is also shared reciprocally, with males exchanging meat for coalitionary support (Mitani and Watts 2001). Additional analyses indicate that males also trade grooming for meat (Mitani 2006b).

Male dominance and territoriality

Male aggression is one of the most conspicuous aspects of wild chimpanzee behavior. Male chimpanzees are much more aggressive than females are (Goodall 1986; Muller 2002). Male aggression occurs primarily in the context of competition for dominance status within communities. Male chimpanzees are inordinately preoccupied with striving for dominance status. They perform elaborate displays to intimidate conspecifics and to maintain or challenge the existing dominance hierarchy. Males frequently enlist the support of others in short-term coalitions and long-term alliances to help them achieve or maintain their dominance status (Riss and Goodall 1977; Nishida 1983; Nishida et al. 1992; Uehara et al. 1994; Nishida and Hosaka 1996). Maintaining dominance has demonstrable costs. High-ranking males possess high levels of cortisol, the so-called "stress" hormone (Muller and Wrangham 2004), and dominance interactions can lead to severe injuries with fatal consequences (Fawcett and Muhumuza 2000; Goodall 1992; Nishida et al. 1995; Nishida 1996; Watts 2004). Despite these costs, high-ranking male chimpanzees derive significant reproductive benefits. Recent studies from Gombe, Taï, and Mahale indicate that high-ranking males father more offspring than do low-ranking males (Constable et al. 2001; Vigilant et al. 2001; Boesch et al. 2006; Inoue et al. 2008).

Male aggression by chimpanzees is also prominently displayed in their territorial behavior. Male chimpanzees defend their territories vigorously against neighbors (Goodall et al. 1979; Boesch and Boesch-Achermann 2000; Watts and Mitani 2001; Williams et al. 2004). Males actively patrol the boundaries of their territory, occasionally making deep incursions into those of neighbors (Goodall et al. 1979; Boesch and Boesch-Achermann 2000; Watts and Mitani 2001; Mitani and Watts 2005b). Encounters take place between members of different communities, either by design during patrols or by chance while chimpanzees feed in the peripheral parts of their territory. During some encounters, male chimpanzees launch communal gang attacks on members of other groups, with such episodes sometimes resulting in fatalities (Goodall et al. 1979; Wilson et al. 2004; Watts et al. 2006). Adult males and infants are frequent victims of these attacks, although females are killed as well (Wilson and Wrangham 2003). For reasons that we do not currently understand, male chimpanzees typically cannibalize infants after killing them. Why male chimpanzees kill individuals from other communities has only been made clear recently. Current evidence is consistent with the hypothesis that chimpanzees employ lethal intercommunity aggression to reduce the coalitionary strength of their neighbors and to expand their territories (Wrangham 1999; Mitani et al. 2010). Territorial expansion in turn increases the availability of food resources and thereby improves female reproduction (Williams et al. 2004). Killing neighboring conspecifics and territory expansion may also facilitate the recruitment of females (Goodall et al. 1979; Nishida et al. 1985).

The preceding review provides a basis for comparing the behavior of chimpanzees and humans. Behavioral similarities can be used to reconstruct the behavior of our common ancestor with chimpanzees, while behavioral differences point to unique changes that occurred during the course of human evolution. In the next section, I review studies that have employed the behavior of chimpanzees to model human behavioral evolution.

CHIMPANZEE MODELS OF HUMAN BEHAVIORAL EVOLUTION

Several attempts have been made to reconstruct the behavior of early humans based on information provided by the behavior of wild chimpanzees (Table 11.4). These attempts employ three distinct strategies. First, some have adopted an explicitly cladistic framework to reconstruct the behavior of the common ancestor of chimpanzees and humans. Second, others have made inferences about early human behavior by focusing on single traits, such as tool use, hunting, meat sharing, and intergroup aggression. Finally, others have compared the

Table 11.4. Chimpanzee models of human behavioral evolution.

Type of model	*Reference*
Cladistic analysis	Ghiglieri 1987; Wrangham 1987; Foley 1989
Trait models	
tool use	Tanner and Zihlman 1976; McGrew 1979; Tanner 1987
hunting	Stanford 1995, 1996, 1998, 1999
meat sharing	Stanford 2001
lethal coalitionary aggression	Wrangham and Peterson 1996; Wrangham 1999
Behavioral reconstructions	Goodall and Hamburg 1975; McGrew 1981; Boesch-Achermann and Boesch 1994; Boesch and Boesch-Achermann 2000

behavior of extant chimpanzees and humans to develop scenarios regarding the behavioral repertoire of early humans.

Cladistic analyses

Wrangham (1987) was the first to employ a cladistic approach to reconstruct the behavior of ancestral humans. Because the phylogenetic relationships between chimpanzees, bonobos, gorillas, and humans had not yet been fully resolved (see above), Wrangham (1987) compared the behavior of all four species to reconstruct the behavior of their common ancestor. To do so, he first summarized behavioral similarities in social organization between these closely related species. Assuming that these similarities represented behavioral homologies, he hypothesized that they would be present in the common ancestor. Based on this analysis, Wrangham (1987) concluded that the common ancestor of chimpanzees, bonobos, gorillas, and humans was likely to have lived in social groups with female exogamy. Social bonds among females were probably only weakly developed, while males pursued matings with multiple females. Intergroup relationships were hostile and dominated by males.

Using the same logic and approach, Ghiglieri (1987) adopted the now generally accepted phylogeny that indicates humans are more closely related to chimpanzees and bonobos than they are to gorillas. Using this phylogeny he made inferences regarding the behavior of the common ancestor of chimpanzees, bonobos, and humans. Ghiglieri (1987) concurred with the conclusions drawn by Wrangham, but suggested that several other traits could be added to the behavioral repertoire of the common ancestor of these three species. These include frequent solitary travel by females; female sociality based on attraction to the same male; male group territoriality; relatively weak male mating competition within communities; and the development of cooperative alliances between males. Ghiglieri (1987) concluded that the retention of males within communities is the driving force that accounts for many of the similarities between humans, chimpanzees, and bonobos. This sets the stage for the evolution of cooperative alliances between male kin, leading to communal reproductive strategies, such as group territoriality.

In a third cladistic analysis, Foley (1989), like Wrangham and Ghiglieri before him, argued that the social organization of the common ancestor of chimpanzees, bonobos, and humans resembled that of modern chimpanzees. Foley (1989) based this claim on the suggestion that strong social bonds between genetically related males and between the sexes represent derived traits among the hominoids, and that the polarity of change indicates that chimpanzees lie at an intermediate state with respect to these characters. If one accepts this scenario at face value, Foley's (1989) reconstruction indicates that the social system of modern chimpanzees is a good model for that of the common ancestor of chimpanzees, bonobos, and humans. As such, it is remarkably congruent with conclusions drawn earlier by Wrangham (1987) and Ghiglieri (1987), which also suggest a chimpanzeelike human ancestor.

Trait models

Tool Use. In a second set of models, some researchers have made inferences about human behavioral evolution by focusing on specific behaviors shared by chimpanzees and humans. Using early observations made at Gombe, McGrew (1979) documented sex differences in chimpanzee tool use and hunting. Female chimpanzees make and use tools extensively to gather insects and do so much more frequently than males. In contrast, male chimpanzees hunt and eat more vertebrate prey than females. Members of both sexes rarely hunt with the aid of tools (Whiten et al. 1999). Based on these observations, McGrew (1979) drew parallels between chimpanzee sex differences in foraging and the division of labor that exist in modern hunter–gatherer societies. He went on to suggest that sex differences in chimpanzee tool use and foraging had two important implications for understanding human evolution. First, he speculated that tool use by humans first emerged via solitary female feeding activities rather than male hunting (cf. Pruetz and Bertolani 2007). Second, McGrew (1979) noted that in chimpanzee society most food is shared between mother and offspring, and because of this, food sharing by females probably preceded the evolution of hunting by human males.

What types of habitats did early hominins occupy? Considerable controversy surrounds this question (see Chapters 1–4, this volume). Most paleoecological reconstructions agree, however, that sometime during the course of our evolution, we made a transition from a forested habitat to a woodland–savanna ecosystem. Regardless of whether this transition took place early or late, a shift in diet is likely to have occurred (but see Grine et al. 2006). Some investigators have suggested that observations of chimpanzee tool use provide a means to understand how early hominins made this dietary transition.

Reacting to early models of human evolution that placed a primacy on hunting by males (Washburn and Lancaster 1968), Tanner and Zihlman (1976) and Tanner (1987) argued that it was equally, if not more important, to focus on the subsistence activities of females. Impressed by observations of frequent tool use by female chimpanzees, they hypothesized that the key behavioral innovation that led to humans involved food gathering with tools practiced by females. According to this view, tool-assisted gathering allowed early humans to adapt to life on the savannas. Yamakoshi (2001) has elaborated this proposal. Chimpanzees are frugivores, who experience seasonal shortfalls in the availability of fruit. In a thorough review, Yamakoshi (2001) noted that tool use by wild chimpanzees permits them to expand their feeding niche in important ways. Chimpanzees utilize tools seasonally to exploit foods that are otherwise inaccessible and whose nutritional payoffs can be enormous. A good example involves nut cracking by West African chimpanzees (see above). In this case, chimpanzees use tools to buffer the effects created by seasonally scarce food resources. Yamakoshi (2001) suggested that tool use by early hominins could have permitted them to expand their feeding niche in a similar way. Because meat or underground storage organs may have been important during the transition to a more open habitat (Peters and O'Brien 1981; Sponheimer and Lee-Thorp 1999; de Heinzelin et al. 1999; Wrangham 2001), the advent of a "hunting" tool or "digging" tool may have facilitated expansion of the early hominin dietary niche (Yamakoshi 2001).

Meat Eating and Sharing. In a series of papers and books, Stanford (1995 1996, 1998, 1999) has used information regarding the hunting behavior of chimpanzees to make inferences about the behavior of early hominins. He suggests that the hunting behavior of early hominins resembled that of modern chimpanzees in several ways. According to his reconstruction, early hominins hunted small- to medium-size prey opportunistically and seasonally within a circumscribed part of their territory. In addition, hunting by early hominins was conducted by males, who shared favored parts, such as marrow and brains, to promote their own social and reproductive interests.

Patterns of meat sharing observed in wild chimpanzees furnish the basis for Stanford's claims regarding early hominin meat-sharing behavior. These same observations have also suggested to him a novel hypothesis concerning its significance during our evolution (Stanford 2001). Early models of human evolution, most notably the "Man the Hunter" hypothesis (Washburn and Lancaster 1968), placed emphasis on the acquisition of meat as the driving force behind the evolution of human behavior. These models have been widely criticized on a variety of conceptual and empirical grounds, and as a consequence, have fallen into disfavor (Tanner and Zihlman 1976; Shipman 1986; Blumenschine and Cavallo 1992; Stanford 1995, 1996). Stanford (2001) has nevertheless continued to argue that meat eating played a critical role in human evolution, stressing that the control and distribution of meat is important, rather than its capture. Recent observations that show male chimpanzees share meat nonrandomly and reciprocally to curry the favor and support of allies suggest that chimpanzees share for social and political purposes (Nishida et al. 1992; Mitani and Watts 2001). Stanford (2001) proposes that the high degree of social strategizing and mental scorekeeping required for such tasks creates a positive selective force for the evolution of intelligence (Humphrey 1976; Byrne 1996; Dunbar 1998).

Lethal Intergroup Aggression. Lethal coalitionary aggression is rare among mammals, taking place in chimpanzees, humans, and some social carnivores (Wrangham 1999). In chimpanzees and humans, such aggression occurs when groups of males make communal gang attacks on neighbors (Wilson and Wrangham 2003). Attacks in both species are often preceded by active searches during territorial boundary "patrols" (Watts and Mitani 2001; Mitani and Watts 2005b). In these patrols, parties of males move together to the periphery of their territory and into those of neighbors. Patrollers move silently and surreptitiously as they seek vulnerable individuals from other groups. If contact is made, patrollers may launch a surprise attack with lethal consequences. The limited taxonomic distribution of lethal intergroup aggression shared by two closely related species, chimpanzees and humans, led Wrangham (1999; Wrangham and Peterson 1996) to propose that its evolutionary roots run deep in both lineages. The pattern of intergroup aggression in chimpanzees and humans combines several unusual features: raids are unprovoked, are conducted surreptitiously, and can lead to fatalities. This unique combination of features suggests that the behaviors are homologous and that lethal intergroup aggression characterized the behavior of the common ancestor of chimpanzees and humans (Wrangham and Peterson 1996).

Behavioral reconstructions of early human behavior

Goodall and Hamburg (1975) reasoned that behavioral similarities between modern chimpanzees and humans were likely to represent homologies and were thus present in the common ancestor as well as the first humans. Drawing largely on Goodall's own studies at Gombe and what was then known about wild chimpanzees, they noted that several similarities characterized the behavior of chimpanzees and humans. These included toolmaking and tool using; hunting, cooperation, and food sharing; a prolonged period of infant dependence on mothers; enduring social bonds between mothers and offspring and between siblings; and some gestures and postures used in nonverbal communication. Despite these similarities, Goodall and Hamburg (1975) pointed out that humans display several behaviors that differentiate us from chimpanzees. Bipedal locomotion, obligatory hunting, continuous female sexual receptivity, and language are uniquely human attributes that appeared sometime during the course of our evolution.

Following the lead of Goodall and Hamburg (1975), McGrew (1981) furnished an update of what was then current knowledge about the behavior of wild chimpanzees. He used this information to suggest a series of behavioral changes that occurred during the early part of human evolution. Based on a comparison of modern chimpanzees and humans, he concluded that behavioral changes occurred in subsistence, technology, mating behavior, and intergroup relations. Key changes involved food sharing, a sexual division of labor with men hunting and women gathering, extensive tool use, sexual bonding between the sexes with accentuated levels of male parental care, and nonaggressive social relations between groups.

Many scenarios propose that ecological change, in the form of a shift from forest to woodland–savanna habitat, led to the evolution of some of key behavioral adaptations that make humans distinct from our ape relatives. Hunting and tool use are two frequently cited examples (Washburn and Lancaster 1968; Tanner and Zihlman 1976; McGrew 1981; Moore 1996; Yamakoshi 2001). Boesch-Achermann and Boesch (1994; Boesch and Boesch-Achermann 2000), however, proposed that some of these behaviors may have had their start in the forest occupied by the common ancestor of chimpanzees and humans. They base this suggestion on their observations of forest-dwelling chimpanzees at Taï. Compared with chimpanzees that live in more open habitats, Taï forest chimpanzees have been reported to "use more tools, make them in more different ways, hunt more frequently and more often in groups, and show more frequent cooperation and food sharing" (Boesch-Achermann and Boesch 1994,

10). In a more recent discussion, Boesch and Boesch-Achermann (2000) argued that tool use and hunting were likely to have persisted as major components in the behavioral repertoire of the earliest humans such as *Ardipithecus* and the australopiths.

CHALLENGES, CONTROVERSIES, AND UNRESOLVED PROBLEMS

While knowledge of chimpanzee behavior can undoubtedly provide insights into the evolution of human behavior, applying this knowledge is not without its difficulties (cf. Sayers and Lovejoy 2008). Previous authors have seldom discussed potential problems, and here I draw attention to some of these that have not received widespread attention or discussion. In a few cases the internal validity of chimpanzee models can be questioned if they do not accord with what we know about the behavior of wild chimpanzees. Intraspecific variation in behavior, difficulties associated with homologizing behavior, and the fact that bonobos are a sister species of chimpanzees create additional problems for chimpanzee models of human behavioral evolution.

Internal validity of models

Ongoing field research continues to alter our understanding of the behavior of chimpanzees, and changes in this understanding may require us to modify reconstructions of early human behavior. For example, Foley (1989, 1999) has been a proponent of a chimpanzeelike common ancestor of humans and chimpanzees. He has used the behavior of chimpanzees to hypothesize that strong kin bonds among males characterized the behavior of the common ancestor. Ghiglieri (1987) has gone further to argue that male kin bonds account for relatively weak male mating competition within communities of ancestral humans.

The idea that male kinship mitigates intracommunity male competition in chimpanzees permeates the literature. Some authors even suggest that chimpanzees as a species are characterized by a relatively low degree of male competition (Plavcan and van Schaik 1997). This view is difficult to reconcile with the severe aggression that commonly takes place between male chimpanzees living in the same community and the fact that these same males will sometimes kill each other (Fawcett and Muhumuza 2000; Goodall 1992; Nishida et al. 1995; Nishida 1996; Watts 2004).

Additional research questions the importance of male kin bonds in chimpanzee society. Because females typically disperse, male chimpanzees in some communities are more closely related to other males than females are to each other

(Inoue et al. 2008; Langergraber et al., unpublished data). Male chimpanzees are not closely related to each other in all communities, however (Vigilant et al. 2001; Lukas et al. 2005). Recently, we have used an array of genetic markers derived from the autosome, Y chromosome, X chromosome, and mtDNA to assay genetic relatedness between pairs of male chimpanzees living in an unusually large community at Ngogo, Kibale National Park, Uganda (Langergraber et al. 2007). Combining these data with behavioral observations confirms the importance of kinship as males cooperate preferentially with their maternal brothers. Nevertheless, paternal brothers fail to preferentially interact and most cooperation occurs between males who are unrelated or distantly related (Langergraber et al. 2007). Thus, kinship has only a limited impact on the lives of male chimpanzees. Examples such as these indicate that reconstructions of human behavioral evolution may require revision as our understanding of the behavior of chimpanzees continues to grow.

The problem of intraspecific variation

Chimpanzees are well known for displaying considerable intraspecific variation in behavior (Whiten et al. 1999; Boesch et al. 2002). Intraspecific variation in behavior represents a problem for cladistic analyses that attempt to reconstruct the behavior of early humans. Reconstructions based on species-typical characteristics of chimpanzees and humans disregard the inherent complexity that exists in the behavior of both species. Two examples highlight the problem. Ignoring for a moment whether it is appropriate to equate sex-biased dispersal in chimpanzees with patrilocal residence in humans (see below), consider the heterogeneous nature of these characteristics in each species. As noted previously, at some field sites most female chimpanzees leave their natal groups and disperse to the territories of neighbors. A large fraction of females, however, fail to disperse at some sites. Although humans are frequently characterized as patrilocal for purposes of phylogenetic reconstructions of human behavior, considerable variability exists in human residence patterns. Based on a careful review of the literature, Alvarez (2004) has recently concluded that the situation is much more complicated and variable. Humans appear to display a pattern of bilocal residence with individuals moving to the homes of either husbands or wives. Those who have applied a cladistic approach to reconstruct the behavior of ancestral humans have realized that variability of this sort creates a problem for analysis (Wrangham 1987; Foley 1989), but these same authors seldom grapple with it in a satisfying way. More careful thought to this issue is clearly needed.

Stanford's (1995, 1996, 1999) efforts to reconstruct the hunting behavior and ecology of early humans furnish a second example of some of the difficulties posed by ignoring intraspecific variation in chimpanzee behavior. Stanford makes a comprehensive attempt to address several questions regarding early human hunting, but his reconstruction is less than compelling because it is based largely on what we know about the hunting behavior of chimpanzees at a single site, namely Gombe, at a single time. For instance, Stanford (1995, 1996, 1999) uses observations of the Gombe chimpanzees to argue that early humans hunted opportunistically. This claim, however, neglects the apparently intentional hunting patrols conducted by chimpanzees at Ngogo and Taï (see above). Stanford's reconstruction is further weakened because it simply transfers the behavior of one population of chimpanzees directly to that of early humans without acknowledging the possibility of any evolutionary change. This raises a question of whether it is even possible to homologize behavior.

Does homology apply to behavior?

Comparisons of living chimpanzees and humans furnish the means to establish behavioral similarities between them. These behavioral similarities are typically assumed to be homologous, which permits them to be used to reconstruct the behavior of our human ancestors. While the evolutionary logic underlying this approach is unquestionable, some investigators have nonetheless argued that behavior cannot be homologized and is therefore of limited phylogenetic utility. According to this view, homology by definition applies only to morphology (Atz 1970; Klopfer 1975; Hodos 1976).

The idea that homology is inexorably tied to morphology and morphology alone is ingrained in the literature dating back to Owen's (1843) original non-evolutionary use of the term. Adhering to a strict morphological definition of homology that excludes behavior, however, is difficult to justify insofar as it suggests that behavior does not evolve. This, of course, is patently false. Behavior is variable, heritable, affected by natural selection, and does change over time. If behavior evolves like any morphological structure, then there is no a priori reason exclude the possibility of homologous behaviors. A persistent view that behavior is too labile and prone to homoplasy has led some to reject the concept of behavioral homology (Atz 1970; Klopfer 1975). This claim is also difficult to defend. Empirical observations reveal that behavior is neither more variable nor more susceptible to convergence than morphology (Sanderson and Donoghue 1989; deQueiroz and Wimberger 1993; Greene 1994). In sum, there is no theoretical or empirical reason to suggest that behavior cannot be homologized.

For purposes of reconstructing the behavior of early humans, a greater concern involves identifying homologous behaviors.

How does one recognize behavioral homology?

Though an infrequent topic of study, investigating behavioral homology is, in principle, no different than investigating homology in anatomical structures (Lauder 1986; Wenzel 1992; Greene 1994). Making an effective case for homology in either situation, however, requires that we are comparing the same thing. Several criteria have been proposed to recognize homologies. A clear consensus on this matter does not exist, but a common theme that emerges in all discussions places significance on similarity in form. Geoffroy Saint-Hilaire (1818) was the first to propose that similarity in structure can be used to recognize homologies. Remane (1952) echoed this view with his frequently cited criterion of "special quality." Here two traits are considered homologous if they can be shown to share a special quality, such as structural similarity. Structural attributes relevant to behavior include the underlying anatomy and neural systems used to perform them (Haas and Simpson 1946; Baerends 1958; Simpson 1958; Pribram 1958). While the criterion of "special quality" is not without problems, it emphasizes that behavioral homologies are fundamentally rooted in similarities in structure.

For purposes of reconstructing the behavior of early humans, behavioral similarities between chimpanzees and humans are often simply assumed to be homologous (see above). This assumption is only occasionally questioned (Ghiglieri 1987), and clearly warrants greater scrutiny, especially in light of the complexity of the behaviors being compared. Can we be assured that tool using and hunting are the same in chimpanzees and humans? Is it fair to equate sex-biased dispersal and mating patterns in chimpanzees with patrilocal residence and marriage systems in humans? Do intergroup aggression in chimpanzees and warfare in humans represent homologous behaviors?

Applying a structural-homology criterion to the hunting behavior of chimpanzees and humans serves to illustrate the problem of assuming homology. Chimpanzees hunt arboreal monkeys while humans hunt terrestrial prey. Human hunting involves planning and cooperation, while predation by chimpanzees is frequently opportunistic. These comparisons suggest that the anatomical and neural structures underlying chimpanzee and human hunting may not be same, forcing one to reevaluate them as behavioral homologies. Additional functional considerations reinforce this conclusion. Chimpanzees differ from humans not only in how they hunt, but why they hunt. While meat eating is obligatory for humans living in some areas, there is no evidence that meat is required for chimpanzees to survive and reproduce. Field studies reveal considerable heterogeneity in the degree to which chimpanzees consume meat in the wild (Mitani and Watts 1999), with some populations of chimpanzees preying on other vertebrates only rarely (Yamakoshi 1998; Newton-Fisher et al. 2002). In sum, it remains an open question whether chimpanzee hunting and human hunting represent the same behavior, and as a result, more prudence may be required before using data on chimpanzee hunting to make inferences about human evolution (cf. Sayers and Lovejoy 2008). Similar care is necessary before assuming homology in other complex behaviors that appear to be shared by chimpanzees and humans (cf. Kelly 2005).

Bonobos are a sister taxon of chimpanzees

The weight of current evidence indicates that chimpanzees and bonobos shared a common ancestor, with humans branching off at some earlier time (Figure 11.1c). Most chimpanzee models of human evolution, however, fail to incorporate the behavior of all three species to reconstruct ancestral states. Why this is a potential problem can be illustrated by considering lethal coalitionary aggression. Intergroup interactions between members of different bonobo communities vary. While some are peaceful (Idani 1990), others are marked by aggression (Kano 1992; personal observation). Despite the variable nature of these encounters, the pattern of coalitionary killing of neighbors reported in chimpanzees has not been observed in bonobos.

Assuming for a moment that lethal coalitionary aggression in humans and chimpanzees represent homologous behaviors, what are we to do with the observation that bonobos do not display this behavior? One reconstruction would assign lethal aggression to the common ancestor of all three species; only later was it lost in the bonobo. In contrast, an alternative hypothesis is that lethal aggression evolved independently in chimpanzees and humans. Clearly, any argument for lethal coalitionary aggression by early humans as a consequence of shared ancestry would be on much firmer ground if it were practiced on a regular and frequent basis by all three species. Alternative efforts to use the bonobo to model human evolution suffer from the same logical problem (Zihlman et al. 1978; Zihlman 1996; de Waal 2001b). Because chimpanzees and bonobos both share a common ancestor with humans and the three species differ in several ways, it is difficult to reconstruct the behavior of early humans based on interspecific comparisons.

Some authors have recognized the problem outlined above and have explicitly addressed it. For instance, Wrangham and Pilbeam (2001) suggested that outgroup comparison with gorillas furnishes a means to establish whether traits that differ between chimpanzees and bonobos represent homologies or homoplasies with respect to humans. They note that traits reliably identified as homologous and shared between gorillas and either species of *Pan* are expected to be present in the common ancestor. Alternate versions of the trait in the other *Pan* would be absent in the ancestor and thus derived. In proposing this research strategy, Wrangham and Pilbeam (2001) recognize that it will be impractical to implement in many situations because of the difficulty of defining traits and recognizing whether they represent homologies or homoplasies. They nevertheless suggest that one clear case exists. Following the work of Shea (1983), Wrangham and Pilbeam (2001) note that the pattern of cranial ontogeny differs considerably between chimpanzees and bonobos. The chimpanzee pattern, however, resembles that displayed by gorillas. Thus, the common ancestor was likely to have been more chimplike than bonobolike with respect to this character. Whether we can generalize from this single example and conclude that the common ancestor resembled modern chimpanzees as a whole is unclear. Following the same logic outlined above, the absence of lethal coalitionary aggression in bonobos and gorillas would seem to imply that its presence in chimpanzees and humans is due to convergence rather than shared ancestry.

CONCLUDING COMMENTS

Chimpanzee models of human behavioral evolution primarily draw upon similarities in the behavior of these two species. Differences as well as similarities, however, play an important role in the reconstruction of human evolution (cf. Sayers and Lovejoy 2008). Insofar as differences in the behavior of chimpanzees and humans represent adaptive responses to specific ecological conditions, they illuminate how we as a species met the challenges of life during the course of evolution. One significant difference bears final comment. As the human population relentlessly grows at an exponential clip, chimpanzees in the wild continue to disappear at an alarming rate. Unless swift and effective action is taken soon, it is unclear whether the next generation of field researchers will be able to the study the behavior of chimpanzees and continue to shed light on our own past.

ACKNOWLEDGMENTS

I thank Matt Sponheimer, Julia Lee-Thorp, Kaye Reed, and Peter Ungar for inviting me to participate in the Pliocene Hominin Paleoecology Workshop and to contribute this chapter. I am grateful to P. Ungar and M. Sponheimer for helpful comments on the manuscript. My field research on chimpanzees has been supported by grants from the Detroit Zoological Institute, L.S.B. Leakey Foundation, NSF (SBR-9253590, BCS-0215622, and IOB-0516644), University of Michigan, and Wenner-Gren Foundation.

REFERENCES

Alvarez, H. 2004. "Residence Groups among Hunter-Gatherers: A View of the Claims and the Evidence for Patrilocal Bands." In *Kinship and Behavior in Primates*, ed. B. Chapais and C. Berman, 420–442. Oxford: Oxford University Press.

Andrews, P., and L. Martin. 1987. "Cladistic Relationships of Extant and Fossil Hominoids." *Journal of Human Evolution* 16 (1): 101–18. http://dx.doi.org/10.1016/0047-2484(87)90062-5.

Atz, J. 1970. "The Application of the Idea of Homology to Behvior." In *Development and Evolution of Behavior*, ed. L. Aronson, E. Tobach, and D. Lehrman, 53–74. San Francisco: W. H. Freeman.

Baerends, G. 1958. "Comparative Methods and the Concept of Homology in the Study of Behavior." *Archives Neerlandaises de Zoologie Supplement* 13: 401–17.

Bermejo, M., and G. Illera. 1999. "Tool-Set for Termite Fishing and Honey Extraction by Wild Chimpanzees in the Lossi Forest, Congo." *Primates* 40 (4): 619–27. http://dx.doi .org/10.1007/BF02574837.

Blumenschine, R. J., and J. A. Cavallo. 1992. "Scavenging and Human Evolution." *Scientific American* 267 (4): 90–6. http://dx.doi.org/10.1038/scientificamerican1092-90. Medline:1411457

Boesch, C. 1994a. "Chimpanzees–Red Colobus Monkeys: A Predator-Prey System." *Animal Behaviour* 47 (5): 1135–48. http://dx.doi.org/10.1006/anbe.1994.1152.

Boesch, C. 1994b. "Cooperative Hunting in Wild Chimpanzees." *Animal Behaviour* 48 (3): 653–67. http://dx.doi.org/10.1006/anbe.1994.1285.

Boesch, C. 1994c. "Hunting Strategies of Gombe and Tai Chimpanzees." In *Chimpanzee Cultures*, ed. R. Wrangham, W. McGrew, F. de Waal, and P. Heltne, 77–92. Cambridge, MA: Harvard University Press.

Boesch, C. 1996. "Social Grouping in Tai Chimpanzees." In *Great Ape Societies*, ed. W. McGrew, L. Marchant, and T. Nishida, 101–113. Cambridge: Cambridge University Press. http://dx.doi.org/10.1017/CBO9780511752414.010

Boesch, C., and H. Boesch. 1981. "Sex Differences in the Use of Natural Hammers by Wild Chimpanzees: A Preliminary Report." *Journal of Human Evolution* 10 (7): 585–93. http:// dx.doi.org/10.1016/S0047-2484(81)80049-8.

Boesch, C., and H. Boesch. 1983. "Optimization of Nut-Cracking with Natural Hammers by Wild Chimpanzees." *Behaviour* 83 (3): 265–86. http://dx.doi.org/10.1163 /156853983X00192.

Boesch, C., and H. Boesch. 1989. "Hunting Behavior of Wild Chimpanzees in the Taï National Park." *American Journal of Physical Anthropology* 78 (4): 547–73. http://dx.doi .org/10.1002/ajpa.1330780410. Medline:2540662

Boesch, C., and H. Boesch-Achermann. 2000. *The Chimpanzees of the Taï Forest*. Oxford: Oxford University Press.

Boesch, C., G. Hohmann, and L. Marchant. 2002. *Behavioral Diversity in Chimpanzees and Bonobos*. Cambridge: Cambridge University Press. http://dx.doi.org/10.1017/CBO9780511606397

Boesch, C., G. Kohou, H. Néné, and L. Vigilant. 2006. "Male Competition and Paternity in Wild Chimpanzees of the Taï Forest." *American Journal of Physical Anthropology* 130 (1): 103–15. http://dx.doi.org/10.1002/ajpa.20341. Medline:16353223

Boesch-Achermann, H., and C. Boesch. 1994. "Hominization in the Rainforest: The Chimpanzee's Piece of the Puzzle." *Evolutionary Anthropology* 3 (1): 9–16. http://dx.doi.org/10.1002/evan.1360030106.

Byrne, R. 1996. "Machiavellian Intelligence." *Evolutionary Anthropology* 5 (5): 172–80. http://dx.doi.org/10.1002/(SICI)1520-6505(1996)5:5<172::AID-EVAN6>3.0.CO;2-H.

Chapman, C., L. Chapman, and R. Wrangham. 1995. "Ecological Constraints on Group Size: An Analysis of Spider Monkey and Chimpanzee Subgroups." *Behavioral Ecology and Sociobiology* 36 (1): 59–70. http://dx.doi.org/10.1007/BF00175729.

Chapman, C., and R. Wrangham. 1993. "Range Use of the Forest Chimpanzees of Kibale: Implications for the Understanding of Chimpanzee Social Organization." *American Journal of Primatology* 31 (4): 263–73. http://dx.doi.org/10.1002/ajp.1350310403.

Chen, F. C., and W.-H. Li. 2001. "Genomic Divergences between Humans and Other Hominoids and the Effective Population Size of the Common Ancestor of Humans and Chimpanzees." *American Journal of Human Genetics* 68 (2): 444–56. http://dx.doi.org/10.1086/318206. Medline:11170892

Chimpanzee Sequence and Analysis Consortium. 2005. "Initial Sequence of the Chimpanzee Genome and Comparison with the Human Genome." *Nature* 437 (7055): 69–87. http://dx.doi.org/10.1038/nature04072. Medline:16136131

Constable, J. L., M. V. Ashley, J. Goodall, and A. E. Pusey. 2001. "Noninvasive Paternity Assignment in Gombe Chimpanzees." *Molecular Ecology* 10 (5): 1279–300. http://dx.doi.org/10.1046/j.1365-294X.2001.01262.x. Medline:11380884

Creel, S. 1997. "Cooperative Hunting and Group Size: Assumptions and Currencies." *Animal Behaviour* 54 (5): 1319–24. http://dx.doi.org/10.1006/anbe.1997.0481. Medline:9398386

de Heinzelin, J., J. D. Clark, T. White, W. Hart, P. Renne, G. WoldeGabriel, Y. Beyene, and E. Vrba. 1999. "Environment and Behavior of 2.5-Million-Year-Old Bouri Hominids." *Science* 284 (5414): 625–9. http://dx.doi.org/10.1126/science.284.5414.625. Medline:10213682

de Queiroz, A., and P. Wimberger. 1993. "The Usefulness of Behavior for Phylogeny Estimation: Levels of Homoplasy in Behavioral and Morphological Characters." *Evolution; International Journal of Organic Evolution* 47 (1): 46–60. http://dx.doi.org/10.2307/2410117.

de Waal, F. 2001a. *The Ape and the Sushi Master*. New York: Basic Books.

de Waal, F. 2001b. "Apes from Venus: Bonobos and Human Social Evolution." In *Tree of Origin*, ed. F. de Waal, 39–68. Cambridge, MA: Harvard University Press.

de Waal, F. B. 2005. "A Century of Getting to Know the Chimpanzee." *Nature* 437 (7055): 56–9. http://dx.doi.org/10.1038/nature03999. Medline:16136128.

Di Fiore, A., and D. Rendall. 1994. "Evolution of Social Organization: A Reappraisal for Primates by Using Phylogenetic Methods." *Proceedings of the National Academy of Sciences of the United States of America* 91 (21): 9941–5. http://dx.doi.org/10.1073/pnas.91.21.9941. Medline:7937922

Dunbar, R. 1998. "The Social Brain Hypothesis." *Evolutionary Anthropology* 6 (5): 178–90. http://dx.doi.org/10.1002/(SICI)1520-6505(1998)6:5<178::AID-EVAN5>3.0.CO;2-8.

Emery Thompson, M., J. H. Jones, A. E. Pusey, S. Brewer-Marsden, J. Goodall, D. Marsden, T. Matsuzawa, T. Nishida, V. Reynolds, Y. Sugiyama, et al. 2007. "Aging and Fertility Patterns in Wild Chimpanzees Provide Insights into the Evolution of Menopause." *Current Biology* 17 (24): 2150–6. http://dx.doi.org/10.1016/j.cub.2007.11.033. Medline:18083515

Fawcett, K., and G. Muhumuza. 2000. "Death of a Wild Chimpanzee Community Member: Possible Outcome of Intense Sexual Competition." *American Journal of Primatology* 51 (4): 243–7. http://dx.doi.org/10.1002/1098-2345(200008)51:4<243::AID-AJP3>3.0.CO;2-P. Medline:10941440

Foley, R. 1989. "The Evolution of Hominid Social Behaviour." In *Comparative Socioecology: The Behavioural Ecology of Humans and Other Mammals*, ed. V. Standen and R. Foley, 473–494. Oxford: Blackwell Scientific Publications.

Foley, R. 1999. "Hominid Behavioural Evolution: Missing Links in Comparative Primate Socioecology." In *Comparative Primate Socioecology*, ed. P. Lee, 363–386. Cambridge: Cambridge University Press. http://dx.doi.org/10.1017/CBO9780511542466.018

Ghiglieri, M. 1987. "Sociobiology of the Great Apes and the Hominid Ancestor." *Journal of Human Evolution* 16 (4): 319–57. http://dx.doi.org/10.1016/0047-2484(87)90065-0.

Gilby, I. 2006. "Meat Sharing among the Gombe Chimpanzees: Harassment and Reciprocal Exchange." *Animal Behaviour* 71 (4): 953–63. http://dx.doi.org/10.1016/j.anbehav.2005.09.009.

Gilby, I., L. Eberly, L. Pintea, and A. Pusey. 2006. "Ecological and Social Influences on the Hunting Behaviour of Wild Chimpanzees, *Pan troglodytes schweinfurthii*." *Animal Behaviour* 72 (1): 169–80. http://dx.doi.org/10.1016/j.anbehav.2006.01.013.

Goodall, J. 1963. "Feeding Behaviour of Wild Chimpanzees: A Preliminary Report." *Symposium of the Zoological Society of London* 10:39–48.

Goodall, J. 1964. "Tool-Using and Aimed Throwing in a Community of Free-Living Chimpanzees." *Nature* 201 (4926): 1264–6. http://dx.doi.org/10.1038/2011264a0. Medline:14151401

Goodall, J. 1986. *The Chimpanzees of Gombe*. Cambridge, MA: Belknap Press.

Goodall, J. 1998. "Learning from the Chimpanzees: A Message Humans Can Understand." *Science* 282 (5397): 2184–5. http://dx.doi.org/10.1126/science.282.5397.2184.

Goodall, J. 1992. "Unusual Violence in the Overthrow of an Alpha Male Chimpanzee at Gombe." In *Topics in Primatology*: Volume 1, *Human Origins*, ed. T. Nishida, W. McGrew, P. Marler, M. Pickford, and F. de Waal, 131–42. Tokyo: University of Tokyo Press.

Goodall, J., A. Bandora, E. Bergmann, C. Busse, H. Matama, E. Mpongo, A. Pierce, and D. Riss. 1979. "Intercommunity Interactions in the Chimpanzee Population of the Gombe National Park." In *The Great Apes*, ed. D. Hamburg and E. McCown, 13–54. Menlo Park, CA: Benjamin/Cummings.

Goodall, J., and D. Hamburg. 1975. "Chimpanzee Behavior as a Model for the Behavior of Early Man: New Evidence of Possible Origins of Human Behavior." *American Handbook of Psychiatry* 6:14–43.

Greene, H. 1994. "Homology and Behavioral Repertoires." In *Homology: The Hierarchical Basis of Comparative Biology*, ed. B. Hall, 369–391. New York: Academic Press.

Grine, F. E., P. S. Ungar, M. F. Teaford, and S. El Zaatari. 2006. "Molar Microwear in *Praeanthropus afarensis*: Evidence for Dietary Stasis through Time and under Diverse Paleoecological Conditions." *Journal of Human Evolution* 51 (3): 297–319. http://dx.doi.org/10.1016/j.jhevol.2006.04.004. Medline:16750841

Günter, M., and C. Boesch. 1993. "Energetic Cost of Nut-Cracking Behaviour in Wild Chimpanzees." In *Hands of Primates*, ed. H. Preuschoft and D. Chivers, 109–129. New York: Springer. http://dx.doi.org/10.1007/978-3-7091-6914-8_8

Haas, O., and G. Simpson. 1946. "Analysis of Some Phylogenetic Terms, with Attempts at Redefinition." *Proceedings of the American Philosophical Society* 90 (5): 319–49. Medline:20282032

Halperin, S. 1979. "Temporary Association Patterns in Free Ranging Chimpanzees: An Assessment of Individual Grouping Preferences." In *The Great Apes*, ed. D. Hamburg and E. McCown, 491–499. Menlo Park, CA: Benjamin/Cummings.

Hasegawa, T. 1990. "Sex Differences in Ranging Patterns." In *The Chimpanzees of the Mahale Mountains*, ed. T. Nishida, 99–114. Tokyo: University of Tokyo Press.

Herbinger, I., C. Boesch, and H. Rothe. 2001. "Territory Characteristics among Three Neighboring Chimpanzee Communities in the Tai National Park, Cote d'Ivoire." *International Journal of Primatology* 22 (2): 143–67. http://dx.doi.org/10.1023/A:1005663212997.

Hill, K., C. Boesch, J. Goodall, A. Pusey, J. Williams, and R. Wrangham. 2001. "Mortality Rates among Wild Chimpanzees." *Journal of Human Evolution* 40 (5): 437–50. http://dx.doi.org/10.1006/jhev.2001.0469. Medline:11322804

Hodos, W. 1976. "The Concept of Homology and the Evolution of Behavior." In *Evolution, Brain, and Behavior: Persistent Problems*, ed. R. Masterton, W. Hodos, and H. Jerison, 153–167. Hillsdale, NJ: John Wiley and Sons.

Hosaka, K., T. Nishida, M. Hamai, A. Matsumoto-Oda, and S. Uehara. 2001. "Predation of Mammals by the Chimpanzees of the Mahale Mountains, Tanzania." In *All Apes Great and Small*, Volume 1: *African Apes*, ed. B. Galdikas, N. Briggs, L. Sheeran, G. Shapiro, and J. Goodall, 107–130. New York: Kluwer Academic Publishers. http://dx.doi.org/10.1007/0-306-47461-1_11

Humphrey, N. 1976. "The Social Function of Intellect." In *Growing Points in Ethology*, ed. P. Bateson and R. Hinde, 303–317. Cambridge: Cambridge University Press.

Huxley, T. H. 1863. *Evidence as to Man's Place in Nature*. New York: D. Appleton.

Idani, G. 1990. "Relations between Unit-Groups of Bonobos at Wamba, Zaire: Encounters and Temporary Fusions." *African Study Monographs* 11:153–86.
Inoue, E., M. Inoue-Murayama, L. Vigilant, O. Takenaka, and T. Nishida. 2008. "Relatedness in Wild Chimpanzees: Influence of Paternity, Male Philopatry, and Demographic Factors." *American Journal of Physical Anthropology* 137 (3): 256–62. http://dx.doi.org/10.1002/ajpa.20865. Medline:18512686
Irwin, R. 1996. "The Phylogenetic Content of Avian Courtship Display and Evolution." In *Phylogenies and the Comparative Method in Animal Behavior*, ed. E. Martins, 234–252. Oxford: Oxford University Press.
Kano, T. 1992. *The Last Ape*. Stanford, CA: Stanford University Press.
Kelly, R. C. 2005. "The Evolution of Lethal Intergroup Violence." *Proceedings of the National Academy of Sciences of the United States of America* 102 (43): 15294–8. http://dx.doi.org/10.1073/pnas.0505955102. Medline:16129826
Klopfer, P. 1975. "Review of Animal Behavior: An Evolutionary Approach by John Alcock." *American Scientist* 63:578–9.
Langergraber, K. E., J. C. Mitani, and L. Vigilant. 2007. "The Limited Impact of Kinship on Cooperation in Wild Chimpanzees." *Proceedings of the National Academy of Sciences of the United States of America* 104 (19): 7786–90. http://dx.doi.org/10.1073/pnas.0611449104. Medline:17456600
Lauder, G. 1986. "Homology, Analogy, and the Evolution of Behavior." In *Evolution of Animal Behavior*, ed. M. Nitecki and J. Kitchell, 9–40. New York: Oxford University Press.
Lonsdorf, E. 2005. "Sex Differences in the Development of Termite-Fishing Skills in the Wild Chimpanzees, *Pan troglodytes schweinfurthii*, of Gombe National Park, Tanzania." *Animal Behaviour* 70 (3): 673–83. http://dx.doi.org/10.1016/j.anbehav.2004.12.014.
Lonsdorf, E. V., L. E. Eberly, and A. E. Pusey. 2004. "Sex Differences in Learning in Chimpanzees." *Nature* 428 (6984): 715–6. http://dx.doi.org/10.1038/428715a. Medline:15085121
Lorenz, K. 1941. "Vergleichende Bewegungsstudien bei Anatiden." *Journal fur Ornithologie* 32:194–294.
Lukas, D., V. Reynolds, C. Boesch, and L. Vigilant. 2005. "To What Extent Does Living in a Group Mean Living with Kin?" *Molecular Ecology* 14 (7): 2181–96. http://dx.doi.org/10.1111/j.1365-294X.2005.02560.x. Medline:15910336
Macedonia, J. M., and K. F. Stanger. 1994. "Phylogeny of the Lemuridae Revisited: Evidence from Communication Signals." *Folia Primatologica* 63 (1): 1–43. http://dx.doi.org/10.1159/000156787. Medline:7813970
Matsumoto-Oda, A., K. Hosaka, M. Huffman, and K. Kawanaka. 1998. "Factors Affecting Party Size in Chimpanzees of the Mahale Mountains." *International Journal of Primatology* 19 (6): 999–1011. http://dx.doi.org/10.1023/A:1020322203166.
McGrew, W. 1979. "Evolutionary Implications of Sex Differences in Chimpanzee Predation and Tool Use." In *The Great Apes*, ed. D. Hamburg and E. McCown, 440–463. Menlo Park, CA: Benjamin-Cummings.

McGrew, W. 1981. "The Female Chimpanzee as a Human Evolutionary Prototype." In *Woman the Gatherer*, ed. F. Dahlberg, 35–73. New Haven, CT: Yale University Press.

McGrew, W. 1992. *Chimpanzee Material Culture*. Cambridge: Cambridge University Press. http://dx.doi.org/10.1017/CBO9780511565519

Mitani, J. C. 2006a. "Demographic Influences on the Behavior of Chimpanzees." *Primates* 47 (1): 6–13. http://dx.doi.org/10.1007/s10329-005-0139-7. Medline:16283424

Mitani, J. C. 2006b. "Reciprocal Exchange in Chimpanzees and Other Primates." In *Cooperation in Primates: Mechanisms and Evolution*, ed. P. Kappeler and C. van Schaik, 107–119. Heidelberg: Springer-Verlag. http://dx.doi.org/10.1007/3-540-28277-7_6

Mitani, J. C., and D. P. Watts. 1999. "Demographic Influences on the Hunting Behavior of Chimpanzees." *American Journal of Physical Anthropology* 109 (4): 439–54. http://dx.doi.org/10.1002/(SICI)1096-8644(199908)109:4<439::AID-AJPA2>3.0.CO;2-3. Medline:10423261

Mitani, J. C., and D. Watts. 2001. "Why Do Chimpanzees Hunt and Share Meat?" *Animal Behaviour* 61 (5): 915–24. http://dx.doi.org/10.1006/anbe.2000.1681.

Mitani, J. C., and D. Watts. 2005a. "Seasonality in Hunting by Nonhuman Primates." In *Primate Seasonality: Studies of Living and Extinct Human and Non-human Primates*, ed. D. Brockman and C. van Schaik, 215–242. Cambridge: Cambridge University Press. http://dx.doi.org/10.1017/CBO9780511542343.009

Mitani, J. C., and D. Watts. 2005b. "Correlates of Territorial Boundary Patrol Behaviour in Wild Chimpanzees." *Animal Behaviour* 70 (5): 1079–86. http://dx.doi.org/10.1016/j.anbehav.2005.02.012.

Mitani, J. C., D. P. Watts, and S. J. Amsler. 2010. "Lethal Intergroup Aggression Leads to Territorial Expansion in Wild Chimpanzees." *Current Biology* 20 (12): R507–8. http://dx.doi.org/10.1016/j.cub.2010.04.021. Medline:20620900

Mitani, J. C., D. Watts, and J. Lwanga. 2002a. "Ecological and Social Correlates of Chimpanzee Party Size and Composition." In *Behavioural Diversity in Chimpanzees and Bonobos*, ed. C. Boesch, G. Hohmann, and L. Marchant, 102–111. Cambridge: Cambridge University Press. http://dx.doi.org/10.1017/CBO9780511606397.011

Mitani, J. C., D. Watts, and M. Muller. 2002b. "Recent Developments in the Study of Wild Chimpanzee Behavior." *Evolutionary Anthropology* 11 (1): 9–25. http://dx.doi.org/10.1002/evan.10008.

Moore, J. 1996. "Savanna Chimpanzees, Referential Models and the Last Common Ancestor." In *Great Ape Societies*, ed. W. McGrew, L. Marchant, and T. Nishida, 275–292. Cambridge: Cambridge University Press. http://dx.doi.org/10.1017/CBO9780511752414.022

Muller, M. 2002. "Agonistic Relations among Kanyawara Chimpanzees." In *Behavioural Diversity in Chimpanzees and Bonobos*, ed. C. Boesch, G. Hohmann, and L. Marchant, 112–124. Cambridge: Cambridge University Press. http://dx.doi.org/10.1017/CBO9780511606397.012

Muller, M., and J. Mitani. 2005. "Conflict and Cooperation in Wild Chimpanzees." *Advances in the Study of Behavior* 35:275–331. http://dx.doi.org/10.1016/S0065-3454(05)35007-8.

Muller, M., and R. Wrangham. 2004. "Dominance, Cortisol and Stress in Wild Chimpanzees (*Pan troglodytes schweinfurthii*)." *Behavioral Ecology and Sociobiology* 55 (4): 332–40. http://dx.doi.org/10.1007/s00265-003-0713-1.
Nakamura, M., and T. Nishida. 2006. "Subtle Behavioral Variation in Wild Chimpanzees, with Special Reference to Imanishi's Concept of Kaluchua." *Primates* 47 (1): 35–42. http://dx.doi.org/10.1007/s10329-005-0142-z. Medline:16132167
Newton-Fisher, N. E. 1999. "The Diet of Chimpanzees in the Budongo Forest Reserve, Uganda." *African Journal of Ecology* 37 (3): 344–54. http://dx.doi.org/10.1046/j.1365-2028.1999.00186.x.
Newton-Fisher, N.E. 2003. "The Home Range of the Sonso Community of Chimpanzees from the Budongo Forest, Uganda." *African Journal of Ecology* 41 (2): 150–6. http://dx.doi.org/10.1046/j.1365-2028.2003.00408.x.
Newton-Fisher, N. E., H. Notman, and V. Reynolds. Sep-Oct 2002. "Hunting of Mammalian Prey by Budongo Forest Chimpanzees." *Folia Primatologica* 73 (5): 281–3. http://dx.doi.org/10.1159/000067454. Medline:12566760
Newton-Fisher, N. E., V. Reynolds, and A. Plumptre. 2000. "Food Supply and Chimpanzee (*Pan troglodytes schweinfurthii*) Party Size in the Budongo Forest Reserve, Uganda." *International Journal of Primatology* 21 (4): 613–28. http://dx.doi.org/10.1023/A:1005561203763.
Nishida, T. 1968. "The Social Group of Wild Chimpanzees in the Mahale Mountains." *Primates* 9 (3): 167–224. http://dx.doi.org/10.1007/BF01730971.
Nishida, T. 1973. "The Ant-Gathering Behaviour by the Use of Tools among Wild Chimpanzees of the Mahali Mountains." *Journal of Human Evolution* 2 (5): 357–70. http://dx.doi.org/10.1016/0047-2484(73)90016-X.
Nishida, T. 1983. "Alpha Status and Agonistic Alliance in Wild Chimpanzees (*Pan troglodytes schweinfurthii*)." *Primates* 24 (3): 318–36. http://dx.doi.org/10.1007/BF02381978.
Nishida, T. 1990. "A Quarter Century of Research in the Mahale Mountains: An Overview." In *The Chimpanzees of the Mahale Mountains: Sexual and Life History Strategies*, ed. T. Nishida, 3–35. Tokyo: University of Tokyo Press.
Nishida, T. 1996. "The Death of Ntologi, the Unparalleled Leader of M Group." *Pan Africa News* 3: 4.
Nishida, T., N. Corp, M. Hamai, T. Hasegawa, M. Hiraiwa-Hasegawa, K. Hosaka, K. D. Hunt, N. Itoh, K. Kawanaka, A. Matsumoto-Oda, et al. 2003. "Demography, Female Life History, and Reproductive Profiles among the Chimpanzees of Mahale." *American Journal of Primatology* 59 (3): 99–121. http://dx.doi.org/10.1002/ajp.10068. Medline:12619045
Nishida, T., T. Hasegawa, H. Hayaki, Y. Takahata, and S. Uehara. 1992. "Meat-Sharing as a Coalition Strategy by an Alpha Male Chimpanzee?" In *Topics in Primatology*, Volume 1: *Human Origins*, ed. T. Nishida, W. McGrew, P. Marler, M. Pickford, and F. de Waal, 159–74. Tokyo: Tokyo University Press.
Nishida, T., and M. Hiraiwa. 1982. "Natural History of a Tool-Using Behavior by Wild Chimpanzees in Feeding upon Wood-Boring Ants." *Journal of Human Evolution* 11 (1): 73–99. http://dx.doi.org/10.1016/S0047-2484(82)80033-X.

Nishida, T., and M. Hiraiwa-Hasegawa. 1985. "Responses to a Stranger Mother-Son Pair in the Wild Chimpanzee: A Case Report." *Primates* 26 (1): 1–13. http://dx.doi.org/10.1007/BF02389043.

Nishida, T., M. Hiraiwa-Hasegawa, T. Hasegawa, and Y. Takahata. 1985. "Group Extinction and Female Transfer in Wild Chimpanzees in the Mahale Mountains National Park, Tanzania." *Zeitschrift für Tierpsychologie* 67:281–301.

Nishida, T., and K. Hosaka. 1996. "Coalition Strategies among Adult Male Chimpanzees of the Mahale Mountains, Tanzania." In *Great Ape Societies*, ed. W. McGrew, L. Marchant, and T. Nishida, 114–134. Cambridge: Cambridge University Press. http://dx.doi.org/10.1017/CBO9780511752414.011

Nishida, T., and S. Uehara. 1983. "Natural Diet of Chimpanzees (*Pan troglodytes schweinfurthii*): Long-term Record from the Mahale Mountains, Tanzania." *African Study Monographs* 3:109–30.

Nishida, T., S. Uehara, and R. Nyundo. 1979. "Predatory Behavior among Wild Chimpanzees of the Mahale Mountains." *Primates* 20 (1): 1–20. http://dx.doi.org/10.1007/BF02373826.

Nishida, T., K. Hosaka, M. Nakamura, and M. Hamai. 1995. "A Within-Group Gang Attack on a Young Adult Male Chimpanzee: Ostracism of an Ill-Mannered Member?" *Primates* 36 (2): 207–11. http://dx.doi.org/10.1007/BF02381346.

Owen, R. 1843. *Lectures on the Comparative Anatomy and Physiology of the Invertebrate Animals*. London: Longman, Brown, Green, and Longmans. http://dx.doi.org/10.5962/bhl.title.22459

Pääbo, S. 2003. "The Mosaic that Is Our Genome." *Nature* 421 (6921): 409–12. http://dx.doi.org/10.1038/nature01400. Medline:12540910

Pepper, J., J. Mitani, and D. Watts. 1999. "General Gregariousness and Specific Social Preferences among Wild Chimpanzees." *International Journal of Primatology* 20 (5): 613–32. http://dx.doi.org/10.1023/A:1020760616641.

Peters, C., and E. O'Brien. 1981. "The Early Hominid Plant-Food Niche: Insights from an Analysis of Plant Food Exploitation by *Homo, Pan*, and *Papio* in Eastern and Southern Africa." *Current Anthropology* 22 (2): 127–40. http://dx.doi.org/10.1086/202631.

Plavcan, J. M., and C. P. van Schaik. 1997. "Interpreting Hominid Behavior on the Basis of Sexual Dimorphism." *Journal of Human Evolution* 32 (4): 345–74. http://dx.doi.org/10.1006/jhev.1996.0096. Medline:9085186

Pribram, K. 1958. "Comparative Neurology and the Evolution of Behavior." In *Behavior and Evolution*, ed. A. Roe and G. Simpson, 140–164. New Haven, CT: Yale University Press.

Proctor, H. 1996. "Behavioral Characters and Homoplasy: Perception versus Practice." In *Homoplasy: The Recurrence of Similarity in Evolution*, ed. M. Sanderson and L. Hufford, 131–149. New York: Academic Press.

Pruetz, J. D., and P. Bertolani. 2007. "Savanna Chimpanzees, *Pan troglodytes verus*, Hunt with Tools." *Current Biology* 17 (5): 412–7. http://dx.doi.org/10.1016/j.cub.2006.12.042. Medline:17320393

Prum, R. 1990. "Phylogenetic Analysis of the Evolution of Display Behavior in the Neotropical Manakins (Aves: Pipridae)." *Ethology* 84 (3): 202–31. http://dx.doi.org/10.1111/j.1439-0310.1990.tb00798.x.

Pusey, A. 2001. "Of Genes and Apes." In *Tree of Origin*, ed. F. de Waal, 9–38. Cambridge, MA: Harvard University Press.

Pusey, A., G. Oehlert, J. Williams, and J. Goodall. 2005. "Influence of Ecological and Social Factors on Body Mass of Wild Chimpanzees." *International Journal of Primatology* 26 (1): 3–31. http://dx.doi.org/10.1007/s10764-005-0721-2.

Pusey, A., J. Williams, and J. Goodall. 1997. "The Influence of Dominance Rank on the Reproductive Success of Female Chimpanzees." *Science* 277 (5327): 828–31. http://dx.doi.org/10.1126/science.277.5327.828. Medline:9242614

Remane, A. 1952. *Die Grundlagen des natürlichen Systems, der vergleichenden Anatomie und der Phylogenetik*. Leipzig: Akad. Verlags.

Rendall, D., and A. Di Fiore. 2007. "Homoplasy, Homology, and the Perceived Special Status of Behavior in Evolution." *Journal of Human Evolution* 52 (5): 504–21. http://dx.doi.org/10.1016/j.jhevol.2006.11.014. Medline:17383711

Reynolds, V. 2005. *The Chimpanzees of the Budongo Forest*. Oxford: Oxford University Press. http://dx.doi.org/10.1093/acprof:oso/9780198515463.001.0001

Riss, D., and J. Goodall. 1977. "The Recent Rise to the Alpha-Rank in a Population of Free-Living Chimpanzees." *Folia Primatologica* 27 (2): 134–51. http://dx.doi.org/10.1159/000155784. Medline:557429

Ruvolo, M. 1997. "Molecular Phylogeny of the Hominoids: Inferences from Multiple Independent DNA Sequence Data Sets." *Molecular Biology and Evolution* 14 (3): 248–65. http://dx.doi.org/10.1093/oxfordjournals.molbev.a025761. Medline:9066793

Saint-Hilaire, E. G. 1881. *Philosophie Anatomique*. Paris: J.B. Baillière.

Sakura, O. 1994. "Factors Affecting Party Size and Composition of Chimpanzees (*Pan troglodytes verus*) at Bossou, Guinea." *International Journal of Primatology* 15 (2): 167–83. http://dx.doi.org/10.1007/BF02735272.

Sanderson, M., B. Baldwin, G. Bharathan, C. Campbell, C. von Dohlen, D. Ferguson, J. Porter, M. Wojciechowski, and M. Donoghue. 1993. "The Growth of Phylogenetic Information and the Need for a Phylogenetic Database." *Systematic Biology* 42 (4): 562–68. http://dx.doi.org/10.1093/sysbio/42.4.562.

Sanderson, M., and M. Donoghue. 1989. "Patterns of Variation in Levels of Homoplasy." *Evolution; International Journal of Organic Evolution* 43 (8): 1781–95. http://dx.doi.org/10.2307/2409392.

Sanz, C., D. Morgan, and S. Gulick. 2004. "New Insights into Chimpanzees, Tools, and Termites from the Congo Basin." *American Naturalist* 164 (5): 567–81. http://dx.doi.org/10.1086/424803. Medline:15540148

Sarich, V. M., and A. C. Wilson. 1967. "Immunological Time Scale for Hominid Evolution." *Science* 158 (3805): 1200–3. http://dx.doi.org/10.1126/science.158.3805.1200. Medline:4964406

Sayers, K., and O. Lovejoy. 2008. "The Chimpanzee Has No Clothes: A Critical Examination of *Pan troglodytes* in Models of Human Evolution." *Current Anthropology* 49 (1): 87–114. http://dx.doi.org/10.1086/523675.

Shea, B. T. 1983. "Paedomorphosis and Neoteny in the Pygmy Chimpanzee." *Science* 222 (4623): 521–2. http://dx.doi.org/10.1126/science.6623093. Medline:6623093

Shipman, P. 1986. "Scavenging or Hunting in Early Hominids: Theoretical Framework and Tests." *American Anthropologist* 88 (1): 27–43. http://dx.doi.org/10.1525/aa.1986.88.1.02a00020.

Sibley, C. G., J. A. Comstock, and J. E. Ahlquist. 1990. "DNA Hybridization Evidence of Hominoid Phylogeny: A Reanalysis of the Data." *Journal of Molecular Evolution* 30 (3): 202–36. http://dx.doi.org/10.1007/BF02099992. Medline:2109085

Simpson, G. 1958. "Behavior and Evolution." In *Behavior and Evolution*, ed. A. Roe and G. Simpson, 507–535. New Haven, CT: Yale University Press.

Smuts, B., D. Cheney, R. Seyfarth, R. Wrangham, and T. Struhsaker. 1987. *Primate Societies*. Chicago: University of Chicago Press.

Sponheimer, M., and J. A. Lee-Thorp. 1999. "Isotopic Evidence for the Diet of an Early Hominid, *Australopithecus africanus*." *Science* 283 (5400): 368–70. http://dx.doi.org/10.1126/science.283.5400.368. Medline:9888848

Stanford, C. B. 1995. "Chimpanzee Hunting Behavior and Human Evolution." *American Scientist* 83:256–61.

Stanford, C. B. 1996. "The Hunting Ecology of Wild Chimpanzees: Implications for the Evolutionary Ecology of Pliocene Hominids." *American Anthropologist* 98 (1): 96–113. http://dx.doi.org/10.1525/aa.1996.98.1.02a00090.

Stanford, C. B. 1998. *Chimpanzee and Red Colobus: The Ecology of Predator and Prey*. Cambridge, MA: Harvard University Press.

Stanford, C. B. 1999. *The Hunting Apes: Meat Eating and the Origins of Human Behavior*. Princeton, NJ: Princeton University Press.

Stanford, C. B. 2001. "The Ape's Gift: Meat-Eating, Meat-Sharing, and Human Evolution." In *Tree of Origin*, ed. F. de Waal, 95–118. Cambridge, MA: Harvard University Press.

Stanford, C. B., J. Wallis, H. Matama, and J. Goodall. 1994a. "Patterns of Predation by Chimpanzees on Red Colobus Monkeys in Gombe National Park, 1982-1991." *American Journal of Physical Anthropology* 94 (2): 213–28. http://dx.doi.org/10.1002/ajpa.1330940206. Medline:8085613.

Stanford, C., J. Wallis, E. Mpongo, and J. Goodall. 1994b. "Hunting Decisions in Wild Chimpanzees." *Behaviour* 131 (1): 1–18. http://dx.doi.org/10.1163/156853994X00181.

Sugiyama, Y. 1994. "Tool Use by Wild Chimpanzees." *Nature* 367 (6461): 327. http://dx.doi.org/10.1038/367327a0. Medline:8114932.

Sugiyama, Y. 2004. "Demographic Parameters and Life History of Chimpanzees at Bossou, Guinea." *American Journal of Physical Anthropology* 124 (2): 154–65. http://dx.doi.org/10.1002/ajpa.10345. Medline:15160368

Sugiyama, Y., and J. Koman. 1979. "Tool-Using and Making Behavior in Wild Chimpanzees at Bossou, Guinea." *Primates* 20 (4): 513–24. http://dx.doi.org/10.1007/BF02373433.

Suzuki, S., S. Kuroda, and T. Nishihara. 1995. "Tool-Set for Termite-Fishing by Chimpanzees in the Ndoki Forest, Congo." *Behaviour* 132 (3): 219–35. http://dx.doi.org/10.1163/156853995X00711.

Tanner, N. 1987. "The Chimpanzee Model Revisited and the Gathering Hypothesis." In *The Evolution of Human Behavior: Primate Models*, ed. W. Kinzey, 3–27. Albany: State University of New York Press.

Tanner, N., and A. Zihlman. 1976. "Women in Evolution, Part I: Innovation and Selection in Human Origins." *Signs* 1 (3): 585–608. http://dx.doi.org/10.1086/493245.

Teleki, G. 1973. *The Predatory Behavior of Wild Chimpanzees*. Lewisburg, PA: Bucknell University Press.

Tinbergen, N. 1960. "Comparative Studies of the Behaviour of Gulls (*Laridae*): A Progress Report." *Behaviour* 15 (1): 1–69. http://dx.doi.org/10.1163/156853960X00098.

Tinbergen, N. 1963. "On the Aims and Methods of Ethology." *Zeitschrift für Tierpsychologie* 20 (4): 410–33. http://dx.doi.org/10.1111/j.1439-0310.1963.tb01161.x.

Uehara, S. 1997. "Predation on Mammals by the Chimpanzee (*Pan troglodytes*)." *Primates* 38 (2): 193–214. http://dx.doi.org/10.1007/BF02382009.

Uehara, S., M. Hiraiwa-Hasegawa, K. Hosaka, and M. Hamai. 1994. "The Fate of Defeated Alpha Male Chimpanzees in Relation to Their Social Networks." *Primates* 35 (1): 49–55. http://dx.doi.org/10.1007/BF02381485.

Uehara, S., T. Nishida, M. Hamai, T. Hasegawa, H. Hayaki, M. Huffman, K. Kawanaka, K. Kobayashi, J. Mitani, Y. Takahata, et al. 1992. "Characteristics of Predation by Chimpanzees in the Mahale Mountains National Park, Tanzania." In Topics in Primatology, Volume 1: Human Origins, ed. T. Nishida, W. McGrew, P. Marler, M. Pickford, and F. de Waal, 143–158. Tokyo: Tokyo University Press.

Vigilant, L., M. Hofreiter, H. Siedel, and C. Boesch. 2001. "Paternity and Relatedness in Wild Chimpanzee Communities." *Proceedings of the National Academy of Sciences of the United States of America* 98 (23): 12890–5. http://dx.doi.org/10.1073/pnas.231320498. Medline:11606765

Wallis, J. 1997. "A Survey of Reproductive Parameters in the Free-Ranging Chimpanzees of Gombe National Park." *Journal of Reproduction and Fertility* 109 (2): 297–307. http://dx.doi.org/10.1530/jrf.0.1090297. Medline:9155740

Washburn, S. 1963. "Behavior and Human Evolution." In *Classification and Human Evolution*, ed. S. Washburn, 190–203. New York: Wenner-Gren Foundation.

Washburn, S., and I. Devore. 1961. "Social Behavior of Baboons and Early Man." In *Social Life of Early Man*, ed. S. Washburn, 91–105. Chicago: Aldine.

Washburn, S., and C. Lancaster. 1968. "The Evolution of Hunting." In *Man the Hunter*, ed. R. Lee and I. DeVore, 293–303. Chicago: Aldine.

Watts, D. 2004. "Intracommunity Coalitionary Killing of an Adult Male Chimpanzee at Ngogo, Kibale National Park, Uganda." *International Journal of Primatology* 25 (3): 507–21. http://dx.doi.org/10.1023/B:IJOP.0000023573.56625.59.

Watts, D., and J. Mitani. 2001. "Boundary Patrols and Intergroup Encounters in Wild Chimpanzees." *Behaviour* 138 (3): 299–327. http://dx.doi.org/10.1163/15685390152032488.

Watts, D., and J. Mitani. 2002. "Hunting Behavior of Chimpanzees at Ngogo, Kibale National Park, Uganda." *International Journal of Primatology* 23 (1): 1–28. http://dx.doi.org/10.1023/A:1013270606320.

Watts, D. P., M. Muller, S. J. Amsler, G. Mbabazi, and J. C. Mitani. 2006. "Lethal Intergroup Aggression by Chimpanzees in Kibale National Park, Uganda." *American Journal of Primatology* 68 (2): 161–80. http://dx.doi.org/10.1002/ajp.20214. Medline:16429415

Wenzel, G. 1992. "Behavioral Homology and Phylogeny." *Annual Review of Ecology and Systematics* 23 (1): 361–81. http://dx.doi.org/10.1146/annurev.es.23.110192.002045.

Whiten, A. 2005. "The Second Inheritance System of Chimpanzees and Humans." *Nature* 437 (7055): 52–5. http://dx.doi.org/10.1038/nature04023. Medline:16136127

Whiten, A., V. Horner, and F. B. de Waal. 2005. "Conformity to Cultural Norms of Tool Use in Chimpanzees." *Nature* 437 (7059): 737–40. http://dx.doi.org/10.1038/nature04047. Medline:16113685

Whiten, A., J. Goodall, W. C. McGrew, T. Nishida, V. Reynolds, Y. Sugiyama, C. E. Tutin, R. W. Wrangham, and C. Boesch. 1999. "Cultures in Chimpanzees." *Nature* 399 (6737): 682–5. http://dx.doi.org/10.1038/21415. Medline:10385119

Wiley, E. O. 1981. *Phylogenetics*. New York: John Wiley.

Williams, J., G. Oehlert, J. Carlis, and A. Pusey. 2004. "Why Do Male Chimpanzees Defend a Group Range?" *Animal Behaviour* 68 (3): 523–32. http://dx.doi.org/10.1016/j.anbehav.2003.09.015.

Williams, J., A. Pusey, J. Carlis, B. Farm, and J. Goodall. 2002. "Female Competition and Male Territorial Behaviour Influence Female Chimpanzee's Ranging Patterns." *Animal Behaviour* 63 (2): 347–60. http://dx.doi.org/10.1006/anbe.2001.1916.

Wilson, M., W. Wallauer, and A. Pusey. 2004. "New Cases of Intergroup Violence among Chimpanzees in Gombe National Park, Tanzania." *International Journal of Primatology* 25 (3): 523–49. http://dx.doi.org/10.1023/B:IJOP.0000023574.38219.92.

Wilson, M., and R. Wrangham. 2003. "Intergroup Relations in Chimpanzees." *Annual Review of Anthropology* 32 (1): 363–92. http://dx.doi.org/10.1146/annurev.anthro.32.061002.120046.

Wrangham, R. 1977. "Feeding Behaviour of Chimpanzees in Gombe National Park, Tanzania." In *Primate Ecology*, ed. T. Clutton-Brock, 503–538. London: Academic Press.

Wrangham, R. 1987. "The Significance of African Apes for Reconstructing Human Social Evolution." In *The Evolution of Human Behavior: Primate Models*, ed. W. Kinzey, 51–71. Albany: State University of New York Press.

Wrangham, R. W. 1999. "Evolution of Coalitionary Killing." *American Journal of Physical Anthropology* 42 (Suppl. 29): 1–30. http://dx.doi.org/10.1002/(SICI)1096-8644(1999)110:29+<1::AID-AJPA2>3.0.CO;2-E. Medline:10601982

Wrangham, R. W. 2000. "Why Are Male Chimpanzees More Gregarious than Mothers? A Scramble Competition Hypothesis." In *Primate Males Causes and Consequences of Variation in Group Composition*, ed. P. Kappeler, 248–258. Cambridge: Cambridge University Press.

Wrangham, R. W. 2001. "Out of the Pan, Into the Fire: How Our Ancestors' Evolution Depended on What They Ate." In *Tree of Origin*, ed. F. de Waal, 119–144. Cambridge, MA: Harvard University Press.

Wrangham, R., and D. Peterson. 1996. *Demonic Males: Apes and the Origins of Human Violence*. New York: Houghton Mifflin.

Wrangham, R., N. Conklin-Brittain, and K. Hunt. 1998. "Dietary Response to Chimpanzees and Cercopithecines to Seasonal Variation in Fruit Abundance. I. Antifeedants." *International Journal of Primatology* 19 (6): 949–70. http://dx.doi.org/10.1023/A:1020318102257.

Wrangham, R., and D. Pilbeam. 2001. "African Apes as Time Machines." In *All Apes Great and Small*, Volume 1: *African Apes*, ed. B. Galdikas, N. Briggs, L. Sheeran, G. Shapiro, and J. Goodall, 5–17. New York: Kluwer Academic Publishers.

Yamakoshi, G. 1998. "Dietary Responses to Fruit Scarcity of Wild Chimpanzees at Bossou, Guinea: Possible Implications for Ecological Importance of Tool Use." *American Journal of Physical Anthropology* 106 (3): 283–95. http://dx.doi.org/10.1002/(SICI)1096-8644(199807)106:3<283::AID-AJPA2>3.0.CO;2-O. Medline:9696145

Yamakoshi, G. 2001. "Ecology of Tool Use in Wild Chimpanzees: Toward a Reconstruction of Early Hominid Evolution." In *Primate Origins of Human Cognition and Behavior*, ed. T. Matsuzawa, 537–614. Tokyo: Springer. http://dx.doi.org/10.1007/978-4-431-09423-4_27

Zihlman, A. 1996. "Reconstructions Reconsidered: Chimpanzee Models and Human Evolution." In *Great Ape Societies*, ed. W. McGrew, L. Marchant, and T. Nishida, 293–304. Cambridge: Cambridge University Press. http://dx.doi.org/10.1017/CBO9780511752414.023

Zihlman, A. L., J. E. Cronin, D. L. Cramer, and V. M. Sarich. 1978. "Pygmy Chimpanzee as a Possible Prototype for the Common Ancestor of Humans, Chimpanzees and Gorillas." *Nature* 275 (5682): 744–6. http://dx.doi.org/10.1038/275744a0. Medline:703839

12

Analogies and Models in the Study of the Early Hominins

Clifford J. Jolly

Paleontologists, including those specializing in the fossil evidence for human evolution, have little prospect of eventually being able to observe a whole, let alone a living, representative of a species represented only by fossil fragments. Especially in vertebrate paleontology, therefore, the task of reconstructing the behavior, ecology, and physiology of "new" forms from fragmentary and indirect evidence is routine and essential, and draws upon advances in neontological biology concerned with function and behavior.

Elton (2006) has recently provided a thorough and thoughtful historical review of works in which anthropologists have attempted to fill the gaps in our knowledge of the behavioral and ecological aspects of human evolution by comparing fossil hominins with living anthropoid primates. As Elton notes, the favored nonhuman primates for comparison have been the great apes, especially the chimpanzees, and the cercopithecoid monkeys. Her account documents the alternating predominance of advocates of great apes and cercopithecoids, respectively, as the most illuminating comparison, or *referent*, against which early hominins are to be matched and interpreted. The argument to be developed here is that although both the chimp-based and monkey-based comparisons are often regarded as exercises in *analogy*, and the referents are labeled *analogs*, they actually represent two quite distinct modes of operation. Most (though not all) studies that draw upon information from chimpanzees to illuminate

DOI: 10.5876/9781607322252:c12

the unobservable aspects of early hominin biology are quite different in logic from those that use cercopithecoid monkeys (or other organisms, primate and nonprimate). It is suggested that while both approaches are vital and complementary sources of insight, only the latter involve *analogies* in the strict sense. In these terms, most studies that have invoked chimpanzees as a referent have used them not as analogs but as "best extant models" (BEM).

ANALOGIES

According to the OED (*Oxford English Dictionary* 1933), one of the earliest (mid-sixteenth century) usages of the term *analogy* is mathematical, referring to the likeness of proportion between two sets of numbers, as in "3 is to 6 as 2 is to 4." From the first, therefore, a *likeness of relations* was a fundamental part of the concept. Only a small step is required from this usage to a slightly more general one, which apparently appeared at almost the same time, and extends the basic idea beyond mathematics. Analogies in this strict sense survived most familiarly in psychological tests, where they are stated in the form: A is to B as C is to (X/Y/Z), as in Foot is to Shoe as Head is to (Hair/Glove/Hat). It is the likeness of relationship (foot in shoe, head in hat) that is important in true analogy, not the resemblance of a member of pair 1 to a member of pair 2. That feet and gloves both have digits, for example, and come in pairs, is irrelevant. This essential concept is retained, even when the analogy is not spelled out as fully, as is illustrated by two nineteenth-century quotations chosen by the OED (both, by curious coincidence, were written by close friends of Charles Darwin: John Tyndall and Sir John Lubbock):

> "The analogy between a river and a glacier moving through a sinuous valley is therefore complete" (Tyndall 1896) [Glac. II. 10. 285].

> "There seem to be three principal types [of ants] offering a curious analogy to the three great phases: the hunting, pastoral, and agricultural stages, in the history of human development" (Lubbock 1879) [Sci. Lect. iv. 137] (OED 1933, 304).

Analogies of this kind are here called *true analogies*. They can easily be stated in a way that formalizes the "likeness of relations." Tyndall's could be stated as: "moving glaciers are to static masses of ice as rivers are to static masses of fresh water."

But it is also obvious that flowing glaciers do not have all the attributes of rivers, even of rivers-as-contrasted-to-lakes. Thus another feature of analogies is that they always include, often implicitly, a phrase that delimits the comparison. In presenting his analogy, Tyndall implies an introductory phrase such as "with respect to the way that they follow the course of the rocky beds in which they lie." Apparently he did not think it necessary to point out that glaciers, unlike rivers, do not house fish and mayflies, and are therefore unprofitable for trout fishing, nor, presumably, did he regard his analogy as less "complete" because glaciers do not share this, or other, attributes of rivers.

Similarly, Lubbock's insight could be accurately, though less elegantly, restated as a three-component analogy: "With respect to their general mode of food acquisition, hunting ants are to aphid-farming ants are to fungus-gardening ants as human hunter-gatherers are to pastoralists are to cultivators." Presumably Sir John thought it irrelevant that the gardeners pruning his roses did not have six legs and an exoskeleton. The many differences between ants and humans in general, and the fact that ants "garden" by using techniques not employed in human horticulture, do not affect Lubbock's analogy, nor do they weaken its insight. Analogy thus implies informative parallels in particular features, not overall similarity or identity.

This implication is seen in pre-Darwinian biology, where *analogy* acquired special significance in the sense of resemblance between organisms that is superimposed upon, different "essences." The OED cites examples in the writings of mid-nineteenth-century natural historians: "Resemblances of form and habits without agreement of structure . . . are termed relations of . . . analogy" (Woodward 1854) [Man. Mollusca 55] and "We understand by analogy those cases in which organs have identity of function, but not identity of essence or origin" (Berkeley 1857) [Cryptog. Bot. §25] (OED 1933, 304).

With the general acceptance of the reality of evolution, *essence* was replaced by the more concrete concept of common ancestry or heritage. *Analogy* then became the opposite of what is now known as *homology*, or similarity that results from heritage. *Homoplasy* rather than *analogy* is the term now more generally used to include all nonhomologous resemblance. It is usually discussed in the context of phylogenetic inference, since homoplasy complicates the search for clades in exactly the same way as—in the terms used by an earlier generation—analogy could confound the quest for shared essence. In recent years, it has become clear that much homoplasy, especially at the molecular genetic level, is attributable to chance alone. Such random homoplasy is indeed of interest only as a hindrance to phylogenetic inference and is not considered here. This chapter focuses on informative analogies, in which comparisons between unrelated organisms with respect to homoplastic traits yield new insights, as in Lubbock's comparison of ant and human subsistence patterns.

Two main examples are used here as illustrations of the way that analogy *sensu stricto* can illuminate unobservable aspects of early hominin biology and evolution. The first is the venerable "seed-eaters" idea, an analogy explaining some derived features of basal hominin craniodental and postcranial structure (Jolly 1970). The second example is provided by recent data on the population structure of terrestrial papionin monkeys and its bearing on problems of hominin evolutionary dynamics and systematics.

THE "SEED-EATERS" ANALOGY

No attempt will be made here to evaluate or update the "seed-eater" concept. It will be discussed in its own, mid-twentieth-century terms—with *Australopithecus africanus* representing the "basal hominin," for example—merely as an example of the use of analogy. The "seed-eater" idea (Jolly 1970, 1973) did not result from an attempt to reconstruct the natural history of the earliest hominins (cf. Bartholomew and Birdsell 1953) or to find the closest ecological match for early hominins among extant primates (cf. Washburn and DeVore 1961). It was prompted by the observation that there appear to be many anatomical parallels between *Theropithecus* and *Australopithecus* that were not covered by the contemporary explanatory framework, which focused largely on tool use and hunting–gathering. If these parallels were due to chance alone, one would expect that an equal number would link *Theropithecus* and *Pan*, which was not the case. There was, therefore, "something-to-be-explained" (an *explanandum*). From the start, therefore, the analysis was based on ideas that could be expressed in "likeness of relationships" terms, as in: "With respect to dental proportions, *Theropithecus* is to *Papio* as *Australopithecus* is to *Pan*." In this statement, the second term of each pair *Theropithecus/Papio* and *Australopithecus/Pan* is the (supposed) sister taxon of the first. To demonstrate the existence of an analogy that is functionally relevant, and thus potentially informative, it is not strictly necessary to know the polarity of the character states. The analogy holds, for instance, whether the dental proportions of the last common ancestor of *Pan* and *Australopithecus* were chimplike or homininlike, or even somewhere between the two. If the analogy is to be used as the basis of an evolutionary scenario, however, which is generally the case, the polarity matters a great deal. Scenarios invoking environmental change, for example, will be very different if a homininlike-to-chimplike rather than a chimp-to-hominin shift is to be "explained." It will almost always be the case, then, that a more useful statement of an evolutionary analogy will make the second term a higher-level clade that includes the first term, and the element of evolutionary change is more explicitly introduced if the second term is expressed as the clade's ancestral morphotype. "With respect to dental proportions, *Theropithecus* is to the *Theropithecus–Papio* morphotype as *Australopithecus* is to the *Pan*–hominin morphotype." (In the original "seed-eater" formulation, the common ancestor of *Theropithecus–Papio* was tacitly assumed to be generally *Papio*-like in teeth and jaws, and that of *Pan*–hominins to be generally chimpanzeelike. Currently, the *Pan*–hominin morphotype, and the *Pan*–hominin relationship itself, are contested, but consideration of the alternatives is beyond the scope of the present essay.) The notion of "parallel change" from ancestral to derived condition may be directly incorporated in the statement of an evolutionary analogy, for example: "With respect to dental proportions, similarities in the derived conditions of *Theropithecus* among the papionins and *Australopithecus* among the hominids are the result of parallel evolutionary change from their respective ancestral morphotypes." Or, simply: "*Theropithecus* **parallels** early hominins in dental proportions." The last statement can become misleading if the likeness-of-relations component is lost by allowing it to morph into: "*Theropithecus* **resembles** early hominins in dental proportions," which has quite a different meaning.

Recognition of an analogy immediately raises questions of *why* these two particular, phylogenetically distant taxa should have followed parallel paths with respect to this particular set of features. These are questions of function and adaptation. The goal is to identify the likely functional correlates of the analogous structural traits, in general, and then in particular. In the "seed-eater" case, we adopt two working postulates: first, that, in general, the size, shape, and proportions of cheek teeth and incisors are likely to respond adaptively to dietary demands, and second, that incisor/cheek-tooth ratios are likely to be related to the relative importance in the diet of foods processed principally by biting, and foods processed principally by crushing and grinding, respectively. (Either of these assumptions could in principle be challenged and refuted on empirical or theoretical grounds.) In the pair of extant baboon genera, the expectation that geladas generally eat smaller, more cheek-tooth-demanding food objects than *Papio* baboons is supported by field observation.

The *explanandum* element is important. The dental proportions and gnathic shape of hominins and geladas are measurable attributes, and the parallelism between them can be disproved. Unless the parallel happened by chance, some element common to the two situations is implicated. Since teeth and jaws are used primarily for feeding, the composition of the diet seems an obvious place to seek an explanation. "Seed-eaters" suggested a property common to the crucial foods of geladas and basal hominins that would explain the parallel. The foods were different, it was suggested, but in both cases the "packets" in which they could be gathered were small (less than a mouthful), and their consistency demanded high occlusal forces and prolonged chewing. Of course, this particular dietary explanation may turn out to be incomplete, imprecise, or just plain wrong. But a refutation of the particular explanation in whole or part, cannot, thereby, abolish the *explanandum*. Unless shown to be based on misinterpretation of the evidence, the structural parallel would still be a fact of nature, and an alternative explanatory hypothesis would have to be found for it.

Such use of analogy in paleontology thus has a simple, three-step structure, in which the analogy is first established (a likeness of relations is demonstrated, and shown to be inexplicable either by pure chance or by common heritage), then interpreted (functional correlates with behavior are distinguished), and finally used predictively to reconstruct this aspect of the fossil form's behavior:

1. "Organism A [extant] parallels Organism B [extinct] with respect to feature X [fossilizable structure]."
2. "Feature X can be reliably linked functionally to feature Y [unfossilizable behavior or physiology]."

therefore

3. "We predict that Organism B paralleled Organism A also in feature Y."

A crucial element here is obviously step 2, the structural–functional link, without which the analogy cannot be used predictively. Establishing this link necessarily involves finding analogous cases—preferably, more than one—among extant forms, in which both structure and behavior (response and outcome) can be observed. A hypothesis of structural–functional relationship can then be advanced—for example, that the way an animal habitually moves will be associated with its limb proportions, and that long legs are advantageous in leaping. One way of testing the hypothesis is statistical: showing, for example, that across a variety of different, extant taxa, relative elongation of the hind limbs is significantly associated with dependence upon hopping and leaping locomotion. A quantitative expression of the same relationship could be achieved by showing a significant correlation between hind-limb length and some measure of the "importance" of hopping/leaping in the locomotor profile. Obviously, the greater the number of cases in which the relationship can be demonstrated, the more reliable the inference will be. The cases must, however, be truly analogous, that is, each representing an independent evolutionary transition, not the homologous possession of an ancestral trait. In recent years, it has become widely accepted that correlations of this kind have often been biased by nonindependence of the data points, due to multiple representation of clades (that is, some of the statistical relationship was due to homology rather than analogy). Independent contrast analysis has been developed to circumvent this statistical double dipping. Since it compares trends of divergence only between pairs of phylogenetically coordinate taxa, its logic is very close to that of analogy.

Though some kind of statistical test (of association or correlation) may be the preferred method of evaluating a hypothesis of structural–functional linkage, such a test may be impossible, because a large-enough sample of independent cases is simply not available, especially after the nonindependent cases are removed. We then have to rely on plausibility to establish the structure–function link. Every schoolchild "knows," for example, that pterosaurs flew, but no human has seen them flying. We simply know that they had structures so closely similar, functionally, to the forelimbs of birds and bats (though unlike them in structural detail), and so difficult to explain mechanically as anything but wings, that the presumption that they used them to fly is almost inescapable. This is, however, an interpretation of function based upon informed common sense rather than a statistically based inference. The pterosaur membrane is so winglike that it is hard to imagine any other functional explanation for it; but the number of independent, analogous cases actually linking wings to true flying in extant vertebrates are only two—the ancestral bird and bat, respectively.

A demonstrated analogy between two or more organisms in one functional system can suggest where to look for parallels in others. For instance, the analogy of birds and bats suggests that dependence upon true, sustained flight in vertebrates requires homeothermy, and strongly favors a circumscribed period of rapid growth and maturation. Evidence for both these features has been sought, and found, in the pterosaurs (de Ricqles et al. 2000). The "seed-eater" model was extended in a similar way. Small-object critical foods (as observably eaten by geladas and inferred, from jaws and teeth, for basal hominins) would strongly favor two-handed gathering. This, in turn, would favor hands adapted for fine manipulation, and positional behaviors in which both hands were free. Postcranial features and manual morphology that meet these criteria are observable in geladas and in early hominins (e.g., Susman 1998). An analogy that appears to be heuristically powerful can also suggest parallels that cannot be directly confirmed by observation but that open up new avenues of speculation. An example is the position of female sexual signals, which, as a derived trait, is pectoral rather than perineal in both geladas and *Homo sapiens* (and thus is displayed when the trunk is erect in two-handed feeding). This implies that the human breast is an ancient hominin trait, but this speculation is unlikely ever to be directly testable. Such speculations on the outer limits of the analogy cannot, of course, be used as support for the original interpretation. If analogous adaptations are identified in different systems (hands, jaws/

teeth, and position of sexual signals, for example) and all can be traced to a common behavioral element (in this case, habitual, two-handed feeding in a trunk-upright position on small, tough or hard objects), they can be described as comprising an *adaptive complex*.

A particular referent species may provide a rich source of informative analogies, but no single referent species can be expected to provide analogies needed to understand all the features of a fossil form, or even all its derived features. For instance, "seed-eaters" suggests that a *Theropithecus* analogy can throw light on the origins of a suite of derived traits in early hominins, all of which can ultimately be linked to a "small-object feeding on the ground" functional complex. A gelada analogy is not expected, however, to throw any light on derived hominin traits that lie outside that adaptive complex. One example of this was cited in "seed-eaters." The height of the molar crowns and the thickness of enamel are *not* paralleled in geladas and hominins. This is predictable, however, because these features are unrelated to small-object feeding per se, but rather to the consistency and composition of the foods processed. *Theropithecus*, a grazer on grasses and low herbs, has, as expected, features of the molar crown analogous to those of other grass-eating mammals, while hominins have thick-enamelled, flat-crowned molars that, in Miocene hominoids, have been likened to those of pigs and bears (Kay 1981) and are also paralleled to some extent in primates such as *Cebus* that specialize in crushing hard-shelled fruits (Kinzey 1987). A still more striking example was pointed out by Martin (1990), who showed that the modest increase in brain size associated with the differentiation of the earliest hominins was not paralleled in *Theropithecus*, which in fact appears to be somewhat small-brained for a cercopithecoid of its body mass, especially as compared with *Papio*. Since greater encephalization was never suggested to be part of the small-hard-object-feeding complex, and there is no reason that it should be, this observation does not invalidate the hypothesis. It does, however, suggest a different analogy. A substantial comparative literature suggests that the modestly enlarged brains of the earliest hominins could have been associated with a foraging strategy concentrated upon high-unit-energy but patchily dispersed resources, spread across comparatively large home ranges (e.g., Milton 1981; Marino 2005; Walker et al. 2006). In all these respects, the econiche of the earliest hominins would have resembled that of the frugivore/generalist *Papio* more closely than the sedentary grazer, *Theropithecus*.

As this example demonstrates, overall reconstruction of the lifeways of an extinct form will need to draw upon functionally interpreted analogies from many different sources in order to understand the implications of all its derived character states. When these different strands of evidence, from a wide range of species, are evaluated and combined, the result will be very close to a "principled" analysis as recommended by Tooby and DeVore (1987).

MONKEY ANALOGIES FOR HOMININ POPULATION STRUCTURE

A second example of analogy, also linking papionin monkeys with hominins (Jolly 2001), appears to be rather different from "seed-eaters," because it involves the structure and history and populations, rather than the functional morphology of bones and teeth. It has, however, a similar logic:

1. With respect to time-since-common-ancestry (ranging from ~10 mya [million years ago] to < 0.5 mya), pairs of taxa among cercopithecoids present parallels to pairs of taxa among the African hominids (gorillas, chimpanzees, and various hominins).

2. Among mammals, time-since-common-ancestry has, in general, a predictable inverse relationship to probability of gene flow between populations.

3. Pairs of African hominid taxa, at any point in time, had a predictable likelihood of exchanging genes by marginal or sympatric gene-flow.

Obviously, the hypothesized analogy can and should be tested at several different points. The dates of particular divergences mentioned in statement 1 are based on a fragile combination of fossil and molecular evidence and are clearly open to revision. The relationship in statement 2, though broadly sustained by evidence to date (Holliday 2003), obviously requires further testing. It also requires underpinning with information about the mechanisms underlying reproductive isolation in mammals, especially higher primates. The occurrence of natural hybridization, now and in the past, needs far more factual documentation. It is also possible that reproductive compatibility between two taxa is better predicted by their degree of morphological divergence than by the time elapsed since cladogenesis (e.g., Harvati et al. 2004).

The conclusion drawn from the analogy (statement 3) is obviously of most value to the paleoanthropologist if it can be applied to particular cases. A full treatment of particular applications is well beyond the scope of this chapter, and in any case would be premature before the basis of the analogy is more fully tested. Some general implications are, however, worth recalling. The first is that taxa whose Last Common Ancestor (LCA) lived less than 2 myr (million years) previously are most unlikely to have been totally reproductively isolated by biological or intrinsic barriers. The separation between Neandertals and *Homo sapiens* (*s.s.*) as it existed in the late Pleistocene would certainly fall within this "grey zone" (Krings et al. 1997; Krings et al. 1999; Ovchinnikov et al. 2000). Even occasional hybridization can permit a population of one species, expanding and displacing a resident population of another, to acquire locally adaptive genes. As previously suggested (Jolly 2001), the ancestors of Europeans and North Asians could thus have acquired light skin pigmentation, adaptive in high latitudes, from the local "archaic" populations they replaced, even if, for cultural or ecological reasons, there was very little genetic interchange between "natives" and "invaders" (Currat and Excoffier 2005). Recent genetic analysis suggests a contribution to the modern human gene-pool from "native archaic" populations (Garrigan et al. 2005).[1]

Similarly, it seems very likely that a lack of complete, intrinsic reproductive isolation also applied to the various named forms of mid-to-late Pleistocene "archaic humans," to species of *Australopithecus*, and even to the stem lineages of hominins and chimpanzees in the period of approximately 7–5 mya (cf. Patterson et al. 2006). For the split between pregorillas and the chimp–human lineage, the period during which cladogenesis was proceeding but some genetic exchange was still possible, can be estimated from about 10 to about 8 mya. Its younger bound thus approaches the current best estimate for the chimpanzee–human split (Raaum et al. 2005). If this is so, it would help to explain the ambiguity of the chimp–human–gorilla phylogenetic pattern, with most loci supporting a chimpanzee–human clade, but minorities suggesting either chimpanzee–gorilla or gorilla–human (Rogers 1994; Samollow et al. 1996). It should also be borne in mind that the "molecular dates" for cladogenetic events are for the most part based upon mitochondrial evidence, and that, given female migration, or "mitochondrial capture" (e.g. Roca, Georgiadis, and O'Brien 2005; Jolly 2006), mitochondria can "flow" between populations just like any other single gene. This effectively extends the "grey area" back in time from the mitochondrially derived date of cladogenesis.

Similarly, the analogy of hybridization between monkey lineages separate for 4–5 myr (Jolly et al. 1997) suggests that sympatric populations of adaptively divergent hominin lineages at about 2 mya—*Paranthropus* and *Homo*—might have been able to produce occasional hybrids, and that these could have been viable and fertile, though probably disadvantaged compared to the parental species. Even if few in number, such occasional hybrids could form a conduit

---

1. Since the above was written, extensive evidence for prolonged hybridization between these diverging hominid lineages has emerged from the comparative genomics of pre-sapiens *Homo* and the African apes. The speculation about Neandertal skin pigmentation has, however, been refuted.

by which locally adaptive genes could pass between populations. Finally, the local context of extensive hybridization between "good" guenon (*Cercopithecus*) species that normally coexist without interbreeding (Detwiler et al. 2005) suggests, by analogy, circumstances under which, rarely, a new hominin species might have arisen by fusion between distinct lineages. Hybridization of this kind might explain some apparently unresolvable homoplasy in early hominin lineages.

The monkey analogy, of course, in no way *proves* that hybridization actually occurred between recognizable hominin or hominid lineages, or, if it did, that it had any perceptible impact on the course of hominin evolution. Its role, like that of all analogy, lies simply in expanding the realm of possibilities from which testable models can be constructed.

THE SOURCE OF ANALOGIES

The analogies used here are all drawn from the cercopithecoid monkeys, the sister taxon of the hominoids. There is no reason, however, to regard analogies found in closely related taxa as inherently more powerful. In fact, the wider the phylogenetic distance between the analogs, the more distinctly the analogous feature and its functional correlate stand out from the background of heritage. This is the case, for example, in Darwin's (1871) analogy that compared relative brain size of humans among mammals with that of ants among insects, arguing that both are products of selection for survival in a complex social setting. On the other hand, in practice, as the distance between analogs increases, the harder it may become to find comparable structures in which detailed morphological parallels can be sought. Coelenterates, for example, provide few analogies for understanding primate dentitions. These considerations suggest that informative analogies are likely to be most numerous in a midrange of phylogenetic distance between analogs. If the distance is too great, parallels may be hard to find; but if close phylogenetic relatives are being compared, it may be hard to distinguish analogous from homologous similarities. For example, similarities exist in the feet of gorillas (especially mountain gorillas) and humans, in which both contrast with chimpanzees (shorter phalanges on digits 2–5, for example) (Schultz 1930, 1934); but there seems no foolproof way to decide whether these resemblances are analogous adaptations to terrestriality, or homologous retentions that have been lost in chimpanzees. By contrast, similarities in the foot linking baboons (compared to more arboreal monkeys) and humans (compared to chimpanzees) are clearly much more likely to be analogous than homologous. The implications of the baboon's toes for the course of human evolution are, therefore, much less ambiguous than those of the gorilla, even though a gorilla's foot is, overall, much more similar to that of a human. Similarly, the African papionin monkeys are a good source of analogies for hominin population structure not because they are catarrhine primates, but because, as medium-sized, social, omnivorous mammals associated with tropical seasonal habitats, their evolutionary history in Africa is likely to have tracked that of the earlier stages of hominin evolution (Foley 1993; Jolly 2001; Elton 2006). Aspects of their population structure, such as the geographical patchwork of morphologically distinct allotaxa, often with marginal gene-flow between them, find analogies in many other taxa, both mammalian and nonmammalian in Africa (e.g., Mackworth-Praed and Grant 1980 [1957]; Kingdon 1997; Grubb 2006) and elsewhere (e.g., Hewitt 2001), and even better analogies may well be found outside the primates. Recently, for example, the genus *Canis* has been used as an informative analog for understanding aspects of the population structure of early *Homo* (Arcadi 2006).

Treated as whole species, chimpanzees, bonobos, and gorillas are poor sources of analogies, but true analogies can be constructed on the basis of their intraspecific variation. For example, several authors (e.g., Kortlandt and Kooij 1963; Kortlandt 1965; Izawa 1970; Jolly 1970; Hunt and McGrew 2000) have suggested that the natural history of chimpanzees living in seasonally deciduous woodland and savanna habitats, if compared with conspecifics living in evergreen forests, might provide informative analogies for understanding aspects of basal hominin ecology. The difference in foot structure between the more arboreal mountain gorilla and the more terrestrial mountain gorilla (Schultz 1934) is another such case. Similarly, interpopulational differences in behavior of chimpanzees and gorillas can be legitimately investigated as sources of analogies for hominins (Whiten et al. 1999; Pilbrow 2006). Stone-tool use by some but not all chimpanzee populations, for example, most likely appeared after the separation of chimpanzee and hominin species, and therefore, by definition, would be analogous.

THE SEARCH FOR "BEST EXTANT MODELS"

True analogies, in the sense adopted here, are explicitly limited by the "with respect to" clause, and do not seek single referents that purport to explain all the derived features of a fossil or other partly known organism. Some analogs, of course, are more broadly informative than others, but even the most apt will inevitably leave a residue of derived features that are unexplained. For a comprehensive picture, the comparative net must be cast widely, to find a variety of partial analogs, each of which is relevant to some of the derived features of an extinct organism.

Both the "likeness of relations" format and the explicit limitation of the scope of comparison to demonstrated convergences distinguish what is here called "true analogy" from another sort of comparison, which is often, confusingly, also called "analogy" or, somewhat disparagingly, "simple analogy" (e.g., various authors in Kinzey 1987). This sense is described by the OED's definition 4: "more vaguely, agreement between things, similarity." A nineteenth-century paleontological example is drawn from yet another member of Darwin's circle: "The trilobites . . . bear so strong an analogy to those described by M. Brongniart" (Murchison 1839) [Silur. Syst. I. xxvii. 358] [OED 1933, 304]. The concept of analogy is thus reduced from "likeness of relationship" to "likeness." This represents a major departure, especially for evolutionary biology, because this definition ignores the distinction between shared ancestral and shared derived traits, and implies a search for a particular living animal—the "best extant (or "living") model" (BEM). In paleontology, choosing referent organisms likely to be "very similar" for comparison with a fossil form is a process quite distinct from true analogy, both in its logic, and in the kind of information it provides. In the case of the early hominins, for example, "true analogy," with its focus on derived characters, attempts to answer the question, "what *new* behaviors led the hominin stem to become physically distinct from the LCA shared with *Pan*?" By contrast, finding a BEM is part of a process aimed at reconstructing the total way of life of the fossil form, not just its newly derived aspects. The BEM is the living species that most closely resembles the fossil in all respects that can be documented, and is therefore presumed, unless there is evidence to the contrary, to resemble it most closely in traits, such as behavior, that cannot be directly observed.

In practice, in most cases, it is most unlikely that the BEM will be a species other than a close phylogenetic relative of the target species, simply because evolution is essentially a conservative process, and the overwhelming majority of traits of any organism are shared with its immediate ancestor. Thus, for the very earliest hominins, it is reasonable to assume that they will resemble the LCA of chimpanzees and humans in all characters that are neither demonstrably different and derived (like teeth, jaws, and pelvis) nor, if invisible in fossils, suspected to be part of a known, derived adaptive complex (like sexual signals). Furthermore, the best living model for the LCA will be the species that has acquired the fewest derived features of its own since the divergence. In the hominin case, this is generally presumed to be *Pan*. An informative complication arises from the fact that there are two living chimpanzee species, each of which has its champions as BEM (Zihlman et al. 1978; Wrangham and Peterson 1996). Insofar as they differ in behavior and structure, neither chimpanzee species can be presumed *a priori* to be the BEM. This is a reminder that a *Best* Extant Model can never be presumed to be a *Perfect* Extant Model. If either *Pan troglodytes* or *P. paniscus* had become extinct, it would have beeen easy to assume that the surviving member of the genus was a complete model for the hominin–chimpanzee LCA, and thus the BEM for a basal hominin. Furthermore, it is quite possible that both *Pan* species have acquired many derived traits, before their divergence from each other, as well as after. In this case, the BEM for the LCA, and, *a fortiori*, for the basal hominin, could well be neither chimpanzee but a less-close but more conservative relative, perhaps *Gorilla*. This simple thought-exercise illustrates the weakness of using the closest sister taxon uncritically as a referent, and indeed leads one to question the value of identifying any one species as the BEM. Although the ancestral morphotype, the LCA, is a mere construct, not a moving, behaving animal that can be watched in the field, it has the great advantage over the BEM that its construction follows well-established rules, it can be modified with new data, and it is not tied to or biased by the survival of any species into the extant fauna. Most importantly, it provides a suite of character states that by definition are an exact complement to the derived features identified in the descendant lineages and interpreted by analogy.

TWO USES OF REFERENT SPECIES: COMPLEMENTARY, NOT CONFLICTING

It is argued here that if living animals are to be used to reconstruct the lifeways of fossil forms in general, and early hominins in particular, the soundest approach is to combine information from two kinds of "referential" analysis—"true analogy" and BEM. Each provides a different kind of information. Analogies are strictly delimited, aim to explain particular, derived, adaptive complexes, and are strengthened by multiple parallels in species that are unrelated phylogenetically. Baboons, among many other species, are a good source for hominin analogies. Identifying a BEM, on the other hand, brings ancestral characters into the picture. Though less rigorous and ultimately less valuable than reconstructing the LCA of the fossil and extant forms, it has the advantage of identifying a species that can actually be observed, and which shares with the extinct form an extensive background of heritage characters. It can be used (with caution) to predict traits of behavior and physiology in the fossil form that are intrinsically unobservable but that are likely to be inherited from the LCA. For the basal hominins, the African apes, and especially *Pan*, are generally identified as BEM.

Although these two formulations are conceptually quite distinct, actual cases may involve elements of both that need to be carefully teased apart to realize their implications. For instance, if chimpanzees living in one particular habitat were found to use hammerstones more often to open nuts, this would provide a possible analogy for the "invention" of tool use in hominins; it would be one of many such parallel cases—in *Cebus* monkeys (Langguth and Alonso 1997), sea-otters (Hall and Schaller 1964), and even Egyptian vultures (Thouless et al. 1989)—and the elements common to the parallel cases (hard and brittle food-objects, for instance) could be informative. On the other hand, the details of chimpanzee tool use might be especially informative about early hominins because of their shared cognitive abilities, hand anatomy, and similarity of habitat—all ancestral features of "general resemblance" that make the chimpanzee a good model.

Because these two kinds of analysis are quite distinct, an analogy cannot be refuted by showing that its referent is a relatively poor model, and *vice versa*. For example, to cite an example much favored by the late Professor Sherwood Washburn, pointing out that a chimpanzee will cross its legs in much the same way as Ms. Dorothy Lamour, while a baboon (presumably) will not, may show that the chimp is a better human model, but it does not refute an analogy between baboon and hominin dentitions. On the other hand, it would obviously be equally mistaken to assume that because some aspects of early hominin ecology may be better matched in baboons than in chimpanzees, baboons are necessarily a better all-round model for early hominins (Strum and Mitchell 1987).

If the interpretation suggested here is correct, the debates and even quarrels over the relative merits of various primate and nonprimate species as referents for early hominins are largely pointless. So long as they are presented explicitly as either analogies or models, all such comparisons are potentially informative.

REFERENCES

Arcadi, A. C. 2006. "Species Resilience in Pleistocene Hominids that Traveled Far and Ate Widely: An Analogy to the Wolf-Like Canids." *Journal of Human Evolution* 51 (4): 383–94. http://dx.doi.org/10.1016/j.jhevol.2006.04.011. Medline:16904731

Bartholomew, G. A., and J. B. Birdsell. 1953. "Ecology and the Protohominids." *American Anthropologist New Series* 55 (4): 481–98. http://dx.doi.org/10.1525/aa.1953.55.4.02a00030.

Berkeley, M. J. 1857. *Introduction to Cryptogamic Botany*. Vol. 12. London: H. Bailliere.

Currat, M., and L. Excoffier. 2004. "Modern Humans Did Not Admix with Neanderthals during Their Range Expansion into Europe." *PLoS Biology* 2 (12): e421. http://dx.doi.org/10.1371/journal.pbio.0020421. Medline:15562317

Darwin, C. 1871. *Descent of Man, and Selection in Relation to Sex*. London: Charles Murray.

de Ricqles, A. J., K. Padian, J. R. Horner, and H. Francillon-Vieillot. 2000. "Paleohistology of the Bones of Pterosaurs (Reptilia: Archosauria): Anatomy, Ontogeny, and Biomechanical Implications." *Zoological Journal of the Linnean Society* 129 (3): 349–85. http://dx.doi.org/10.1111/j.1096-3642.2000.tb00016.x.

Detwiler, K. M., A. S. Burrell, and C. J. Jolly. 2005. "Conservation Implications of Hybridization in African Cercopithecine Monkeys." *International Journal of Primatology* 26 (3): 661–84. http://dx.doi.org/10.1007/s10764-005-4372-0.

Elton, S. 2006. "Forty Years on and Still Going Strong: The Use of Hominin-Cercopithecid Comparisons in Palaeoanthropology." *Journal of the Royal Anthropological Institute* 12 (1): 19–38. http://dx.doi.org/10.1111/j.1467-9655.2006.00279.x.

Foley, R. A. 1993. "African Terrestrial Primates: The Comparative Evolutionary Biology of Theropithecus and the Hominidae." In *Theropithecus: The Rise and Fall of a Primate Genus*, ed. N. G. Jablonski, 245–70. Cambridge: Cambridge University Press. http://dx.doi.org/10.1017/CBO9780511565540.010

Garrigan, D., Z. Mobasher, S. B. Kingan, J. A. Wilder, and M. F. Hammer. 2005. "Deep Haplotype Divergence and Long-Range Linkage Disequilibrium at Xp21.1 Provide Evidence that Humans Descend from a Structured Ancestral Population." *Genetics* 170 (4): 1849–56. http://dx.doi.org/10.1534/genetics.105.041095. Medline:15937130

Grubb, P. 2006. "Geospecies and Superspecies in the African Primate Fauna." *Primate Conservation* 20:75–8. http://dx.doi.org/10.1896/0898-6207.20.1.75.

Hall, K.R.L., and G. B. Schaller. 1964. "Tool-Using Behavior of the California Sea Otter." *Journal of Mammalogy* 45 (2): 287–98. http://dx.doi.org/10.2307/1376994.

Harvati, K., S. R. Frost, and K. P. McNulty. 2004. "Neanderthal Taxonomy Reconsidered: Implications of 3D Primate Models of Intra- and Interspecific Differences." *Proceedings of the National Academy of Sciences of the United States of America* 101 (5): 1147–52. http://dx.doi.org/10.1073/pnas.0308085100. Medline:14745010

Hewitt, G. M. 2001. "Speciation, Hybrid Zones and Phylogeography—or Seeing Genes in Space and Time." *Molecular Ecology* 10 (3): 537–49. http://dx.doi.org/10.1046/j.1365-294x.2001.01202.x. Medline:11298967

Holliday, T. W. 2003. "Species Concepts, Reticulation, and Human Evolution." *Current Anthropology* 44 (5): 653–73. http://dx.doi.org/10.1086/377663.

Hunt, K. D., and W. C. McGrew. 2000. "Chimpanzees in Dry Habitats at Mount Assirik, Senegal and at the Semliki-Toro Wildlife Reserve, Uganda." In *Behavioural Diversity in Chimpanzees and Bonobos*, ed. C. Boesch, G. Hohmann, and L. F. Marchant, 35–51. Cambridge: Cambridge University Press.

Izawa, K. 1970. "Unit Groups of Chimpanzees and Their Nomadism in the Savanna Woodland." *Primates* 11 (1): 1–45. http://dx.doi.org/10.1007/BF01730674.

Jolly, C. J. 1970. "The Seed Eaters: A New Model of Hominid Differentiation Based on a Baboon Analogy." *Man* 5 (1): 5–26. http://dx.doi.org/10.2307/2798801.
Jolly, C. J. 1973. "Changing Views of Hominid Origins." *Yearbook of Physical Anthropology* 16:1–17.
Jolly, C. J. 2001. "A Proper Study for Mankind: Analogies from the Papionin Monkeys and their Implications for Human Evolution." *Yearbook of Physical Anthropology* 44 (Suppl. 33): 177–204. http://dx.doi.org/10.1002/ajpa.10021. Medline:11786995
Jolly, C. J. 2006. "Mitochondrial Capture in the Evolution of Baboon and Human Allotaxa." *American Journal of Physical Anthropology* (Suppl. 42): 110.
Jolly, C. J., T. Woolley-Barker, S. Beyene, T. R. Disotell, and J. E. Phillips-Conroy. 1997. "Intergeneric Hybrid Baboons." *International Journal of Primatology* 18 (4): 597–627. http://dx.doi.org/10.1023/A:1026367307470.
Kay, R. F. 1981. "The Nut-Crackers—A New Theory of the Adaptations of the Ramapithecinae." *American Journal of Physical Anthropology* 55 (2): 141–51. http://dx.doi.org/10.1002/ajpa.1330550202.
Kingdon, J. 1997. *The Kingdon Field Guide to African Mammals*, 464. San Diego, CA: Academic Press.
Kinzey, W. G. 1987. "A Primate Model for Human Mating Systems." In *The Evolution of Human Behavior: Primate Models*, ed. W. G. Kinzey, 105–114. Albany: State University of New York Press.
Kortlandt, A. 1965. "How Do Chimpanzees Use Weapons When Fighting Leopards?" *Year Book - American Philosophical Society* 1965:327–32.
Kortlandt, A., and M. Kooij. 1963. "Protohominid Behaviour in Primates (Preliminary Communication)." *Symposium of the Zoological Society of London* 10:61–88.
Krings, M., H. Geisert, R. W. Schmitz, H. Krainitzki, and S. Pääbo. 1999. "DNA Sequence of the Mitochondrial Hypervariable Region II from the Neandertal Type Specimen." *Proceedings of the National Academy of Sciences of the United States of America* 96 (10): 5581–5. http://dx.doi.org/10.1073/pnas.96.10.5581. Medline:10318927
Krings, M., A. Stone, R. W. Schmitz, H. Krainitzki, M. Stoneking, and S. Pääbo. 1997. "Neandertal DNA Sequences and the Origin of Modern Humans." *Cell* 90 (1): 19–30. http://dx.doi.org/10.1016/S0092-8674(00)80310-4. Medline:9230299
Langguth, A., and C. Alonso. 1997. "Capuchin Monkeys in the Caatinga: Tool Use and Food Habits during Drought." *Neotropical Primates* 5 (3): 77–8.
Lubbock, J. 1879. *Transactions of the Royal Entomological Society of London* 27: xli–xlv. doi: 10.1111/j.1365-2311.1879.tb02008.x.
Mackworth-Praed, C. W., and C.H.B. Grant. 1980 [1957]. *Birds of Eastern and North Eastern Africa*. London: Longman.
Marino, L. 2005. "Big Brains Do Matter in New Environments." *Proceedings of the National Academy of Sciences of the United States of America* 102 (15): 5306–7. http://dx.doi.org/10.1073/pnas.0501695102. Medline:15811939
Martin, R. D. 1990. *Primate Origins and Evolution: A Phylogenetic Reconstruction*. London: Chapman and Hall.

Milton, K. 1981. "Distribution Patterns of Tropical Plants as Evolutionary Stimulus to Primate Mental Development." *American Anthropologist* 83 (3): 534–48. http://dx.doi.org/10.1525/aa.1981.83.3.02a00020.

Murchison, R. I. 1839. *The Silurian System, Founded on a Series of Geological Researches in the Counties of Salop, Hereford, Radnor, Montgomery, Caermarthen, Brecon, Pembroke, Monmouth, Gloucester, Worcester, and Stafford; With Descriptions of the Coalfields and Overlying Formations*. London: John Murray.

Ovchinnikov, I. V., A. Götherström, G. P. Romanova, V. M. Kharitonov, K. Lidén, and W. Goodwin. 2000. "Molecular Analysis of Neanderthal DNA from the Northern Caucasus." *Nature* 404 (6777): 490–3. http://dx.doi.org/10.1038/35006625. Medline:10761915.

Oxford English Dictionary. 1933. "Analogy," p. 304. Oxford University Press, Oxford.

Patterson, N., D. J. Richter, S. Gnerre, E. S. Lander, and D. Reich. 2006. "Genetic Evidence for Complex Speciation of Humans and Chimpanzees." *Nature* 441 (7097): 1103–8. http://dx.doi.org/10.1038/nature04789. Medline:16710306

Pilbrow, V. 2006. "Population Systematics of Chimpanzees Using Molar Morphometrics." *Journal of Human Evolution* 51 (6): 646–62. http://dx.doi.org/10.1016/j.jhevol.2006.07.008. Medline:16965803

Raaum, R. L., K. N. Sterner, C. M. Noviello, C. B. Stewart, and T. R. Disotell. 2005. "Catarrhine Primate Divergence Dates Estimated from Complete Mitochondrial Genomes: Concordance with Fossil and Nuclear DNA Evidence." *Journal of Human Evolution* 48 (3): 237–57. http://dx.doi.org/10.1016/j.jhevol.2004.11.007. Medline:15737392

Roca, A. L., N. Georgiadis, and S. J. O'Brien. 2005. "Cytonuclear Genomic Dissociation in African Elephant Species." *Nature Genetics* 37 (1): 96–100. Medline:15592471.

Rogers, J. 1994. "Levels of the Genealogical Hierarchy and the Problem of Hominoid Phylogeny." *American Journal of Physical Anthropology* 94 (1): 81–8. http://dx.doi.org/10.1002/ajpa.1330940107. Medline:8042707

Samollow, P. B., L. M. Cherry, S. M. Witte, and J. Rogers. 1996. "Interspecific Variation at the Y-Linked RPS4Y Locus in Hominoids: Implications for Phylogeny." *American Journal of Physical Anthropology* 101 (3): 333–43. http://dx.doi.org/10.1002/(SICI)1096-8644(199611)101:3<333::AID-AJPA3>3.0.CO;2-#. Medline:8922180

Schultz, A. H. 1930. "The Skeleton of the Trunk and Limbs of the Higher Primates." *Human Biology* 2:303–438.

Schultz, A. H. 1934. "Some Distinguishing Characters of the Mountain Gorilla." *Journal of Mammalogy* 15 (1): 51–61. http://dx.doi.org/10.2307/1373897.

Stanford, C. B., and J. S. Allen. 1991. "On Strategic Storytelling: Current Models of Human Behavioral Evolution." *Current Anthropology* 32 (1): 58–61. http://dx.doi.org/10.1086/203914.

Strum, S. C., and W. Mitchell. 1987. "Baboon Models and Muddles." In *The Evolution of Human Behavior: Primate Models*, ed. W. G. Kinzey, 87–104. Albany: State University of New York Press.

Susman, R. L. 1998. "Hand Function and Tool Behavior in Early Hominids." *Journal of Human Evolution* 35 (1): 23–46. http://dx.doi.org/10.1006/jhev.1998.0220. Medline:9680465

Thouless, C. R., J. H. Fanshawe, and B.C.R. Bertram. 1989. "Egyptian Vultures *Neophron percnopterus* and Ostrich *Struthio camelus* Eggs: The Origins of Stone-Throwing Behaviour." *Ibis* 131 (1): 9–15. http://dx.doi.org/10.1111/j.1474-919X.1989.tb02737.x.

Tooby, J., and I. DeVore. 1987. "The Reconstruction of Hominid Behavioral Evolution through Strategic Modelling." In *The Evolution of Human Behavior: Primate Models*, ed. W. G. Kinzey, 183–237. Albany: State University of New York Press.

Tyndall, John. 1896. *The Glaciers of the Alps: Being a Narrative of Excursions and Ascents, an Account of the Origin and Phenomena of Glaciers, and an Exposition of the Physical Principles to Which They Are Related.* London: Longmans, Green, and Co.

Walker, R., O. Burger, J. Wagner, and C. R. Von Rueden. 2006. "Evolution of Brain Size and Juvenile Periods in Primates." *Journal of Human Evolution* 51 (5): 480–9. http://dx.doi.org/10.1016/j.jhevol.2006.06.002. Medline:16890272.

Washburn, S. L., and I. DeVore. 1961. "The Social Life of Baboons." *Scientific American* 204 (6): 62–71. http://dx.doi.org/10.1038/scientificamerican0661-62.

Whiten, A., J. Goodall, W. C. McGrew, T. Nishida, V. Reynolds, Y. Sugiyama, C.E.G. Tutin, R .W. Wrangham, and C. Boesch. 1999. "Cultures in Chimpanzees." *Nature* 399 (6737): 682–5. http://dx.doi.org/10.1038/21415. Medline:10385119

Woodward, S. P. 1854. *A Manual of the Mollusca; or, Rudimentary Treatise of Recent and Fossil Shells.* London: John Weale.

Wrangham, R., and D. Peterson. 1996. *Demonic Males: Apes and the Origins of Human Violence.* Boston: Houghton Mifflin.

Zihlman, A. L., J. E. Cronin, D. L. Cramer, and V. M. Sarich. 1978. "Pygmy Chimpanzee as a Possible Prototype for the Common Ancestor of Humans, Chimpanzees and Gorillas." *Nature* 275 (5682): 744–6. http://dx.doi.org/10.1038/275744a0. Medline:703839

Contributors

David R. Braun, Department of Archaeology, University of Cape Town, Cape Town, South Africa

Beth Christensen, Environmental Studies Program, Adelphi University, Garden City, NY

David J. Daegling, Department of Anthropology, University of Florida, Gainesville, FL

Craig S. Feibel, Department of Earth and Planetary Sciences, Rutgers University, New Brunswick, NJ

Frederick E. Grine, Department of Anthropology, SUNY Stony Brook, Stony Brook, NY

Clifford J. Jolly, Department of Anthropology, New York University, New York, NY

Julia A. Lee-Thorp, Research Lab for Archaeology, University of Oxford, Oxford, United Kingdom

Naomi E. Levin, Department of Earth and Planetary Sciences, Johns Hopkins University, Baltimore, MD

Mark A. Maslin, Environmental Change Research Centre, Department of Geography, University College London, UK

John C. Mitani, Department of Anthropology, University of Michigan, Ann Arbor, MI

Jay Quade, Department of Geosciences, University of Arizona, Tucson, AZ

Amy L. Rector, Anthropology, School of World Studies, Virginia Commonwealth University, RuchmandRichmond, VA

Kaye E. Reed, Institute of Human Origins, School of Human Evolution and Social Change, Arizona State University, Tempe, AZ

Jeanne Sept, Department of Anthropology, Indiana University, Bloomington, IN

Lillian M. Spencer, Department of Social Sciences, Glendale Community College, Glendale, AZ

Matt Sponheimer, Department of Anthropology, University of Colorado at Boulder, Boulder CO

Mark F. Teaford, School of Health Sciences, High Point University, High Point, NC

Peter S. Ungar, Department of Anthropology, University of Arkansas, Fayetteville, AR

Carol V. Ward, Department of Pathology and Anatomical Science, University of Missouri School of Medicine, Columbia, MO

Katy E. Wilson, Department of Earth Sciences, University College London, UK

Index

Page numbers in italics indicate illustrations.